Smart Materials for Science and Engineering

Scrivener Publishing
100 Cummings Center, Suite 541J
Beverly, MA 01915-6106

Publishers at Scrivener
Martin Scrivener (martin@scrivenerpublishing.com)
Phillip Carmical (pcarmical@scrivenerpublishing.com)

Smart Materials for Science and Engineering

Edited by
Upendra Kumar
and
Piyush Kumar Sonkar

Scrivener Publishing

WILEY

This edition first published 2024 by John Wiley & Sons, Inc., 111 River Street, Hoboken, NJ 07030, USA and Scrivener Publishing LLC, 100 Cummings Center, Suite 541J, Beverly, MA 01915, USA

For more information about Scrivener publications please visit www.scrivenerpublishing.com.

Wiley Global Headquarters
111 River Street, Hoboken, NJ 07030, USA

For details of our global editorial offices, customer services, and more information about Wiley products visit us at www.wiley.com.

Library of Congress Cataloging-in-Publication Data

ISBN 978-1-394-18581-8

Cover images: Solar Panel: Wave Break Media Ltd | Dreamstime.com, Black & White Abstract: Kateryna Linnik | Dreamstime.com, Fuel Cell: Luchschen | Dreamstime.com, Dental Implants: Bymandesigns Dreamstime.com
Cover design and drawings: Kris Hackerott

Set in size of 11pt and Minion Pro by Manila Typesetting Company, Makati, Philippines

Printed in the USA

10 9 8 7 6 5 4 3 2 1

Dedicated
to ours
Authors, Collaborators
&
Beloved Parents

Contents

Preface

Advanced and creative materials are other names for smart materials. They can be described as advanced materials that react intuitively to environmental changes or as materials that can return to their original shape in response to certain stimuli. Based on their characteristics, such as active or passive response, smart materials are classified. There are two types of active materials; the first kind cannot change its characteristics when subjected to outside stimuli, such as photochromatic spectacles, which only alter their colour when exposed to sunlight. The other kinds, like piezoelectric materials, can change one sort of energy (thermal, electrical, chemical, mechanical, and optical) into another. When subjected to external pressure, it can generate an electric charge. As an example, optical fibers can transmit electromagnetic waves. In contrast, passive smart materials can transmit a specific sort of energy. They have some amazing qualities that set them apart from other materials, such as transiency—they can react to different kinds of external stimuli, immediacy—the response time is much shorter, self-actuation—the capacity to change their appearance and shape, selectivity—the response is divided and expected, directness—the response is limited to the activating event, shape-changing—the material can change its shape to external stimuli, self-diagnostic—their ability to determine their own health, and self-healing - their ability to recover and fixed issue by themselves.

The ability to synthesize novel materials has substantially progressed thanks to science and technology over the past 20 years. They fall mostly into the following four categories: polymers, ceramics, metals, and smart materials. Among these, smart materials are gaining in popularity since they have more uses than conventional materials. Smart materials are unusual substances that have the ability to alter their properties, such as those that can immediately change their phase when placed near a magnet or their shape simply by applying heat. The human race will be significantly impacted by this new era of smart materials. For instance, some of them can adapt their properties to the environment, some have sensory capabilities, some can repair themselves automatically, and some can degrade themselves. These extraordinary properties of smart materials will have an effect on all facets of civilization. There are many different types of intelligent materials, including magnetorheological materials, electro-rheostat materials, shape memory alloys, piezoelectric materials, and more. This book describes many forms of smart materials and their possible uses in various fields. Thus, this book titled "*Smart Materials for Science and Engineering*" cover most of the significant areas of smart materials and useful for the readers.

Acknowledgements

- First of all, we are thankful to our contributing authors for their valuable contribution to this book.
- We are thankful to Prof. B. N. Dwivedi (Department of Physics, IIT BHU) Prof. Devendra Kumar (Department of Ceramic Engineering, IIT BHU) for their continuous help, support and motivation.
- We are thankful to Dr. Shail Upadhyay (Department of Physics, IIT BHU) and Prof. V. Ganesan (Department of Chemistry, Institute of Science, BHU) for their sincere effort and dedication to contribute in this book.
- We are thankful to our research group members Dr. Satyam Kumar, Dr. Varsha Yadav, Mr. Narvadeshwar, Amit Kumar Verma, Ms. Vedika, Harshpreet, Manisha, Mr. Ramsundar, and Raj Kumar for their help and support.
- We are thankful to Mr. Angesh Kumar Maurya and Ms. Kavita Sonkar for helping in development of book cover page and index, respectively.
- The authors, Dr. Upendra Kumar acknowledges funding support from SERB, Govt. of India for EEQ project and Indian Institute of Information Technology Allahabad for SEED Grant; Dr. Piyush Kumar Sonkar acknowledges funding support from Institute of Eminence (IoE), Banaras Hindu University (BHU), Varanasi, India for Seed Grant and Trans-disciplinary Research Project.
- We are thankful to our colleagues, collaborators, friends and well-wishers to provide the moral support for this book.
- We are thankful to our family members and friends for providing continuous moral support in this book.

Scope of the Book

The present book gives a complete description about the different types of smart materials. In this book, a literature survey discusses the different types of smart materials such as based ceramics, polymers, organic compounds, etc., and their need, advantages, disadvantages, and applications will be comprehensively discussed. In this book, the discussion about the well investigated smart materials including piezoelectric, Magnetostrictive, shape memory alloys, electro-rheological fluid, and magnetorheological fluid will be discussed with their present prospects and current literature survey.

This book covers the various aspects of the smart materials. Chapter 1 describes the detailed introduction of the smart materials, historical overview and future prospective. Fabrication and characterization of the smart materials are discussed in Chapter-2. Chapter-3 and Chapter-4 provides the details about medical application in dental science and tissue engineering. The various application and preparation strategies of the smart materials are discussed in the Chapter-5 to Chapter-8. The energy storage applications of the smart materials are discussed in Chapter-9 to Chapter-12. The fuel cells and biofuel cell applications are discussed in Chapter-13 and Chapter-14. The CO_2 reduction, capture and semiconductor application from the smart materials are discussed in Chapter-15 and Chapter-16, respectively. Further, the futuristic microelectronics from the smart materials are discussed in Chapter-17. Hence, this book is quite beneficial for undergraduate, post-graduate, Ph.D. scholar, Post-Doc fellow, faculty and scientist working in interdisciplinary areas to understand fundamentals of smart materials, their advantages, disadvantages, and applications in various societal and smart city problems.

1

Introduction: Historical Overview, Current and Future Perspective

Unni kisan*, **R.R. Awashthi and Sanjeev Kumar Trivedi**

Faculty of Engineering and Technology, KMC Language University, Lucknow, India

Abstract

Human civilization was dependent on the use of materials in the ancient era as it is today. Humans started using stone before progressing to functional nanomaterials; today, humans benefit from using new advanced materials named smart materials. In the present scenario, humans develop different types of smart materials like piezoelectric materials, magnetostrictive materials, dielectrics, thermoelectric materials, nano-medicines, shape-memory alloys, and rheological fluids. These types of smart materials are applicable for sensing devices, data storage, fast commutation, cloud computing, and different engineering as well as medical tools and equipment. Biodegradable and low-cost smart materials will develop due to the synthesis of different amalgamation of materials in the future. The newly developed smart materials may be rapidly used in the engineering, medical, and the information technology sectors.

This book chapter aims to promote awareness on smart materials for the extensive research and knowledge enhancement for new applications in the future. Historical views and modern and future perspectives of smart materials are discussed in this chapter.

Keywords: Historical overview, smart materials, current prospective, future prospective, application of smart materials

1.1 Introduction

In ancient times, humans used different materials for various purposes, due to which there was an enhancement in their living standards. Civilizations were categorized on the basis of their invention of material; the primary age was the Stone Age. Bronze Age was the most radical and was more sustainable. The development of bronze signified the start of a new metallurgical age, which saw the synthesis of numerous materials. Engineering and technology have made significant advancements in the manufacture of novel materials during the last two decades. They could be subdivided primarily into four groups: polymers, ceramics, metals, and smart materials. Because they have more applications than traditional materials, smart materials are among them and are growing in popularity. Smart materials

**Corresponding author*: krishnanswami1996@gmail.com

Upendra Kumar and Piyush Kumar Sonkar (eds.) Smart Materials for Science and Engineering, (1–16) © 2024 Scrivener Publishing LLC

are unique materials with the capability of changing their properties, such as substances that may instantly change their phase when placed near a magnet or their shape by simply reheating. These unexpected abilities of advanced material will have an effect on every aspect of civilization [1].

The terms "smart," "intelligent," and "adaptive" were first used to describe the newly developing field of research that involved incorporating electro-active efficient materials in massive structures as in actuators and *in situ* sensors in the beginning of the 1980s. Tiny and microstructure transducers and precise mechatronics (mechanical + electrical) controllers constituted the only applications for electro active materials in the past [2].

Mecha, which signifies mechanical, and tronics are terms from the fields of electrical and electronic engineering, etc. Also included is digital engineering. In the other meaning, as produced items and technologies are advanced, it will become gradually difficult to distinguish the electronics between how deeply and naturally they are incorporated into processes. When seen in the context of systems design, the field of mechatronics could be characterized as the intersection of these three main domains rather than just the total of the three. Figure 1.1 [3] depicts mechatronics' multidisciplinary approach. "The effective integration of electronic control, systems thinking, and precision mechanical engineering in the design of goods and industrial processes" [4].

During the 19th century, mechanical engineering as a popular discipline saw a burst in growth as it established the groundwork of successful and quick advancement for the revolution of the industry. Mechanical, electrical, civil, and chemical engineering were the four main engineering disciplines of the 20th century. These disciplines still have their own bodies of knowledge, textbooks, and professional journals since they are thought to have separate intellectual and professional domains. Entrants might evaluate their unique intellectual abilities and select one of the fields as a career. The information revolution is a current scientific and social change that we are currently witnessing, and oddly, engineering expertise appears to be both concentrating and diversifying. The advancement of engineering electronics that has sparked communication and information that revolutionized people is what gave rise to this modern revolution. One of the newest and most fascinating areas of engineering is mechatronics, which incorporates elements of more established disciplines

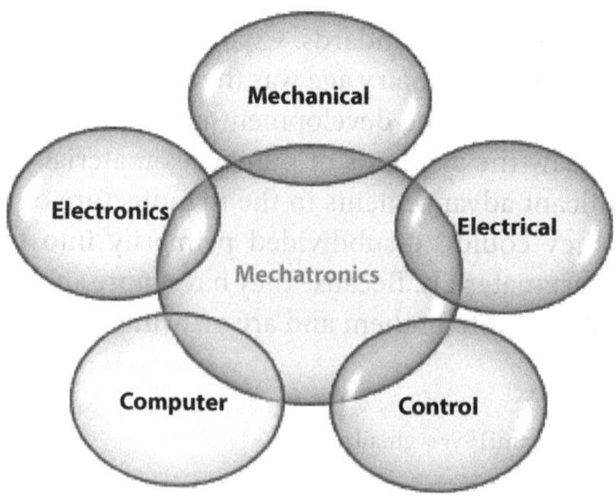

Figure 1.1 Mechatronics: a multi-disciplinary approach [4].

and necessitates a more comprehensive approach to the design of what we can properly refer to as mechatronic systems. So, exactly what is mechatronics? The word "mechatronics" refers to a multidisciplinary subject of engineering that is currently in rapid development. The term "mechatronics" was first used in Japan in the late 1960s; it later gained popularity in Europe and is now widely used in the US. Mechatronic system design primarily draws on the fields of mechanics, semiconductors, controls, and computer engineering.

An engineer of mechatronic systems must be capable of construction and choose digital and analogical circuitry systems, micro-processor-based elements, mechanical components, actuators, and sensors so that the finished result meets the necessary objectives. Smart devices are another name for mechatronic systems. Although a specific definition of the word "smart" is elusive, in the context of engineering, we refer to the incorporation of aspects like computing, logic, and feedback system that are combined in a complex design, which may appear to emulate human thought procedures. The engineering of mechatronic systems requires knowledge from numerous domains, making it difficult to encapsulate within a traditional field of engineering. The designer of mechatronic systems needs to be a generalist who is eager to learn from a variety of sources and apply it to their work. The learner may initially feel intimidated by this, yet it has many advantages for originality and lifelong learning. Nowadays, almost all mechanical machines come equipped with electrical parts and some kind of computer monitoring or control. As a result, a widespread array of system and components fall under the mechatronic systems. Microcontrollers are being included into electromechanical devices increasingly frequently, giving system designers far greater flexibility and control. All the components of an engineering mechatronic system are shown in Figure 1.2. The interface circuit between the input/output and control circuits are controlled by the digital devices [3].

The engineering disciplines that deal with the design of controlled electromechanical systems are currently in a process of evolutionary transformation. A mechanical system that is computer controlled is referred to as mechatronic. Control decisions are frequently made by an embedded computer rather than a general-purpose computer. Yaskawa Electric Company engineers originally used the term "mechatronics" Nowadays, an embedded computer controller is almost built into every electromechanical system. As a result, concerns with computer hardware and software are included in the discipline of mechatronics when applied to the control of electromechanical systems. The field of mechatronics as we know it today would not exist if cheap microcontrollers were not widely available for the

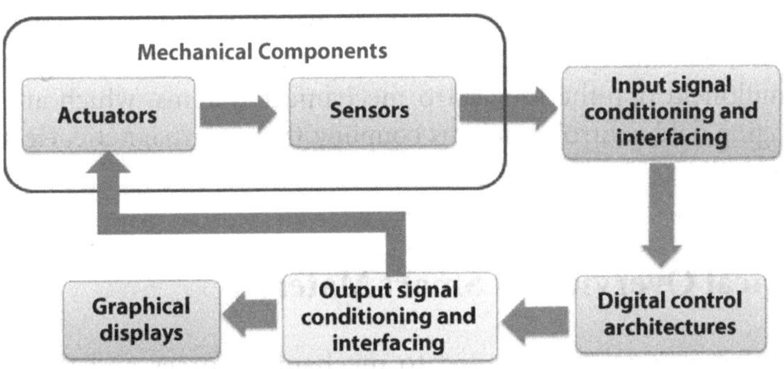

Figure 1.2 Mechanical components in mechatronics system.

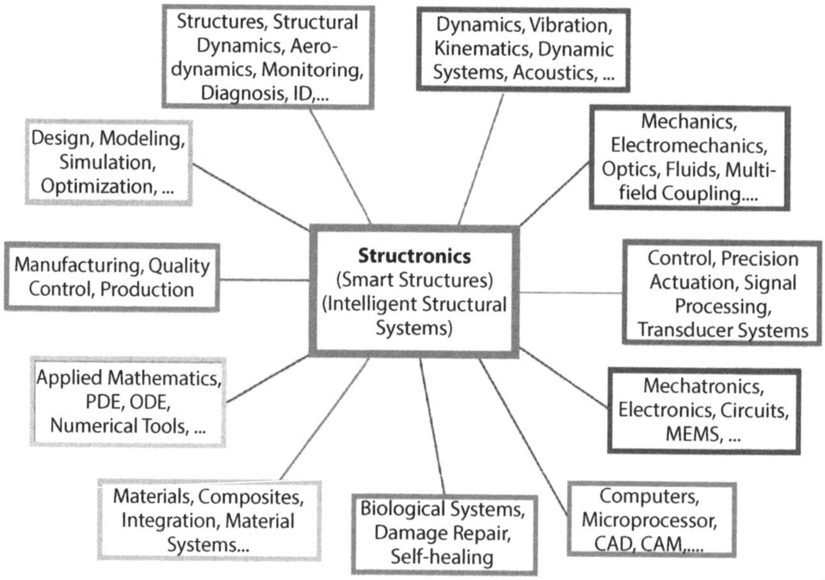

Figure 1.3 Structronics—A multidiscipline integration.

mainstream market. The application of computer control in countless consumer products is made possible by the accessibility of embedded microprocessors for the mass market at continually decreasing cost and rising performance. The preceding model [3] is the outdated model for an electromechanical product design team.

(i) Engineers who design a manufacturer's mechanical parts;
(ii) Engineers who create a product's electrical parts, including actuators, sensors, amplifiers, and other devices, as well as the control logic and algorithms;
(iii) Computer hardware and software developers who design the real-time controls for the product.

Adaptive materials or structures, and smart and intelligent materials are generally thought to have the capacity to be sophisticated, trendy, and active. In Figure 1.3, the structronic system is depicted. As a consequence, the objective of this study is to discuss the fundamental properties, design concepts, and real-world uses of the major smart materials listed in Table 1.1, the smart materials that have been studied. Also covered are the specifications for multifield optothermoelectro mechanical systems, which are used to solve various challenging field control problems coupling thermal, magnetic, electric, magnetic, and optical interactions [2].

1.2 Historical Overview of Smart Material

In 1880, quartz crystals were subjected to mechanical forces and the Curie brothers (Jacques and Pierre) detected the creation of electric fields on the crystals (the Greek

Table.1.1 Different eras smart materials [2].

Smart materials	Time periods
Pyroelectrics	315 B.C.
Piezoelctric	1880
Electro and Magneto-rheological Fluids	1784/1947
Superconductors	1911
Electro and Magneto-strictive Materials	1954/1840
Shape-memory Material	1932
Polyelectrolyte Gels (pH muscles)	1949
Photostrictive Material	1974

word piezo means "push"). When the crystal was exposed to electric fields and they also saw strain formation. In general, the piezo-electricity is an electro-mechanical phenomenon that couples the electric (static coupling) and elastic (dynamic coupling) fields. The direct piezoelectric effect is a piezo-electric material that responds to stresses or pressures of mechanical property by producing voltages. The inverse piezoelectric effect is the ability of electric charges or fields to cause mechanical stresses or strains in a material.

Chang and Read discovered shape-memory-like behavior in a gold cadmium (AuCd) sample for the first time in 1932. Later on, in 1938, brass underwent this shape change, and in 1951, an AuCd bend bar did as well. Buehler, Gilfrich, and Wiley did not discover the whole shape-memory effect in various nickel–titanium alloys until 1962. The most popular shape-memory alloy (SMA) is commercially manufactured as nitinol (NiTi). Military high-performance hydraulic systems used the Cryofit hydraulic pipe coupling, which Raychem successfully manufactured in 1969. In Japan, demand for SMA goods significantly rose during the 1970s.

The ER feature was first noted by Winkler in 1784. The Winslow effect is occasionally used to refer to the ER phenomenon because Winslow continued to rigorously study ER property variations in 1947. Colloidal suspensions known as ER fluids undergo significant property changes in response to an appropriate electrical field.

Polymeric solutions known as ionic polymeric gels have the capacity to change their physical characteristics in response to external environmental factors like pH, electrical charge and field, and so on. A paper on the ionization-induced swelling and contraction of polymeric acids is published in 1949 by Katchalsky and Kuhn. Although many novel polymers were found between 1988 and 1991, little work was started until the early 1990s theoretically. To ascertain characteristics of several of the older polymers, tests had to be conducted. Some gel qualities still need to be determined experimentally because not all gels are now covered by theories that can predict polymer bf.

Michael Faraday demonstrated in 1845 that polarized light's polarization angles may be altered by passing through a thick glass sheet fastened to the poles of a strong magnet. The magnetooptical effect is sometimes known as the Faraday effect.

The disappearance of mercury wire's resistance as the temperature approaches zero degrees Fahrenheit was originally noticed by Kamerlingh Onnes in 1911. A new era of superconductivity began as a result of this observation. At room temperature, ion oscillations within the lattice structure of metallic conductors and material impurities, flaws, and imperfections cause electrical resistance. The resistance brought on by the vibrating ions, however, reduces as the temperature of some materials rises. When the temperature falls below a specific level, known as the critical temperature. Various nonmagnetic elements, alloys, and compounds transform into a superconductive state [2].

1.3 About Smart Materials

Any new material that can be controlled and experiences a macroscopic change in one of its physical and mechanical properties as a result of a non- mechanical external stimuli is described as a smart material [5]. Advanced or intelligent materials are other names for smart materials. They can be described as sophisticated materials that react intelligently to environmental changes or as materials that can return to their basic shape in response to certain stimuli. Intelligent materials are categorized into two parts: active and passive smart materials. Active smart materials can transport electromagnetic waves; examples of passive smart materials are fiber optics. Additionally, there are two categories for active materials. For example, photochromatic spectacles only change colour when placed in sunlight; they cannot modify their attributes when subjected to external stimuli. The capability of intelligent material to change their shape in response to external stimuli is self-diagnostic; they are capable of automatically detecting cracks on their surfaces, and are self-healing when damaged [1].

Table 1.2 Type of smart materials [5].

Stimulus	Active materials	Semi-active materials
Thermal Field	Shape Memory Alloys (SMAs) Shape Memory Plymers (SMPs);	• Magnetorheological Fluid (MRF) • Electrorheological Fluid (ERF) • Magnetorheological Elastomers (MRE)
Magnetic Field	Magnetic Shape Memory Alloys (MSMAs); Magnetostrictive Materials;	
Electric Field	Electrostrictive Materials; Piezoelectric Materials; Dielectric Materials; Photomechanical Materials;	

Another important consideration is how controllable the materials are, as it is critical to comprehend how external stimuli might alter these materials. The sensor and the feedback control system are responsible for direct physical changes. It is useful to distinguish between the semi-active materials, which have controllable property but cannot add energy to the system, and the active materials, which can generate mechanical work. The most popular classes of smart materials can be split into groups based on their external activation stimuli and active and semi-active activities. The stimuli of intelligent materials are included in the table on the basis of active and semi-active materials. Table 1.2 suggests one possible classification for these materials [5].

1.3.1 Analysis of Some Active and Semi-Active Material

1.3.1.1 Shape-Memory Alloy Analysis

Shape-memory alloy (SMA) is a widely used type of intelligent material. These intelligent materials can take on any shape as a result of temperature changes, but they differ from other materials in that they can recall their previous shape in response to stress or other external stimuli. Additionally, they are known as intelligent materials. Several phase changes between the martensite and austenite take place. The martensite phase is stable at lower temperature and austenite is stable at higher temperatures. These thermomechanical capabilities of SMAs result from a reversible solid–solid phase change controlled by mechanical stress and temperature. Regarding the phase transformation that occurs in steels, the phases are austenite and martensite. The material's ability to recover from enormous elastic strains (up to 8%) and to withstand significant cyclic deformations with no residual strains is known as superelasticity. The intelligent memory alloy such as nickel-based titanium alloy has 45 to 50% titanium. The nickel–titanium alloys that are sold commercially go by the trade name Nitinol. In contrast to the mechanical and shape-memory qualities of nickel–titanium alloys, they perform less well. The pseudo elasticity of SMAs has two effects that SMAs, a type of new materials, exhibit. The phase transition between the martensite and austenite phases of an SMA is what causes a widespread trend, also known as pseudoelasticity, which has elasticity property response to applied stress. Contrary to the form memory effect, which is shown when the temperature changes, faux elasticity is seen when the temperature stays the same. When SMA is removed at a temperature higher than the finishing austenite temperature, stress-induced strain recovery occurs. When the temperature is really high and the austenite structure is stable, superelasticity arises. The austenite phase, also known as "stress-induced martensite," can be forced to change into the martensite phase when an external load is acted. This strain is held by the local areas' related shear. The austenite phase recovers after removing force and the created strain is completely restored. This is substantially more than what would typically be attainable during standard elastic deformation (up to 0.5%) and can be accomplished without being plastically deformed. This aspect, however, can only be realized if martensitic phase changes have produced all of the deformations [1].

1.3.1.2 Piezo-Electric Material Analysis

The word "piezo-electricity" is a combination of the Greek words "piezo", meaning pressure, and electricity", which refers to voltage or electrical charges. Piezo-electric materials alter mechanical energy into electrical energy and vice versa by applying stress or strain. Similar to this, piezoelectric actuators translate electrical information into mechanical motion that is used to adjust mirrors, lenses, and other car components. The performance of current items can be enhanced using piezoelectric materials. The direct piezo-electric effect and the reverse piezo-electric effect are the two primary manifestations of the piezo-electric effect that occurs in several directions.

1.3.1.2.1 The Direct Piezo-Electric Effect

According to the direct piezo-electric effect, anytime mechanical stress of forces is applied to piezo-electric-based materials, the electric charges are generated. The effect of generators and sensor effect are other names for the direct piezoelectric effect. Sensors are the major application for piezoelectric direct effect technology.

1.3.1.2.2 The Inverse Piezo-Electric Effect

The ability of piezo-electric-based materials to experience changes or strain in electric field is known as the inverse piezoelectric effect. The minor effect or actuator effect are other names for inverse piezo-electric effect. Utilization in piezo-electric actuators is the opposite effect. Piezo-crystals are the most exact tool for positioning things; thus, they are used as actuators. At very high electric intensities, the crystals' width can change in very small ways, even with a precision higher than a micrometer. The following equation can be used to explain the inverse piezoelectric effect:

$$\mathbf{S=S^{E}\times T+d\times E}$$

where S = Strain;

S^E= Coefficient of elasticity;

T = Stresses;

d = Constant of piezo-electric charge; and

E = electrical fields.

1.3.1.3 Magneto-Rheological Fluids Analysis

When a magnetic field is applied, magnetorheological fluids (MRFs) will modify their rheological characteristics, such as viscosity and stresses. Smart materials that are magneto-sensitive are also known as MRFs. A soft particle that resembles rubber called a magneto-rheological elastomer (MRE) can change mechanical properties when a magnetic field is applied. The non-magnetic elastic matrices, magnetic particles, and additives are the main parts of MREs. These materials' viscous characteristics can be changed by adjusting magnetic fields. MRFs are employed in various components such as the clutch, buffer, and damper. MREs are frequently utilized in soak absorbers that are tuned adaptively. Because carbonyl iron has a high saturation magnetization, it is utilized to prepare MRFs. MRFs have a number of characteristics, including viscoelasticity, magnetic properties, light weight, tunable modulus, and good sound absorption.

1.3.1.4 Magneto-Strictive Materials Analysis

Materials that are magnetically restrictive can get magnetized while under stress or deform when exposed to a magnetic field. The transducer is a term they use frequently. Magnetization may be the cause of the length variations. These substances could be categorized as either positive or negative magneto-strictive. By altering the magnetic field, they can choose to unwind. Since these materials include nonlinear properties and frequency-dependent hysteresis, it can be difficult to adequately capture complicated behavior. For magneto-strictive uses in sensors and actuators, the use of cobalt ferrites primarily occurs. Its high saturation magneto-striction is the cause of this. Due to the absence of rare-earth elements, it is an effective Terfenol-D alternative. Its magneto-strictive characteristics can be adjusted via magnetic uni-axial anisotropy. This can be accomplished by uni-axial pressure-assisted compaction, magnetic annealing, or magnetic field annealing.

1.3.1.5 Electro-Rheological Fluids Analysis

When an electric field is generated, very minute particles suspended in electrically insulating types of fluid quickly solidify in the field's direction, creating an electro-rheological fluid (ERF). They will change back from one state to another, such as from a gel to a liquid, and vice versa when electrical field is applied. ERFs have numerous uses in the automobile sector, such as shock absorbers, clutches, and vibration isolators. The interfacial bond strength, stiffness, high dielectrical constant, unique rheology, and dielectric in nature are some of characteristics of electro-rheological fluids.

1.3.1.6 Optical Fiber Analysis

The optical-based fiber is a transparent and flexible material that is produced by the drawing process of glass or plastic. The optical fiber is widely used in transmission lines based on light. Civil structures, mechanical blades, and shafts are a few of the structures that need to be checked for strain fluctuations caused by fractures and damage. Optical fiber sensors and active piezo-electric sensors are the two intelligent materials that are most frequently utilized for structural health monitoring (SHM). In the past 20 years, the two optical fiber sensors that received the greatest research attention are the fiber Bragg grating (FBG) and the optical fiber polarimetric sensor (FOPS). For measuring local strain, FBGs are employed. For monitoring the structural health of composite constructions, FBG sensors are the ones that can be easily rooted. Recently, the interior strain field can be easily measured by FEG sensors. FOPS measures the strain along a structure's whole length. Because of the more effort required regardless of where the damage is in the structure, a damaged or cracked area causes a rise in strain, which is detected by FOPS. The FOPS keeps an eye on damage to numerous structures worldwide. FOPS, however, is unable to identify the damage's exact location. The electricity can be transmitted using optical fiber and a solar cell. Medical and other applications can make use of optical fibers as light sources. Optical fiber lights can be used to illuminate decorative applications such as signs, toys, etc.

1.3.1.7 Dielectric Elastomers

The dielectric elastomers are essentially compliant capacitors that lack inherent intelligence. Two acquiescent electrodes are deposited on their faces, which are typically formed of silicone or acrylic polymers, to create a capacitor-like structure. The elastomer layer, which is relatively acquiescent in plane, is squeezed when an electrical charge is applied by the electro-static pressure generated by the Coulomb forces.

1.3.1.8 Photo-Mechanical Materials

In essence, photo-mechanical materials are those materials that deform slightly when exposed to light (photons). Uchino demonstrated how they might be applied to the creation of a modulated light-powered walking device. Nematic liquid crystal elastomers can be photoisomerized to produce a reversible form change (LCEs). The substitution of LCE has also been seen responding to light by swimming on the plane of the water. They are not used in engineering like other materials, but the scientific community is looking at their characteristics.

1.3.1.9 Magnetorheological Fluids (MRF)

It is an intelligent material made of fluid, often an oil, in which micron-sized ferromagnetic particles are dispersed along with various thixotropic additives. The fluid's apparent viscosity increases significantly in the presence of a magnetic field, reaching levels where it resembles a visco-plastic solid. This behavior happens as a result of the magnetic particles pulling together into chains in the field's direction. The "MRFs" section provides a more thorough explanation of material along with a thorough orientation to the technical research that has been done. The ERFs are comparable to the MRF, except in this instance, electric fields are used to change the viscosity characteristics. The distribution of the particles is uniform and random in the neutral state. They have little technical uses because the useable yield stress is smaller and a high magnetic field is considerably simpler to manage and produce than a strong electric field [5].

1.3.1.10 Magneto-Rheological Elastomers (MRE)

The MREs are created using an elastomeric foundation, a few additives, and iron particles that are smaller than a micron. Although the basic elastomers might change, polydimethylsiloxane is frequently employed. Chains are formed when magnetic poles begin to attract one another in the direction of an outside activated magnetic field [5].

1.4 Current and Future Perspectives of Smart Materials

There are wide potential applications of smart materials at present for modern devices. It is also used in Industry 4.0 [7]. Future applications of smart materials will include some "smart behavior" that is intended. However, intelligent behavior was discovered in some materials by accident, and it can take some time for the significance of the unique character to be understood as having any practical value [8]. Smart materials are those with the ability to perceive changes in their surroundings and external stimuli and to adjust their functions accordingly [9]. Hydrogels are a good illustration of a smart material since they are three-dimensional polymer networks that can hold a lot of water when they are swollen [10].

1.4.1 Smart Composite

Any novel composite intelligent material's ultimate goals are to combine more than two intelligent materials in stoichiometry ratio. Because of this, "intelligent composite-based materials" come extremely close to meeting all of the aforementioned requirements. Their benefits and flexibility in meeting the aforementioned design criteria have resulted in an abundance of innovative products. Essentially, there are two kinds:

- A composite substance manufactured entirely by humans. This substance is meant to strengthen or stiffen anything.
- A blend of single or composite materials and fiber- or polymer-reinforced materials (FRPs). The adaptability of the many design combinations is another important benefit. A beautiful and especially economical smart composite will be produced if fiber optic sensors are integrated with FRP [6].

1.4.2 Smart Structures

As previously mentioned, a crystal structure has the specific quality to enhance sensing and actuating characteristics due to its small surface-to-volume ratio. The five fundamental parts of a smart structure can be summed up as follows (Figure 1.4):

1.4.2.1 Data Acquisition (Tactile Sensing)

This section is generally used for sensing and monitoring made from intelligent materials.

1.4.2.2 Data Transmission (Sensory Nerves)

This section's goal is to transmit the fresh data to the regional or centralized control and command centers.

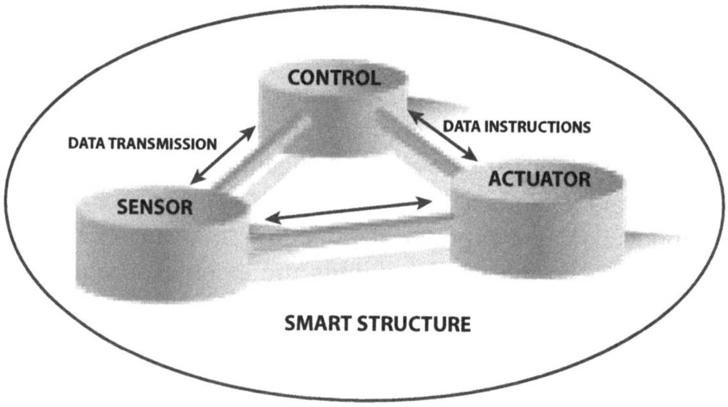

Figure 1.4 Basic component of smart structure.

1.4.2.3 Command and Control Unit (Brain)

This section works as the central processing unit, whose responsibility is to supervise and regulate the whole system by collecting and evaluating data that need to be done.

1.4.2.4 Data Instructions (Motor Nerves)

This component's role is to relay the pronouncements and related commands back to the structure's constituent parts.

1.4.2.5 Action Devices (Muscles)

This section's purpose is to initiate an activity by activating the controlling elements or units.

1.4.3 Fiber Optic Sensing

Sensors that mimic the operation of traditional strain gauges are made of fiber optics. They react when the amount of transmitted light changes. This variation could be in wavelength, mode, polarization, frequency, intensity, or phase. They function very well since they are exceedingly sensitive and can spot even the smallest deviations. There are three different kinds of sensors:

- The intensiometric type adapts to any disturbance that modifies the intensity of the transmitted light, such as bending or twisting. (Benefit: Relatively easy.)
- The interferometric type, which reacts to spectrometric and phase variations. These strain-sensitive phase sensors are available in a variety of designs.

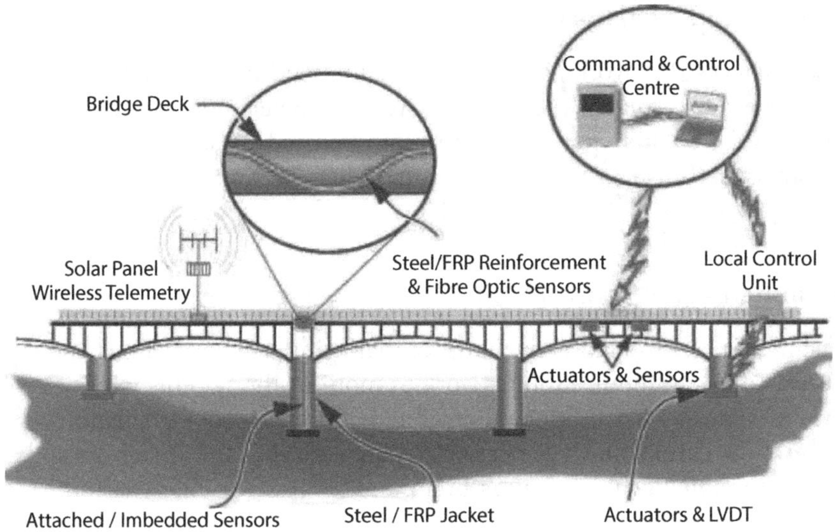

Figure 1.5 presents, in a schematic way, a section of a smart bridge [6].

- The spectroscopy based on variation of refractive index of the materials. The benefit of this type is that the sensed data immediately recorded the parameter based on the inverse of frequency that is unaffected by fluctuation. In order to achieve wavelength division multiplexing, every grating element is manufactured at a slightly various inverse of wavelengths. This results in the formation of several gratings along a single fiber, each with a slightly different resonance frequency.

Although they are more expensive than Fabry Perot sensors, Bragg grating sensors have excellent performance. Several applications in civil engineering already employ this technology. For instance, deformation and vibration of a broken bridge in Berlin, Germany, are being measured. Dams in Germany and Switzerland have joints between concrete panels that are shifting. A few bridges in the US and Canada are being monitored for vibration and corrosion. In Swiss pilings, tunnels, and anchoring cables, rock slides are being identified. In Japan, numerous structures and bridges are also under surveillance and section of a smart bridge shows in the Figure 1.5 [6].

1.4.4 Actuating Component

The design of the component of actuation, like those of the sensing component, depends on the type of actuation that is needed, such as physio-chemical factors. Environmental factors like temperature, magneto-electrical phenomena, rust, and the dimension and mechanical characteristics of the interface must all be taken into account in this situation. Actuators, pumps, heaters, dampers, and other regulating devices, as well as a number of novel materials like those previously mentioned, can all create movement [6].

1.4.5 Sensors and Actuators

Sound-based devices that work as acoustic actuators and microphones serve as acoustic sensors in active noise control applications. Accelerometers and the two most common types of actuators are additional devices that can be used. When little frequency, a large amount of stress, and displacements are required, hydraulic and pneumatic actuators are used [6].

1.4.6 Command and Control Unit

The command and control unit is in charge of managing daily operations and is in charge of using a real-time communication network to monitor the system's integrity and overall health. The unit manages a collection of integrated non-destructive evaluation instruments, operates operational and control devices, and manages optical fiber sensors and actuators. This serves as the intelligent structure's brain and serves two different and fundamental purposes.

1.4.6.1 The Processing Function

The processing function depends on physical parameters such as wavelength, frequency, and its surroundings where data are received. It is possible to apply specialized algorithms to regulate behavior or find damage. To effectively process the raw data, pattern recognition algorithms and neural networks with fuzzy logic might be used. Expert systems can also organize, classify, and store the data after they have been retrieved.

1.4.6.2 The Analysis Function

This work includes carefully and intelligently reviewing the raw data. It will use the outcomes of the analysis described above to evaluate the state of the structure. In this study, specific variables, items, or traits are localized and identified in comparison to predetermined threshold levels. Action controlling devices identify the suitable correlation between raw data and response. These functions are operated using specialized algorithms.

1.4.7 Automobile Sector

1.4.7.1 Noise Reduction in Vehicles

In conventional fabric or material production, filaments of piezoelectric ceramic fiber are fashioned into different shapes to reduce vibrational property in helicopter rotor blades and vehicle base, as well as noise in vehicles.

1.4.8 Military Areas

There are several specific military uses of intelligent materials and intelligent systems:

1.4.8.1 Smart Skin

A T-shirt constructed of a specific tactic material that can pick up various frequencies from the human body during combat, such as the recognition of bullet strikes, might be worn by soldiers. Then, it may signal the type of wound or injury, assess its severity, choose whether to act urgently, and even take action to stabilize the wound.

1.4.8.2 Autonomous Smart Systems

Future fights will involve intelligent vehicles on land, at sea, or in space. Intelligent materials will increase observation, target identification, and battlefield awareness thanks to their sensors, actuators, and sophisticated controls.

1.4.8.3 Stealth Applications

The B-2 stealth bomber and the F-117 stealth fighter are excellent examples of novel machinery in which the smart vehicles indicated above can be used for defense purposes. These systems are also required for the rapid and accurate recognition of underwater or space stealth targets [6].

References

1. Shashi Bahal.; Himanshu Nagar.; Inderpreeet Singh.; Shankar Sehgal.; Smart materials types, properties and application : A review. *Mater. Today: Proc.* 2020, doi: 10.1016/j.matpr.2020.04.505.
2. H.S. Tzou.; H-J. Lee.; S.M. Arnold, A.; Smart materials, precision Sensors/ Actuators, Smart Structures, and Structronic System. *Mech. Adv. Mater. Struct.* 2010, doi: 10.1080/15376490490451552.
3. Sabri Centinkunt.; *Mechatronics with experiments*; Wiley, 2015, pp. 1-3.
4. Robert H. Bshop.; *The Mechatronics Hand Book*; ISA, 2002, pp. 1-25.
5. A Spaggiari.; D Castagnetti.; N Golinelli.; E Dragoni and G Scire Mammano.; Smart materials: Properties, design and mechatronic application. *J. Mater.: Design Appl.*. 2016, doi:10.1177/1464420716673671.
6. Georges Akhras.; Smart material and smart system for the futrue. *Can. Mil. J.*. 2000.
7. Pai ZHENG.;Honghui Wang.; Zhiqian Sang.; Ray Y.Zhong; Yongkui LIU.; Chao LIU.; Khamdi Mubarok.; Shiqian YU.; Xun Xu.; *Samrt manufacturing systems for industry 4.0: Conceptual framework, scenarios, and future perspectrives*. Springer. 2018, doi:10.1007/s11465-018-0499-5.
8. JF McCabe.; Z Yan.; OT Al Naimi, G Mahmoud.; SL Rolland.; Smart materials in dentistry. *Aust. Dental J.* 2011, doi: 10.1111/j.1834-7819.2010.01291.x.
9. Hyun Chan Kim.; Seongcheol Mun.; Hyun-U Ko.; Lindong Zhai.; Abdullahil Kafy and Jaehwan Kim.; Renewable smart materials. *IOP Publishing*. 2016, doi:10.1088/0964-1726/25/7/073001.
10. Sytze J. Buwalda.; Kristel W.M. Boere.; Pieter J.; Dijkstra.; jan Feijen.; Tina Vermonden.; WimE. Hennink.; hydrogels in a histrorical perspective: From simple networks to smart materials. *Elsevier J. Control. Release*. 2014, doi:10.1016/j.jconrel.2014.03.052.

2

Fabrication and Characterization Tools for Organic Semiconductors as Smart Materials in Optoelectronic Device Applications

Minakshi Sharma[1]*, Chandra Mohan Singh Negi[2], Parvez Ahmed Alvi[2] and Saral Kumar Gupta[2]

[1]Department of Physics, Shri Khushal Das University, Hanumangarh, Rajasthan, India
[2]Department of Physical Sciences, Banasthali Vidyapith, Banasthali, India

Abstract

Global warming continues to build up as the evidence clarifies that we will have to find techniques to produce electricity without the expulsion of carbon dioxide and all other types of greenhouse gases. Fortunately, there are numerous ways to produce renewable energy sources without endangering the environment. In comparison to silicon and other inorganic material-based devices, organic electronics have the potential to produce electronic devices that are more energy-efficient and environmentally friendly, which would help to create more efficient and sustainable electronic world. Modern electronic and optoelectronic devices including photodiodes, organic light-emitting diodes (OLEDs), organic solar cells (OSCs), and organic field effect transistors (OFETs) are made mostly from organic materials, primarily conjugated polymers (CPs). The benefits of facile solution processing, strong conductivity, great mechanical flexibility, and superior environmental stability are all provided by CPs. However, due to its weak electrical conductivity and persistent stability concerns, the promise of providing inexpensive, large-area, comparably efficient electronic devices by organic/polymer materials has been hampered. By mixing polymers with other species, such as inorganic oxides and compounds, their material characteristics can be dramatically improved. The conjugation system of carbon-based compounds, such as carbon nanotubes (CNTs), which also have a large conjugating system with graphene-based material and oxides, has increased electrical conductivity, impact resistance, mechanical strength, and heat resistance. Also, there are several possibilities to modify the low-band gap (LBG) polymer or conjugated polymers to enhance the optoelectronic properties. Using the LBG polymer with fullerene and its derivative in the active layer of the devices leads to improved optoelectronic properties. The aim is to impart a good knowledge of the materials used for the optoelectronic applications. After a general overview of materials, this chapter describes tools for compound synthesis, solution preparation, and device fabrication. We then described detailed methods for fabricating substrates and devices using various techniques and tools. Finally, we described the characterization tools used to study different properties of the samples.

***Keywords*:** Organic semiconductors, photoactive layer, conjugating polymers, fabrication, electrical properties

**Corresponding author*: minakshisharma9869@gmail.com

Upendra Kumar and Piyush Kumar Sonkar (eds.) Smart Materials for Science and Engineering, (17–38) © 2024 Scrivener Publishing LLC

2.1 Introduction

The increasing effect of global warming leads scientists to demonstrate that we must find ways to generate electricity without generating carbon dioxide or any other greenhouse gases [1, 2]. Luckily, there are many ways to generate renewable energy sources that do not harm the environment [3]. One of this century's most severe technological challenges is the development of sparingly practical and maintainable energy capitals [4]. Currently, the world's energy needs are mainly fulfilled by the energy supply from fossil fuels, including coal, oil, and natural gas. According to the U.S. Energy Information Administration, fossil fuels account for roughly 83% of the world's energy consumption, with petroleum, coal, and natural gas making up about 37%, 20%, and 26% of that total, respectively [5].

A further obstacle to the worldwide goal of lowering carbon dioxide emissions and global warming is the substantial amount of greenhouse gases emitted during the burning of fossil fuels [6]. Solar energy is undoubtedly one of the most promising alternatives to traditional energy resources in this context. Photovoltaic (PV) systems convert sunlight into electricity and can be used to harness solar energy. Despite this, the contribution of inorganic solar cell technology to the total world energy supply remains below 0.7% because the cost of producing PV power is still higher than the cost of producing electricity from conventional sources [7]. Despite their poorer power conversion efficiency, organic devices have the potential to be a low-cost, easy-to-process alternative to the present, primarily crystalline silicon-based, inorganic semiconductor industry [8].

2.2 Overview of Organic Semiconductors

The primary difference between inorganic and organic semiconductors is that organic semiconductors have carbon-based materials in their structure, and inorganic semiconductors are non-carbon-based materials such as silicon and germanium. Organic semiconductors are used in compound or oxide form in optoelectronic devices like CdS, SiO_2, InP, etc. The classification of smart semiconductor materials depends on the structure, as shown in Figure 2.1. Organic semiconductors distinguish from inorganic semiconductors in those

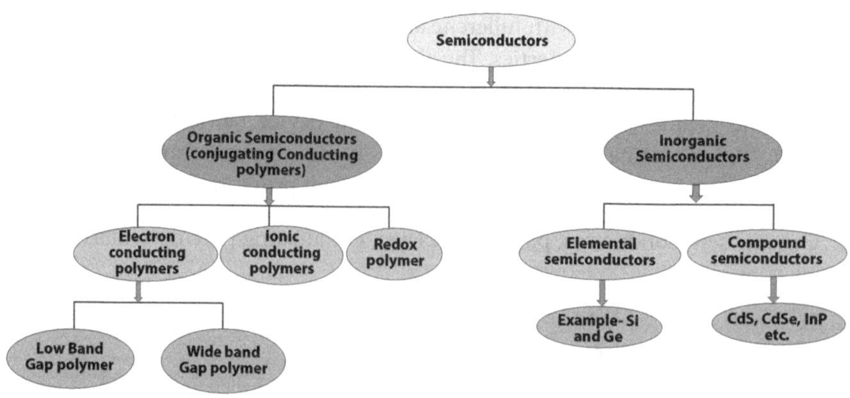

Figure 2.1 Different classifications of semiconductors depend on the structure.

organic molecules form solids by van der Waals interactions. In contrast, inorganic solids have covalent molecular bonds [19]. Organic semiconductors are a type of unsaturated hydrocarbon that may be readily divided into two groups: small conjugated molecules and conjugated polymers [9]. Long chains of alternating single and double bonds with overlapping p-orbitals are known as conjugated polymers. Conducting polymers are frequently referred to as quasi-one-dimensional metals due to weak van der Waals-type intrachain coupling interactions and the solid intrachain contacts (strong covalent bonding along the chain), which cause electrons to delocalize along the polymer chain [10].

2.3 Optoelectronic Properties of Conjugated Polymers

Organic semiconductors have a dielectric constant of approximately 3, which is significantly less than conventional inorganic semiconductors. Due to low dielectric constants, organic semiconductors' charge carriers have a low lifetime, which causes the small diffusion length of the charge carriers [11]. Due to electrostatic coulombic attraction, oppositely charged carriers electrons and holes are bound together, and this bounded pair of bound electron–hole is known as an exciton. Semiconductors and insulators both have electrically neutral quasi-particle excitons. Excitons are usually produced by charge carriers' injection or generation (photo or thermal excitation) [12]. Exciton energy is slightly less than the unbound electron and hole due to other electrons and holes in the surroundings.

When inorganic semiconductors are exposed to light, photons get absorbed in the material, and electrons, by gaining energy, excite from the valence band to the conduction band, leaving behind the corresponding holes in the valence band [13]. These photogenerated electrons and holes then diffuse or drift to the appropriate electrodes, where they are eventually removed to provide the photocurrent [14]. Due to their opposite charge on electrons and holes, some recombine, leading to a loss in the photocurrent. An organic semiconductor-based optoelectronic device possesses donor and acceptor material in the active layer sandwiched between the two electrodes as anode and cathode. The photogenerated charge carriers move by the drift and diffusion process as the concentration gradient causes diffusion, and drift is supported by the built-in electric field between the electrodes [15].

2.4 Optoelectronic Devices

The structure of a typical optoelectronic device is shown in Figure 2.2a. Firstly, an indium tin oxide (ITO)- or fluoride tin oxide (FTO)-coated glass substrate is used to fabricate the device. Above this layer, a buffer layer hole transport layer (HTL) is usually coated, which should have good hole-transport properties and a work function higher than ITO. Poly(ethylene-dioxythiophene):polystyrene sulfonate (PEDOT: PSS), MoO_3, V_2O_5, and NiO are some materials used for this purpose. PEDOT:PSS polymer possesses high conductivity and work function [16]. Using a suitable method, the light-absorbing active layers of organic semiconductors are deposited on HTL. A spin coating process coats a combination of donor and acceptor polymer over the HTL active layer. An electron transporting layer (ETL) is coated on the active layer using the spin coating technique. E-beam or thermal evaporation in a vacuum coats the active layer with an electron-collecting top

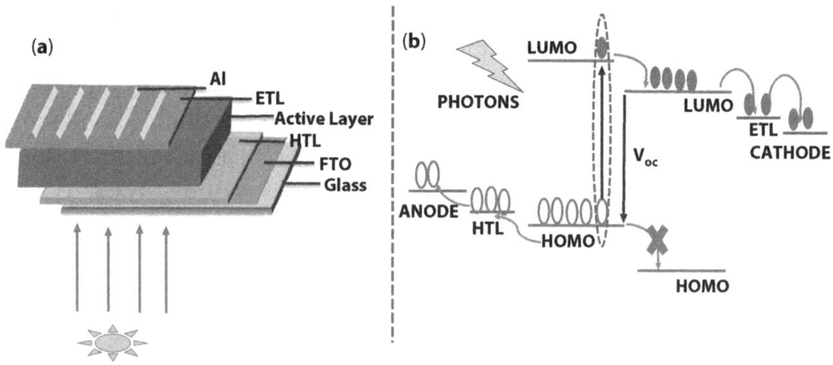

Figure 2.2 (a) Device structure and (b) the device's working principle in the illumination condition.

electrode (cathode) [17]. The open circuit voltage (V_{oc}) is the energy gap between the highest occupied molecular orbital (HOMO) of the acceptor and the lowest unoccupied molecular orbital (LUMO) of the donor material. Energy conversion by an optoelectronic device can be broken down into four critical processes as shown in Figure 2.2b: (1) light absorption induced exciton generation, (2) the diffusion of excitons to the junction (interface), (3) dissociation of the excitons at the interface by the built-in field or an applied field, and (4) transportation of charge carrier and collection to their respective electrodes. Therefore, a clear fundamental understanding of these processes is necessary to design and develop an efficient and stable optoelectronic device.

2.4.1 Organic Light-Emitting Diodes (OLEDs)

The layer-to-layer deposition of organic or hybrid material can be carried out between two electrodes to fabricate the OLED. OLEDs use less energy compared to backlight-dependent display technologies [18]. Due to their numerous advantages, such as rapid response, large viewing angle, flexibility, and low operating voltages, they are becoming a hot research topic. Due to their outstanding features, they are used in various applications, together with flat panel displays, solid-state lighting, etc. In the early 1990s, researchers demonstrated the study of graphene-based polymer heterostructures for application in photodetectors' top electrodes. They showed that the easily synthesized solution-processed polymer might be used to make large-area organic light-emitting displays. This was quickly followed by developing OLED utilizing a soluble PPV derivative. Novel television displays produced using polymer-printing techniques are now thoroughly commercially available, based on these initial experiments on polymer OLEDs [19].

2.4.2 Organic Photodetectors (OPDs)

Organic photodetectors (OPDs) are becoming more popular in light-sensing applications due to their broad absorption spectrum, high photo-generation yield, inexpensive fabrication costs, lightweight, and flexibility [1]. OPDs provide several critical advantages over amorphous silicon detectors. Most notably, its compatibility with the thermomechanical characteristics of plastic allows solution processing of large-area image sensors on various flexible substrates. Whole absorbing OPD arrays with semitransparent top electrodes provide more excellent resolution than the imaging systems built on glass substrates, where light refraction and accompanying dispersion result in more optical cross-talk between single pixels [20]. Light is detected in OPDs by absorbing incident photons with an energy equal to or greater than the photoactive material's optical bandgap.

The active layer of OPDs, like that of many organic photovoltaic (OPV) devices, is based on a bulk heterojunction (BHJ) structure, having amalgamated percolating networks of phases of electron donor and acceptor materials, and having a large interface area that facilitates the dissociation of the exciton. Photodetectors are mainly used for security in conjunction with other optical devices [21]. The p-n photodiode, PIN photodiode, avalanche photodiode, and phototransistor are examples of photodetectors. A photodiode is a photodetector that is made up of semiconductors as p-type and n-type that are joined together to form a junction. It depletes the charge carriers, resulting in a high impedance zone. It is primarily used for reverse-biased conditions, and as it allows current to flow through the circuit even in reverse bias when illuminated by light, it can be used as a switch or relay. An inherent layer of high resistivity is inserted between the p-type and n-type materials to boost response speed; these photodiodes are known as PIN photodiodes. Numerous organic and organic/inorganic hybrid photodetectors with excellent linearity of the output current for incident light, rapid response, extensive spectrum range, and lower temperature sensitivity have been developed in recent years [22].

2.4.3 Organic Solar Cells (OSCs)

Solar energy harvesting depends on the chemical and physical interactions at the interfaces of the devices, which are used to harvest the light and transport the charge carriers [23]. Organic–organic and organic–inorganic interactions are both possible. OSCs have shown promises of low cost, conformability, lightweight, and low-temperature simple processing. It is now becoming the most thrilling application of organic materials. However, it is still a challenge to build a comprehensive understanding of polymers' physical and chemical properties. Also, numerous variables can influence the device's performance, such as polymers' physical properties and the active layer morphology. Continuous efforts are required to overcome the challenges and make a product commercially available.

Traditionally, p-n junction solar cells have shown their potential to convert solar energy into electricity directly. These devices can offer high power conversion efficiency (PCE). However, the production cost is very high. More recently, organic solar cells have accompanied two types of architecture. The first consists of multilayered donor–acceptor materials.

The other is based on the composites of the donor–acceptor materials, also called BHJ; the schematic diagram of OSC with BHJ as active layer material is shown in Figure 2.3. More electrons are present in the donor than in the acceptor. As light excites the donor, the acceptor becomes negatively charged while the donor becomes positively charged. Via charge transfer between molecules, positive charges travel by hopping holes, whereas negative charges travel by hopping electrons. Due to the movement of charges in the biological layers, the electrodes must also shift charges.

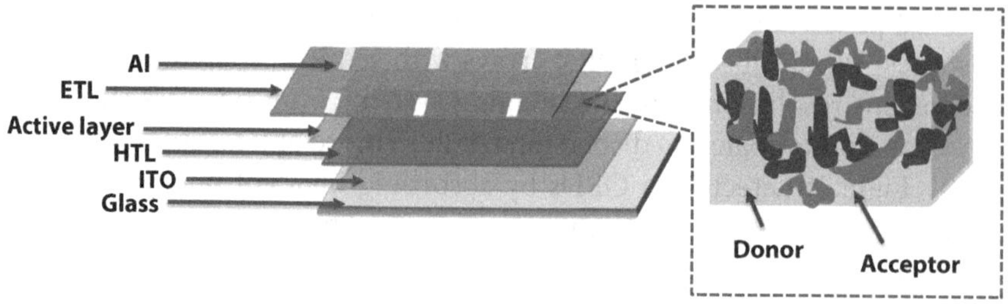

Figure 2.3 Schematic diagram of a bulk heterojunction-based organic solar cell (OSC).

2.5 Overview of Smart Materials

This section discusses the various smart materials used to fabricate optoelectronic devices. The objective is to provide sufficient knowledge of the smart materials being used. Following a general description of materials, the chapter describes the instruments synthesizing the composites, preparing the solution, and fabricating the devices. Next, we consider the methodology to prepare the substrates and devices. Finally, the characterization instruments used to investigate the various properties of the samples have been discussed.

2.5.1 Specification of Smart Materials

Table 2.1 lists the raw materials and their specifications, which have been used to carry out the research work. Some other high-grade chemicals, reagents, and solvents that have been utilized are 2-Propanol (C_2H_8O), ethanol (C_2H_5OH), acetone (C_3H_6O), ortho dichlorobenzene ($C_6H_4Cl_2$), 1,8 di-iodooctane ($C_8H_{16}I_2$), dimethyl sulfoxide (C_2H_6OS), chlorobenzene (C_6H_4Cl), deionized water, acetonitrile (C_2H_3N), dimethyl fluoride (DMF), zinc dust, etc.

2.5.2 Carbon Nanotubes (CNTs)

The rolled graphene sheet with sp^2 hybridization in tubular form is known as a carbon nanotube. Based on their chirality, they can be semiconductors or metallic [24]. In the case of semiconducting CNTs, the electric current density is approximately 1000 times higher than that of metallic CNTs. Due to the presence of carbon atoms, they are lightweight and chemically stable. CNTs have been successfully applied in electrochemical sensors, supercapacitors, Li-ion batteries, and energy storage devices. In optoelectronic devices like solar cells, photodiodes, and photodetectors, CNTs can be used as hole and electron-transporting layers.

They can also be used in the active layer as a composite to increase absorption, strength, and conductivity. CNTs can have various forms, depending on size and diameter, like single-wall (SWCNTs), multiwall (MWCNTs), and double-wall carbon nanotubes (DWCNTs). Multiwalled carbon nanotubes are zero-band gap metals. They are repeatedly considered an infinite long sheet roll, or many single-wall tubes are framed concentric. They are incorporated into the polymer matrix to form the composites and significantly improve the

Table 2.1 Few smart materials for optoelectronic applications.

S. no.	Raw materials	Purity	Manufacturer
1	SWCNT	>99%	Sigma-Aldrich
2	MWCNT	>99%	Sigma-Aldrich
3	rGO	>99%	Technistro
4	P3HT	>99%	Sigma-Aldrich
5	PEDOT:PSS	>99%	Sigma-Aldrich
6	MoO_3	>99%	Sigma-Aldrich
7	C_{60}	>99%	Sigma-Aldrich
8	ZnO	>95%	Sigma-Aldrich
9	PANI-ES	>99%	Sigma-Aldrich
10	ICBA	>99	Sigma-Aldrich
11	PCBM	>99	Sigma-Aldrich

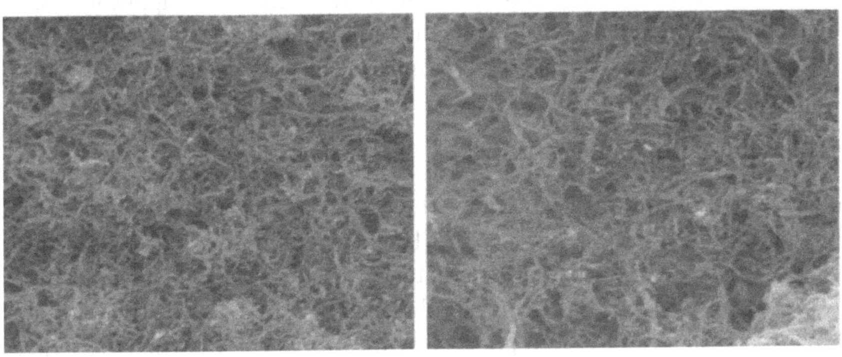

Figure 2.4 FESEM image of dispersed (a) SWCNTs and (b) MWCNTs.

conductivity of polymers [25]. The dispersion of CNTs can be observed using the FESEM image, as shown in Figure 2.4.

2.5.3 Reduced Graphene Oxide (rGO)

Reduced graphene oxide is in the form of graphene oxide in which the oxygen content is decreased chemically to improve the optical and electrical properties [26]. The dispersibility of rGO is lower than that of graphene oxide and all the functionalized graphene-like (amine and phosphorus-doped) graphene. The polar solvents, such as DMF and water, are the most suitable medium to disperse the rGO. We obtained the homogeneous solution of rGO as shown in Figure 2.5. We dissolved 0.1 mg/ml rGO in DMF or added 3:2 of Propan-2-ol and ethylene glycol and placed it in the probe sonicator for several hours [27].

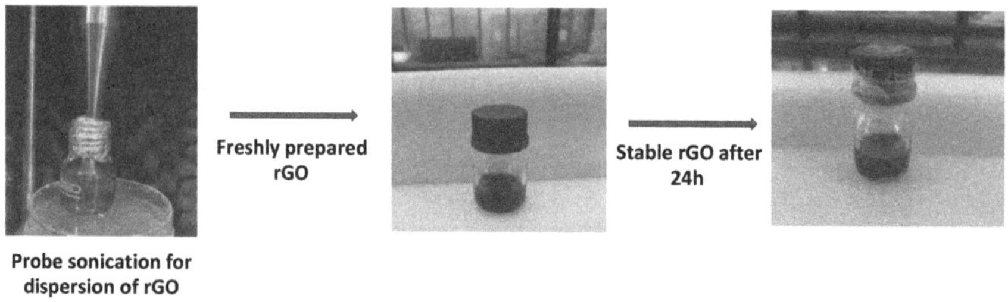

Figure 2.5 The dispersion of rGO using a probe sonicator.

2.5.4 Poly(3-Hexylthiophene) (P3HT)

The P3HT is the Π-conjugated polymer with a polycrystalline structure and broad optical absorption in the visible band. It is soluble in various organic solvents; therefore, it can be used as an active p-type material for organic-based optoelectronics such as photovoltaic cells [28]. P3HT has become necessary because of its solubility, good conductivity, and tunable photoluminescence. Among the conjugated polymers, poly (3-hexylthiophene) (P3HT) is a suitable donor material, having the highest hole charge carrier mobility with a band gap of 1.9 eV. The P3HT:CNTs hetero-structures have been intensively studied for organic photovoltaic applications. The P3HT-CNT composites exhibit attractive physical, optical, and electrical properties, making them the material of choice for various electronic applications. The blending of P3HT with fullerene derivative forms the bulk heterojunctions, which offer enhanced exciton dissociation, leading to substantially improved electrical and optical properties [29].

2.5.5 Fullerenes and Its Derivative

Fullerene C_{60} is one of the most widely used electron acceptor materials in organic solar cells (OSCs). It has been used as an acceptor layer in the bilayer solar cells, where a P3HT or another polymer, having a p-type character, is used as a donor layer, whereas in the bulk-heterojunction OSCs, it has been blended with P3HT or any other donor. It exhibits a well-symmetrical structure and good electron mobility [30]. The emergence of ICBA (indene-C_{60}-bisadduct), $PC_{60}BM$, and its corresponding C_{70} derivative ($PC_{70}BM$) as acceptors in OPVs has been seen in the past decade. The $PC_{70}BM$ exhibits more excellent absorption than the $PC_{60}BM$; therefore, it has recently attracted much interest.

Nevertheless, it involves more tedious synthesis methods than C60, making it more costly and restricting its use. The fullerene derivatives' electrochemical properties and energy level have greater importance in OSCs. For instance, V_{oc} is dictated by the maximum separation between fullerene acceptors' LUMO and polymer donors' HOMO. Hence, the LUMO energy level of the fullerene derivatives can be the critical parameter in selecting the accepter material to match with donor materials to offer unprecedented performance [31].

2.5.6 Polyaniline (PANI)

Another conducting polymer we have used in this research is PANI. It possesses many impressive properties, including facile synthesis, lightweight, cost-effectiveness, tunable properties, high conductivity, and environmental stability, and, therefore, has become one

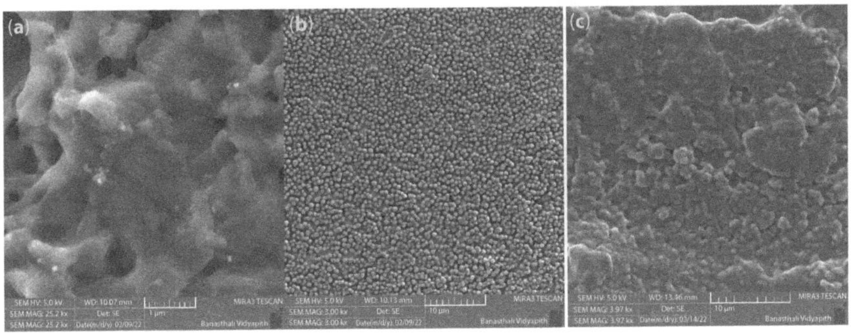

Figure 2.6 (a) FESEM image of PANI-ES. (b) Pure ICBA. (c) Composite of PANI-ES: ICBA-based thin films.

of the most attractive conducting polymers. Therefore, it has many applications, such as chemical sensing, actuators, fuel cells, and optoelectronic energy storage devices.

The PANI-based device has recently demonstrated a high specific capacitance than the electrochemical double-layer supercapacitors. Depending on the pH value, PANI is available in two forms; PANI under acidic conditions (pH less than 4) exists as polyaniline emeraldine salt form (PANI-ES) and converts to polyaniline emeraldine base (PANI-EB) if pH is greater than 4 [32]. In this section, the FESEM image of PANI-ES, a P-type conducting polymer, pristine ICBA, and a composite of both are shown in Figure 2.6.

2.6 Methods and Techniques

This section presents the numerous methods used to prepare various polymer and composite solutions and deposit the thin films with the corresponding solutions.

2.6.1 Sonication Process

Sonication is when a sound wave is used to perturb the particles in the solution to convert electrical energy into vibration to break the substances in the solution. The ultrasonication process works on a frequency greater than 20 kHz, and this energy passes to the particles present in the solution for occurring chemical reactions. The ultrasonication process can be used to clean the samples or the substrates, and the probe sonicator is used to mix or break the molecules in the solution. The presence of intense sound waves generates alternate low-pressure (rarefaction) cycles in which bubbles form voids or small vacuum in the liquid solution, which absorbs energy, or at high pressure (compression), in which the bubbles collapse and no longer absorb any energy.

There are two types of sonication processes: direct and indirect sonication. Figure 2.7a and b shows the Probe sonicator and Ultrasonic cleaner photograph, available at Banasthali Vidyapith. In the probe sonication process, the probe is directly contacting the sample. It is applicable for small volumes (~10 ml). In this process, only a tiny amount of energy is required to agitate the particles in the solution. The indirect sonication process includes the ultrasonic cleaner in which the sample is not in direct contact with the sound waves. However, both sonication processes are beneficial when it is impossible to stir the sample and are in an inaudible frequency range [33].

Figure 2.7 Photograph of (a) Probe sonicator. (b) Ultrasonic Cleaner in Banasthali Vidyapeeth.

2.6.2 Centrifugation

Centrifugation is the process that helps to separate the mixture from the immiscible liquid solution by spinning, occurring due to the centrifugal force. The centrifuge consists of a rotor, a refrigerated chamber, glass or plastic tubes in which solution is placed, and an electric motor. The separation rate of the mixture from the liquid solution depends on the particles' shape, size, density, and viscosity [34]. The centrifugation rate is expressed as the angular velocity or the revolution per minute (RPM), and g shows the respective centrifuge force. The RPM is proportional to r, the radius from the center of the rotor to the sample tube. Centrifugation is used to separate the mixture and has many applications in the hydrodynamics of macromolecules. The RPM is proportional to r, the radius from the center of the rotor to the sample tube. This relation is expressed as

$$RPM = \sqrt{\frac{g}{r}}$$

2.6.3 Spin Coating Unit

Many techniques like doctor blade, drop-casting, spray printing, dip coating, spin coating, and roll-to-roll printing exist to deposit the thin film of solution on a glass substrate. Spin coating provides the uniform thin film of the solution on a substrate with the desired thickness. The spin coating unit used in this work is displayed in Figure 2.8a. As the name illustrates, the coating solution is first placed on the substrate using the pipette and deposited on the prescribed rpm and time duration to obtain a particular thickness.

Figure 2.8b shows the sub-steps for the solution deposition. The acceleration time shows a very small or negligible effect on the thickness of the dispersed solution, but if it has a higher value, it affects the radial uniformity of the solution. Further, a centrifugal force throws the liquid radially outward. The solvent evaporation process increases the viscosity of the film. To control the flow of dispersed solution, a balance between centrifugal and viscous force is required. The rotational speed or the RPM affects the thickness of the deposited film. Generally, a reduced thickness value can be obtained at higher RPM values for a particular acceleration time with a slight variation in the thickness [35].

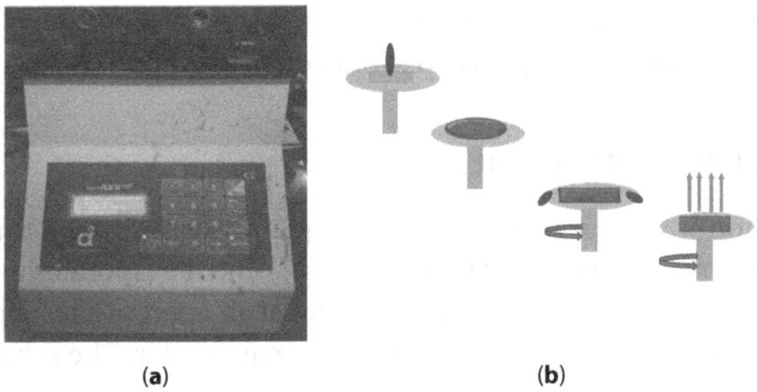

Figure 2.8 (a) Photograph of spin coating. (b) Process for sample preparation using spin coating technique.

2.6.4 Glove Box

A glove box comprises an enormous container, also known as a dry box, with long arm-length gloves, which help to work inside the inert atmosphere. The glove box has a transparent chamber with two ports and a window. The glove box is usually filled with the gases like nitrogen, although argon and helium gases can also be used if that particular material reacts with the nitrogen gas. It is used in research work to mix materials or isolate the deposition of materials on a specific substrate to avoid oxidation or contamination. To prevent the presence of oxygen inside the chamber, regular purging should be done. The pressure inside the glove box is relatively higher than the surroundings, so any tiny leak causes the leakage of inert gas instead of letting air inside the box as shown in Figure 2.9. Most commonly, two types of glove boxes exist. One is used to keep hazardous materials like highly radioactive

Figure 2.9 Glove box photograph in Banasthali Vidyapith.

materials that must be placed in isolation, and the other is for the specific material that requires purity and an inert atmosphere to avoid contamination [36].

2.6.5 Thermal Evaporation Unit

Thermal evaporation is a commonly used technique for depositing a thin film on a solid substrate. When the material is heated to its melting point, it evaporates under the high vacuum in the chamber and is deposited on the samples at an appropriate height. The thickness of the deposited material varies from millimeter to nanometer. The schematic diagram of the vacuum chamber is shown in Figure 2.10.

This chamber consists of a clamping system to hold the sample holder on which samples are placed for deposition. Also, tungsten filaments are used to place the target material [37]. A source of low tension is used to heat the material, which is connected through the filament. After placing the sample, the chamber door is closed, and firstly, roughing is used to pump the system or to eject all the impurities from the chamber. If the foreign particle comes in contact with the evaporated atoms, it may react if oxygen is present in the chamber, and the evaporated aluminum will become aluminum oxide. That is why a significantly high vacuum of 10^{-6} bar is required for the deposition of the thin film.

To create the vacuum, two types of pumps are used. Generally, the backing pump evolved in the vacuum system; the diffusion pump is coupled with the oil-sealed rotary pump. Using a rotary device, the mechanical rotary pump is used to evacuate the air from the chamber. At atmospheric pressure, the air in the chamber is dragged out initially with the help of the rotary pump. The first step is to exhaust the air using the rotary pump as the diffusion pump cannot work at atmospheric pressure, so it discharges to the rotary pump and finally discharges to the atmosphere [38]. The photograph of the thermal Evaporation Unit is displayed in Figure 2.10.

The diffusion pump starts when the pressure reaches 10^{-2} bar using a rotary pump as they are connected in series. Therefore, rotary pumps are also known as fire pumps, roughing pumps, or backing pumps. After that, the diffusion pump helps to attain a high vacuum in the chamber, less than or equal to 10^{-6} bar. Two gauges, the Pirani gauge and the Penning

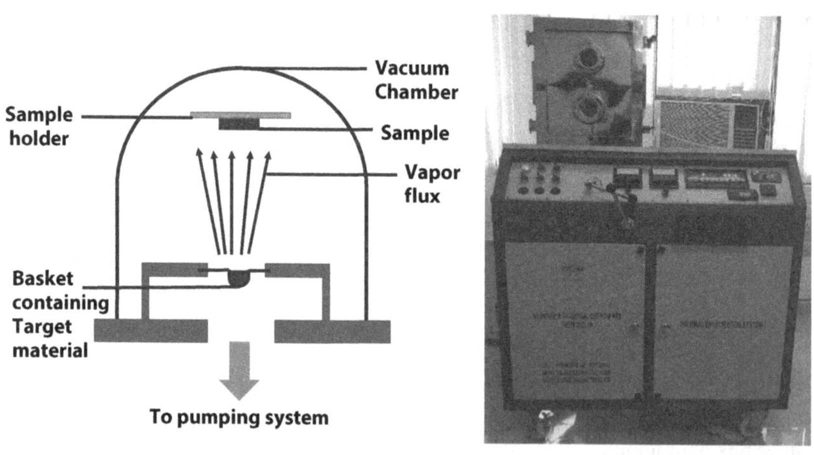

Figure 2.10 Diagram of the vacuum chamber in the thermal evaporation unit in Banasthali Vidyapeeth.

gauge, are used to monitor the pressure in the pumps' reference. Once the required pressure is achieved in the high vacuum, the low-tension meter is switched on, and the material evaporates. After reaching the melting point, the material evaporates and is deposited on the substrate.

2.6.6 Sample Preparation Processes

The step-by-step procedure shown in Figure 2.11 was followed to fabricate the thin film optoelectronic devices. Active layers consist of a donor, acceptor, or donor–acceptor blends, where light absorption and charge carriers are generated. In the present work, we have investigated the various inorganic and organic materials as the device's active layer, like CNTs: rGO composites, P3HT:rGO, and PANI, and studied their various properties.

2.7 Methodology

2.7.1 Substrate Preparation and Device Fabrication

Firstly, the etching of ITO substrates was done using zinc dust and 1 M HCl, and the substrate was subsequently washed with the leveling soap solution. These substrates were dipped in deionized water and cleaned with an ultra-bath sonicator for 30 min, followed by a 30-min ultrasonic bath with acetone and 2-propanol, respectively. In step (a) the bare ITO of size 2×2 is taken firstly, then in step (b) the etching of the ITO is done, in step (c) deposition of various layer is done then step (d) is done to sketch the edge part of the ITO, in step (e) the deposition of electrode is done using the mask shown in (f) of size 2×2 to form the

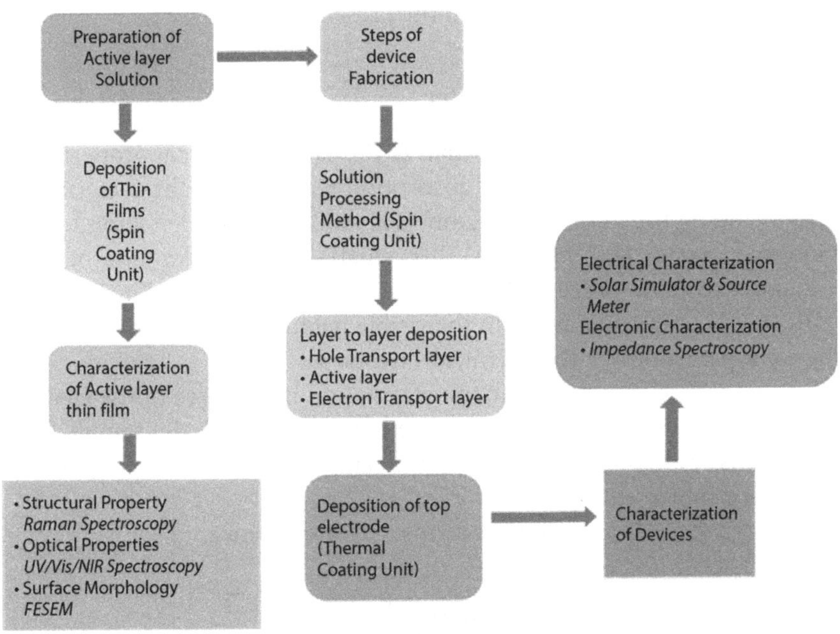

Figure 2.11 Steps followed for the fabrication and characterization of the device.

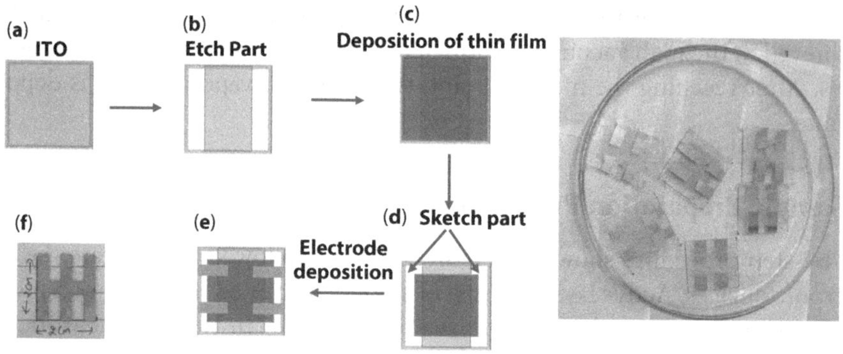

Figure 2.12 Substrate preparation is followed by the device fabrication and photograph of the final fabricated devices.

final device. The ITO was treated with UV ozone plasma for 20 minutes. The hole transport layer was spin-coated on the ITO substrate at a specific rpm for a specific time to obtain the desired thickness and annealed at 120°C for 20 minutes. After that, the active layer solution was poured over the HTL and spin-coated to get the desired thickness. Afterward, spin coating deposited the ETL with the appropriate material. Eventually, electrode (Al) deposition was finally done using a thermal evaporation unit. Figure 2.12 illustrates the block diagram of the steps to fabricate the device.

2.8 Characterization Techniques

The structural, morphological, optical, and electrical characterization techniques have been discussed briefly in this section and used in the present study.

2.8.1 UV/Vis/NIR Spectrophotometer

A UV-Vis spectrophotometer is used to analyze the samples' optical properties. The technique is mainly used for examining the absorption spectrum of the material in different wavelengths, including the ultraviolet region (200–400 nm), visible region (400–800 nm), and near infrared region (800–1100 nm). It evaluates the concentration of the materials respective to the absorption range [39]. Beer–Lambert's law is the working principle behind UV/VIS/NIR spectroscopy operation. In the visible field, absorption or reflectance directly impacts the color perception of the chemicals involved.

The molecules undergo an electronic transformation in this area of the electromagnetic spectrum. It measures the light intensity before it passes through the sample (I) and compares it to its intensity after it passes through the sample (I_o). The ratio (I/I_o) is called transmittance expressed in percentage (%T).

However, absorbance (A) is defined in terms of transmittance.

$$A = \varepsilon \times c \times \ell = \log 10\left(\frac{I}{I_o}\right) = -\log\left(\frac{T}{100}\right)$$

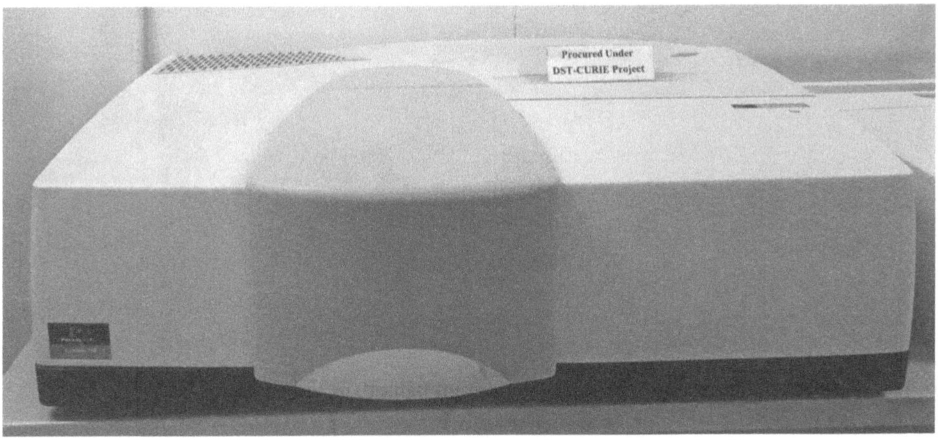

Figure 2.13 Photograph of the UV/Vis/NIR spectrophotometer (Perkin Elmer Lambda) at Banasthali Vidyapith.

Reflectance can also be measured with a UV-Vis spectrometer. The spectrophotometer is made up of four main components: (i) light source (ii) detector, (iii) sample holder, and (iv) monochromator diffraction grating. The radiation source is operated in continuous mode for the visible wavelengths. The Perkin Elmer Lambda 750 UV/Vis/NIR Spectrophotometer utilized for the experiment is shown in Figure 2.13.

2.8.2 Raman Spectroscopy

Raman spectroscopy can be used to characterize the phase, structure, and crystallinity and study an amalgamation of the dopants and lattice ailment of the sample. Inelastic scattering of monochromatic light, usually from a laser source to the material, is the basic process of this technique. A Raman spectrophotometer has four main components: (i) excitation source (LASER); (ii) sample illumination device; (iii) light collecting optics, wavelength selector (FILTER, spectrophotometer); and (iv) detector [photodiode collection, CCD, or photo-multiplier-tube (PMT)]. The sample is frequently illuminated with a laser beam in the ultraviolet (UV), visible (Vis), or near-infrared (NIR) range, as available with a spectrometer. The scattered light is collected by a lens and sent through an interference filter to a Raman spectrometer, where the Raman spectrum of the investigating samples is captured [40].

A photograph of the DXRxi Raman Imaging Microscope is shown in Figure 2.14. The frequency of the incident light varies because of inelastic interaction. It may be shifted to a higher frequency and a lower frequency as the monochromatic beam is incident on the sample. When the scattered beam has the same frequency as a source, it is known as Rayleigh radiation; the scattered frequency with higher and lower frequencies is known as anti-Stokes radiations and Stokes radiations, respectively. The variability in the frequencies provides information about the modes of the material. This effect is known as the Raman Effect. The samples are in different forms like liquids, gels, powders, thin films, etc., and can be easily characterized. In this technique, the molecule undergoes a momentary distortion

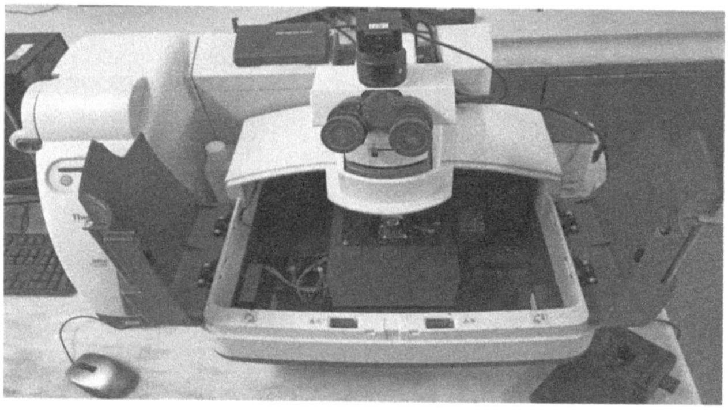

Figure 2.14 Photograph of a Raman spectrophotometer at Banasthali Vidyapith.

of electrons, scattered around a bond by the incoming photon from the laser, resulting in an induced dipole moment in the molecule, accompanied by photon re-emission as it relaxes. As a molecule relaxes, the frequency of specific photons changes (Raman scattering), and they do not return to their original state. Instead, they transition to a new energy state called the Raman shift. Raman shift can be treated as a fingerprint of the materials [41].

2.8.3 Field Emission Scanning Electron Microscopy (FESEM)

FESEM is used to analyze the sample's surface morphology and elemental detail and generate an image at a particular value of magnification ranging from 10× to 300,000×. The field emission cathode in a scanning electron microscope electron gun produces a narrow beam at low and high electron energy, improving spatial resolution and reducing sample charging and damage. The electrons were generated by a field gradient using a field emission source in a vacuum. The beam passing through the electromagnetic lenses was focused on the sample, resulting in various types of electrons from the sample. The detector obtained secondary electrons, and a further image of the sample's surface was created by comparing the strength of the secondary electron beam to the scanning primary electron beam. The resolution achieved by FESEM is ~2 nm and permits the examination of tiny microstructural details, morphological analysis like particle size and shape, interface behavior, fracture studies, quantitative and qualitative elemental analysis, grain orientation, texture, and phase identification of any material.

To analyze the elemental compositional analysis, a facility is attached with the SEM and TEM microscopy, generally known as energy dispersive X-ray analysis (EDAX or EDX). It uses X-ray analysis to investigate the elemental composition of any material. The information regarding elements of the specimen can be obtained based on the energy of this emitted X-ray. EDX provides qualitative and semi-quantitative information on elements in any composition [42]. In the present section, FESEM and EDS investigations of synthesized samples have been done with MIRA 3 TESCAN EDS with Z-Tech, respectively attached with FESEM, represented in Figure 2.15.

Figure 2.15 Photograph of FESEM at Banasthali Vidyapith.

2.8.4 Current–Voltage (I–V) Measurements

A solar simulator measures the I–V characteristics and studies the optoelectronic properties. The device's current and voltage (I–V) characteristics in dark and illumination regions are typically measured for the standard reference conditions [43]. The device's spectrum, intensity, temperature, and area define the conditions. Several factors influence I–V measurements, including a light source, voltage sweep rate, direction, junction temperature, contact to the metallization, instrumentation, and intensity. Various types of simulators in the photovoltaic community approximate a standard reference solar spectrum, and the user compares the intensity of the ASTM process to determine the simulator class. The Sun's radiation is divided into two types: (I) direct radiation (which comes directly from the Sun) and (II) scattered radiation (which comes from the rest of the sky that includes the portion of the radiation that is reflected from the ground). For several conditions, the solar simulator offers the spectral distribution of sunlight. The solar simulator's experimental setup and the Keithley source meter are shown in Figure 2.16a.

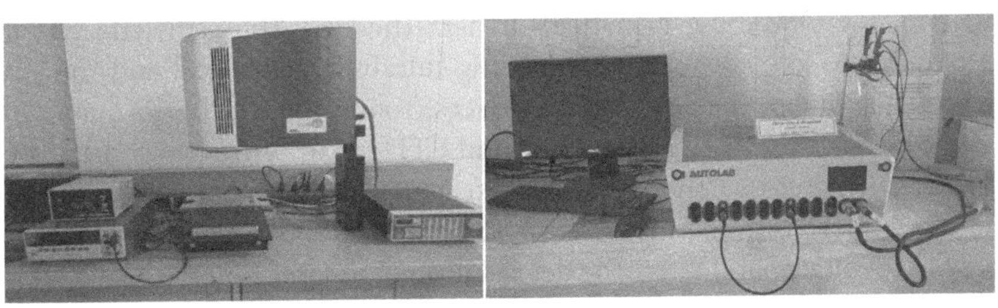

(a) (b)

Figure 2.16 (a) Photograph of the solar simulator and (b) impedance analyzer at Banasthali Vidyapith.

2.8.5 Electrochemical Impedance Spectroscopy (EIS)

The opposition experienced by the circuit to a current during the application of a voltage is called electrical impedance. It represents a more generalized concept compared to resistance. It can be described as the complex ratio of the voltage to the current in the alternating current (AC) circuit. The base-level impedance spectroscopy experiment applies an electrical stimulus to the electrodes and observes the response as the frequency function [44]. Therefore, impedance is frequency-dependent; if the impedance spectrum is depicted as the fundamental part vs. the imaginary part of the impedance, the diagram is known as the Nyquist plot, in which each point represents a frequency. In other graphical representations, the depiction is done as frequency vs. magnitude of Z, known as Bode modulus, and frequency vs. phase of Z, known as Bode phase. The method characterizing the electrical properties of materials and their interfaces by the electronic means of electrodes is called impedance spectroscopy (IS), and the setup is shown in Figure 2.16b. The intention is that the data of experimental impedance is approximately similar to the impedance of the equivalent circuit formed through ideal resistors, capacitors, and inductors. In a typical EIS measurement unit, the sine waves with user-defined frequency and small amplitude are generated by the frequency response analyzer (FRA) module. The obtained circuit parameters also provide information about the interfacial properties of the device. The various circuit parameters can also be correlated with the device's performance; it can help redesign the device to improve performance.

2.9 Conclusion and Future Work

Organic semiconductors have attained a unique role among several materials utilized in modern energy or optoelectronic technology owing to their fascinating optical, chemical, physical, electronic, and mechanical properties. Latest innovations in OSCs have raised hopes for developing more powerful and long-lasting devices at a low cost. In this study, we attempted to improve the performance of devices fabricated using various narrow bandgap smart materials that lead to a more effective and sustainable electronic world than their organic counterparts. The low bandgap material in the active layer with the non-fullerene acceptors in the active layer of the organic optoelectronic devices has been investigated recently to enhance OSCs' performance further. The PET (polyethylene terephthalate) substrates offer some key advantages, including durability, chemical and moisture resistance, and heat stability for the flexibility of devices. To improve the performance of organic optoelectronic devices, the synthesis of HTL and ETL can improve transportation, and the photocurrent also increases. The interface defect density between active layer/ETL and active layer/HTL interfaces determines the recombination and leakage current that directly impacts the photovoltaic parameter. The degradation of organic optoelectronic devices over time can be investigated. We believe that combining improved morphology and control of interfacial properties will bring novel, cost-effective, environment-friendly organic/polymer devices.

References

1. Fuentes-Hernandez, C. *Charge Transport and Photogeneration in Organic Semiconductors: Photorefractives and Beyond*; 2016; Vol. 240. https://doi.org/10.1007/978-3-319-29334-9_2.
2. Chandrakar, A. K. Energy Resources : Indian Scenario An Assignment On. 2015, No. May.
3. Qiu, Z.; Deconinck, G.; Belmans, R. A Literature Survey of Optimal Power Flow Problems in the Electricity Market Context. *2009 IEEE/PES Power Systems Conference and Exposition, PSCE 2009*, 2009, 1–6. https://doi.org/10.1109/PSCE.2009.4840099.
4. Of, N. S. Renewable Sources of Energy. *Renewable sources Energy*, 1987, 96–110.
5. Gielen, D.; Boshell, F.; Saygin, D.; Bazilian, M. D.; Wagner, N.; Gorini, R. The Role of Renewable Energy in the Global Energy Transformation. *Energy Strategy Rev.*, 2019, *24* (June 2018), 38–50. https://doi.org/10.1016/j.esr.2019.01.006.
6. Ratier, B.; Nunzi, J. M.; Aldissi, M.; Kraft, T. M.; Buncel, E. Organic Solar Cell Materials and Active Layer Designs-Improvements with Carbon Nanotubes: A Review. *Polym. Int.*, 2012, *61* (3), 342–354. https://doi.org/10.1002/pi.3233.
7. Abdulrazzaq, O. A.; Saini, V.; Bourdo, S.; Dervishi, E.; Biris, A. S. Organic Solar Cells: A Review of Materials, Limitations, and Possibilities for Improvement. *Part. Sci. Technol.*, 2013, *31* (5), 427–442. https://doi.org/10.1080/02726351.2013.769470.
8. Pathipati, S. R.; Pavlica, E.; Bratina, G. Enhanced Photoconductivity of Semiconductor Polymers Using Graphene Nanoflakes. *Mater. Today: Proc.*, 2019, *19* (xxxx), 2591–2595. https:// doi.org/10.1016/j.matpr.2019.10.092.
9. Wetzelaer, G. J. A. H.; Blom, P. W. M. Diffusion-Driven Currents in Organic-Semiconductor Diodes. *NPG Asia Mater.*, 2014, *6* (7), e110-13. https://doi.org/10.1038/am.2014.41.
10. Yu, Y.; Luszczynska, B.; Szymanski, M. Z. Organic Bulk Heterojunction Photodetectors Based on Polymer Blends with Unbalanced Mobilities. *Semicond. Sci. Technol.*, 2020, *35* (10). https://doi.org/10.1088/1361-6641/abaaeb.
11. Gebhardt, R. S.; Du, P.; Peer, A.; Rock, M.; Kessler, M. R.; Biswas, R.; Ganapathysubramanian, B.; Chaudhary, S. Utilizing Wide Band Gap, High Dielectric Constant Nanoparticles as Additives in Organic Solar Cells. *J. Phys. Chem. C*, 2015, *119* (42), 23883–23889. https:// doi.org/10.1021/acs.jpcc.5b08581.
12. Fu, Z.; Zhang, X.; Zhang, H.; Li, Y.; Zhou, H.; Zhang, Y. On the Understandings of Dielectric Constant and Its Impacts on the Photovoltaic Efficiency in Organic Solar Cells. *Chin. J. Chem.*, 2021, *39* (2), 381–390. https://doi.org/10.1002/cjoc.202000289.
13. Laquai, F.; Andrienko, D.; Deibel, C.; Neher, D. Charge Carrier Generation, Recombination, and Extraction in Polymer–Fullerene Bulk Heterojunction Organic Solar Cells. *Adv. Polym. Sci.*, 2017, 272, 267–291. https://doi.org/10.1007/978-3-319-28338-8_11.
14. Candiotto, G.; Torres, A.; Mazon, K. T.; Rego, L. G. C. Charge Generation in Organic Solar Cells: Interplay of Quantum Dynamics, Decoherence, and Recombination. *J. Phys. Chem. C*, 2017, *121* (42), 23276–23286. https://doi.org/10.1021/acs.jpcc.7b07165.
15. Shoaee, S.; Armin, A.; Stolterfoht, M.; Hosseini, S. M.; Kurpiers, J.; Neher, D. Decoding Charge Recombination through Charge Generation in Organic Solar Cells. *Solar RRL*, 2019, *3* (11). https://doi.org/10.1002/solr.201900184.
16. Wade, J.; Wood, S.; Beatrup, D.; Hurhangee, M.; Bronstein, H.; McCulloch, I.; Durrant, J. R.; Kim, J. S. Operational Electrochemical Stability of Thiophene-Thiazole Copolymers Probed by Resonant Raman Spectroscopy. *J. Chem. Phys.*, 2015, *142* (24). https://doi.org/10.1063/1.4923197.
17. Ren, J.; Kan, Z. *Chapter 5: Charge Recombination in Organic Solar Cells*; AIP Publishing LLC, 2020.

18. Juhasz, P.; Nevrela, J.; Micjan, M.; Novota, M.; Uhrik, J.; Stuchlikova, L.; Jakabovic, J.; Harmatha, L.; Weis, M. Charge Injection and Transport Properties of an Organic Light- Emitting Diode. *Beilstein J. Nanotechnol.*, 2016, *7* (1), 47–52. https://doi.org/10.3762/ bjnano.7.5.
19. Arredondo, B.; De Dios, C.; Vergaz, R.; Del Pozo, G.; Romero, B. High-Bandwidth Organic Photodetector Analyzed by Impedance Spectroscopy. *IEEE Photon. Technol. Lett.*, 2012, *24* (20), 1868–1871. https://doi.org/10.1109/LPT.2012.2217488.
20. Bag, M.; Narayan, K. S. *Studies of Long-Lived Photogenerated Carriers in Low Band Gap Polymer Photodiodes*, 2015, 1–6.
21. Wu, Z.; Yao, W.; London, A. E.; Azoulay, J. D.; Ng, T. N. Temperature-Dependent Detectivity of near-Infrared Organic Bulk Heterojunction Photodiodes. *ACS Appl. Mater. Interfaces*, 2017, *9* (2), 1654–1660. https://doi.org/10.1021/acsami.6b12162.
22. Yazmaciyan, A.; Meredith, P.; Armin, A. Cavity Enhanced Organic Photodiodes with Charge Collection Narrowing. *Adv. Opt. Mater.*, 2019, *7* (8). https://doi.org/10.1002/ adom.201801543.
23. Wei, Z.; Zhang, J.; He, X.; Ma, S.; Zhang, A.; Cui, W.; Li, Y.; Cai, X.; Zhang, S.; Fan, J. Preparation of Graphene-Multi-Walled Carbon Nanotube Composite for Quantitive Determination of 2-Hydroxy-3-Methylanthraquinone in Hedyotis Diffusa. *Int. J. Electrochem. Sci.*, 2017, *12* (1), 629–638. https://doi.org/10.20964/2017.01.49.
24. Liao, K. S.; Yambem, S. D.; Haldar, A.; Alley, N. J.; Curran, S. A. Designs and Architectures for the next Generation of Organic Solar Cells. *Energies*, 2010, *3* (6), 1212–1250. https://doi.org/10.3390/en3061212.
25. Kumar, U.; Sikarwar, S.; Sonker, R. K.; Yadav, B. C. Carbon Nanotube: Synthesis and Application in Solar Cell. *J. Inorg. Organomet. Polym. Mater.*, 2016, *26* (6), 1231–1242. https://doi.org/10.1007/s10904-016-0401-z.
26. Sharma, M.; Sharma, N.; Alvi, P. A.; Gupta, S. K.; Negi, C. M. S. P3HT-RGO Composites for High-Performance Optoelectronic Devices. *Opt. Mater.*, 2022, *127* (April). https://doi.org/10.1016/j.optmat.2022.112326.
27. Sharma, M.; Alvi, P. A.; Gupta, S. K.; Negi, C. M. S. The Optoelectronic Behavior of Reduce Graphene Oxide-Carbon Nanotube Nanocomposites. *Synth. Met.*, 2021, *281* (August), 116892. https://doi.org/10.1016/j.synthmet.2021.116892.
28. Sharma, N.; Singh Negi, C. M.; Sharma, M.; SinghVerma, A.; Gupta, S. K. A Comparative Analysis of the Optoelectronic Performance of Conventional and Inverted Design Organic Photodetectors. *Opt. Mater.*, 2019, *95* (July), 109273. https://doi.org/10.1016/j.optmat.2019.109273.
29. Zang, Y.; Xin, Q.; Zhao, J.; Lin, J. Effect of Active Layer Thickness on the Performance of Polymer Solar Cells Based on a Highly Efficient Donor Material of PTB7-Th. *J. Phys. Chem. C*, 2018, *122* (29), 16532–16539. https://doi.org/10.1021/acs.jpcc.8b03132.
30. A.Jenekhe. Chem. Mater. 2012, 24, 1995–2001.Pdf. *Chem. Mater.* 2012, *24*, 1995.
31. Benavides, C. M.; Murto, P.; Chochos, C. L.; Gregoriou, V. G.; Avgeropoulos, A.; Xu, X.; Bini, K.; Sharma, A.; Andersson, M. R.; Schmidt, O.; Brabec, C. J.; Wang, E.; Tedde, S. F. High-Performance Organic Photodetectors from a High-Bandgap Indacenodithiophene-Based π-Conjugated Donor-Acceptor Polymer. *ACS Appl. Mater. Interfaces*, 2018, *10* (15),12937–12946. https://doi.org/10.1021/acsami.8b03824.
32. Andreas, R.; Lesbani, A.; Akhmad Yusuf, F. The Character Istics (Compositions, Morphological, and Structure) of Nanocomposites Polyaniline (PANI)/ZnO. *IOP Conference Series: Materials Science and Engineering*, 2019, *509* (1). https://doi.org/10.1088/1757-899X/509/1/012126.
33. Hafeez, H. Y.; Iro, Z. S.; Adam, B. I.; Mohammed, J. Fabrication and Characterization of Organic Photovoltaic Cell Using Keithley 2400 SMU for Efficient Solar Cell. *J. Physics: Conf. Ser.*, 2018, *1000* (1). https://doi.org/10.1088/1742-6596/1000/1/012124.

34. Kidowaki, H.; Oku, T.; Akiyama, T.; Suzuki, A.; Jeyadevan, B.; Cuya, J. Fabrication and Characterization of CuO-Based Solar Cells. *J. Mater. Sci. Res.*, 2011, *1* (1). https://doi.org/10.5539/jmsr.v1n1p138.
35. Zheng, J.; Zhang, M.; Lau, C. F. J.; Deng, X.; Kim, J.; Ma, Q.; Chen, C.; Green, M. A.; Huang, S.; Ho-Baillie, A. W. Y. Spin-Coating Free Fabrication for Highly Efficient Perovskite Solar Cells. *Sol. Energy Mater. Sol. Cells*, 2017, *168* (April), 165–171. https://doi.org/10.1016/j.solmat.2017.04.029.
36. Corporation, S. Application News Measurement of Electrolyte Solutions for Lithium-Ion Secondary Batteries with IRSpirit Glove Box System.
37. Pavlovic, A.; Fragassa, C.; Bertoldi, M.; Mikhnych, V. Thermal Behavior of Monocrystalline Silicon Solar Cells: A Numerical and Experimental Investigation on the Module Encapsulation Materials. *J. Appl. Comput. Mech.*, 2021, *7* (3), 1847–1855. https://doi. org/10.22055/jacm.2021.37852.3101.
38. Nakamura, S.; Nishioka, K. Application News Measurement of Electrolyte Solutions for Lithium-Ion Secondary Batteries with IRSpirit Glove Box System. *MATEC Web Conf.*, 2016, *65*. https://doi.org/10.1051/matecconf/20166504001.
39. Pamain, A.; Pogrebnaya, T.; King'ondu, C. K. Natural Dyes for Solar Cell Application: UV-Visible Spectra and Outdoor Photovoltaic Performance. *Res. J. Eng. Appl. Sci.*, 2014, *3* (5), 332–336.
40. Sarau, G.; Bochmann, A.; Lewandowska, R.; Christianse, S. From Micro– to Macro–Raman Spectroscopy: Solar Silicon for a Case Study. *Adv. Aspects Spectrosc.*, 2012. https:// doi.org/10.5772/48143.
41. Beinert, A. J.; Büchler, A.; Romer, P.; Heinrich, M.; Schubert, M. C.; Aktaa, J.; Eitner, U. Stress Mapping by Confocal Raman Spectroscopy on Solar Cells and Modules. *2018 IEEE 7th World Conference on Photovoltaic Energy Conversion, WCPEC 2018 - A Joint Conference of 45th IEEE PVSC, 28th PVSEC and 34th EU PVSEC*, 2018, 3613–3617. https://doi.org/10.1109/PVSC.2018.8547729.
42. Khalili, S.; Chenari, H. M.; Yıldırım, F.; Orhan, Z.; Aydogan, S. Highly Sensitive, Self-Powered Photodetector Based on Reduced Graphene Oxide- Polyvinyl Pyrrolidone Fibers (Fs)/p-Si Heterojunction. *J. Alloys Compd.*, 2022, 889, 161647. https://doi.org/10.1016/j.jallcom.2021.161647.
43. Salam, R. A.; Munir, M. M.; Warsahemas, T.; Saputra, C.; Latief, H.; Khairurrijal, K. A Simple Solar Simulator with Highly Stable Controlled Irradiance for Solar Panel Characterization. *Meas. Control (U. K.)*, 2019, *52* (3–4), 159–168. https://doi. org/10.1177/0020294019827327.
44. Lasia, A. Electrochemical Impedance Spectroscopy and Its Applications. *Electrochemical Impedance Spectroscopy and Its Applications*, 2014, *9781461489* (July 2014), 1–367. https://doi.org/10.1007/978-1-4614-8933-7.

3

Smart Scaffold Constructs for Regenerative Medicine and Tissue Engineering

Princy Choudhary[1], Ayushi Gupta[1], Saurabh Kumar Gupta[2], Shrey Dwivedi[1] and Sangeeta Singh[1]*

[1]Department of Applied Science, Indian Institute of Information Technology, Allahabad, U.P., India
[2]Department of Biotechnology and Medical Engineering, National Institute of Technology, Rourkela, Odisha, India

Abstract
Tissue engineering represents a prominent tool in regenerative medicine and has been an active domain of scientific research for the last few decades. It is an alluring approach to restore, retain, and revitalize lost or damaged tissues and organs by combining the understanding and inventions from developmental biology, stem cells, and materials science. Basically, the approach incorporates three significant modules: (i) a scaffold that creates a microenvironment facilitating cell adhesion for tissue development; (ii) a suitable cell type; and (iii) biomolecules that direct cell behavior and function. With the advancement of new technologies and novel therapeutic agents, smart approaches like the development of smart scaffolds to deliver the requirements in an efficient and controllable manner become inevitable. Smart scaffolds seek to imitate the dynamic behavior of native extracellular matrix of tissues. The multi-functionality and smartness of the material enable them to execute multiple crucial tasks simultaneously, viz., delivery of essential bioactive agents, cell growth regulation, differentiation, and direct stem cell functioning. In this chapter, the characteristics and significance of such scaffold constructs will be discussed, focusing on the different application areas of regenerative medicine like bone, muscle, skin, cardiovascular, and neural tissue engineering along with recent efforts through techniques in improving efficiency, controlling the smartness of the scaffolds, and translating the *in vitro* smartness to successful *in vivo* tissue repair and regeneration.

Keywords: Smart scaffold, tissue engineering, regenerative medicine, tissue repair, stem cell differentiation, shape memory polymer, 4D bioprinting, smart material

3.1 Introduction

Tissue or organ loss due to injury or illness is a major global concern. Standard therapy involves tissue or organ transplantation, but this is severely bounded by the shortage of donors. The other available treatment possibilities include drug therapy, surgical reconstruction, synthetic prostheses, or medical devices. However, in addition to availability

**Corresponding author*: sangeeta@iiita.ac.in

Upendra Kumar and Piyush Kumar Sonkar (eds.) Smart Materials for Science and Engineering, (39–74) © 2024 Scrivener Publishing LLC

constraints, they also carry the drawback of not being able to completely replace the function of damaged tissue/organ. The advent of tissue engineering and regenerative medicine emerged as an alternative approach to replacing and restoring damaged tissue/organs and offers hope in combating these drawbacks of fully restoring their functions (Atala, 2004; G. Chen *et al.*, 2002). Tissue engineering incorporates cells, scaffolds, and growth factors to replace damaged/diseased tissues. At the same time, regenerative medicine integrates tissue engineering with other approaches, such as gene therapy, cellular therapy, and immunomodulation techniques to regenerate tissue/organ *in vivo* (Han *et al.*, 2020).

Tissue engineering is one of the essential aspects of regenerative medicine and is currently an active research domain for scientists. The working principle of tissue engineering involves an interdisciplinary approach to replace or restore lost or damaged tissues and organs, employing current understanding and modern inventions in developmental biology, stem cells, and materials science (Kaliva *et al.*, n.d.). In general, two approaches are followed: first, acellular matrices are used, which rely on the body's native potential to regenerate into new tissue, and second, cell–matrix constructs are used (Atala, 2004). Acellular matrices are prepared either by artificial scaffold or by removing cellular components from tissues through chemical and physical modifications for producing collagen-rich matrices. For the second approach, a scaffold is used, which provides an adhesive substrate for the implanted cells and mechanical assistance to direct the new tissue/organ formation (Ahadian & Khademhosseini, 2018). In both ways, a scaffold is an inevitable component, forming the base of the tissue engineering approach.

A scaffold is a 3D (three-dimensional) structure that serves as a template for tissue formation and is seeded with cells, sometimes growth factors, or subjected to biophysical stimuli through a bioreactor, a device/system that provides various kinds of chemical and mechanical stimuli to the cells (O'Brien, 2011). It mimics the ECM for cellular adhesion, migration, proliferation, and tissue regeneration in three dimensions. A scaffold should possess several characteristics to be ideal for use in tissue engineering. An appropriate scaffold should repair the body tissues with minor requirements for cell growth, proliferation, vascularization, and host incorporation, and lastly, the material should be degraded during or after healing (Eltom *et al.*, 2019; Roseti *et al.*, 2017). The scaffold design should possess high porosity, high surface area, a thoroughly interconnected framework, structural durability, and a three-dimensional structure that necessitates cell adhesion, proliferation, host incorporation, and differentiation while retaining the cell functions. They should be biocompatible as neither the biomaterial nor its by-product should eventually cause any toxicity and should be biodegradable to get eventually eliminated (Carletti *et al.*, 2011; G. Chen *et al.*, 2002; Eltom *et al.*, 2019; Roseti *et al.*, 2017). Hence, the main challenge lies in the selection of biomaterial, design, and fabrication of the scaffold as it must address physical, mechanical, and chemical properties of the tissue in consideration (G. Chen *et al.*, 2002). Different materials ranging from natural biomass to synthetic biopolymers have been explored and implemented in tissue engineering scaffold fabrication. They have some remarkable features like good flexibility and elasticity, commendable biocompatibility, and biodegradability. Natural biopolymers are typically produced during the growth cycles of all species and are widely available (Atala, 2004; Choudhary *et al.*, 2020; X. Zhang *et al.*, 2021). Most of these polymers were recognized even before tissue engineering was introduced into the research field. This category comprises proteins like fibrin and collagen; polysaccharides such as alginate, chitosan, and hyaluronic acid (HA); and some ECM derivatives. Synthetic biopolymers could also be used in scaffold fabrication. PLA (polylactic acid), an aliphatic polyester, is one such polymer that could degrade within human body and form lactic acid, which can be easily removed from the body. Other similar materials are PCL (polycaprolactone), poly (lactic-co-glycolic acid) (PLGA), polyurethane (PU), and PGA (polyglycolic acid). These materials are considered safe for various applications and have a long history of usage in the biomedical domain. Nonetheless, there are still several disadvantages including their bioinert properties, poor hydrophilicity, and showing inadequacy of biological signals in ECM, which substantially reduces their efficiency in tissue engineering applications (Atala, 2004; X. Zhang *et al.*, 2021). The implanted cells adhere to the temporary scaffold in 3D, proliferate, release their own ECM (extracellular matrix), and substitute the scaffold. Cells and growth factors in natural tissues are immersed in a spatially defined and diverse extracellular matrix (ECM). Along with tissue's physical and biological properties, ECM regulates other aspects like cell dynamics, nutrients, soluble factors, and excretory products within tissues (Ahadian & Khademhosseini, 2018). It is said to be a dynamic matrix that constantly varies in concentration and composition depending on cell development, aging, remodeling, and repair. One of the arduous tasks in scaffold design is to recapitulate the dynamic nature of ECM (Furth *et al.*, 2007).

Smart scaffolds have emerged as a next-generation "Smart" biomaterial due to advancements in polymer synthesis, molecular self-assembly, peptide engineering, and biofabrication technologies that fulfill the above needs by stipulating bio-responsive and structurally tunable properties (Ahadian & Khademhosseini, 2018; Kowalski *et al.*, 2018). Smart biomaterial differs from traditional biomaterials in the way of responding to internal/external stimuli; hence, they can also be termed Smart stimuli-responsive biomaterials where the term "smart" refers to a biomaterial's capacity to influence tissues in a stimulating or inducing manner in response to the stimuli. Smart biomaterials can escalate therapeutic efficacy

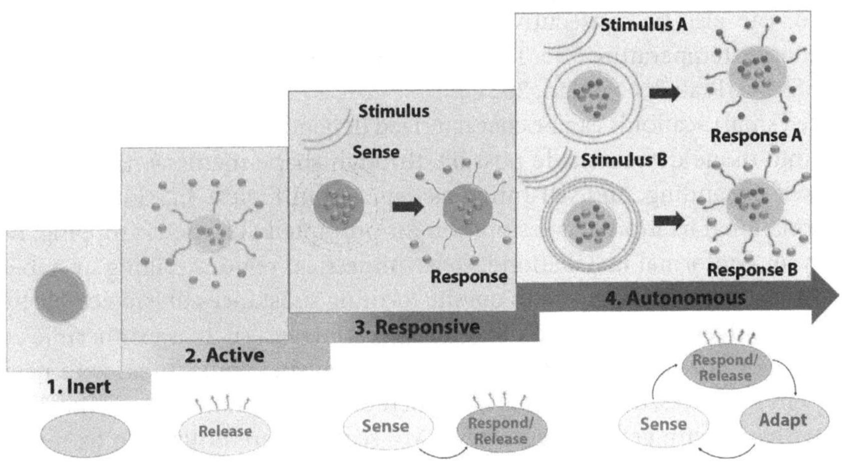

Figure 3.1 The four ascending stages of biomaterial smartness: (1) Inert, (2) Active, (3) Responsive, and (4) Autonomous. Inert signifies the biocompatibility of the biomaterial or does no harm. Active biomaterials provide a one-way unrestricted release of remedies. Responsive biomaterial has the potential to sense specific signals or stimuli and then respond accordingly. Autonomous biomaterials may perceive a signal, deliver a definite dosage, and alter their features in response to changing environmental factors to continue imparting new, improved, or alternative therapies.

and reduce undesirable side effects by quickly detecting and reacting to disease, delivering therapeutic effects while maintaining physiologically healthy cells and tissues (Wei *et al.*, 2022). "Smart biomaterial" has been redefined to acknowledge four grades of smartness, i.e., *inert, active, responsive,* and autonomous. These four grades of smartness have been explained in Figure 3.1. Based on their ability to adapt their physical and chemical properties in response to variations in physiological factors and exogenous stimuli, smart biomaterials can be categorized into smart materials that can be triggered by external stimuli (pH, enzymes, magnetic field, and piezoelectricity) or smart materials that can be triggered by internal material properties (mechanical features, topography, scaffold chemistry, and surface charge) (Montoya *et al.*, 2021; Wei *et al.*, 2022).

Smart scaffolds follow a multifunctional approach, i.e., they can execute multiple crucial tasks simultaneously, delivering essential bioactive agents, cell growth regulation, differentiation, and direct stem cell functioning. Appealingly, some smart scaffolds can regulate the host tissue response and enhance the therapeutic competence of the scaffold *in vivo.* Smart scaffolds are typically thermoresponsive, i.e., shape recovery is temperature responsive, or hydroresponsive, i.e., shape recovery is hydration responsive. Smart scaffolds based on shape memory polymer (SMP) are progressively being explored in the field of tissue engineering. They have unique shape-actuating capabilities and could deliver biomolecules in a precisely programmed way (Ahadian & Khademhosseini, 2018; Kaliva *et al.*, n.d.; Pfau & Grunlan, 2021). SMPs or shape memory polymers refer to intelligent materials that can respond towards extrinsic stimulants through shape change. This reaction comprises structure constancy and shape retrieval, i.e., deformation followed by fixing into a transient form and restoration of the original/permanent shape. Regarding thermoresponsive SMPs, thermoresponsive temperature (T_{trans}) linked to SMP shifting section can be T_g, i.e., glass transition temperature, or T_m, i.e., melting transition temperature. At temperature higher than thermoresponsive temperature ($T > T_{trans}$), the material deformed into a temporary shape that could be recovered by cooling to temperature less than thermoresponsive temperature ($T < T_{trans}$) or can also be consecutively restored by raising the temperature higher than thermoresponsive temperature ($T > T_{trans}$) too. Entropy determines the shape remembrance outcome in SMPs (Pfau & Grunlan, 2021).

SMP-based smart scaffolds can be characterized depending on introduction of functional substances into tissue defects made possible through shape memory nature, such as self-deploying, self-expanding, and self-fitting as depicted in Figure 3.2. Such scaffolds utilize their conformational fixation and restoration for prompted volumetric topping, non-invasive delivery, or conformal installation into asymmetrical regions. Filling unstable regions is a characteristic commonly linked to on-site forming substances like injectable hydrogels, bone cement, etc. However, they have several disadvantages, such as exothermic cures, low setting time, poor pore connectivity, and shrinkage causing contact loss with neighboring tissues. In some instances, the perpetual shape of the scaffold is tailored to fit the tissue defect. On the other hand, with generic geometry, SMP scaffolds are utilized to fix numerous yet irregular deformities. *Self-deploying smart material polymer scaffolds* are usually heated to their T_{trans}, enabling them to pleat into a tight conformation and then cool for securing the structure, allowing filling into a needle or catheter. Hence, they can be delivered in the least invasive way. Such scaffolds have T_{trans} almost similar to body temperature (37°C); therefore, as delivered in the body, it is prompted to recover its shape. Also, post-delivery hydration triggers shape recovery. *Self-expanding scaffolds* also experience conformational recovery

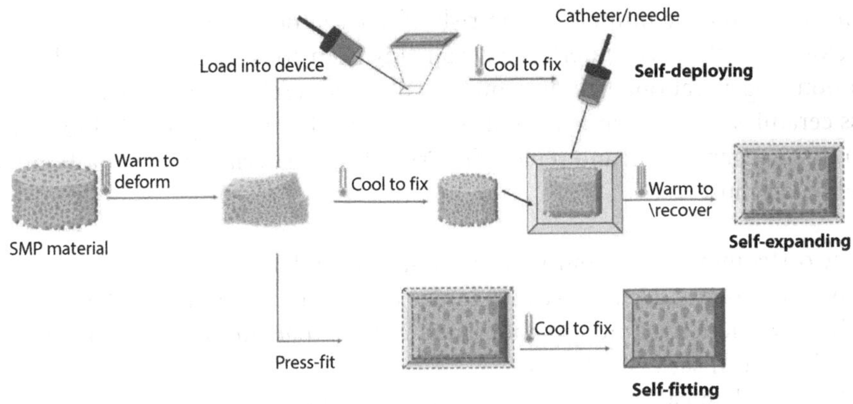

Figure 3.2 Categories of shape memory polymer scaffolds on the basis of delivery mode into tissue defects.

after implanting at tissue distortions despite not being administered through a minimally invasive approach. The scaffold, proportionately constricted formerly, is promoted by shape repair to stretch among the required site. Moreover, the shape recovery can also be actuated by raising the temperature to 37°C, i.e., T_{body} (body temperature) or through hydration after scaffold administration. *Self-fitting scaffolds* often belong to the SMP category possessing T_{trans} somewhat greater than T_{body}. They can be warmed to slacken, enabling the press-fit of the construct into the requisite tissue space. Subsequently, the scaffold expanded via shape recovery within tissue defects comprising those with uneven shapes. As the temperature lowers to body temperature, the scaffold fits in this new shape (Pfau & Grunlan, 2021).

The manufacturing of smart scaffolds using smart biomaterials has attracted research interests in recent years. During the last few years, there has been a rapid advancement in the fabrication of these scaffolds, from external stimulation to enhance pharmacological outcomes for smart reaction towards modifications in the inner milieu to rational combinatorial treatment, which combines different techniques to ameliorate remedial potency. This chapter mainly focuses on the different application areas of regenerative medicines like bone, cartilage muscle, skin, cardiovascular, and neural tissue engineering along with the recent efforts through the techniques in improving the efficiency, controlling smartness of the scaffolds, and successfully converting *in vitro* intelligence into *in vivo* rejuvenation and repair of tissues.

3.2 Applications of Smart Scaffolds in Different Areas

3.2.1 Bone Tissue Engineering

The quest for a suitable material for bone regeneration is still ongoing. At the microscopic level, bone can heal itself. However, significantly larger defects require clinical intervention. Autologous bone implants comprising the highest biocompatibility, osteoinductive, osteoconductive, and osteogenic properties are the gold standard for bone regeneration. These implants osseointegrate and are vascularized, with the bone having nearly zero risk of infection and immune rejection, although they also have drawbacks like donor-site morbidity, limited availability, and graft harvesting time (Rogers & Greene, 2012). In comparison, xenografts are primarily limited by the risk of allergic or immune response and disease transmission. A plethora of materials have already been considered or are under consideration for bone regeneration, including metals, alloys, several natural to synthetic polymers, and glass ceramics. A new generation of "smart" materials is being studied for their application in tissue regeneration and repair. The "smart" biomaterial suitable for bone regeneration and repair must

1. be ossteointegrative (ability to communicate with host tissue),
2. be osteoconductive (ability to colonize bone cell and blood vessels),
3. be osteoinductive (induces differentiation of progenitor cells into bone cells) (Mojtaba Ansari, 2019), and
4. have mechanical competence compared to bone.

Over the past five years, thousands of research articles have been published employing smart/intelligent biomaterials that could positively influence tissue regeneration by responding to various physical, chemical, and biological stimulation. Adding growth factor, extracellular proteins, or nanoparticles can achieve this. Multiple approaches, including bioprinting, gas foaming, salt leaching, freeze-drying, etc., were employed to develop scaffolds, tissue construct, or graft-supporting tissue regeneration.

The most common functionality of materials is shape memory under the physiological change in the surroundings. Scientists have studied the bone regeneration capability of shape-changing porous foam applying hydrophilic polyethylene glycol in conjugation with degradable polycaprolactone and calcium citrate or amorphous calcium phosphate (CA) (L. Wang *et al.*, 2022). The prepared scaffolds have shown up to 87% of shape recovery after four cycles at 50°C with varying amounts of CA particles. These scaffolds have shown short shape recovery time in an aqueous environment. Incorporating CA particles in these scaffolds have persuaded mesenchymal stem cells to differentiate in 7 days, showing excellent biocompatibility and mechanical properties.

The scaffold or graft degradation should meet the rate of bone regeneration. A smart scaffold can be utilized to monitor cell–material interaction. An electrochemically responsive scaffold using carbon nanotubes and bone morphogenic protein loaded in a chitosan-based hydrogel system was studied to monitor bone regeneration. Carbon nanotube (CNT) incorporation has increased 7-fold in mechanical properties with a sustained degradation. The added bone morphogenic proteins (BMPs) in the presence of CNT have synergistically induced human adipose-derived stem cells towards osteogenic differentiation. Electrochemical cyclic voltammetry and electrochemical spectroscopy performance of the scaffold in cranial defect repair in the mouse model showed good sensitivity in monitoring early osteogenic differentiation (Huang *et al.*, 2023). It has been recognized that bone undergoes remodeling stimulated by electrochemical cues. The property of mechanoelectrical transduction (tensile/compressive forces converted into an electrical signal) of piezoelectric materials can be incorporated into a scaffold to induce bone formation smartly and naturally. Several piezoelectric polymers and ceramics listed in Figure 3.3 have shown capability in bone tissue engineering (Jacob *et al.*, 2018a).

The constructs have been designed to combine bone regeneration and therapy for bone defects caused by tumor excision, severe infection, and other bone diseases like osteoporosis. Either external stimuli or tissue microenvironment acts upon these materials, smartly

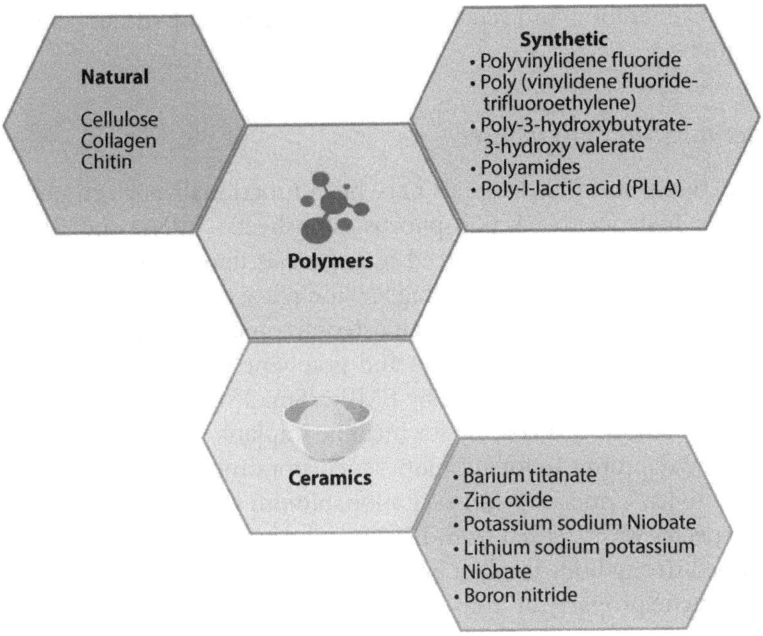

Figure 3.3 Prospective biomaterials in the field of bone tissue engineering.

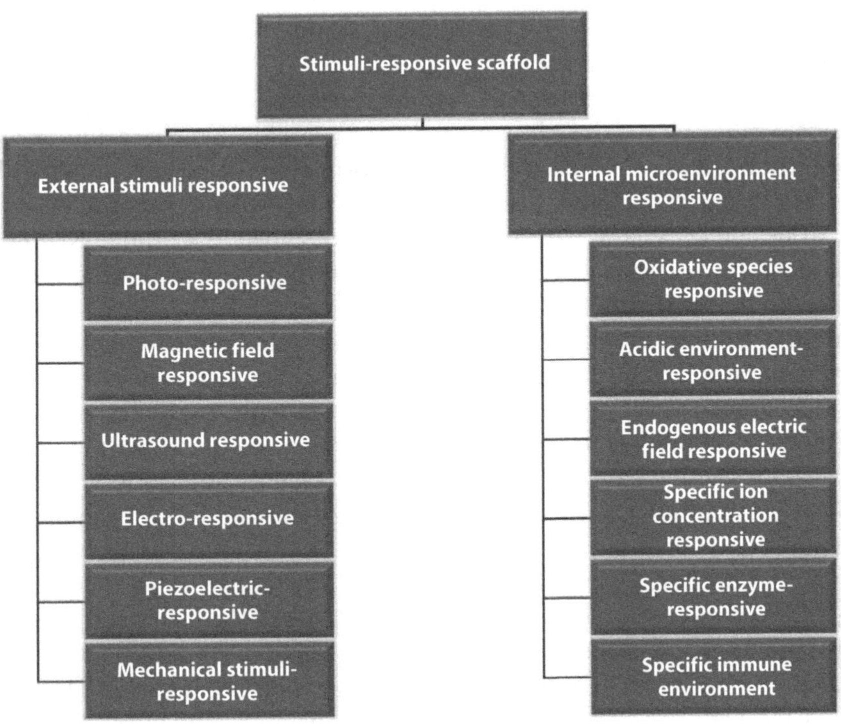

Figure 3.4 Categories of stimuli-responsive scaffold.

activating bone regeneration and repair. Different categories of stimuli-sensitive scaffolds are listed in Figure 3.4.

3.2.1.1 External Stimuli Responses

Recently, conventional scaffold materials have been functionalized with photo-responsive nanoparticles. For instance, black phosphorus nanosheets (BPNs) and BPNs conjugated with $SrCl_2$ loaded in PLGA have enhanced osteogenesis upon near-infrared excitation by heat shock protein expression and triggering Sr^{2+} ion release, respectively (Tong *et al.*, 2019; X. Wang *et al.*, 2018). The ablation of human osteosarcoma cells was achieved by functionalizing chitosan with nano-hydroxyapatite and graphene oxide upon 808-nm laser excitation and enhancing formation of bone by BMP-2/Smad signaling mechanism in hMSCs (L. Ma *et al.*, 2020). Similarly, a coating for metallic implants was developed using polydopamine, IR820, and daptomycin for antibiotics incorporating antibiotic, photothermal, and photodynamic activity. Under 808-nm excitation, biofilm was eradicated by hyperthermia, singlet oxygen species was released from IR820, and daptomycin ruptured bacterial membrane and attained strong bone implant bond (Zeng *et al.*, 2020).

Using the magnetic property of iron-based NPs in the tissue construct may be helpful in bone therapy and regeneration under an external magnetic field. A multifunctional mesoporous bioactive glass system was developed by Zhu *et al.* incorporating iron. The designed scaffold has shown better mechanical durability, elucidating apatite formation potential and prolonging drug-release profile. Iron incorporation resulted in slower dissolution and better osteoblast activity, and showed alternating magnetic field-induced hyperthermia. Such materials can be functionalized in critical-sized defects caused by tumor excision, where a magnetic field synergizes hyperthermia with the released drug that targets the cancerous cell and enhanced bone regeneration (Y. Zhu *et al.*, 2013). A 3D-printed akermanite containing calcium peroxide and iron oxide has the capability of magnetic hyperthermia via tumor-oxidative therapy and the release of hydrogen peroxide in tumor acidic environment and subsequent release of Ca^{2+} ion-induced bone formation (Dong *et al.*, 2020). A biphasic injectable bone cement containing calcium sulfate-coated nanocomposite of iron oxide/graphene oxide and tricalcium phosphate have been designed that promotes tumor necrosis by magnetic field-assisted hyperthermia and inducing osteogenic differentiation and bone formation *in vivo* (F. Yan *et al.*, 2019).

Another non-invasive way of delivering bone therapy and inducing bone regeneration is employing ultrasound. Non-thermogenic clinical ultrasounds were used to release bone morphogenic protein (BMP-2), given as nanocomplex with liposomes enhancing local bone regeneration (Crasto *et al.*, 2016). A research group modified the titanium implant surface by sulfur-causing oxygen deficiency and achieved 99.99% antibacterial efficiency against *Staphylococcus aureus* when exposed to ultrasound and NIR in combined therapy for 15 min (Su *et al.*, 2020).

Electrical stimuli have been clinically established to enhance bone regeneration by inducing differentiation in MSCs and upregulating BMP-2. The localized treatment using electrical stimulus can be achieved by employing electroactive materials (GO, CNT, etc.) and conductive biocompatible polymers (polypyrrole, polyaniline, etc.). Hydroxyapatite NPs conjugated with conductive polypyrrole/polydopamine NPs were electrodeposited on the titanium implant. Osteogenic differentiation was achieved by HA and the electrical stimulation that activated Ca^{2+} ion transduction pathways (Zhou *et al.*, 2019). The silver-loaded poly(amine-amide) dendrimer was doped in PANI and coated the Ti surface to simultaneously achieve antibiotic and osteogenic therapy under electrical stimulation (B. Zhu *et al.*, 2019). Under combined electrical and magnetic stimulation, the nanocomposite of PLGA/poly(3-hexylthiophene) functionalized with $GdPO_4.H_2O$ nanobundles emerged with the capability of osteogenic differentiation and implant traceability because of Gd incorporation (H. Yan *et al.*, 2019).

The mechanical stress exerted on the bone tissues is converted into electrical energy that induces bone growth and remodeling (Ribeiro *et al.*, 2017). The polarization of the composite of HA and barium titanate after cyclic loading facilitated the proliferation of bone cell lineage and enhanced interaction with HA (Tang *et al.*, 2017). Incorporating piezo-ceramic $BaTiO_3$ NPs with piezo-polymer poly(vinylidene fluoride-tri fluoroethylene) (PVFT) matrix has achieved higher osteoinductivity and neovascularization *in vivo* (Bai *et al.*, 2019). The synergistic effect of bioactive glass microparticles and PVFT in nanofibrous scaffold was proclaimed to significantly improve the proliferation, migration, and differentiation of BMMSCs (F. Zhao *et al.*, 2020). This work has shown the biomimetic effect of the PVFT, where the negative pole of PVFT attracted positively charged calcium, thus effectively showing potential in treating critical-sized defects.

3.2.1.2 Internal Microenvironment-Responsive Smart Materials

External stimuli response of biomaterial in bone regeneration and therapy are limited by tissue thickness and damage to healthy tissue. In contrast, biomaterial responsive to the microenvironment provides an advantage of faster response, regeneration, and therapy.

Certain pathological conditions, such as a tumor, injury, inflammation, etc., cause an increase in reactive oxygen species (H_2O_2, OH radicals, etc.). Microenvironments containing ROS can stimulate biomaterial response, eradicating ROS and inducing bone formation. H_2O_2 mediated antibacterial and antitumor effects in butyrate-inserted Ni-Ti layered double hydroxide film (D. Wang *et al.*, 2017). A composite of iron (Fe) and calcium silicate ($CaSiO_3$) fabricated in a scaffold by 3D printing having high mechanical strength showed dual-function ROS-mediated tumor and photothermal therapy. The $CaSiO_3$ presents enhanced proliferation and differentiation of hMSCs into osteoblast lineage *in vivo* (H. Ma *et al.*, 2018).

A slightly alkaline tissue pH decreases and becomes acidic under conditions such as chronic inflammation, infection, or tumor niche. A layer of silver NPs sandwiched between polydopamine layers deposited over a 3D-printed polyether-ether ketone scaffold was triggered by acidic pH liberating silver NPs, thus killing microbes. The coating was loaded with an additional apatite layer inducing osteogenesis (Y. Deng *et al.*, 2020). The enhanced osteoclast activity, as in osteoporosis, creates an acidic microenvironment. Lin *et al.* formulated a tetracycline functionalized nanoliposome loaded with sodium bicarbonate that responded to an acidic environment by forming an alkaline layer at the bone surface. Such therapy reduces osteoclast activity by stimulating their apoptosis cascade, reducing bone destruction, and promoting osteogenesis (Lin *et al.*, 2020).

Bone defect causes changes in the naturally existing endogenous electric field, which can be endowed in scaffold material to repair bone tissue efficiently. A nanofilm has been fabricated using an electropositive ferroelectric $BiFeO_3$ that counters electronegative bone defects triggering osteointegration. The osteogenesis of stem cells is mediated by the Ca^{2+} ion and PI3K/AKT signaling mechanism (Y. Liu *et al.*, 2017). An imbalance in the ionic concentration has been used to repair a bone fracture or drug-induced tumor necrosis. The high Ca^{2+} ion concentration in bone fracture triggered higher drug release in chitosan/hyaluronic acid polyelectrolyte gel formulation loaded with bovine serum albumin, which can be utilized in rapid healing (Krishna *et al.*, 2015). A mechanized Zr-metal organic framework showed high drug encapsulation and release of antitumor drug 5-fluorouracil induced by the lower pH and higher concentration of Ca^{2+} ion around the bone tumor cells (L. L. Tan *et al.*, 2015). The event of matrix metalloproteinases in the microenvironment marks the formation of tumors and metastasis. They are overexpressed in bone and cartilage cells in pathological conditions such as rheumatoid arthritis and osteoporosis. Researchers have designed enzyme-based responsive smart materials. An enzyme-responsive platform integrates Ag NPs with mesoporous silica NPs and assembles them layer by layer using poly-L-glutamic acid. The glutamyl endonuclease secreted by *S. aureus* disrupts the assembly, thus releasing Ag NPs and delivering antimicrobial activity, and regeneration is enhanced by poly-L-glutamic acid (Y. Ding *et al.*, 2020). *In situ* polymerized MMP-reactive cross-linker and 2-(methacryloloxy) ethyl phosphorylcholine were linked to the BMP-2 nanocapsule surface. This assembly is responsive to higher MMP concentration at the fracture site, which causes the degradation of the cross-linker, thus disrupting the polymer shell and release of BMP-2, enhancing bone regeneration (Qi *et al.*, 2019).

In the case of bone-related disorder or material introduced for bone regeneration, an immune response is mediated by various cells and cytokines, and they also induce the expression of multiple factors influencing bone repair. Osteoimmunomodulatory biomaterials have come up to enhance bone repair. Self-assembly of heparin-modified gelatin nanofibers linking interleukin-4 was obtained as an injectable microsphere modulated the activity of the macrophages (M1→M2) promoting the healing process (Hu *et al.*, 2018).

3.2.2 Cartilage, Muscle, and Skin Tissue Engineering

Like bone tissue, the cartilage tissues dynamically remodel themselves based on functional requirements—the electrical and chemical communication between the cells and the ECM. However, disruption in these signal pathways occurs due to critical-sized defects due to injury, disease, etc., eventually affecting the natural repair. A tissue engineering approach using biomaterial that changes their structure and property in certain conditions/stimuli can be practical in therapy-based regeneration, mimicking natural healing.

Piezoelectric materials having the intrinsic property that produces electricity in response to mechanical stresses are utilized to fabricate electroactive scaffolds in the field of cartilage tissue regeneration too. Poly(3-hydroxybutyrate-co-3hydroxyvalerate) incorporating piezoelectric $BaTiO_3$ nanoparticles was developed into cartilage-mimicking nanofibrous scaffolds by electrospinning. The incorporation of 20% $BaTiO_3$ improved the mechanical property, and the electrically polarized (increased surface charge density) scaffold has shown better cell viability and adhesion of bone marrow-derived chondrocytes (Jacob *et al.*, 2019).

Attempting to repair articulate cartilage, a hydrogel-based system was developed by photo-cross-linking gelatin methacryloyl (GelMa) and modified by TGF-β1 affinity peptide (HSNGLPL). *In vitro* and *in vivo* studies showed that the heptapeptide sequence increased the autonomous TGF-β1 recruitment to the hydrogel. TGF-β1 induces chondrogenic differentiation of human umbilical cord-derived MSCs *in vitro*, as confirmed by expression of Col II/I/X and aggrecan. The peptide-loaded scaffold efficiently repaired the articular cartilage in the rabbit model (Ju *et al.*, 2022).

With the increasing incidence of accidents, the number of people relying on repairing tendons and ligament tissue to restore daily bodily functions has drastically increased over the years. Of all the injuries, the major ones involve the tear of ligaments and tendons. Featuring the property of high elasticity and locomotor efficiency in tendon fibrous tissues, finding their supplement biomaterial is quite tedious. The current techniques of implanting collagen fibers and other biocompatible scaffolds tend only to subside the issue and leave a massive chance that the tendon may be subjected to re-injury. A method to overcome such shortcomings, hybrid material scaffolds are being preferred now. Such materials are both biocompatible and bioactive in nature, which corresponds to the native characteristics of native tendon tissues. One example is electrochemically aligned Collagen fibers that represent a 3D bio-textile property with robust and porous features like tendon tissues. Thus, making them an elite choice in cases of tendon replacements (Beldjilali-Labro *et al.*, 2018).

Another example of the implantation of smart materials as a method to attempt the regeneration of smooth muscle cells is the employment of nanotubes based on boron nitride. These nanotubes exhibit illustrious piezoelectric properties that help transit an electrical stimulation with the help of a mechanical source (wireless in nature, like ultrasound waves) throughout the length of the applied tissue (Figure 3.5). Further functionalization of the boron nitride nanotubes is favorable for implantation and proliferation of smooth muscle cells, chondrocytes and fibroblasts, and other types of myogenic proliferation (Jacob *et al.*, 2018b). Table 3.1 shows the applications of various smart polymers in the area of muscle tissue engineering.

The advent of skin scaffolds through tissue engineering has become reasonably necessary to facilitate their incorporation into the wound care and dressing field. With the percentage of aging and diabetic population on the rise, along with increasing incidence and prevalence of ulcers and diabetic wound ulcers, it has become of utmost importance that wound care and healing can take the help of the modern approach of tissue engineering and scaffold implantation. Skin is the largest sensory organ, and has a very diverse range

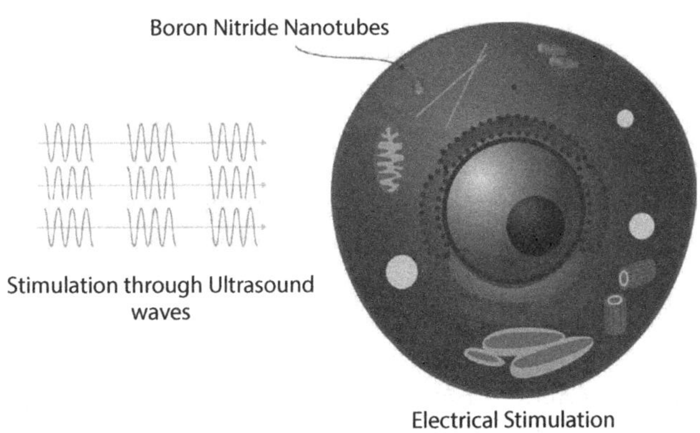

Figure 3.5 Application of boron nitride nanotubes for functional electrical stimulation through ultrasonic waves.

Table 3.1 Applications of smart polymers in muscle tissue engineering.

Polymer	Application	Results	References
β-tricalcium phosphate (β-TCP) bioceramic scaffolds with graphene oxide	For repair and regeneration of skeletal muscles	The results showed successful regeneration of skeletal muscles with antibacterial activity endowed by the scaffold itself.	(Palmieri *et al.*, 2020)
Poly lactic co glycolic acid with micro-grooved patterned structure	The scaffold was implemented for myogenic differentiation of c2c12 cells	The scaffold acted as a bionic structure for the skeletal muscle cell rejuvenation.	(Gao *et al.*, 2021)

of functions, including thermoregulation, balancing of the bodily fluid, sensation of touch and pain, residing location for sensors associated with heat and cold sensation, production of vitamin D, and forming the initial barrier and the first line of defense for the immune system. Dermis, epidermis, and hypodermis comprise the three subcutaneous layers that include the skin; the healing of a particular would involve all three.

The recuperation process for wounds is divided into multiple phases, keeping the wound clean and moist as one of the necessary elements throughout. The traditional way of doing this is by applying daily dressings with the help of gauge and tulle. However, their removal causes disruption and pain. This is because of the adherence of the dressing material to the wound bed. Modern methods of bioactive or interactive materials are being used to shield the wound from getting gouged again. These may include semipermeable films and foams made up of polymers like polyurethane or have specific structural properties like those contained in hydrogels and hydrocolloids, alginates, collagen, and hydrofibers (Talikowska *et al.*, 2019). The most prevalent biocompatible polymers include hyaluronic acid, chitosan, and elastin.

However, most of these polymers fail to effectively act as a matrix to guide the cells towards differentiation and proliferation and provide appropriate mechanical support to the growing tissues, thus maintaining a structure that is possibly the same as ECMs. One of the answers to such problems is the advent of conducting polymers. Conducting polymers facilitate an increased electrical conductivity that can be applied to stimulate recovery in the wound area. They can also help maintain a controlled release of pharmaceutical and biological vehicles near the desired location in a predetermined fashion (Korupalli *et al.*, 2021; Talikowska *et al.*, 2019).

The main structure of the conducting polymers exerts the doping of the non-conducting ones. Here, the term doping implies a redox reaction; p-doping refers to the oxidation of the polymers, while n-doping refers to the reduction of the polymers. In general, conducting polymers have a structural backbone that happens to be conjugated. As the structure comprises alternate single and double bonds, it leaves the π-electrons of the p-orbital to move

Table 3.2 Applications of conducting polymers for wound healing.

Polymer	Application	Results	References
Functionalized polyaniline (fPANI)	For the inhibition of the growth of wild-type *E. coli*, *P. aeruginosa*, *S. aureus*	The fPANI causes oxidative stress, causing ulterior hindrance towards forming cell walls in resistant strains.	(Gizdavic-Nikolaidis *et al.*, 2011)
Nanocomposites of polypyrrole/dextrin that are activated in acidic medium	Antibacterial activity against *P. aeruginosa* and *S. aureus*, covering both gram-positive and gram-negative strains	The composites turned out to be biodegradable with increased antibacterial activity	(Nazarzadeh Zare *et al.*, 2014)
Polypyrrole composite with silver nanoparticles	Antimicrobial activity against both gram-negative and gram-positive bacteria	—	(Chowdhury & Al-Jumaily, 2016; Maráková *et al.*, 2017)
Poly(3, 4-ethylenedioxythiophene) (PEDOT) nanotubes, applied as transdermal patch	Electrically controlled release of insulin	The drug insulin was attached within the polymer nanotubes as a counter ion to its backbone, which was released according to the potential pulse given	(Nguyen *et al.*, 2014)
Polypyrrole electrically active polymer	Electrically controlled release of analgesic and anti-inflammatory drug ibuprofen	The polymer can be easily switched between its oxidation state and the surface charge accumulated that ultimately enhances the ease of loading and release of the drug	(Justin *et al.*, 2012)
Thermo-sensitive hydrogel membranes made up of sodium alginate, poloxamer 407, pluronic F127, and polyvinyl alcohol	The polymeric scaffold was used for hastened wound relieving with the encapsulation of antibacterial amikacin	Hydrogel showed high efficiency regarding wound recovery along with bacterial growth inhibition	(Abbasi *et al.*, 2020)

(Continued)

Table 3.2 Applications of conducting polymers for wound healing. (*Continued*)

Polymer	Application	Results	References
Bilayered thiolated alginate/ PEG diacrylate (BSSPD) hydrogels	Applied for rapid and scarless wound healing through consecutive discharge of tiny extracellular vesicles	The scaffold facilitated a more uniform formation of vascularization and almost scarless healing	(Shen *et al.*, 2021)
Silk nanofiber hydrogels laden with MSCs	Applied for scarless wound healing	The hydrogel resulted in the formation of hair follicles at the site of the wound	(Zheng *et al.*, 2020)
Hydrogel combined with light-sensitive tetrapodal zinc oxide microparticles	Wound healing along with angiogenesis	The hydrogel enabled spatiotemporal controlled release of the growth factor VEGF	(Klimek & Ginalska, 2020)

freely and cause its delocalization. This delocalization causes the polymer to support the flow of electric current (K & Rout, 2021).

Many applications have been implemented for wound healing as far as conducting polymers are concerned. Some examples are listed in Table 3.2.

Along with these, there are several other applications of conducting polymers in skin regeneration, including wound relieving with the help of PANI doped with usnic acid as a method for interactive recovery, combined with polyurethane (dos Santos *et al.*, 2018). Conducting polymers have also been implemented in skin tissue engineering-related areas, as cell growth and proliferation are supported better in cases where the polymer has an applied bipolar or, in fact, multipolar direction (B. K. Gu *et al.*, 2018). There have been many incidences where the polypyrrole-based polymer, doped with the desired agent, is employed to form an absorbable scaffold for the regeneration of fibroblast cells and Schwann cells associated with the dermis (Lee & Schmidt, 2015).

3.2.3 Cardiac Tissue Engineering

The primary aim of any scaffold engineered for cardiovascular implementation is to mimic the properties of the residing myocardium cells, their mechanical support, and their bioelectrical activity. Moreover, given the heart's changing rhythm with the need of the body, it is essential that the scaffold implanted is of tunable nature. As cardiac cells tend to lose their regenerative properties as the person matures and reaches adulthood, a substitute must become readily available when an impactful number of cells lose their functionality after a myocardial infarction. Therefore, the native properties of myocardium cells, like bioelectrical conductivity, structure, and stiffness, along with a degree of flexibility, and a certain level of perfusion for the enablement of the blood vessel genesis need to be attained through these smart material scaffolds.

Taking the electrical aspect of the cardiac muscles into account, materials forming medical devices like the multi-electrode arrays or some similar kind of biomedical device that may support the generation and conduction of an impulse form the basis of such studies. Such devices with bioelectric interface function through the support of intermittent immediacy of cardiac scaffold patches. The compatible cardiac scaffolds can be adjusted in the form of bioelectronic scaffolds. Such bioelectronic scaffolds are made up of materials like the ubiquitous polymer of SU8 that is subjected to the addition of a multi-electrode array to make it suitable as a cardiac scaffold. This composite scaffold is programmed to stimulate a synchronized release of an electric impulse to help regulate cardiac contractions. Another aspect of such electrode array scaffold is the activity of controlled release of growth factors embedded essential for the regeneration of cardiac muscle cells.

Similar endeavors include the advent of hybrid patches used for electrical conductivity within the scaffold, the formation of artificial blood vessels made up of 3D folded bioelectronic scaffolds, suitable for embodiment within the heart walls, and other organ-chip configured devices for smoother cell–cell communication. Another aspect of bioelectronic scaffold is its support and migration properties to the seeded human mesenchymal cells endowed towards cardiac cell differentiation. The native architecture and morphology of the scaffolds enable the said cells to grow and proliferate at a favorable rate, leading to the formation of 3D organoids (Bolonduro *et al.*, 2020).

Apart from scaffolds that generated electric impulses with the help of electrode arrays, there is a range of biomaterials that are electroconductive in nature. Materials like polythiophene, graphene and CNTs, polypyrrole, polyaniline, and MXenes comprise the group of electroconductive biomaterials. Apart from the role that these polymers play in advancing skin tissue engineering and wound care, they also play a stint in the field of cardiac tissue engineering. Due to their ability to withstand and direct the influx of ions, these electromechanically active scaffolds can be employed as actuators that help make the scaffold placed in the heart electrically conductive (Solazzo *et al.*, 2019).

Another smart polymer that has begun to find increasing applications in medical devices and recently into tissue engineering is shape memory polymer. These polymers' ability to switch between shapes according to the external stimuli helps create various clinical applications around them. Cardiovascular stents are among the most desirable applications for these SMPs. With their properties of low cost and shape reconfiguration, it has become relatively easy to apply them during angioplasty (Santo *et al.*, 2020; Serrano & Ameer, 2012). Various types of shape memory polymers, along with their properties, are being used for developing cardiovascular stents that are more and more advanced in the effectiveness and feasibility of the application. These range from thermo-induced ones to bio-absorbable ones (Mehdi Ansari *et al.*, 2019; Yeazel & Becker, 2020).

Apart from stents, smart materials that are also being adopted for tissue engineering are the heart valves so that one may replace the faulty ones without any disadvantages that the current animal-procured ones or natural polymers have. In this scenario, melt electrowriting employed functional scaffolds that are developed and that are architected in such a way that they mimic the microstructure of the valves and have the property of bearing the load of the blood flow passing through it at every second. The scaffolds for valve formation are designed in a J-shaped manner to pose in the leaflet frame of the valve. The leaflet design of the scaffold enables the hydrodynamic capacity and the architecture suitable for the heart

Table 3.3 Polymer products used for cardiovascular applications in the biomedical field.

Polymer/ commercialized material	Application	Results	References
Shape memory polymer–polyurethane-coated nitinol	Used as cardiovascular stents	They are corrosion-resistant and function as a self-expanding stent	(Holman *et al.*, 2021)
Evolut R—self-expanding, repositionable transcatheter heart valve	Used in surgeries for transcatheter aortic valve replacement	The clinical study showed favorable and accurate valve placement through the repositioning of the valve	(Manoharan *et al.*, 2018)
Sparrow® drug-eluting stent (DES)	Employed in percutaneous coronary angioplasty.	Sparrow® DES caused a remarkable depletion at in-stent late lumen loss, ultimately reducing restenosis (decrease in the lumen diameter of the vessels)	(Oliveira *et al.*, 2013)
Self-apposing STENTYS stents	The STENTYS stents act on lesions of bifurcated coronary arteries	The trials have proved favorable in left main coronary artery diseases and saphenous vein graft lesions.	(Lu *et al.*, 2018)
Bioresorbable Igaki-Tamai stents composed of poly(L-lactide) (PLLA)	The stent was employed in patients with an adverse cardiac event probability.	The stent was discovered to have been completely dissolved in about three years, as it degraded into lactic acid, which was used up in Kreb's cycle	(Yeazel & Becker, 2020)
Absorb™ is made up of PLLA coated with PDLA (poly(D-Lactide) with everolimus embedded	The stent was used for restoring the vasomation with the release of the embedded drug.	The drug was observed to be released within 30 days, and the stent was dissolved entirely in about three years	(Yeazel & Becker, 2020)

valves (Saidy *et al.*, 2019). Some smart polymers and commercialized products used for different cardiovascular applications in the biomedical field are presented in Table 3.3.

3.2.4 Neural Tissue Engineering

Our nervous system is responsible for the outcomes of all other systems in the human body. Being a controlling entity of the human body, any disability related to cognitive or motor functions can prompt a significant loss in the lifestyle attributes for the patient. Such occurrences of neurodegenerative diseases have been on the rise for the past couple of decades. Current effective treatment options rely on diminishing the symptoms, leaving the underlying cause as it is (Choudhary, Gupta, & Singh, 2021; Choudhary, Gupta, Shukla, *et al.*, 2021; Dwivedi *et al.*, 2023; Gupta & Singh, 2022; Pahal *et al.*, 2021; Shukla *et al.*, 2022). On this front, tissue engineering presents a prospective answer for regenerating the diseased or injured areas of the PNS (peripheral nervous system) and the CNS (central nervous system). As a way to facilitate the regeneration capacity of the stem cells used in these treatments, scaffolds come in quite handy. With advancements in technology and the evolution of smart materials, their application in the nerve tissue engineering domain has dramatically widened the horizons of the available treatment options for neurodegenerative diseases. The prospective application of biomaterials as 3D tools for *in vitro* neurogenesis of tissues in both physiological and pathological states establishes the most effective and appropriate neural repair strategies. The ongoing studies aim to optimize various biomaterials that can possibly be utilized as scaffold constructs to support neural 3D culture or brain-region-specific organoids. Such neural cultures can be achieved by using customized scaffolds, carrying specific features of human brains so that it would be possible to easily replicate biochemical, physical, and topological features. Using the new smart biomaterials, it is now feasible to anticipate and program progenitor cell adhesion, proliferation, and differentiation *in vitro*, imparting the interaction among neurons and glial cells and neural circuit development (Yuan *et al.*, 2018; Yuan & Xing, 2019).

Considering the functional aspect of neurons, nerves can be depicted as wires that conduct the action potential throughout our body. The action potential of the body follows the rhythm of depolarization, repolarization, and hyperpolarization. Polymers like polyvinylidene fluoride and its modified state with trifluoroethylene are some of the polymers with native piezoelectric characteristics, and their scaffolds help stimulate the number, length, and connections made through the dendrites (Cheng *et al.*, 2020; Zaszczynska *et al.*, 2020). Piezoelectric materials have also been used to power patches as nerve regeneration scaffolds. The electrical signals serve as cues for releasing cytokines and other biological chemicals as inducers for neuronal regeneration. One such patch is made up of polyvinyldiene fluoride with reduced graphene oxide doped with methacrylate gelatin gel. The patch was installed to release chemokine CLCXL12 at electrically generated signals, enhancing nerve regeneration (M. Tan *et al.*, 2022). Piezoelectric scaffolds can develop electrical potential by applying mechanical force in a non-invasive manner independent of any external power supply (Steel & Sundararaghavan, 2016). Concerning neurons, such scaffolds have been derived for optimal performance with pore sizes ranging between 10 and 100 μm and a porosity of 90% via the electrospinning technique (Miguel *et al.*, 2019). In a study, ultrasound exposure was used as a mechanical force to instigate electrical stimulus under *in vitro* culture conditions on piezoelectric scaffolds having poly(vinylidene fluoride-trifluoroethylene P(VDF-TrFE)) and P(VDF-TrFE)/BTNP separately. The SH-SY5Y neuroblastoma cells in culture exhibited appreciable neurite extension and maturation levels in the latter (Genchi *et al.*, 2016). Later, researches built upon this approach and extended it using polylactic acid (PLA) nanofibers/potassium sodium niobate (K0.5Na0.5NbO3, KNN) and nerve guidance conduits. Neuronal differentiation, neural cell multiplication, migration, and motor function rescue were reported under *in vivo* conditions, underlining the future scope of piezoelectric scaffolds for neuroregenerative therapy (P. Chen *et al.*, 2022; Y. Ma *et al.*, 2023b; Wu *et al.*, 2022). Most studies on piezoelectric scaffolds have depended on mechanical input on the protrusion and contraction of adherent cells, which may cause temporary deformation in culture. However, researchers opine that a genuine piezoelectric effect necessitates an actual mechanical input *in vitro*, equating to the *in vivo* deformations on a macro- and microscopic scale. Polymers like polyvinylidene fluoride (PVDF) have peak piezoelectric coefficients but are non-biodegradable. These can be replaced with biodegradable substitutes like composite scaffold of PANi, an electrical stimuli-responsive polymer, and a piezoelectric polymer to amplify the piezoelectric response (Zaszczynska *et al.*, 2020).

Graphene addition to scaffolds also positively affects neural cell proliferation ability, adherence, stemness preservation, and differentiation (Guo *et al.*, 2021). Nanofibrous scaffolds have significantly benefitted from incorporating graphene with existing biomaterials for efficient controlled drug delivery, neural cell migration, and multiplication (Choudhary Princy, 2021; Heidari *et al.*, 2019). The graphene present is known to project the stimulation to produce NGF (nerve growth factor), ultimately inducing neuronal differentiation (Tupone *et al.*, 2021). A graphene-embedded electrospun PCL/gelatin nanofibrous scaffold was evaluated for proliferation on PC12 cells. Its morphological and other characterizations concluded it as an electrically conductive scaffold and a propitious material for applying neural tissue engineering (Heidari *et al.*, 2019).

Another aspect of nerve regeneration is that it is gradual and time-consuming. Traditional scaffolds cannot support the growing cells for an extended period. To overcome such a burden, the advent of shape memory polymers was done in a study in the form of a nerve conduit. The shape memory polymers drive the plasticizing effect required by the cells to overcome injury and form a nerve connection. Different *rac* lactide-to-glycolide ratio polymers were used to develop the conduit. The shape of such polymers is based on their sensitivity towards water, ultimately forming a successful trisegmented smart nerve conduit (C. Chen *et al.*, 2016).

An injectable composite hydrogel consisting of a gelatin matrix accompanying SMP fibers was prepared to treat spinal cord injury. The gelatin matrix provided a suitable microenvironment for cell assemblage, and SMP fibers preserved the microstructures even after significant distortion brought on by the injection process. Besides, fibers were employed to facilitate and direct the motor neuronal differentiation. The composite hydrogel loaded with embryonic stem cells (ESCs) was injected in mice, and improved tissue regeneration with motor function recovery was observed (C. Wang *et al.*, 2018). Inspired by SMP's smart shape responsive property, a self-producing and multichannel nerve-directing conduit with topographical cues was prepared by a degradable PLA-trimethylene carbonate copolymer. This study presented the conduit as a prospective candidate for peripheral nerve regeneration (Jing Wang *et al.*, 2020).

A hydroresponsive scaffold derived from CNT-doped silkworm sericin was fabricated with an asymmetrical shape in an ischemic stroke mouse model. It was administered through a minimal invasive method after pre-seeding with bone marrow-derived mesenchymal stem cells (BMMSCs) and was proven to enhance neuronal development (Pfau & Grunlan, 2021). 4D implementation of shape memory polymers is also used to serve as a guidance conduit for nerve regeneration (Z. Zhang *et al.*, 2019). Some recent studies on scaffolds for neural tissue engineering are mentioned in Table 3.4.

Since smart scaffolds have emerged in recent years, they are currently at their juvenile phase. Some commercially available products that are being used for neural regeneration purposes are shown in Table 3.5.

Table 3.4 Types of scaffolds for neural tissue engineering.

Scaffold type	Biomaterial	Injury/ disorder	Components-molecular	cellular	Result(s)	Reference
Hydrogel-based	3D-extrusion printed hyaluronic acid-glycidyl methacrylate	TBI (*in vitro*, *in vivo* via TBI mice model)	BDNF, GDNF	-	• Negligible *in vitro* cytotoxicity. • Enhanced extension of neurites via BDNF. • GDNF scaffolds elevated neuron–glial interactions, which makes it more efficient for *in vivo* functional and morphological recovery.	(Mishchenko *et al.*, n.d.)
	Cross-linked gelatin	TBI (*in vitro*, *in vivo* via TBI mice model)	HRP and ChOx	Murine BMSC	• Inhibition of inflammation and apoptosis. • Better neural differentiation, neurotrophin secretion	(J. Li *et al.*, 2021)
	Shape memory CNTs-SS	Preclinical ischemic stroke model ((MCAO) mouse)	-	Mouse BMSCs	• Neuronal differentiation ability. • Suitable for implantation in ischemic stroke-targeted therapy	(Jian Wang *et al.*, 2021)
	Chitosan-g-oligo(L,L-lactide) copolymer	Brain injury (*in vitro*)	-	Neonatal rat cortical cells	• Neuro-compatible, non-cytotoxic. • It suported cellular metabolic reaction to glutamate excitotoxicity.	(Grebenik *et al.*, 2020)
	CAGNF /alginate nano-composite	Nerve damage (*in vitro*, *in vivo* study)	CA	PC12 cells	• Biocompatible. • Promotion of biological activity via electrical contact and cell–cell signaling. • Significant cellular proliferation and differentiation	(Homaeigohar *et al.*, 2019)
	Sodium alginate, hyaluronic acid	TBI rat model	-	hUCMSCs	• Improved recovery in terms of memory, learning abilities, exercise, and relocation of loaded stem cells to the lesion spot.	(K. Zhang *et al.*, 2018)

(*Continued*)

Table 3.4 Types of scaffolds for neural tissue engineering. (*Continued*)

Scaffold type	Biomaterial	Injury/ disorder	Components–molecular cellular		Result(s)	Reference
Graphene	3D-(GOSA), RGOSA	-	-	DPSCs	• Smaller pore size and PLL coating imparted greater cell viability, cell adherence, and migration. • Suited for serum-free media due to non-cytotoxicity.	(Mansouri *et al.*, 2021)
	3D bacterial cellulose-graphene foam	-	-	NSCs	• Neuro-compatible. • Suited as a conductive substrate due to improved NSC multiplication, maintenance of NSC adherence, and stemness. • Rapid induction of 3D-neural networks from NSCs	(Guo *et al.*, 2021)
	Electrospun graphene-incorporated PCL/gelatin	-	TCH	PC12 cells	• Biocompatible. • Superior level of controlled drug release. • Enhanced neural cell adherence, migration, and multiplication.	(Heidari *et al.*, 2019)
Piezoelectric	Multichannel cryogel NGC	PNI (*in vitro*, *in vivo* study)	Silk fibroin, PEDOT, PVDF/ PLCL	PC12 cells	• Stimulation of neuronal fate induction, Schwann cell multiplication, growth, and myelination. • Improved *in vivo* neural regeneration and functional rescue due to piezoelectric and conductive stimuli.	(Y. Ma *et al.*, 2023a)
	PENG	PNI (sciatic nerve injury model)	-	-	• *In vivo* electrical stimulation via US exposure positively impacted the regrowth of nerves.	(Wu *et al.*, 2022)

(*Continued*)

Table 3.4 Types of scaffolds for neural tissue engineering. (*Continued*)

Scaffold type	Biomaterial	Injury/ disorder	Components-molecular	cellular	Result(s)	Reference
	3D-PLA nanofiber/ potassium sodium niobate ($K_{0.5}Na_{0.5}NbO_3$, KNN) composite	SCI (rat model)	-	-	• *In vivo*, US-mediated electrical cues stimulated NSC differentiation, nerve repair, and motor function recovery.	(P. Chen *et al.*, 2022)
	P(VDF-TrFE)) and P(VDF-TrFE)/ BTNP	*In vitro* study	$BaTiO_3$	SH-SY5Y neuroblastoma cells	• US-piezoelectric stimuli combination elicited neuronal activation, higher neurite extension, and maturation *in vitro*.	(Genchi *et al.*, 2016)
3D Bioprinted	Fibrin-based bioink	PD	SB, LDN, puro, FGF8, BFGF, BDNF	ATMSCs	• Prominent cell viability. • Neural tissue with DN-specific markers' expression.	(Restan Perez *et al.*, 2021)
Bio-scaffold	NNT	SCI (*in vitro*, *in vivo* rat model)	NT-3	TrkC-overexpressing NSCs	• NNT formation via controlled release of NT-3 *in vitro*. • Integration with host neural network *in vivo*. • Motor function restoration post-engraftment.	(G. Li *et al.*, 2021)

Abbreviation List for Table 3.4: GDNF—glial cell line-derived neurotrophic factor, BDNF—brain-derived neurotrophic factor, PC12 cells—pheochromocytoma cells, hUCMSCs—human umbilical cord mesenchymal stem cells, BMSCs—bone marrow mesenchymal stem cells, ATMSCs—adipose tissue-derived mesenchymal stem cells, TBI—traumatic brain injury, PNI—peripheral nerve injury, SCI—spinal cord injury, PD—Parkinson's disease, HRP—horseradish peroxidase, ChOx—choline oxidase, CA—citric acid, CAGNF—citric acid (CA) functionalized graphite nanofilament, CNTs-SS—carbon nanotubes (CNTs)-doped silkworm sericin scaffold, MCAO—middle cerebral artery occlusion, TCH—tetracycline hydroxide, NSCs—neural stem cells, PCL—poly ε-caprolactone, GOSA—graphene oxide/sodium alginate, RGOSA—reduced graphene oxide/sodium alginate, PLL—poly-L-lysine, PENG—piezoelectric nanogenerator, US—ultrasound, NGC—nerve guidance conduit, PEDOT—poly(3,4-ethylenedioxythiophene), PVDF—polyvinylidene fluoride, PLCL—poly(l-lactic acid-co-caprolactone), P(VDF-TrFE)—poly(vinylidene fluoride-trifluoroethylene, P(VDF-TrFE)/BTNP—poly(vinylidene fluoride-trifluoroethylene/barium titanate nanoparticle, PLA—polylactic acid, NNT—neural network tissue, TrkC—tropomyosin receptor kinase C, NT-3—neurotrophin-3, DN—dopaminergic neurons, SB—SB431542, LDN—LDN-193189, puro—purmorphamine, FGF8—fibroblast growth factor 8, bFGF—fibroblast growth factor-basic, BDNF—brain-derived neurotrophic factor.

Table 3.5 Commercial products available for neural tissue engineering.

Biomaterial	Product	Brand	References
Collagen Type 1	NeuraWrap™	Integra Life Sciences Corp. Plainsboro, NJ	(Du *et al.*, 2018; X. Gu *et al.*, 2014)
	NeuraGen™	Integra Life Sciences, Corp. Plainsboro, NJ	(Du *et al.*, 2018; X. Gu *et al.*, 2014)
	NeuroMend™	Collagen Matrix Inc. Oakland, NJ	(Du *et al.*, 2018)
	NeuraGen™ 3D	-	(Du *et al.*, 2018)
PGA	Neurotube™	Synovis Micro Companies Alliance Birmingham, AL	(Du *et al.*, 2018)
	Nerbridge™	Toyobo Co., Ltd.	(Du *et al.*, 2018)
PLCL	Neurolac™	Polyganics The Netherlands	(Du *et al.*, 2018)

3.3 Future Advancements and Techniques to Improve Efficiency of Scaffolds

3.3.1 4D Bioprinting

3D bioprinting has successfully mimicked the ECM's geometrical environment, though it failed to reproduce the dynamic environment. The 4D technology has been developed to fulfill the need for stimulus-responsive geometry. It is the futuristic form of 3D printing technology where advanced material/shape-morphing materials transform their shape over time in a simulated setting. In the presence of numerous various external stimuli, 4D printing transforms these intelligent materials into novel dimensions. These external stimuli may be physical (light, temperature, magnetic and electric field, and ultrasound), chemical (moisture and pH), or biological (biomolecule, enzymes, and cell traction force). 4D-printed materials exhibit five smart characteristics, i.e., shape memory, self-actuating, self-assembly, self-sensing, and self-healing. Some 4D printing-assisted scaffolds are mentioned in Table 3.6.

Table 3.6 4D printing-assisted materials/scaffolds in recent studies.

S. no.	Material/scaffold	Application	Effect	Stimuli	References
1.	Nano-hydroxyapatite incorporated chemically cross-linked PCL scaffold loaded with BMP2	Mandibular bone defect repair	Shape memory	Temperature	(X. Liu *et al.*, 2014)

(Continued)

Table 3.6 4D printing-assisted materials/scaffolds in recent studies. (*Continued*)

S. no.	Material/scaffold	Application	Effect	Stimuli	References
2.	Soybean oil epoxidized acrylate	Wound dressing	Shape memory	Thermal	(Miao *et al.*, 2016)
3.	Stereolithography of hydroxybutyl methacrylate-chitosan		Change in pore characteristics by thermal cross-linking of polymer	Temperature	(Seo *et al.*, 2020)
4.	Black phosphorus nanosheets and osteogenic peptide incorporated β-tricalcium phosphate, poly (lactic acid-co-trimethylene carbonate) nanocomposite scaffolds	Bone tissue engineering	Shape reconfiguration	Photothermal	(C. Wang *et al.*, 2020)
5.	Cell-laden methacrylated alginate (MA), methacrylated gelatin (GelMA), and PEGDA incorporating ultraviolet (UV) absorber and photo-initiator (PI)	Biosensing and bioactuation	UV-induced cross-linking	UV light	(A. Ding *et al.*, 2022)
6.	Limonene and β-mycrene-derived oligomeric photopolymer	Biomedical scaffold	Thermomechanical	UV irradiation	(Constant *et al.*, 2022)
7.	Gelatin methacrylate-based 3D micro-scaled hollow tubules (MHTs)	Cardiac tissue engineering	Self-molding	Water	(L. Zhang *et al.*, 2020)
8.	Glycidyl methacrylate and carbon nanotubes nanocomposite cryogel	Wound healing and hemostasis	Shape recovery	Blood-triggered	(X. Zhao *et al.*, 2018)
9.	PCL and amino-capped aniline trimers	Myogenic differentiation	Stretchable; electroactive	Temperature	(Z. Deng *et al.*, 2016)
10.	Polyurethane (PU) elastomers, citric acid-PCL, dopamine, and aniline hexamer composite		Stretchable, bioactive, electroactive	Temperature	(X. Zhao *et al.*, 2017)

3.3.2 MD Simulation

As composite biocompatible scaffolds are increasingly used in modern tissue engineering, there is a rising need for obtaining the finest or perfect biocompatibility that promotes optimal cell attachment and growth. Scaffolds made up of polymeric composites including different kind of polymers, like hydrophobic and hydrophilic ones, are popular these days because they provide better control over scaffold properties. Although using different proportions of each polymer imparts distinct features, identifying the correct optimal ratio of polymers to deliver the best-desired features remains a challenging concern for modern tissue engineering. Despite the availability of experimental assays to evaluate the scaffold characteristics, it is not feasible to perform them for each proposed structure due to time and financial constraints. In such a case, the computational approach, offering a fair assessment of cell adhesion on the constructs, can be a considerable technique in screening the samples, accelerating the stage of optimization. Remarkably, this technique is practicably beneficial for the composite constructs/scaffolds, i.e., those made up of different polymers and also for preparing novel polymer constructs (Abpeikar *et al.*, 2022; Sarmadi *et al.*, 2017).

It is widely accepted that the success or non-success of scaffolds is strongly determined by their capacity to specifically adsorb target proteins, specifically those constituting the ECM because they play a key responsibility in early cell adhesion and growth (Rabe *et al.*, 2011; Roach *et al.*, 2005; Sarmadi *et al.*, 2017). Moreover, the unwanted non-specific blood protein adsorption in medical implants may raise the risk of blood clotting. In addition to precisely accommodating the adsorption procedure, a prudent computational method, competent in estimating protein adsorption level, should be capable of anticipating the cell's adherence ability to the substrate (Shamloo *et al.*, 2015). In recent years, molecular dynamics (MD) simulation turned up as an effective tool for simulating various biotechnological phenomena at the nanoscale level (Shamloo *et al.*, 2016). It can explicitly model completely atomistic interactions among biomolecules and surfaces in nanoscales.

MD simulation has been utilized for a variety of surfaces, such as structured surfaces, carbon nanoconstructs, or polymeric amorphous surfaces (Abpeikar *et al.*, 2022; Sarmadi *et al.*, 2017). Protein adsorption on inorganic surfaces is a significant concern for biomedical fields (Abpeikar *et al.*, 2022; Gray, 2004). With a good understanding of scaffolds and MD simulation, the best protein selection can be made, and further, the gene of the same protein can be utilized for gene delivery purposes, contributing to the upliftment of the tissue engineering approach. Also, the computational method can help select the best scaffold for efficient protein adsorption at the surface by evaluating the protein adsorption with different scaffolds (Abpeikar *et al.*, 2022).

A research group has used MD simulations to model protein adsorption on heterogenous polymeric surfaces with different proportions of PVA and PCL in a water-like medium. Two main ECM proteins, i.e., collagen type I and fibronectin, were taken for the simulation setting. In addition to simulation results, the experimental study showed that surface hydrophobicity is directly proportional to the initial cell adherence and proliferation, which was particularly prevalent in samples with a PCL greater than 70% (Sarmadi *et al.*, 2017). MD simulations have also been used to study the dynamic behavior of bone morphogenetic protein-2 (BMP-2) on a flat gold substrate commonly used in medical devices. Protein adsorption was evaluated at three different temperatures, viz., room temperature (293 K), *in vivo* environment (310 K), and boiling point of water (373 K). The analysis of

BMP-2 interaction at various physiological conditions depicts various biochemical parameters for tissue engineering (Marquetti & Desai, 2018). Moreover, the initial phase of BMP-2 adsorption while approaching two implant surfaces, hydrophobic graphite and hydrophilic titanium dioxide rutile, was examined. This was done using classical molecular dynamics (MD) accompanied by a hybrid model consisting of steered molecular dynamics (SMD) and MD simulations (Utesch *et al.*, 2011). SMD simulation was found to accelerate protein adsorption and desorption on material surfaces. Many researchers are combining MD and SMD simulations to understand the interaction between biomolecules and biomaterials. For instance, MD and SMD simulation were merged to explore the mechanical behavior of osteopontin residues on hydroxyapatite surfaces.

MD has always been a unique technique for determining biological mechanisms. As MD simulation can practically reduce the cost and time and aids in finding new composites, the judicious use of MD simulations in combination with simultaneous experimental techniques could be an efficient way to advance tissue engineering and other related fields (Abpeikar *et al.*, 2022; Mitchell *et al.*, 2022).

3.4 Conclusion

Smart scaffolds usher the tissue engineering field in a new dawn in the era of smart technologies. With their limitless possibilities, smart materials have brought hope for treating several life-threatening diseases requiring organ replacement. With the myriad of properties like shape memory feature, external and internal stimuli responses are continuously being explored in different biomedical aspects and further for new prospective applications. This chapter has discussed the properties of smart responsive materials and recent studies that utilize those properties. Despite the encouraging results of existing literature, smart materials are still in their infancy, with various obstacles and issues that must be confronted in future studies. Their immune response, metabolic pathways, pharmacological acquisition, and long-term toxicity have not yet been explored. Meanwhile, clinical translation is still a hurdle, resulting in the non-availability of smart scaffolds in the market currently. Given the current pace of research in the field, smart scaffolds are anticipated to have significant biomedical applications very soon.

References

Abbasi, A. R., Sohail, M., Minhas, M. U., Khaliq, T., Kousar, M., Khan, S., Hussain, Z., & Munir, A. (2020). Bioinspired sodium alginate based thermosensitive hydrogel membranes for accelerated wound healing. *Int. J. Biol. Macromol.*, *155*, 751–765. https:// doi.org/10.1016/j.ijbiomac.2020.03.248

Abpeikar, Z., Alizadeh, A. A., Ahmadyousefi, Y., Najafi, A. A., & Safaei, M. (2022). Engineered cells along with smart scaffolds: critical factors for improving tissue engineering approaches. *Regener. Med.*, *17*(11), 855–876. https://doi.org/10.2217/rme-2022-0059

Ahadian, S., & Khademhosseini, A. (2018). Smart scaffolds in tissue regeneration. *Regener. Biomater.*, *5*(3), 125–128. https://doi.org/10.1093/rb/rby007

Ansari, Mehdi, Golzar, M., Baghani, M., Abbasishirsavar, M., & Taghavimehr, M. (2019). Force recovery evaluation of thermo-induced shape-memory polymer stent: material, process and thermo-viscoelastic characterization. *Smart Mater. Struct.*, *28*(9), 095022. https:// doi.org/10.1088/1361-665X/ab28fc

Ansari, Mojtaba. (2019). Bone tissue regeneration: biology, strategies and interface studies. *Prog. Biomater.*, *8*(4), 223–237. https://doi.org/10.1007/s40204-019-00125-z

Atala, A. (2004). Tissue Engineering and Regenerative Medicine: Concepts for Clinical Application. *Rejuvenation Res.*, *7*(1), 15–31. https://doi.org/10.1089/1549168043231O5053

Bai, Y., Dai, X., Yin, Y., Wang, J., Sun, X., Liang, W., Li, Y., Deng, X., & Zhang, X. (2019). Biomimetic piezoelectric nanocomposite membranes synergistically enhance osteogenesis of deproteinized bovine bone grafts. *Int. J. Nanomed.*, *14*, 3015–3026. https://doi. org/10.2147/IJN.S197824

Beldjilali-Labro, M., Garcia Garcia, A., Farhat, F., Bedoui, F., Grosset, J.-F., Dufresne, M., & Legallais, C. (2018). Biomaterials in Tendon and Skeletal Muscle Tissue Engineering: Current Trends and Challenges. *Materials*, *11*(7), 1116. https://doi.org/10.3390/ma11071116

Bolonduro, O. A., Duffy, B. M., Rao, A. A., Black, L. D., & Timko, B. P. (2020). From biomimicry to bioelectronics: Smart materials for cardiac tissue engineering. *Nano Res.*, *13*(5), 1253–1267. https://doi.org/10.1007/s12274-020-2682-3

Carletti, E., Motta, A., & Migliaresi, C. (2011). *Scaffolds for Tissue Engineering and 3D Cell Culture* (pp. 17–39). https://doi.org/10.1007/978-1-60761-984-0_2

Chen, C., Hu, J., Huang, H., Zhu, Y., & Qin, T. (2016). Design of a Smart Nerve Conduit Based on a Shape-Memory Polymer. *Adv. Mater. Technol.*, *1*(4), 1600015. https://doi. org/10.1002/admt.201600015

Chen, G., Ushida, T., & Tateishi, T. (2002). Scaffold Design for Tissue Engineering. *Macromol. Biosci.*, *2*(2), 67–77. https://doi.org/10.1002/1616-5195(20020201)2:2<67::AID-MABI67>3.0.CO;2-F

Chen, P., Xu, C., Wu, P., Liu, K., Chen, F., Chen, Y., Dai, H., & Luo, Z. (2022). Wirelessly Powered Electrical-Stimulation Based on Biodegradable 3D Piezoelectric Scaffolds Promotes the Spinal Cord Injury Repair. *ACS Nano*, *16*(10), 16513–16528. https://doi.org/10.1021/acsnano.2c05818

Cheng, Y., Xu, Y., Qian, Y., Chen, X., Ouyang, Y., & Yuan, W.-E. (2020). 3D structured self-powered PVDF/PCL scaffolds for peripheral nerve regeneration. *Nano Energy*, *69*, 104411. https://doi.org/10.1016/j.nanoen.2019.104411

Choudhary, P., Gupta, A., & Singh, S. (2021). Therapeutic Advancement in Neuronal Transdifferentiation of Mesenchymal Stromal Cells for Neurological Disorders. *J. Mol. Neurosci.*, *71*(5), 889–901. https://doi.org/10.1007/s12031-020-01714-5

Choudhary, P., Gupta, S., Shukla, R., Gupta, A., Pahal, S., & Singh, S. (2021). Regulation of neuronal repair and regeneration through inhibition of oligodendrocyte myelin glycoprotein (OMgp). *J. Biomol. Struct. Dyn.*, 1–17. https://doi.org/10.1080/07391102.2021.1997820

Choudhary, P., Jaiswal, A., Singh, S., & Gupta, S. K. (2020). Bacterial cellulose based composites: Preparation and characterization. *Mater. Sci. Forum*, *978 MSF*, 183–190. https://doi.org/10.4028/www.scientific.net/MSF.978.183

Choudhary Princy, S. S. (2021). Targeted Drug Delivery. In K. M. Rajesh K. Kesharwani (Ed.), *Biotechnology in the Modern Medicinal System* (pp. 55–74). Apple Academic Press. https://doi.org/10.1201/9781003129783

Chowdhury, N. A., & Al-Jumaily, A. M. (2016). Regenerated cellulose/polypyrrole/silver nanoparticles/ionic liquid composite films for potential wound healing applications. *Wound Med.*, *14*, 16–18. https://doi.org/10.1016/j.wndm.2016.07.001

Constant, E., King, O., & Weems, A. C. (2022). Bioderived 4D Printable Terpene Photopolymers from Limonene and β-Myrcene. *Biomacromolecules*, *23*(6), 2342–2352. https://doi.org/10.1021/acs.biomac.2c00085

Crasto, G. J., Kartner, N., Reznik, N., Spatafora, M. V., Chen, H., Williams, R., Burns, P. N., Clokie, C., Manolson, M. F., & Peel, S. A. F. (2016). Controlled bone formation using ultrasound-triggered release of BMP-2 from liposomes. *J. Controlled Release, 243*, 99–108. https:// doi.org/10.1016/J.JCONREL.2016.09.032

Deng, Y., Shi, X., Chen, Y., Yang, W., Ma, Y., Shi, X. L., Song, P., Dargusch, M. S., & Chen, Z. G. (2020). Bacteria-Triggered pH-Responsive Osteopotentiating Coating on 3D-Printed Polyetheretherketone Scaffolds for Infective Bone Defect Repair. *Ind. Eng. Chem. Res.*, *59*(26), 12123–12135. https://doi.org/10.1021/ACS.IECR.0C02107/ASSET/IMAGES/LARGE/IE0C02107_0008.JPEG

Deng, Z., Guo, Y., Zhao, X., Li, L., Dong, R., Guo, B., & Ma, P. X. (2016). Stretchable degradable and electroactive shape memory copolymers with tunable recovery temperature enhance myogenic differentiation. *Acta Biomater.*, *46*, 234–244. https://doi.org/10.1016/J.ACTBIO.2016.09.019

Ding, A., Lee, S. J., Ayyagari, S., Tang, R., Huynh, C. T., & Alsberg, E. (2022). 4D biofabrication via instantly generated graded hydrogel scaffolds. *Bioactive Materials*, *7*, 324–332. https://doi.org/10.1016/j.bioactmat.2021.05.021

Ding, Y., Hao, Y., Yuan, Z., Tao, B., Chen, M., Lin, C., Liu, P., & Cai, K. (2020). A dual-functional implant with an enzyme-responsive effect for bacterial infection therapy and tissue regeneration. *Biomater. Sci.*, *8*(7), 1840–1854. https://doi.org/10.1039/C9BM01924C

Dong, S., Chen, Y., Yu, L., Lin, K., & Wang, X. (2020). Magnetic Hyperthermia–Synergistic H2O2 Self-Sufficient Catalytic Suppression of Osteosarcoma with Enhanced Bone-Regeneration Bioactivity by 3D-Printing Composite Scaffolds. *Adv. Funct. Mater.*, *30*(4), 1907071. https://doi.org/10.1002/ADFM.201907071

dos Santos, M. R., Alcaraz-Espinoza, J. J., da Costa, M. M., & de Oliveira, H. P. (2018). Usnic acid-loaded polyaniline/polyurethane foam wound dressing: preparation and bactericidal activity. *Mater. Sci. Eng.: C*, *89*, 33–40. https://doi.org/10.1016/j.msec.2018.03.019

Du, J., Chen, H., Qing, L., Yang, X., & Jia, X. (2018). Biomimetic neural scaffolds: a crucial step towards optimal peripheral nerve regeneration. *Biomater. Sci.*, *6*(6), 1299–1311.

Dwivedi, S., Choudhary, P., Gupta, A., & Singh, S. (2023). Therapeutical growth in oligodendroglial fate induction via transdifferentiation of stem cells for neuroregenerative therapy. *Biochimie*, *211*, 35–56. https://doi.org/10.1016/j.biochi.2023.02.012

Eltom, A., Zhong, G., & Muhammad, A. (2019). Scaffold Techniques and Designs in Tissue Engineering Functions and Purposes: A Review. *Adv. Mater. Sci. Eng.*, *2019*, 1–13. https://doi.org/10.1155/2019/3429527

Furth, M. E., Atala, A., & Van Dyke, M. E. (2007). Smart biomaterials design for tissue engineering and regenerative medicine. *Biomaterials*, *28*(34), 5068–5073. https://doi.org/10.1016/j.biomaterials.2007.07.042

Gao, H., Xiao, J., Wei, Y., Wang, H., Wan, H., & Liu, S. (2021). Regulation of Myogenic Differentiation by Topologically Microgrooved Surfaces for Skeletal Muscle Tissue Engineering. *ACS Omega*, *6*(32), 20931–20940. https://doi.org/10.1021/acsomega.1c02347

Genchi, G. G., Ceseracciu, L., Marino, A., Labardi, M., Marras, S., Pignatelli, F., Bruschini, L., Mattoli, V., & Ciofani, G. (2016). P (VDF-TrFE)/BaTiO3 nanoparticle composite films mediate piezo-electric stimulation and promote differentiation of SH-SY5Y neuroblastoma cells. *Adv. Healthcare Mater.*, *5*(14), 1808–1820.

Gizdavic-Nikolaidis, M. R., Bennett, J. R., Swift, S., Easteal, A. J., & Ambrose, M. (2011). Broad spectrum antimicrobial activity of functionalized polyanilines. *Acta Biomater.*, *7*(12), 4204–4209. https://doi.org/10.1016/j.actbio.2011.07.018

Gray, J. J. (2004). The interaction of proteins with solid surfaces. *Curr. Opin. Struct. Biol.*, *14*(1), 110–115. https://doi.org/10.1016/j.sbi.2003.12.001

Grebenik, E. A., Surin, A. M., Bardakova, K. N., Demina, T. S., Minaev, N. V, Veryasova, N. N., Artyukhova, M. A., Krasilnikova, I. A., Bakaeva, Z. V, & Sorokina, E. G. (2020). Chitosan-g-oligo (L, L-lactide) copolymer hydrogel for nervous tissue regeneration in glutamate excitotoxicity: *in vitro* feasibility evaluation. *Biomed. Mater.*, *15*(1), 15011.

Gu, B. K., Park, S. J., & Kim, C.-H. (2018). Beneficial effect of aligned nanofiber scaffolds with electrical conductivity for the directional guide of cells. *J. Biomater. Sci. Polym. Ed.*, *29*(7–9), 1053–1065. https://doi.org/10.1080/09205063.2017.1364097

Gu, X., Ding, F., & Williams, D. F. (2014). Neural tissue engineering options for peripheral nerve regeneration. *Biomaterials*, *35*(24), 6143–6156.

Guo, R., Li, J., Chen, C., Xiao, M., Liao, M., Hu, Y., Liu, Y., Li, D., Zou, J., & Sun, D. (2021). Biomimetic 3D bacterial cellulose-graphene foam hybrid scaffold regulates neural stem cell proliferation and differentiation. *Colloids Surf., B*, *200*, 111590.

Gupta, A., & Singh, S. (2022). Potential Role of Growth Factors Controlled Release in Achieving Enhanced Neuronal Trans-differentiation from Mesenchymal Stem Cells for Neural Tissue Repair and Regeneration. *Mol. Neurobiol.*, *59*(2), 983–1001. https://doi.org/10.1007/s12035-021-02646-w

Han, F., Wang, J., Ding, L., Hu, Y., Li, W., Yuan, Z., Guo, Q., Zhu, C., Yu, L., Wang, H., Zhao, Z., Jia, L., Li, J., Yu, Y., Zhang, W., Chu, G., Chen, S., & Li, B. (2020). Tissue Engineering and Regenerative Medicine: Achievements, Future, and Sustainability in Asia. *Front. Bioeng. Biotechnol.*, *8*. https://doi.org/10.3389/fbioe.2020.00083

Heidari, M., Bahrami, S. H., Ranjbar-Mohammadi, M., & Milan, P. B. (2019). Smart electrospun nanofibers containing PCL/gelatin/graphene oxide for application in nerve tissue engineering. *Mater. Sci. Eng., C*, *103*, 109768. https://doi.org/10.1016/j.msec.2019.109768

Holman, H., Kavarana, M. N., & Rajab, T. K. (2021). Smart materials in cardiovascular implants: Shape memory alloys and shape memory polymers. *Artif. Organs*, *45*(5), 454–463. https:// doi.org/10.1111/aor.13851

Homaeigohar, S., Tsai, T.-Y., Young, T.-H., Yang, H. J., & Ji, Y.-R. (2019). An electroactive alginate hydrogel nanocomposite reinforced by functionalized graphite nanofilaments for neural tissue 45.engineering. *Carbohydr. Polym.*, *224*, 115112.

Hu, Z., Ma, C., Rong, X., Zou, S., & Liu, X. (2018). Immunomodulatory ECM-like Microspheres for Accelerated Bone Regeneration in Diabetes Mellitus. *ACS Appl. Mater. Interfaces*, *10*(3), 2377–2390. https://doi.org/10.1021/ACSAMI.7B18458

Huang, Y., Zhang, L., Ji, Y., Deng, H., Long, M., Ge, S., Su, Y., Chan, S. Y., Loh, X. J., Zhuang, A., & Ruan, J. (2023). A non-invasive smart scaffold for bone repair and monitoring. *Bioact. Mater.*, *19*, 499–510. https://doi.org/10.1016/J.BIOACTMAT.2022.04.034

Jacob, J., More, N., Kalia, K., & Kapusetti, G. (2018a). Piezoelectric smart biomaterials for bone and cartilage tissue engineering. *Inflammation and Regeneration*, *38*(1), 1–11. https://doi.org/10.1186/s41232-018-0059-8

Jacob, J., More, N., Kalia, K., & Kapusetti, G. (2018b). Piezoelectric smart biomaterials for bone and cartilage tissue engineering. *Inflammation and Regeneration*, *38*(1), 2. https://doi.org/10.1186/s41232-018-0059-8

Jacob, J., More, N., Mounika, C., Gondaliya, P., Kalia, K., & Kapusetti, G. (2019). Smart Piezoelectric Nanohybrid of Poly(3-hydroxybutyrate- co-3-hydroxyvalerate) and Barium Titanate for Stimulated Cartilage Regeneration. *ACS Appl. Bio Mater.*, *2*(11), 4922–4931. https://doi.org/10.1021/acsabm.9b00667

Ju, X., Liu, X., Zhang, Y., Chen, X., Chen, M., Shen, H., Feng, Y., Liu, J., Wang, M., & Shi, Q. (2022). A photo-crosslinked proteinogenic hydrogel enabling self-recruitment of endogenous TGF-β1 for cartilage regeneration. *Smart Mater. Med.*, *3*(October 2021), 85–93. https://doi. org/10.1016/j. smaim.2021.12.002

Justin, G. A., Zhu, S., Nicholson, T. R., Maskrod, J., Mbugua, J., Chase, M., June-ho Jung, & Mercado, R. M. L. (2012). On-demand controlled release of anti-inflammatory and analgesic drugs from conducting polymer films to aid in wound healing. *2012 Annual International Conference of the IEEE Engineering in Medicine and Biology Society*, 1206–1209. https://doi.org/10.1109/ EMBC.2012.6346153

K, N., & Rout, C. S. (2021). Conducting polymers: a comprehensive review on recent advances in synthesis, properties and applications. *RSC Adv.*, *11*(10), 5659–5697. https://doi.org/10.1039/ D0RA07800J

Kaliva, M., Chatzinikolaidou, M., & Vamvakaki, M. (n.d.). *CHAPTER 1. Applications of Smart Multifunctional Tissue Engineering Scaffolds* (pp. 1–38). https://doi.org/10.1039/978178 8010542-00001

Klimek, K., & Ginalska, G. (2020). Proteins and Peptides as Important Modifiers of the Polymer Scaffolds for Tissue Engineering Applications—A Review. *Polymers*, *12*(4), 844. https://doi. org/10.3390/polym12040844

Korupalli, C., Li, H., Nguyen, N., Mi, F., Chang, Y., Lin, Y., & Sung, H. (2021). Conductive Materials for Healing Wounds: Their Incorporation in Electroactive Wound Dressings, Characterization, and Perspectives. *Adv. Healthcare Mater.*, *10*(6), 2001384. https://doi.org/10.1002/ adhm.202001384

Kowalski, P. S., Bhattacharya, C., Afewerki, S., & Langer, R. (2018). Smart Biomaterials: Recent Advances and Future Directions. *ACS Biomater. Sci. Eng.*, *4*(11), 3809–3817. https://doi. org/10.1021/acsbiomaterials.8b00889

Krishna, A. S., Radhakumary, C., & Sreenivasan, K. (2015). Calcium ion modulates protein release from chitosan-hyaluronic acid poly electrolyte gel. *Polym. Eng. Sci.*, *55*(9), 2089–2097. https:// doi.org/10.1002/pen.24050

Lee, J. Y., & Schmidt, C. E. (2015). Amine-functionalized polypyrrole: Inherently cell adhesive conducting polymer. *J. Biomed. Mater. Res. Part A*, *103*(6), 2126–2132. https:// doi.org/10.1002/ jbm.a.35344

Li, G., Zhang, B., Sun, J., Shi, L., Huang, M., Huang, L., Lin, Z., Lin, Q., Lai, B., & Ma, Y. (2021). An NT-3-releasing bioscaffold supports the formation of TrkC-modified neural stem cell-derived neural network tissue with efficacy in repairing spinal cord injury. *Bioact. Mater.*, *6*(11),3766–3781.

Li, J., Zhang, D., Guo, S., Zhao, C., Wang, L., Ma, S., Guan, F., & Yao, M. (2021). Dual-enzymatically cross-linked gelatin hydrogel promotes neural differentiation and neurotrophin secretion of bone marrow-derived mesenchymal stem cells for treatment of moderate traumatic brain injury. *Int. J. Biol. Macromol.*, *187*, 200–213.

Lin, X., Wang, Q., Gu, C., Li, M., Chen, K., Chen, P., Tang, Z., Liu, X., Pan, H., Liu, Z., Tang, R., & Fan, S. (2020). Smart Nanosacrificial Layer on the Bone Surface Prevents Osteoporosis through Acid-Base Neutralization Regulated Biocascade Effects. *J. Am. Chem. Soc.*, *142*(41), 17543–17556. https://doi.org/10.1021/JACS.0C07309/ASSET/IMAGES/LARGE/JA0C07309_0006.JPEG

Liu, X., Zhao, K., Gong, T., Song, J., Bao, C., Luo, E., Weng, J., & Zhou, S. (2014). Delivery of growth factors using a smart porous nanocomposite scaffold to repair a mandibular bone defect. *Biomacromolecules*, *15*(3), 1019–1030. https://doi.org/10.1021/bm401911p

Liu, Y., Zhang, X., Cao, C., Zhang, Y., Wei, J., Li, Y. jun, Liang, W., Hu, Z., Zhang, J., Wei, Y., & Deng, X. (2017). Built-In Electric Fields Dramatically Induce Enhancement of Osseointegration. *Adv. Funct. Mater.*, *27*(47), 1703771. https://doi.org/10.1002/adfm.201703771

Lu, H., De Winter, R. J., & Koch, K. T. (2018). The STENTYS self-apposing stent technology in coronary artery disease: literature review and future directions. *Expert Rev. Med. Devices*, *15*(7), 479–487. https://doi.org/10.1080/17434440.2018.1491305

Ma, H., Li, T., Huan, Z., Zhang, M., Yang, Z., Wang, J., Chang, J., & Wu, C. (2018). 3D printing of high-strength bioscaffolds for the synergistic treatment of bone cancer. *NPG Asia Mater.*, *10*(4), 31–44. https://doi.org/10.1038/s41427-018-0015-8

Ma, L., Feng, X., Liang, H., Wang, K., Song, Y., Tan, L., Wang, B., Luo, R., Liao, Z., Li, G., Liu, X., Wu, S., & Yang, C. (2020). A novel photothermally controlled multifunctional scaffold for clinical treatment of osteosarcoma and tissue regeneration. *Mater. Today*, *36*, 48–62. https://doi.org/10.1016/J.MATTOD.2019.12.005

Ma, Y., Wang, H., Wang, Q., Cao, X., & Gao, H. (2023a). Piezoelectric conduit combined with multi-channel conductive scaffold for peripheral nerve regeneration. *Chem. Eng. J.*, *452*, 139424.

Ma, Y., Wang, H., Wang, Q., Cao, X., & Gao, H. (2023b). Piezoelectric conduit combined with multi-channel conductive scaffold for peripheral nerve regeneration. *Chem. Eng. J.*, *452*, 139424. https://doi.org/10.1016/j.cej.2022.139424

Manoharan, G., Van Mieghem, N. M., Windecker, S., Bosmans, J., Bleiziffer, S., Modine, T., Linke, A., Scholtz, W., Chevalier, B., Gooley, R., Zeng, C., Oh, J. K., & Grube, E. (2018). 1-Year Outcomes With the Evolut R Self-Expanding Transcatheter Aortic Valve. *JACC Cardiovasc. Interv.*, *11*(22), 2326–2334. https://doi.org/10.1016/j.jcin.2018.07.032

Mansouri, N., Al-Sarawi, S., Losic, D., Mazumdar, J., Clark, J., Gronthos, S., & O'Hare Doig, R. (2021). Biodegradable and biocompatible graphene-based scaffolds for functional neural tissue engineering: A strategy approach using dental pulp stem cells and biomaterials. *Biotechnol. Bioeng.*, *118*(11), 4217–4230.

Maráková, N., Humpolíček, P., Kašpárková, V., Capáková, Z., Martinková, L., Bober, P., Trchová, M., & Stejskal, J. (2017). Antimicrobial activity and cytotoxicity of cotton fabric coated with conducting polymers, polyaniline or polypyrrole, and with deposited silver nanoparticles. *Appl. Surf. Sci.*, *396*, 169–176. https://doi.org/10.1016/j.apsusc.2016.11.024

Marquetti, I., & Desai, S. (2018). Molecular modeling of bone morphogenetic protein for tissue engineering applications. *IISE Annual Conference and Expo 2018, May*, 1759–1764.

Miao, S., Zhu, W., Castro, N. J., Nowicki, M., Zhou, X., Cui, H., Fisher, J. P., & Zhang, L. G. (2016). 4D printing smart biomedical scaffolds with novel soybean oil epoxidized acrylate. *Sci. Rep.*, *6*(1), 27226. https://doi.org/10.1038/srep27226

Miguel, S. P., Sequeira, R. S., Moreira, A. F., Cabral, C. S. D., Mendonça, A. G., Ferreira, P., & Correia, I. J. (2019). An overview of electrospun membranes loaded with bioactive molecules for improving the wound healing process. *Eur. J. Pharm. Biopharm.*, *139*, 1–22. https://doi.org/10.1016/j.ejpb.2019.03.010

Mishchenko, T. A., Novozhilova, M. O., Kuznetsova, A. I., Savelyev, A. G., Sochilina, A. V, Mariyanats, A. O., Popov, V. K., Khaydukov, E. V, Zvyagin, A. V, & Vedunova, M. V. (n.d.). 3D-printed hyaluronic acid hydrogel scaffolds impregnated with neurotrophic factors (BDNF, GDNF) for post-traumatic brain tissue reconstruction. *Front. Bioeng. Biotechnol.*, 1344.

Mitchell, S. M., Pajovich, H. T., Broas, S. M., Hugo, M. M., & Banerjee, I. A. (2022). Molecular dynamics simulations and *in vitro* studies of hybrid decellularized leaf-peptide-polypyrrole composites for potential tissue engineering applications. *J. Biomol. Struct. Dyn.*, 1–16. https://doi.org/10.1080/07391102.2021.2023643

Montoya, C., Du, Y., Gianforcaro, A. L., Orrego, S., Yang, M., & Lelkes, P. I. (2021). On the road to smart biomaterials for bone research: definitions, concepts, advances, and outlook. *Bone Res.*, *9*(1), 12. https://doi.org/10.1038/s41413-020-00131-z

Nazarzadeh Zare, E., Mansour Lakouraj, M., & Mohseni, M. (2014). Biodegradable polypyrrole/dextrin conductive nanocomposite: Synthesis, characterization, antioxidant and antibacterial activity. *Synth. Met.*, *187*, 9–16. https://doi.org/10.1016/j.synthmet.2013.09.045

Nguyen, T. M., Lee, S., & Lee, S. B. (2014). Conductive polymer nanotube patch for fast and controlled *ex vivo* transdermal drug delivery. *Nanomedicine*, *9*(15), 2263–2272. https://doi.org/10.2217/nnm.13.153

O'Brien, F. J. (2011). Biomaterials & scaffolds for tissue engineering. *Mater. Today*, *14*(3),88–95. https://doi.org/10.1016/S1369-7021(11)70058-X

Oliveira, F. R. A. de, Mattos, L. A. P. e, Abizaid, A., Abizaid, A. S., Costa, J. R., Costa, R., Staico, R., Botelho, R., Sousa, J. E., & Sousa, A. (2013). Miniaturized Self-Expanding Drug-Eluting Stent in Small Coronary Arteries: Late Effectiveness. *Arq. Bras. Cardiol.*, https://doi. org/10.5935/abc.20130199

Pahal, S., Gupta, A., Choudhary, P., Chaudhary, A., & Singh, S. (2021). Network pharmacological evaluation of Withania somnifera bioactive phytochemicals for identifying novel potential inhibitors against neurodegenerative disorder. *J. Biomol. Struct. Dyn.*, 1–12. https://doi.org/10.1080/07391102.2021.1951355

Palmieri, V., Sciandra, F., Bozzi, M., De Spirito, M., & Papi, M. (2020). 3D Graphene Scaffolds for Skeletal Muscle Regeneration: Future Perspectives. *Front. Bioeng. Biotechnol.*, *8*. https://doi.org/10.3389/fbioe.2020.00383

Pfau, M. R., & Grunlan, M. A. (2021). Smart scaffolds: shape memory polymers (SMPs) in tissue engineering. *J. Mater. Chem. B*, *9*(21), 4287–4297. https://doi.org/10.1039/ D1TB00607J

Qi, H., Yang, L., Li, X., Sun, X., Zhao, J., Hou, X., Li, Z., Yuan, X., Cui, Z., & Yang, X. (2019). Systemic administration of enzyme-responsive growth factor nanocapsules for promoting bone repair. *Biomater. Sci.*, *7*(4), 1675–1685. https://doi.org/10.1039/C8BM01632A

Rabe, M., Verdes, D., & Seeger, S. (2011). Understanding protein adsorption phenomena at solid surfaces. *Adv. Colloid Interface Sci.*, *162*(1–2), 87–106. https://doi.org/10.1016/j. cis.2010.12.007

Restan Perez, M., Sharma, R., Masri, N. Z., & Willerth, S. M. (2021). 3D Bioprinting mesenchymal stem cell-derived neural tissues using a fibrin-based bioink. *Biomolecules*, *11*(8), 1250.

Ribeiro, C., Correia, D. M., Rodrigues, I., Guardão, L., Guimarães, S., Soares, R., & Lanceros-Méndez, S. (2017). *In vivo* demonstration of the suitability of piezoelectric stimuli for bone reparation. *Mater. Lett.*, *209*, 118–121. https://doi.org/10.1016/J.MATLET.2017.07.099

Roach, P., Farrar, D., & Perry, C. C. (2005). Interpretation of Protein Adsorption: Surface-Induced Conformational Changes. *J. Am. Chem. Soc.*, *127*(22), 8168–8173. https://doi.org/10.1021/ja042898o

Rogers, G. F., & Greene, A. K. (2012). Autogenous bone graft: basic science and clinical implications. *J. Craniofac. Surg.*, *23*(1), 323–327. https://doi.org/10.1097/SCS.0B013E318241DCBA

Roseti, L., Parisi, V., Petretta, M., Cavallo, C., Desando, G., Bartolotti, I., & Grigolo, B. (2017). Scaffolds for Bone Tissue Engineering: State of the art and new perspectives. *Mater. Sci. Eng., C*, *78*, 1246–1262. https://doi.org/10.1016/j.msec.2017.05.017

Saidy, N. T., Wolf, F., Bas, O., Keijdener, H., Hutmacher, D. W., Mela, P., & De-Juan-Pardo, E. M. (2019). Biologically Inspired Scaffolds for Heart Valve Tissue Engineering via Melt Electrowriting. *Small*, *15*(24), 1900873. https://doi.org/10.1002/smll.201900873

Santo, L., Quadrini, F., Bellisario, D., & Iorio, L. (2020). *Applications of Shape-Memory Polymers, and Their Blends and Composites*, (pp. 311–329). https://doi.org/10.1007/978-981-13-8574-2_13

Sarmadi, M., Shamloo, A., & Mohseni, M. (2017). Utilization of Molecular Dynamics Simulation Coupled with Experimental Assays to Optimize Biocompatibility of an Electrospun PCL/PVA Scaffold. *PLoS One*, *12*(1), e0169451. https://doi.org/10.1371/journal.pone.0169451

Seo, J. W., Shin, S. R., Park, Y. J., & Bae, H. (2020). Hydrogel Production Platform with Dynamic Movement Using Photo-Crosslinkable/Temperature Reversible Chitosan Polymer and Stereolithography 4D Printing Technology. *Tissue Eng. Regener. Med.*, *17*(4), 423–431. https://doi.org/10.1007/s13770-020-00264-6

Serrano, M. C., & Ameer, G. A. (2012). Recent Insights Into the Biomedical Applications of Shape- memory Polymers. *Macromol. Biosci.*, *12*(9), 1156–1171. https://doi.org/10.1002/mabi.201200097

Shamloo, A., Mohammadaliha, N., & Mohseni, M. (2015). Integrative Utilization of Microenvironments, Biomaterials and Computational Techniques for Advanced Tissue Engineering. *J. Biotechnol.*, *212*, 71–89. https://doi.org/10.1016/j.jbiotec.2015.08.005

Shamloo, A., Pedram, M. Z., Heidari, H., & Alasty, A. (2016). Computing the blood brain barrier (BBB) diffusion coefficient: A molecular dynamics approach. *J. Magn. Magn. Mater.*, *410*, 187–197. https://doi.org/10.1016/j.jmmm.2016.03.030

Shen, Y., Xu, G., Huang, H., Wang, K., Wang, H., Lang, M., Gao, H., & Zhao, S. (2021). Sequential Release of Small Extracellular Vesicles from Bilayered Thiolated Alginate/Polyethylene Glycol Diacrylate Hydrogels for Scarless Wound Healing. *ACS Nano*, *15*(4), 6352–6368. https://doi.org/10.1021/acsnano.0c07714

Shukla, R., Pahal, S., Gupta, A., Choudhary, P., Misra, K., & Singh, S. (2022). Modulation of GPCR receptors common to gut inflammatory diseases and neuronal disorders, Alzheimer's and Parkinson's diseases as druggable targets through Withania somnifera bioactives: an in silico study. *J. Biomol. Struct. Dyn.*, 1–19. https://doi.org/10.1080/07391102.2022.2068072

Solazzo, M., O'Brien, F. J., Nicolosi, V., & Monaghan, M. G. (2019). The rationale and emergence of electroconductive biomaterial scaffolds in cardiac tissue engineering. *APL Bioeng.*, *3*(4),041501. https://doi.org/10.1063/1.5116579

Steel, E. M., & Sundararaghavan, H. G. (2016). Electrically Conductive Materials for Nerve Regeneration. In *Neural Eng.* (pp. 145–179). Springer International Publishing. https://doi.org/10.1007/978-3-319-31433-4_5

Su, K., Tan, L., Liu, X., Cui, Z., Zheng, Y., Li, B., Han, Y., Li, Z., Zhu, S., Liang, Y., Feng, X., Wang, X.,& Wu, S. (2020). Rapid Photo-Sonotherapy for Clinical Treatment of Bacterial Infected Bone Implants by Creating Oxygen Deficiency Using Sulfur Doping. *ACS Nano*, *14*(2), 2077–2089. https://doi.org/10.1021/ACSNANO.9B08686/ASSET/IMAGES/MEDIUM/NN9B08686_M001.GIF

Talikowska, M., Fu, X., & Lisak, G. (2019). Application of conducting polymers to wound care and skin tissue engineering: A review. *Biosens. Bioelectron.*, *135*, 50–63. https://doi. org/10.1016/j.bios.2019.04.001

Tan, L. L., Song, N., Zhang, S. X. A., Li, H., Wang, B., & Yang, Y. W. (2015). Ca2+, pH and thermo triple-responsive mechanized Zr-based MOFs for on-command drug release in bone diseases. *J. Mater. Chem. B*, *4*(1), 135–140. https://doi.org/10.1039/C5TB01789K

Tan, M., Xu, X., Yuan, T., Hou, X., Wang, J., Jiang, Z., & Peng, L. (2022). Self-powered smart patch promotes skin nerve regeneration and sensation restoration by delivering biological-electrical signals in program. *Biomaterials*, *283*, 121413. https://doi.org/10.1016/j.biomaterials.2022.121413

Tang, Y., Wu, C., Wu, Z., Hu, L., Zhang, W., & Zhao, K. (2017). Fabrication and *in vitro* biological properties of piezoelectric bioceramics for bone regeneration. *Sci. Rep.*, *7*(1), 43360. https://doi.org/10.1038/srep43360

Tong, L., Liao, Q., Zhao, Y., Huang, H., Gao, A., Zhang, W., Gao, X., Wei, W., Guan, M., Chu, P. K., & Wang, H. (2019). Near-infrared light control of bone regeneration with biodegradable photothermal osteoimplant. *Biomaterials*, *193*, 1–11. https://doi.org/10.1016/J.BIOMATERIALS.2018.12.008

Tupone, M. G., Panella, G., D'Angelo, M., Castelli, V., Caioni, G., Catanesi, M., Benedetti, E., & Cimini, A. (2021). An Update on Graphene-Based Nanomaterials for Neural Growth and Central Nervous System Regeneration. *Int. J. Mol. Sci.*, *22*(23), 13047. https://doi.org/10.3390/ijms222313047

Utesch, T., Daminelli, G., & Mroginski, M. A. (2011). Molecular Dynamics Simulations of theAdsorption of Bone Morphogenetic Protein-2 on Surfaces with Medical Relevance. *Langmuir*, *27*(21), 13144–13153. https://doi.org/10.1021/la202489w

Wang, C., Yue, H., Feng, Q., Xu, B., Bian, L., & Shi, P. (2018). Injectable Nanoreinforced Shape-Memory Hydrogel System for Regenerating Spinal Cord Tissue from Traumatic Injury. *ACS Appl. Mater. Interfaces*, *10*(35), 29299–29307. https://doi.org/10.1021/acsami.8b08929

Wang, C., Yue, H., Liu, J., Zhao, Q., He, Z., Li, K., Lu, B., Huang, W., Wei, Y., Tang, Y., & Wang, M. (2020).Advanced reconfigurable scaffolds fabricated by 4D printing for treating critical-size bone defects of irregular shapes. *Biofabrication*, *12*(4), 045025. https://doi.org/10.1088/1758-5090/ ABAB5B

Wang, D., Peng, F., Li, J., Qiao, Y., Li, Q., & Liu, X. (2017). Butyrate-inserted Ni–Ti layered double hydroxide film for H2O2-mediated tumor and bacteria killing. *Mater. Today*, *20*(5), 238–257. https://doi.org/10.1016/J.MATTOD.2017.05.001

Wang, Jian, Li, X., Song, Y., Su, Q., Xiaohalati, X., Yang, W., Xu, L., Cai, B., Wang, G., & Wang, Z. (2021). Injectable silk sericin scaffolds with programmable shape-memory property and neuro-differentiation-promoting activity for individualized brain repair of severe ischemic stroke. *Bioact. Mater.*, *6*(7), 1988–1999.

Wang, Jing, Xiong, H., Zhu, T., Liu, Y., Pan, H., Fan, C., Zhao, X., & Lu, W. W. (2020). Bioinspired Multichannel Nerve Guidance Conduit Based on Shape Memory Nanofibers for Potential Application in Peripheral Nerve Repair. *ACS Nano*, *14*(10), 12579–12595. https://doi.org/10.1021/acsnano.0c03570

Wang, L., Chen, X., Zeng, X., Luo, K., Zhou, S., Zhang, P., & Li, J. (2022). Degradable smart composite foams for bone regeneration. *Compos. Sci. Technol.*, *217*, 109124. https://doi. org/10.1016/J.COMPSCITECH.2021.109124

Wang, X., Shao, J., Abd El Raouf, M., Xie, H., Huang, H., Wang, H., Chu, P. K., Yu, X. F., Yang, Y., AbdEl-Aal, A. B. M., Mekkawy, N. H. M., Miron, R. J., & Zhang, Y. (2018). Near-infrared light-triggered drug delivery system based on black phosphorus for *in vivo* bone regeneration. *Biomaterials*, *179*, 164–174. https://doi.org/10.1016/J.BIOMATERIALS.2018.06.039

Wei, H., Cui, J., Lin, K., Xie, J., & Wang, X. (2022). Recent advances in smart stimuli-responsive biomaterials for bone therapeutics and regeneration. *Bone Res.*, *10*(1), 17. https://doi. org/10.1038/s41413-021-00180-y

Wu, P., Chen, P., Xu, C., Wang, Q., Zhang, F., Yang, K., Jiang, W., Feng, J., & Luo, Z. (2022). Ultrasound-driven *in vivo* electrical stimulation based on biodegradable piezoelectric nanogenerators for enhancing and monitoring the nerve tissue repair. *Nano Energy*, *102*, 107707. https://doi.org/10.1016/j.nanoen.2022.107707

Yan, F., Liu, Z., Zhang, T., Zhang, Q., Chen, Y., Xie, Y., Lei, J., & Cai, L. (2019). Biphasic Injectable Bone Cement with Fe3O4/GO Nanocomposites for the Minimally Invasive Treatment of Tumor-Induced Bone Destruction. *ACS Biomater. Sci. Eng.*, *5*(11), 5833–5843. https://doi.org/10.1021/ACSBIOMATERIALS.9B00472/SUPPL_FILE/AB9B00472_SI_001.PDF

Yan, H., Li, L., Wang, Y., Huang, J., Wang, Z., Shi, X., & Zhang, P. (2019). An electrically and magnetically responsive nanocomposite of GdPO 4 ·H 2 O/P3HT/PLGA with electrical stimulation for synergistically enhancing the proliferation and differentiation of pre-osteoblasts. *New J. Chem.*, *43*(44), 17315–17326. https://doi.org/10.1039/C9NJ04167B

Yeazel, T. R., & Becker, M. L. (2020). Advancing Toward 3D Printing of Bioresorbable Shape Memory Polymer Stents. *Biomacromolecules*, *21*(10), 3957–3965. https://doi.org/10.1021/acs.biomac.0c01082

Yuan, H., Xing, K., & Hsu, H.-Y. (2018). Trinity of Three-Dimensional (3D) Scaffold, Vibration, and3D Printing on Cell Culture Application: A Systematic Review and Indicating Future Direction.*Bioengineering*, *5*(3), 57. https://doi.org/10.3390/bioengineering5030057

Yuan, & Xing. (2019). A Systematic Study on Design Initiation of Conceptual 3DPVS. *Biomimetics*, *4*(2), 31. https://doi.org/10.3390/biomimetics4020031

Zaszczynska, A., Sajkiewicz, P., & Gradys, A. (2020). Piezoelectric Scaffolds as Smart Materials for Neural Tissue Engineering. *Polymers*, *12*(1), 161. https://doi.org/10.3390/polym12010161

Zeng, J., Wang, Y., Sun, Z., Chang, H., Cao, M., Zhao, J., Lin, K., & Xie, Y. (2020). A novel biocompatible PDA/IR820/DAP coating for antibiotic/photodynamic/photothermal triple therapy to inhibit and eliminate Staphylococcus aureus biofilm. *Chem. Eng. J.*, *394*, 125017. https://doi.org/10.1016/J.CEJ.2020.125017

Zhang, K., Shi, Z., Zhou, J., Xing, Q., Ma, S., Li, Q., Zhang, Y., Yao, M., Wang, X., & Li, Q. (2018). Potential application of an injectable hydrogel scaffold loaded with mesenchymal stem cells for treating traumatic brain injury. *J. Mater. Chem. B*, *6*(19), 2982–2992.

Zhang, L., Xiang, Y., Zhang, H., Cheng, L., Mao, X., An, N., Zhang, L., Zhou, J., Deng, L., Zhang, Y., Sun, X., Santos, H. A., & Cui, W. (2020). A Biomimetic 3D-Self-Forming Approach for Microvascular Scaffolds. *Adv. Sci.*, *7*(9), 1903553. https://doi.org/10.1002/advs.201903553

Zhang, X., Li, Z., Yang, P., Duan, G., Liu, X., Gu, Z., & Li, Y. (2021). Polyphenol scaffolds in tissue engineering. *Mater. Horiz.*, *8*(1), 145–167. https://doi.org/10.1039/D0MH01317J

Zhang, Z., Demir, K. G., & Gu, G. X. (2019). Developments in 4D-printing: a review on current smart materials, technologies, and applications. *Int. J. Smart Nano Mater.*, *10*(3), 205–224. https://doi.org/10.1080/19475411.2019.1591541

Zhao, F., Zhang, C., Liu, J., Liu, L., Cao, X., Chen, X., Lei, B., & Shao, L. (2020). Periosteum structure/function-mimicking bioactive scaffolds with piezoelectric/chem/nano signals for criticalsized bone regeneration. *Chem. Eng. J.*, *402*. https://doi.org/10.1016/J. CEJ.2020.126203

Zhao, X., Dong, R., Guo, B., & Ma, P. X. (2017). Dopamine-Incorporated Dual Bioactive Electroactive Shape Memory Polyurethane Elastomers with Physiological Shape Recovery Temperature, High Stretchability, and Enhanced C2C12 Myogenic Differentiation. *ACS Appl. Mater. Interfaces*, *9*(35), 29595–29611. https://doi.org/10.1021/ACSAMI.7B10583/SUPPL_FILE/AM7B10583_SI_001.PDF

Zhao, X., Guo, B., Wu, H., Liang, Y., & Ma, P. X. (2018). Injectable antibacterial conductive nano-composite cryogels with rapid shape recovery for noncompressible hemorrhage and wound healing. *Nat. Commun.*, *9*(1), 2784. https://doi.org/10.1038/s41467-018-04998-9

Zheng, X., Ding, Z., Cheng, W., Lu, Q., Kong, X., Zhou, X., Lu, G., & Kaplan, D. L. (2020). Microskin-Inspired Injectable MSC-Laden Hydrogels for Scarless Wound Healing with Hair Follicles. *Adv. Healthcare Mater.*, *9*(10), 2000041. https://doi.org/10.1002/adhm.202000041

Zhou, T., Yan, L., Xie, C., Li, P., Jiang, L., Fang, J., Zhao, C., Ren, F., Wang, K., Wang, Y., Zhang, H., Guo, T., & Lu, X. (2019). A Mussel-Inspired Persistent ROS-Scavenging, Electroactive, and Osteoinductive Scaffold Based on Electrochemical-Driven *In Situ* Nanoassembly. *Small (Weinheim an Der Bergstrasse, Germany)*, *15*(25). https://doi.org/10.1002/SMLL.201805440

Zhu, B., Li, Y., Huang, F., Chen, Z., Xie, J., Ding, C., & Li, J. (2019). Promotion of the osteogenic activity of an antibacterial polyaniline coating by electrical stimulation. *Biomater. Sci.*, *7*(11),4730–4737. https://doi.org/10.1039/C9BM01203F

Zhu, Y., Shang, F., Li, B., Dong, Y., Liu, Y., Lohe, M. R., Hanagata, N., & Kaskel, S. (2013). Magnetic mesoporous bioactive glass scaffolds: Preparation, physicochemistry and biological properties. *J. Mater. Chem. B*, 1(9), 1279–1288. https://doi.org/10.1039/c2tb00262k

4

Application of Smart Materials in Dental Sciences

Ruqaiya Saleem[1], Amaresh Kumar Sahoo[2] and Shalini Gupta[1*]

[1]King George's Medical University, Lucknow, Uttar Pradesh, India
[2]Department of Applied Sciences, Indian Institute of Information Technology Allahabad, Devghat, Prayagraj, Uttar Pradesh, India

Abstract

Development of dental materials is an ever-evolving science, which includes generation of advanced formulation and improvement in existing materials related to various branches of dentistry. The latest trend in the field of dental material provides the opportunity to use smart active materials with dynamic qualities that change according to the external environment instead of passive materials whose properties remain fixed for lifetime. Moreover, the advent of nanotechnology offers a huge scope of development of smart nanocomposite materials in restorative dentistry. Likewise, from zirconia in prosthodontics to the use of nickel-titanium (Ni-Ti) wires with shape memory in orthodontics, every field is now actively taking advantage of these advanced materials to successfully treat patients in clinics with more ease in less time. Oral cavity is a very diverse environment where pH, moisture levels, and the oral microbiome continue to change frequently, which demands the use of such smart materials that possess the capacity to sense the stimulus from the surrounding environment and respond accordingly in a useful manner. Therefore, there is a progressive inclination towards formulation of advanced material to develop "active" rather than "passive" materials to cater real-life applications.

One of the most popular and widely explored materials is metal and metal alloy-based nanocomposite materials in restorative dentistry, such as nickel-titanium (Ni-Ti) alloy, which comes under the category of shape memory alloys (SMAs) and is being successfully used to manufacture super-elastic endodontic files used in root canal treatment (RCT) procedures. In orthodontics, Ni-Ti arch wires are now preferred instead of stainless-steel wires to provide low sustainable forces for a longer span of time. Recent studies showed that smart self-healing nanocomposites were prepared for dental caries restorations using an epoxy system that contains microcapsules filled with healing agents, e.g., a visible light cured model resin consisting of triethylene glycol dimethacrylate (TEGMA): urethane dimethacrylate (UDMA): bisphenol A glycidyl methacrylate (BisGMA) (1: 1: 1) at 45% w/w with silane glass with a self-healing material consisting of dicyclopentadiene microcapsules and Grubbs catalyst. Likewise, transformation-toughened zirconia (TTZ) now widely used in prosthetic crowns and bridges involve a mechanism that stabilizes the tetragonal phase of zirconium oxide at room temperature. In endodontics, the C Point system is a root canal filling technique that consists of premade hydrophilic endodontic points to enhance the sealing ability of the root canal filling material, thus reducing the chances of failure and potentiating the long-term success of root canal treatment.

**Corresponding author*: sgmds2002@yahoo.co.in

Upendra Kumar and Piyush Kumar Sonkar (eds.) Smart Materials for Science and Engineering, (75–88) © 2024 Scrivener Publishing LLC

Keywords: Dental materials, smart nano-composites, NiTi wires, NiTi files, glass ionomer cement, self-healing composites, smart polymers, monolithic zirconia

4.1 Introduction

Development of dental materials is an ever-evolving science, which includes generation of advanced formulations and improvement in existing materials related to various branches of dentistry. The latest trend in the field of dental material provides the opportunity to use smart active materials with dynamic qualities that change according to the external environment instead of passive materials whose properties remain fixed for lifetime. Moreover, the use of nanotechnology offers a huge scope for the development of multifunctional active smart nanocomposite materials in almost all fields of dental sciences (Table 4.1). From the use of tetragonal zirconia polycrystal (TZP) in prosthodontics to the nickel-titanium (NiTi) wires with shape memory in orthodontics, every field is now actively taking advantage of these advanced materials to successfully treat patients in clinics with more ease in less time. Smart files made of nickel-titanium (NiTi) have become a game changer in the field of endodontics, allowing accurate preparation of root canals without risk of transportations or ledging, which were common with the conventional file system manufactured from stainless steel.

Oral cavity offers a dynamic environment where pH, moisture levels, and the oral microbiome continue to change with time, which demands the use of such materials that possess the capacity to sense the stimulus from the surrounding environment and respond accordingly in a useful manner [1]. Initially, it was believed that to make materials long lasting, they need to be in an inert or passive state like those of conventional composites, amalgam, and resin cements. However, according to the new criteria, materials more suitable for use in dental restoration are those that can modify themselves according to the

Table 4.1 Types of smart materials used in different branches of dentistry.

Conservative dentistry and endodontics	✓ Smart NiTi files ✓ SmartSeal obturation system ✓ Glass ionomer cement ✓ Smart composites ✓ Self-healing composites
Prosthodontics	✓ Smart denture base polymer ✓ Smart ceramics ✓ Smart impression materials
Periodontics	✓ Smart brush ✓ Smart dental implant coatings
Orthodontics	✓ Smart polymers ✓ Shape memory NiTi wires

external environment, making them resistant to the degrading forces that lead to the failure of the treatment [2, 3]. Therefore, there is a progressive inclination towards formulation of advanced smart materials that are "active" rather than "passive" in nature that would make dental treatment patient-friendly and reduce the risk of treatment failures.

4.2 Clinical Applications of Smart Materials in Various Branches of Dentistry

4.2.1 Smart Files

NiTi alloys were first manufactured by Naval Ordinance Laboratory, Maryland, USA, and are made up of approximately 55% nickel and 45% titanium. NiTi instruments are well known in the field of endodontics for their easy adaptation and efficiency in cleaning and shaping of root canals in root canal procedure (RCT) [4]. RCT is done in inflamed or infected tooth where infection has spread in pulp of the infected tooth. Post preparation of access cavity and removal of inflamed tissue, canals are cleaned, irrigated, and shaped with metallic files to receive an inert filling material. This prevents further spread of infection and saves the tooth. Initially, endodontic files were made of stainless steel as they have high fracture resistance and cutting efficiency. However, due to their rigidity, they have a tendency to straighten in curved canals and cause transportation and ledging causing iatrogenic perforations therefore they are needed to be pre-curved before insertion in the root canals [5]. Canal transportation refers to the canal wall structure removal apart from the natural curve of the canal due to the file tendency to straighten in the root canal. With the advent of smart NiTi files, the field of endodontics has been revolutionized as NiTi files are superelastic and flexible, which allows them to take the shape of the root canal in which they are used, thus eliminating the need to pre-curve to avoid ledging [6]. This reduces the incidence of ledging and perforations, which are common with stainless-steel files. Their unique crystalline structure makes them the material of choice also in terms of greater fatigue resistance.

The smart property of NiTi material is basically the function of temperature and stress. At higher temperatures, it exists as a body-centered austenite form, which transforms to monoclinic martensite form when cooled or any kind of stress is induced. On cooling below a critical point called transformation temperature range (TTR), there is a phase transformation from austenitic into monoclinic martensitic due to linear thermal contraction. TTR for the alloy varies anywhere between −50°C and 100°C depending on the composition of the nitinol alloy [31]. This limit is martensite start (Ms) where crystal transformation has just begun. Beyond this limit, there is further acceleration until monoclinic phase is reached, i.e., martensite finish (Mf). An intermediate phase called R-phase, which is the rhombohedral phase, is formed before martensite formation is completed (Figure 4.1). On reheating, this process is reversed and previous austenite phase begins to reform at a particular temperature, i.e., austenite start (As), and the body-centered structure is completed at austenite finish (Af).

Stress-induced transformation. The stress-induced transformation from austenite to martensite is utilized in endodontics during root canal preparation where the file undergoes

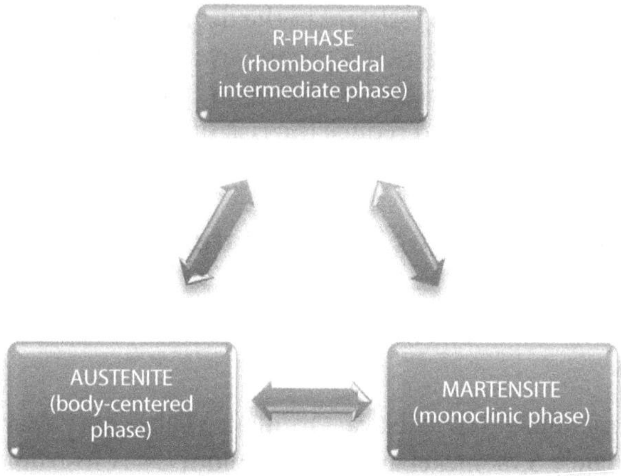

Figure 4.1 Phases formed in the NiTi lattice structure.

a considerable amount of stress while adapting to the root canal curvature when inserted inside the root canal. The austenite phase is converted to martensite phase when the material is subjected to stress. This stress-induced transition of crystal form allows the material to accommodate greater stress without increase in the strain up to approximately 8% unlike the stainless-steel files that exhibit permanent deformation over 1% strain [7, 8]. On releasing the stress, the martensite converts back to its parent phase, i.e., austenite. This property is known as superelasticity and is utilized in endodontics where stressed-induced martensitic conversion takes place, allowing the file to bend according to the shape of the canal without increasing in strain. This increases the life span of the files by preventing fatigue buildup.

Temperature-induced transformation. The smart property of temperature-induced transformation from martensite to austenite is utilized in orthodontic wires, which are discussed in detail further in the chapter (refer to section 2.10)

4.2.1.1 Thermally Treated NiTi Wires

1) M-wire technology: M-wire, which was introduced in 2007 by Tulsa Dental, is thermomechanically processed NiTi alloy that allows martensitic phase to remain stable at room temperature. It has improved fatigue resistance and flexibility [9, 10]. Examples of m-wire in the market currently available are ProfileGT, Protaper Next, Vortex files, WaveOne, and Reciproc.
2) Twisted file: These files are manufactured by twisting raw NiTi wires and subjecting it to a cycle of heating and cooling, causing it to convert into R-phase, which is the intermediate phase that occurs along a narrow temperature range between austenite and martensite phase. These files also have greater flexibility and fatigue resistance than conventional NiTi files.
3) Controlled memory alloys: CM wire was introduced in 2010 and is also an example of heat-treated NiTi wires with stable martensite at room temperature and improved properties. HyflexCM and HyflexEDM are available file systems that use CM wire technology.

4.2.1.2 *Surface-Treated NiTi Files*

Various surface treatments have been attempted to further improve the functional properties of NiTi files. Methods like electropolishing and electroplating are done to reduce surface irregularities in NiTi files and provide a smooth finish, which further assists in preventing crack propagation. This is done by immersing the file in electrolytic solution and passing direct current. EndoSequence and RaCe are examples of file systems prepared through the electrochemical process. Titanium nitride coating can be done to improve the cutting efficiency of the endodontic file, which makes it easier to shape the root canal during root canal procedures [7].

4.2.2 Smart Obturation Systems

Obturation is defined as sealing of cleaned and prepared root canal with a suitable material, which would prevent ingress of micro-organisms, saliva, or blood to prevent any periapical infections post root canal procedure. The most commonly used material for canal obturation is Gutta Percha cones, which are available in various consistencies to be used with different techniques of canal obturation. As gutta percha is a passive material and does not bind with walls of root canal, a sealer is commonly applied on the dentine wall before inserting any obturation material to establish a proper seal. However, many *in vitro* studies show that there are high chances of leakage between GP–sealer and sealer–dentin interface leading to failure of RCT [11, 12].

In search of a smart material that could tightly bind against the root canal wall, locking it into a sealed cavity SmartSeal system (C-Point system) was introduced. This system consists of two parts comprising a smart Propoint (C-Point) and a hydrophilic smart sealer called SmartPaste/SmartPaste Bio [13].

The Propoints also known as C points consist of a central core and an outer hydrophilic layer that swells up in response to any moisture present in the root canal and seals the remaining voids or any lateral canal if present [2]. The central core is made up of nylon polymers, Trogamid CX and Trogamid T [3]. The polymeric core provides the needed flexibility and adaptability to the Propoints, allowing them to pass effortlessly in any curved canal and take its shape. The outer layer consists of cross-linked hydrophilic polymer of vinyl pyrrolidone and acrylonitrile. The outer covering helps the SmartSeal system to swell only sideways with no axial swelling. It also exhibits a controlled expansion mechanism that permits only a certain amount of swelling, allowing the lateral forces to be less than the tensile strength of dentine.

The SmartPaste is a hydrophilic resin-based sealer that also shows expansion in response to moisture filling any leftover voids in the canal. SmartPaste Bio is yet another improved version of smart sealer available that incorporates the use of bio-ceramic along with the resin part. The manufacturer claims that the addition of a bio-ceramic provides additional dimensional stability and makes it non resorbable also. There is also the production of hydroxyapatite and calcium hydroxide as by-products of the reaction, thus making it antibacterial and increasing biocompatibility.

4.2.3 Glass Ionomer Cement (GIC)

The most commonly used material in conservative dentistry and endodontics to show smart behavior is glass ionomer materials, which show remineralizing behavior due to fluoride ion release. In search of a material that could also provide remineralization capacity to the tooth material apart from just passively lying in the cavity as previously available materials like amalgam and conventional composites, GIC was introduced in 1960s by Wilson and colleagues. These materials were the first in dentistry to show smart behavior in terms of remineralization potential. GIC is still the material of choice for restoring primary tooth in case of rampant caries [14].

GIC is prepared when fluoroaluminosilicate glass is reacted with polyacrylic acid in an acid–base reaction, which sets with a hard glass-like finish (Figure 4.2). Due to the presence of CaF2, sustained release of fluoride ions is maintained even when pH falls below the critical level, thus providing anticariogenic properties to the material and at the same time increasing biocompatibility. However, conventional GIC does not show adequate strength to be used as a permanent restorative material. In a further attempt to modify and improve the mechanical properties, i.e., flexural and compressive strength RMGIC (resin-modified glass ionomer cement) was introduced by adding methacrylate to polyacrylic acid [15]. RMGIC shows increased strength without compromising the fluoride releasing capability of the material. Other modifications done with glass ionomer materials are compomers and giomers. Recently, GIC has also been modified by adding CPP-ACP (Casein Phosphopeptide-Amorphous Calcium Phosphate), resulting in enhanced remineralization capability and flexural strength [16].

Another behavior that can be termed as smart is the capacity of dimensional change in GIC with respect to outside environment, which is almost similar to natural dentine. As the coefficient of thermal expansion of GIC is similar to dentine, any dimensional change in tooth dentine due to fluctuation in temperature will also cause similar changes in GIC, thus preventing marginal failure and better marginal adaptation.

4.2.4 Smart Composites

Dental composites are the most widely used restorative material due to their strength and color compatibility. However, there are high chances of marginal failure and secondary caries as they are passive in nature and exhibit a considerable amount of polymerization shrinkage. Thus, smart composites that possess the capability of remineralizing the defect caused

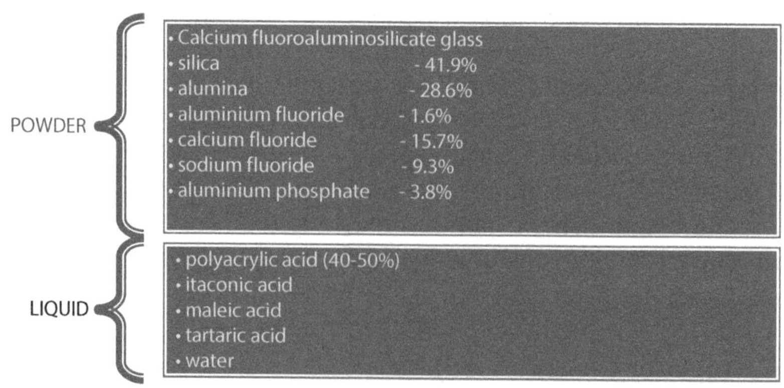

Figure 4.2 Composition of glass ionomer cement (GIC) powder and liquid.

by shrinkage or bacterial invasions are necessary for a successful and long-term restoration [17]. Currently available smart composites contain amorphous calcium phosphate (ACP) as a filler material along with polymer matrix. Amorphous calcium phosphate acts as a reservoir of calcium and phosphate ions and is also a prerequisite for the formation of hydroxyapatite crystals, which, in turn, form the building blocks of hard tissue of tooth [18].

ACP fillers remain in their original form until pH falls below the critical level. As soon as oral pH falls below 5.5, calcium and phosphate ions in ACP merge to form a gel consistency that readily converts into amorphous crystals, thus remineralizing the tooth structure. Due to this property, ACP fillers are now being used in various dental applications, be it pit and fissure sealants, endodontic sealers, or as bases under permanent restorations [19]. Addition of silver nanoparticles (NAg) along with quaternary ammonium methacrylates (QAMs) is also successfully tried in dental restorative polymers to provide antimicrobial property to the restorations and actively prevent secondary caries. Due to their large surface area-to-mass ratio, the addition of even a small amount is sufficient to induce high-quality antimicrobial resistance [20].

4.2.5 Self-Healing Composites

Conventional composite materials gradually age and degrade due to works of erosion, marginal dissolution, fracture, and corrosion. Keeping in mind the dynamic environmental conditions of oral cavity, researchers are now focusing on incorporating new self-healing properties to composite materials, which can easily prevent incidence of marginal failure, thus increasing life of restoration [21]. Self healing composites can be classified according to the mechanism of healing utilised i.e. intrinsic or extrinsic mechanism as shown in Table 4.2.

One such system is the epoxy-based system. It contains microencapsulated dicyclopentadiene (DCPD) and Grubb's catalyst [21]. When fracture occurs, some of the microcapsules present near the crack rupture and release the healing material, which fills in the defect and prevents crack propagation (Figure 4.3). Another study tested TEGDMA-DHEPT with BPO as self-healing material in resin composite [22]. A self-healing composite material was formulated using bis-GMA:TEGDMA in 1:1 composition with benzoyl peroxide (BPO) catalyst and around 20 weight % of SiO_2 nanoparticles as fillers and microcapsules containing polyurea-formaldehyde encapsulating TEGDMA (triethylene glycol dimethacrylate) monomer and *N*,*N*-dihydroxyethyl-*p*-toluidine (DHEPT) amine as accelerators.

However, these technologies are still not commercialized and are under clinical investigation. Future studies should be done on incorporating self-healing property in new materials that are introduced in clinics. Table 4.3 summarizes all the different types of self-healing mechanisms used over the years.

4.2.6 Smart Denture Base Polymers

The most commonly used denture base material is PMMA (polymethyl methacrylate) due to its cost-effectiveness and ease of manipulation and easy repair. Moreover, its optical properties are very much similar to natural human gingiva, which provides a more lifelike appearance.

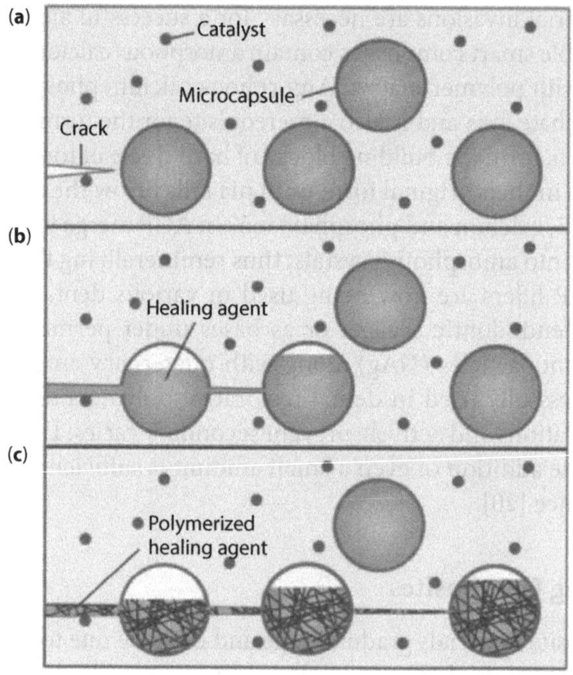

Figure 4.3 Mechanism of self-healing mechanism in dental polymers. Source [23].

Table 4.2 Types of self-healing composites.

▪ **Extrinsic self-healing mechanism**	➢ **Microcapsules mechanism** Catalyst with healing material is encapsulated in small polymeric microcapsules. ➢ **Vascular network self-healing mechanism**
▪ **Intrinsic self-healing mechanism**	➢ Healing agent and catalyst is not needed. ➢ Changes in temperature itself causes changes in molecular orientations and recombination leading to healing mechanism.

PMMA is a synthetic resin derived through polymerization of methyl methacrylate ($C_5O_2H_8$) and polymethylmethacrylate $(C_5O_2H_8)_n$. Polymerization reaction takes place under the influence of an initiator and free radicals helps in the reaction activation. Activators can be chemical based, light based, or heat based. On the basis of type of activation energy used, denture base resins can be classified into light cure, heat cure, or cold cure resin. The heat-based activation system uses benzoyl peroxide as an initiator and hydroquinone as an inhibitor.

However, conventionally used denture base materials, i.e., PMMA, can be the cause of various intraoral diseases, be it fungal, bacterial, or allergy. The rough denture surface adjacent to palatal mucosa can act as a breeding ground for various fungal and bacterial

Table 4.3 Examples of self-healing composite mechanisms tried in recent years [24].

Reaction mechanism	Materials used	
Ring opening metathesis polymerization (ROMP)	Dichloropentadiene (DCPD) and Grubbs catalyst	White *et al.* (2014), [25] Jason *et al.* (2007) [26]
Polycondensation reaction	Polysiloxane catalyst	Cho *et al.* (2006) [27]
Epoxy-based system	Epoxy resin and hardener encapsulated separately	Yin *et al.* (2009) [28]
Curing mechanism	Epoxides	Yuan *et al.* (2008), [29] Henghua *et al.* (2012) [30]
Hydrogen bonding mechanism	Polyacrylic acid cross-linked nanocomposite gel	Xiu *et al.* (2018) [31]
3D printing mechanism	Polymeric resin with embedded microcapsules	Sanders *et al.* (2019) [32]

population, most common being *Candida albicans* causing candidiasis in denture wearers [33]. Chronic erythematous candidiasis, also known as denture stomatitis, causes inflammation, burning, and erythematous changes in the palatal mucosa. Advanced cases may show palatal mucosal hyperplasia as well. To combat these issues, there was a need to incorporate materials in the denture base resins that could show smart behavior and inherently resist the fungal or bacterial growth. Denture base material made of PMMA modified by adding 2-methacryloyloxyethyl phosphorylcholine (MPC), a protein repellent, helps make these dentures fungal resistant by inhibiting *Candida albicans* growth on the surface, thus showing smart behavior. Other materials that can be incorporated to provide additional antibacterial properties are silver nanoparticles, graphene silver nanoparticles (G-AgNP), pre-reacted glass ionomer (PRG), and TiO_2 [34].

4.2.7 Smart Impression Materials

Dental impression materials require to show certain specific properties such as the following:

- Dimensional stability under specific temperature changes with respect to oral and environmental conditions.
- Adequate tear resistance to prevent breaking or distortion during removal from mouth.
- Hydrophilic to properly coat the oral mucosa and avoid formation of voids for smooth surface impression.
- Ease of manipulation and handling properties.
- Adequate working and setting time for the convenience of clinician and patient, respectively.

The most commonly used materials for impression making are impression compound, alginate, and elastomers. Elastomers especially the addition of silicon are the most popular among this category as they show the most accurate surface details. However, they frequently result in surface voids due to their inherent hydrophobic behavior. Polyether is the only one with hydrophilic properties. One of the latest impression materials to show smart behavior by combining the properties of both addition silicones and polyether is vinylsiloxanether, which results in smooth surface details [35]. Other smart impression materials are those with a short setting time like fast-set polyether and fast-set addition silicon to help reduce chairside time and prevent gag reflex. Some of the commercially available materials are Imprint-TM 3VPS, AquasilUltra, and Impregim-TM.

4.2.8 Smart Ceramics

Ceramics in different forms and combinations are widely used in prosthetic dentistry be it single tooth crown or short or long span bridges. One of the common materials used in the clinical scenario is the PFM (porcelain fused to metal) crown, which is a ceramic structure fabricated onto a metal substructure/core. However, the use of metal makes it less aesthetically appealing due to the grayish metallic hue below the tooth colored porcelain jacket. To overcome this shortcoming of PFM crown, more tooth material needs to be removed in order to provide porcelain coating with sufficient thickness to prevent chipping and also gives it a more aesthetic finish.

Monolithic zirconia was introduced in prosthodontics at around 2010, and since then, it has become the most popular material to be used for crown restorations. It offers a less-invasive alternative with no risk of veneer chipping. Moreover, zirconia crown also allows easy cementation, which can be done by a simple one-step procedure using resin cement or GIC unlike other ceramic crowns that needs multi-step adhesive techniques. Zirconia microstructure is made up of atoms arranged in an orderly manner unlike the irregular structure of glass. This makes zirconia an exceptionally strong and durable material that stands the test of time even in harsh conditions. The flexural strength and fracture toughness of transformation-toughened zirconia are considerably higher than other ceramics. Zirconia exhibits phase transformation with change in temperature with three polymorphic forms. At room temperature, it has a monoclinic crystal structure that changes to tetragonal when temperature is raised at about 1000 to 1100°C up to 2000°C. Above 2000°C, an FCC (face-centred cubic) structure is formed (Figure 4.4).

As monoclinic crystal structure occupies about 4.4% more volume than a tetragonal one, and this phenomenon could lead to material crumbling on cooling to room temperature after firing. To avoid this from happening, materials scientists introduced a stabilized form of zirconia by the addition of small amounts of calcium and most recently yttrium, which could retain its tetragonal structure even at normal room temperature. This tetragonal structure is only stable until any kind of external stress is applied, under which the tetragonal structure

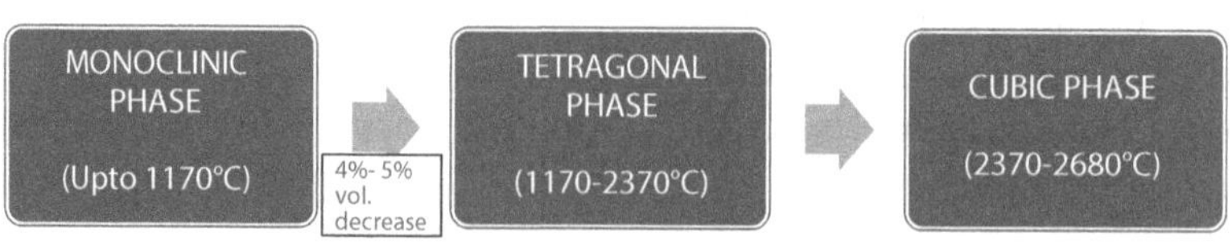

Figure 4.4 Phase transformations of zirconia at various temperatures (monoclinic crystal structure occupies 4.4% more volume at low temperatures).

is again restored to its original form, i.e., monoclinic crystal structure. This phase change under stress proves very beneficial for stopping crack propagation. As soon as crack appears, the highly localized stress in the area of crack triggers the tetragonal phase change to monoclinic, increasing the crystal volume by 4.4%, which stops crack from further propagation, making the material fracture resistant [36, 37]. Materials added as zirconia stabilizers are oxides of yttrium, magnesium, cerium, or aluminum in small molecular percentages like 3 mol%, 4 mol%, or 8 mol%. The most commonly used composition in dental applications is 3Y-TZP, i.e., tetragonal zirconia polycrystal with 3 mol% of yttria as stabilizer.

4.2.9 Smart Polymers

Smart materials that exhibit shape memory are most useful in the field of orthodontics, which include alignment of irregularly placed teeth in the dental arch. Smart wires made of NiTi and smart polymers has made orthodontic treatment much more patient-friendly and convenient for clinicians. Shape memory polymer that has been tried for dental alignment is made of polyurethane (PU) copolymer manufactured via a two-step process using 4,4 methylene bis phenyl isocyanate, poly(ε-caprolactone), and 1,4 butanediol. It consists of hard and soft segments with the soft segment providing shape memory property and the hard segment acting as a backbone forming the cross-links for stability [38].

4.2.10 Shape Memory NiTi Wires

NiTi wires are widely used in orthodontics, with their shape memory property making them resistant to deformation and the function of temperature-induced transformation [39, 40]. It was introduced in orthodontics by Andreasen *et al.* in 1971 as Nitinol (Nickel Titanium Naval Ordinance Laboratory), which was a stabilized form of NiTi alloy with 55% nickel and 45% titanium. These wires are less stiff than stainless-steel wires and could exert constant low forces for a longer period of time without undergoing permanent deformation, which are very beneficial in orthodontic applications. NiTi wires are stabilized in martensitic phase and placed in oral cavity for arch alignment. Transformation temperature range (TTR) for orthodontic purposes is kept generally lower than the temperatures of oral cavity. When the wire is placed in the oral cavity and it gets exposed to higher temperatures, memory-based temperature-induced austenite transformation of the deformed martensite takes place, thus regaining its previous original shape and providing constant force needed for tooth alignment.

4.3 Conclusion

Dental materials are the backbone of almost all various branches of dental sciences. With the evolution of conventional passive materials to active smart materials that adapt their properties according to the changes in the environment, the clinical treatment procedures have undergone a complete revolution that have made dental treatment more comfortable and time saving. Moreover, continuous advancements in the field of nanotechnology and bioactive materials have made development of multifunctional dental materials possible. For example, it is now possible to restore a carious lesion with a composite that will not only restore the previous tooth structure to help resume the normal masticatory function but also prevent future microbial attacks by incorporating nanoparticles like NAg that infers antimicrobial properties to the nanocomposite material. Temperature- and pressure-responsive alloys have taken orthodontic treatment to another level. Zirconia crowns in prosthodontics and self-healing composites have made long-term restorations possible by decreasing the failure rates due to micro-cracks and chipping off. Overall, the advent of smart materials in dentistry has expediated the advancements towards a successful and patient-friendly dental treatment.

Future Prospects

As far as we have advanced in the field of dental nanocomposite materials, their biocompatibility plays an important role in deciding whether the material is a fit for clinical scenario. Nanoparticles present in dental composites can pose cytotoxic and neurotoxic threats in view of their ability to cross the BBB (blood–brain barrier); therefore, future research should focus on modification of these nanoparticles to make them safe for long-term use in the human body [41]. A lot more remains to be done in the field of smart dental materials as a number of materials like smart polymers for denture base materials and antibacterial or regenerative composite resins have only been tried successfully *in vitro* in most of the cases and still need to be tested in clinical scenarios. More clinical studies are needed to confirm their clinical significance.

References

1. Yin Y, Rogers JA. Introduction: Smart Materials. *Chem. Rev.*, 2022 Mar 9;122(5):4885-4886. doi:10.1021/acs.chemrev.2c00074. PMID: 35259868.
2. McCabe JF, Yan Z, Al Naimi OT, Mahmoud G, Rolland SL. Smart materials in dentistry—future prospects. *Dent. Mater. J.*, 2009 Jan;28(1):37-43. doi: 10.4012/dmj.28.37. PMID: 19280966.
3. Badami V, Ahuja B. Biosmart materials: breaking new ground in dentistry. ScientificWorld Journal. 2014 Feb 2;2014:986912. doi: 10.1155/2014/986912. *Erratum in: Sci. World J.*, 2020 Mar 15;2020:7291341. PMID: 24672407; PMCID: PMC3929572.
4. Dagna A. Nickel-Titanium Single-file System in Endodontics. *J. Contemp. Dent. Pract.*, 2015 Oct 1;16(10):834-9. doi: 10.5005/jp-journals-10024-1766. PMID: 26581466.
5. Del Fabbro M, Afrashtehfar KI, Corbella S, El-Kabbaney A, Perondi I, Taschieri S. *In Vivo* and *In Vitro* Effectiveness of Rotary Nickel-Titanium vs Manual Stainless Steel Instruments for Root Canal Therapy: Systematic Review and Meta-analysis. *J. Evid. Based Dent. Pract.*, 2018 Mar;18(1):59-69. doi: 10.1016/j.jebdp.2017.08.001. Epub 2017 Aug 19. PMID: 29478682.
6. Gagliani MA, Cerutti A, Bondesan A, Colombo M, Godio E, Giacomelli G. A 24-month survey on root canal treatment performed by NiTi engine driven files and warm gutta-percha filling associated system. *Minerva Stomatol.*, 2004 Oct;53(10):543-54. English, Italian. PMID: 15531870.
7. Chan WS, Gulati K, Peters OA. *Advancing NiTinol*: From heat treatment to surface functionalization for nickel-titanium (NiTi) instruments in endodontics. *Bioact. Mater.*, 2022 Sep 27;22:91-111. doi: 10.1016/j.bioactmat.2022.09.008. PMID: 36203965; PMCID: PMC9520078.
8. Maloo LM, Patel A, Toshniwal SH, Bagde AD. Smart Materials Leading to Restorative Dentistry: An Overview. *Cureus*, 2022 Oct 28;14(10):e30789. doi: 10.7759/cureus.30789. PMID: 36457623; PMCID: PMC9705074.

9. Liang Y, Yue L. Evolution and development: engine-driven endodontic rotary nickel-titanium instruments. *Int. J. Oral Sci.*, 2022 Feb 18;14(1):12. doi: 10.1038/s41368-021-00154-0. PMID: 35181648; PMCID: PMC8857196.
10. Bhatt A, Rajkumar B. A comparative evaluation of cyclic fatigue resistance for different endodontic NiTi rotary files: An *in-vitro* study. *J. Oral. Biol. Craniofac. Res.*, 2019 Apr-Jun;9(2):119-121. doi: 10.1016/j.jobcr.2018.12.003. Epub 2019 Jan 2. PMID: 30671337; PMCID: PMC6327874.
11. Arora S, Hegde V. Comparative evaluation of a novel smart-seal obturating system and its homogeneity of using cone beam computed tomography: *In vitro* simulated lateral canal study. *J. Conserv. Dent.*, 2014 Jul;17(4):364-8. doi: 10.4103/0972-0707.136512. PMID: 25125851; PMCID: PMC4127697.
12. Hegde V, Murkey LS. Microgap Evaluation of Novel Hydrophilic and Hydrophobic Obturating System: A Scanning Electron Microscope Study. *J. Clin. Diagn. Res.*, 2017 May;11(5):ZC75-ZC78. doi: 10.7860/JCDR/2017/27458.9923. Epub 2017 May 1. PMID: 28658913; PMCID: PMC5 483815.
13. Hegde V, Arora S. Sealing ability of a novel hydrophilic vs. conventional hydrophobic obturation systems: A bacterial leakage study. *J. Conserv. Dent.*, 2015 Jan-Feb;18(1):62-5. doi: 10.4103/0972-0707.148898. PMID: 25657530; PMCID: PMC4313482.
14. Dias AGA, Magno MB, Delbem ACB, Cunha RF, Maia LC, Pessan JP. Clinical performance of glass ionomer cement and composite resin in Class II restorations in primary teeth: A systematic review and meta-analysis. *J. Dent.*, 2018 Jun;73:1-13. doi: 10.1016/j.jdent.2018.04.004. Epub 2018 Apr 9. PMID: 29649506.
15. Uno S, Finger WJ, Fritz U. Long-term mechanical characteristics of resin-modified glass ionomer restorative materials. *Dent. Mater.*, 1996 Jan;12(1):64-9. doi: 10.1016/S0109-5641(96)80066-2. PMID: 8598253.
16. Khoroushi M, Keshani F. A review of glass-ionomers: From conventional glassionomer to bioactive glass-ionomer. *Dent. Res. J.* (Isfahan), 2013 Jul;10(4):411-20. PMID: 24130573; PMCID: PMC3793401.
17. Montoya C, Roldan L, Yu M, Valliani S, T C, Yang M, Orrego S. Smart dental materials for antimicrobial applications. *Bioact. Mater.*, 2022 Dec 9;24:1-19. doi: 10.1016/j.bioactmat.2022.12.002. PMID: 36582351; PMCID: PMC9763696
18. Skrtic D, Antonucci JM, Eanes ED. Amorphous Calcium Phosphate-Based Bioactive Polymeric Composites for Mineralized Tissue Regeneration. *J. Res. Natl. Inst. Stand. Technol.*, 2003 Jun 1;108(3):167-82. doi: 10.6028/jres.108.017. PMID: 27413603; PMCID: PMC4844509.
19. Skrtic D, Antonucci JM. Polymeric dental composites based on remineralizing amorphous calcium phosphate fillers. *Curr. Trends Polym. Sci.*, 2016;17:1-31. PMID: 29599572; PMCID: PMC5870908.
20. Cheng L, Zhang K, Weir MD, Melo MA, Zhou X, Xu HH. Nanotechnology strategies for antibacterial and remineralizing composites and adhesives to tackle dental caries. *Nanomedicine (Lond).*, 2015 Mar;10(4):627-41. doi: 10.2217/nnm.14.191. PMID: 25723095; PMCID: PMC4347904.
21. Lee MW, An S, Yoon SS, Yarin AL. Advances in self-healing materials based on vascular networks with mechanical self-repair characteristics. *Adv. Colloid Interface Sci.*, 2018 Feb;252:21-37. doi: 10.1016/j.cis.2017.12.010. Epub 2017 Dec 24. PMID: 29329666.
22. Wu J, Weir MD, Melo MA, Xu HH. Development of novel self-healing and antibacterial dental composite containing calcium phosphate nanoparticles. *J. Dent.*, 2015 Mar;43(3):317-26. doi:10.1016/j.jdent.2015.01.009. Epub 2015 Jan 24. PMID: 25625674; PMCID: PMC4559273.
23. Yongjing Wang, Duc Truong Pham & Chunqian Ji | Eileen Harkin-Jones (Reviewing Editor) (2015) Self-healing composites: A review, *Cogent Eng.*, 2:1, DOI: 10.1080/23311916.2015.1075686
24. Shankar Kadam *et al.* 2021 *Mater. Res. Express,* **8** 052001**DOI** 10.1088/2053-1591/abfba5

25. White S, Moore J, Sottos N, Krull B, Cruz W and Gergely R. Restoration of large damage volumes in polymers. *Science,* 2014 344: 620–623.
26. Kamphaus Jason M, Rule Joseph D, Moore Jeffrey S, Sottos Nancy R and White Scott R. A new self-healing epoxy with tungsten (VI) chloride catalyst. *J. R. Soc. Interface,* 2008 595–103.
27. Cho S, Andersson H, White S, Sottos N and Braun P. Polydimethylsiloxanebased self-healing materials. *Adv. Mater.* 2006 18: 997–1000.
28. Yin T, Rong M Z, Zhang M Q and Zhao J Q. Durability of self-healing woven glass fabric/epoxy composites. *Smart Mater. Struct.,* (2009) 18: 1–7.
29. Yuan Y, Rong M, Zhang M, Chen J, Yang G and Li X. Self-healing polymeric materials using epoxy/mercaptan as the healant. *Macromolecules,* 2008 41: 5197–5202.
30. Henghua J, Chris L, Dylan S, Jeffrey S, Nancy R and Scott R. Self-healing thermoset using encapsulated epoxy-amine healing chemistry. *Polymer,* (2012) 53: 581–587.
31. Xiu H, Xuanyu M, Ziheng W, Zhenghong W, Xiaole Q, Xiu H, Xuanyu M, Ziheng W, Zhenghong W and Xiaole Q. Dynamic imine bond cross-linked self-healing thermosensitive hydrogels for sustained anticancer therapy via intratumoral injection. *Mater. Sci. & Eng. C,* 2018 93: 1064–1072.
32. Sanders P, Young A, Qin Y, Fancey K, Reithofer M, Guillet-Nicolas R, Kleitz F and Pamme N C J. Stereolithographic 3D printing of extrinsically self-healing composites. *Sci. Rep.,* 2019 9: 388.
33. Webb BC, Thomas CJ, Willcox MD, Harty DW, Knox KW. Candida-associated denture stomatitis. Aetiology and management: a review. Part 2. Oral diseases caused by Candida species. *Aust. Dent. J.,* 1998 Jun;43(3):160-6. doi: 10.1111/j.1834-7819.1998.tb00157.x. PMID: 9707778.
34. Mohd Farid DA, Zahari NAH, Said Z, Ghazali MIM, Hao-Ern L, Mohamad Zol S, Aldhuwayhi S, Alauddin MS. Modification of Polymer Based Dentures on Biological Properties: Current Update, Status, and Findings. *Int. J. Mol. Sci.,* 2022 Sep 9;23(18):10426. doi: 10.3390/ijms231810426. PMID: 36142344; PMCID: PMC9499318.
35. Rose S, Aravindakshan S, Mohamed Usman JA, Mohamed R, Menon S, Shafiullah RS, Salloum MG. Comparative Evaluation of Surface Detail Reproduction and Dimensional Stability of Poly Ether, Vinyl Siloxane, and Vinyl Siloxane Ether Impression Materials: An *In vitro* Study. *J. Pharm. Bioallied Sci.,* 2021 Jun;13(Suppl 1):S851-S856. doi: 10.4103/jpbs.JPBS_819_20. Epub 2021 Jun 5. PMID: 34447214; PMCID: PMC8375845.
36. https://doi.org/10.1016/j.dental.2007.05.005 (https://www.sciencedirect.com/science/article/pii/S0109564107001121)
37. Iftikhar S, Jahanzeb N, Saleem M, Ur Rehman S, Matinlinna JP, Khan AS. The trends of dental biomaterials research and future directions: A mapping review. *Saudi Dent. J.,* 2021 Jul;33(5):229-238. doi: 10.1016/j.sdentj.2021.01.002. Epub 2021 Jan 14. PMID: 34194185; PMCID: PMC8236547.
38. Hassan R, Aslam Khan MU, Abdullah AM, Abd Razak SI. A Review on Current Trends of Polymers in Orthodontics: BPA-Free and Smart Materials. *Polymers (Basel),* 2021 Apr 27;13(9):1409. doi: 10.3390/polym13091409. PMID: 33925332; PMCID: PMC8123702.
39. Fernandes DJ, Peres RV, Mendes AM, Elias CN. Understanding the shape-memory alloys used in orthodontics. *ISRN Dent.,* 2011;2011:132408. doi: 10.5402/2011/132408. Epub 2011 Oct 3. PMID: 21991455; PMCID: PMC3185255.
40. Kapila S, Sachdeva R. Mechanical properties and clinical applications of orthodontic wires. *Am. J. Orthod. Dentofacial Orthop.,* 1989 Aug;96(2):100-9. doi: 10.1016/0889-5406(89)90251-5. PMID: 2667330.
41. Feng X, Chen A, Zhang Y, Wang J, Shao L, Wei L. Application of dental nanomaterials: potential toxicity to the central nervous system. *Int. J. Nanomedicine.,* 2015 May 14;10:3547-65. Doi:10.2147/IJN.S79892. PMID: 25999717; PM

5

Graphene-Related Smart Material (GRSM): Synthesis, Characterization, and Application in Optoelectronics Devices

Varsha Yadav[1]*, Rahul Bhatnagar[2] and Saral Kumar Gupta[3]

[1]*School of Applied Sciences, Shri Venkateshwara University, Gajraula (Amroha), U.P., India*
[2]*School of Engineering and Technology, Shri Venkateshwara University, Gajraula (Amroha), U.P., India*
[3]*Department of Physical Sciences, Banasthali Vidyapith, Banasthali, India*

Abstract
This chapter aims to introduce the emerging technologies of graphene-related smart material (GRSM) such as graphene oxide (GO) in the field of optoelectronics energy storage devices. The graphene oxide (GO) paste is prepared by the well-known Hummers method and GO films are prepared by the blade technique on fluorine-doped tin oxide (FTO) glass substrates to be validated as an efficient electrode material in optoelectronics devices. The prepared powder form of GO is verified and characterized by X-ray diffraction, scanning electron microscopy, EDX, and Raman spectroscopy. The coated graphene oxide films are characterized by X-ray diffraction, scanning electron microscopy, and Raman spectroscopy, and the catalytic properties of films are investigated by cyclic voltammetry. We note that graphene oxide-based electrodes can be considered as a promising electrode in optoelectronics devices, with further optimization, which clearly has advantages in terms of its abundance and low-cost processing towards industrial prospects.

***Keywords*:** Graphene-related smart material (GRSM), efficient electrode, Hummers method, graphene oxide thin film and optoelectronics devices

5.1 Introduction

Today, energy is the main input for practically all economic activity, and it has become essential for improving people's living conditions. In addition, the issue of producing clean energy for an ever-expanding and evolving civilization is emerging in the international community [1]. One of the most important and pressing concerns facing humanity today is the development of low-cost, highly scalable, and emission-free energy [2]. When we look at global energy demand, we can split it into five categories: oil, natural gas, nuclear, coal, and renewable energy.

**Corresponding author*: varshayadav616@gmail.com

Upendra Kumar and Piyush Kumar Sonkar (eds.) Smart Materials for Science and Engineering, (89–104) © 2024 Scrivener Publishing LLC

Currently, fossil fuels such as natural gas and other non-renewable energy sources such as oil, coal, and nuclear energy provide more than half of the world's energy demand [3, 4]. The usage of these conventional sources has a lot of problems, including restricted supplies, pollution, and price increases [5]. These sources are modest and have a number of disadvantages. In order to meet the world's energy needs, fossil fuel usage should be decreased while the technology and development of renewable energy sources, such as solar cells, wind turbines, and fuel cells, should be accelerated [6, 7]. Therefore, switching to alternative, unconventional, and renewable energy sources is essential for society's long-term survival. Many different types of renewable energy are used as alternative energy sources [2]. Geothermal, wind, biomass, hydroelectric, solar, and other renewable energy sources will all be utilized to provide clean energy in the future because they do not harm the environment and are always available [8, 9].

The renewable energy systems such as photovoltaic generation systems have received tremendous attention, because they are a potential solution to environmental problems, in research among the prospective methods [10]. Solar energy is the most attractive renewable energy source due to its inherent abundance and environmental benefits, as it involves the supply of clean energy generated by the Sun [11].

The synthesis of 3rd-generation optoelectronics devices, such as perovskite solar cells, dye-sensitized solar cells (DSSCs), and organic solar cells (OSCs), has been the focus of current research in this area in an effort to address the shortcomings of the 1st- and 2nd-generation solar cells [12]. We believe that dye-sensitized smart material solar cells are an ideal system to provide such a solution. The claim that photovoltaic cells provide an ideal material system for creating a mechanism suitable for collecting solar energy over a large area is strongly supported by a combination of very low material costs and a low-cost manufacturing facility suitable for solution processing, which allows for simple fabrication method and low environmental cost on both energies for production and green materials [13]. Dye-sensitized solar cells are a cost-effective and environmentally friendly substitute to traditional Si solid-state devices (DSSCs) [14]. Direct sun energy is transformed into electric energy using DSSCs, a particular type of photo electrochemical cell [15]. The fact that it is easy to make and that the floats' adaptable thin-film architectures are well suited to automated manufacturing suggests that it might be scaled up without much difficulty. Fast scale-up, low production costs, great performance in dim/dispersive light, conformability with flexible substrates, and compatibility with building window glass are only a few of its remarkable features [16]. The DSSC consists of an electrolyte, a large band gap semiconductor surface coated with dye molecules, and a platinum-coated conductive glass substrate (redox couple) [17]. When DSSCs are exposed to solar radiation, an incoming photon causes a bonded electron–hole pair to form. In the dye sensitizer, a significant fraction of solar light is converted into electron–hole pairs that attach to the photoanode of materials with a large band gap (nanoparticles). As electron–hole pairs occur in dye solutions of molecules, they are easily split at the microscopic level due to the difference in energy levels (picoseconds scale) [18]. The glass-coated transparent conduction oxide (TCO) layer receives electrons from the dye that are pushed into the photoanodes' conduction band. The difference between the semiconductor film's Fermi energy and the electrolyte's redox potential in this configuration determines the maximum output voltage [19]. As illustrated in Figure 5.1, in this case, the sandwich format of DSSCs was designed (a). Using a transparent conductive substrate as a working electrode, a nanocrystalline semiconductor is placed on it.

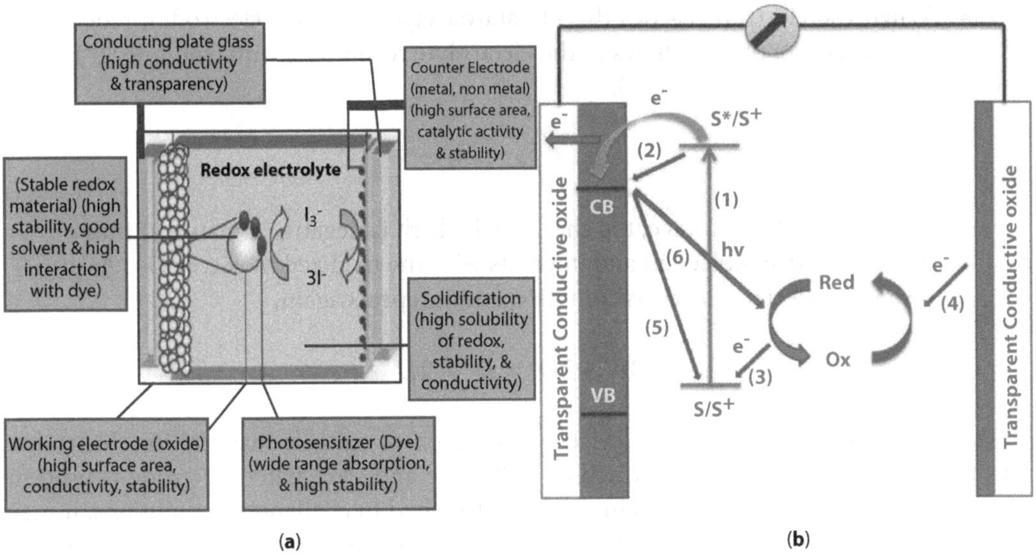

Figure 5.1 The working theory of DSSCs is depicted in (a) the schematic diagram and (b) working diagram of DSSCs.

Unlike conventional systems where semiconductors perform both charge carrier transport and light absorption, in this system, the two tasks are separated [20]. A wide band gap semiconductor's surface is coated with a sensitizer, which absorbs light. TiO2, SnO_2, and ZnO are the most often utilized substances. Due to the dye's photo-induced electron injection into the solid's conduction band, charge separation happens at the interface between the two materials. From the semiconductor's conduction band, carriers are transported to the charge collector, which is typically the TCO [21]. The nanocrystalline film's mesoscopic pores are filled with a mediator, such as an electrolyte (redox couple). The photosensitive dye is chemisorbed into the semiconductor surface to absorb light. A thin spacer separates the counter electrode (CE) from the photoanode (working electrode). TCO substrate with a Pt coating serves as the counter electrode (CE) [22].

A DSSC operates using four fundamental steps: electron injection, light absorption, current selection, and carrier transfer [23]. Figure 5.1b shows the conversion of photons into current via following steps.

- Firstly, a dye molecule-produced electron photon, such as in photosynthesis. A dye molecule is excited by the absorption of photon (*h*), whereas an electron is excited from HOMO-D into LUMO-D*.

$$D + \text{h}\nu = D^*$$

- The free electron is then inserted in the conduction band of TiO2 [24] (with duration in the nanosecond range) and leaves the oxidized dye molecule D+.

$$D^* = D^+ e^-$$

- When the electron reaches the Pt catalyst layer (counter electrode), redox processes take place as holes in the electrolyte recombine with triiodide (I^{3-}), reducing it to iodide ion (I^-).

$$I^{3-} + 2e^- = 3I^-$$

- However, in the last stage, the charge of I which is negative in nature diffuses back to the dye molecules and interacts with the oxidized molecule D+. As a result, the electrical cycle is complete, and it repeats again.

$$3I^- + 2D^+ = I^{3-} + 2D$$

The key component for such a device are selection of suitable material for electrodes. This requires a vital component, which is: A flexible and stretchable electrode is used in the DSSC devices, and even after being bent or stretched in challenging conditions, it may retain its original electrical properties [25]. However, compared to silicon-based solar cells that have already achieved commercial viability, the long-term stability and power conversion efficiency (PCE) of DSSC devices are still not suitable for practical applications. Various DSSC parts, such as electrolyte, counter electrode, and photoanode, are made of a variety of traditional materials that are frequently modified or replaced [26]. As a result, more research is still being done, for example, to improve photon harvesting, mobility and charge carrier generation while suppressing carrier recombination and minimizing leakages and electrolyte evaporation. Numerous research efforts have been performed employing different nanostructure metal oxides (TiO2, ZnO, SnO_2, etc.) and nanocomposites (carbon nanotubes and graphene) as photoelectordes to address the problems with DSSC and increase the efficiency of power conversion [27]. Additionally, numerous attempts have been made to reduce the cost of DSSC using a range of carbon-based nanomaterials, including as carbon nanotubes (CNTs), carbon black, and graphite as a replacement to Pt (due to its high cost). However, by utilizing the special qualities of carbon nanostructured materials including graphene-related materials (GRM), graphene oxide, graphene quantum dots, graphitic carbon nitride, etc. [28], we have developed a novel strategy to get over the challenges in DSSC. Therefore, carbon-based materials, like carbon nanotubes (CNTs), and, because of their low cost, non-toxicity, competitive electrical conductivity, ubiquity, and high optical transparency, carbon nanorods, carbon nanoribbons, and graphene have been extensively studied as potential replacements or additions to existing materials used for fabrication of various DSSC components [29].

The advantage of graphene, a novel two-dimensional monolayer graphite with a hexagonal lattice structure, is its chemical stability, thermal, and high mechanical, large specific surface area, wide spectral absorption range, light weight, flexibility, and large specific surface, all of which are necessary to produce DSSCs [30]. Pure graphene is not suitable for obtaining solutions due to its chemical inertness, which makes it insoluble in organic solvents [31]. Graphene derivatives such as oxygen-functionalized graphene sheets, graphene oxide (GO), and reduced graphene oxide (rGO) have been extensively studied due to their high solubility in organic solvents, making them suitable for facile solution synthesis on a large scale [29].

Although GO has an insulating property, removing the oxygen functional groups through chemical or thermal reduction helps re-establish delocalized electron, improving carrier transport and transformation of electrically insulating GO into conductive graphene sheets [32]. Due to their capacity for large-scale synthesis, the reduction of GO and CVD are the procedures for producing graphene that are most frequently utilized [33]. Several studies have investigated recent developments in the use of graphene-based materials for DSSC components, which, if realized, are expected to lead to cheaper and more reliable carbon-based electronics. Although the PCE of DSSCs is as accurate as possible, adjusting the

Table 5.1 Performances of graphene and its material-based DSSCs in photovoltaic applications.

S. no.	Material	J_{SC} (mAcm^{-2})	V_{OC} (V)	FF (%)	η (%)	Ref.
1	rGO	14.32	0.56	25.0	2.0	[36]
2	rGO	15.57	0.69	52.0	5.6	[37]
3	rGO	12.82	0.78	72.0	7.2	[38]
4	N + GO	13.14	0.71	48.0	4.5	[39]
5	rGO/CNT	12.86	0.78	61.0	6.2	[40]
6	Graphene	14.3	0.54	65.3	5.7	[41]
7	Graphene	12.08	0.60	39.0	2.8	[42]
8	Graphene nanosheet	12.66	0.75	57.20	5.4	[43]
9	Rgo foam	14.34	0.75	45.6	4.8	[44]
10	Graphene nanosheet	17.0	0.75	53.6	6.8	[45]
11	Pt + GO	10.08	0.67	60.0	4.0	[46]
12	GO +CNT +N	16.23	0.77	54.0	6.7	[47]
13	CVD graphene	12.34	0.72	64.33	5.7	[48]
14	Functionalized graphene	13.16	0.64	60.0	5.0	[49]
15	Ru + GO	16.13	0.77	67.0	8.3	[50]
16	Co3O4/rGO	15.70	0.76	67.7	8.1	[51]
17	CNT/GP	14.5	0.68	72.5	7.2	[52]
18	Graphene +Ag	18.27	1.01	77.0	14.3	[53]
19	rGO	3.02	0.76	0.36	0.8	[54]
20	MoS2/rGO	17.20	0.66	60.0	6.8	[55]

various DSSC while keeping the cost still cheap for actual applications new materials, such graphene-based components, has shown to be a desirable choice for materials [34]. Some research literature based on graphene-based material components for optoelectronics applications are summarized in Table 5.1. It is clear that CEs for DSSC can be made of functionalized graphene, graphene composites, and graphene. Higher power conversion efficiency was demonstrated by the state-of-the-art graphene CE-based DSSC [29]. The production of graphene-based CEs uses a variety of processes including sputtering, electro-spray, drop casting, spin casting, electrophoretic deposition doctor blade, and screen printing. In comparison to pristine graphene CE, the photo conversion efficiencies of DSSCs made with graphene oxide CE are much lower [35]. In this study, we describe the creation of GO paste through the use of -terpineol to ground GO powder and ethyl cellulose. In dye-sensitive solar cells, the GO films are produced for use as a counter electrode.

5.2 Experimental Methods and Materials

5.2.1 Synthesis of Graphene Oxide Powder

Hummers technique is a chemical procedure that produces graphene oxide by mixing potassium permanganate into a mixture of graphite, sulfuric acid, and sodium nitrate [56]. The material known as graphene oxide can be modified to produce a one-molecule-thick form of it. In a typical technique, 50 ml of concentrated sulfuric acid was heated to 0°C and then mixed with 1 g of sodium nitrate and 2 g of graphite powder. With vigorous mechanical stirring, 8 g of potassium permanganate ($KMnO_4$) was gradually added to the aforementioned solutions. Stirring was continued at 0°C for 1.5 h and then at 35°C for 2 h, then 100 ml of deionized water was added and the temperature was raised to 90°C and held for 15 min. The reaction was then stopped by adding deionized water (1 L) and about 10 ml of 30% hydrogen peroxide. Finally, after filtering the resulting suspension and washing with 500 ml of 10% HCl and deionized water, the product was held under vacuum at 40°C for 24 hours [57].

5.2.2 Fabrication of Graphene Oxide Electrode

This step was performed in three steps for each semiconductor material as active layer; the first one is (i) synthesis of GO paste, (ii) doctor blade method, and (iii) annealing.

5.2.2.1 Synthesis of TiO_2 Paste

The nanocrystalline GO paste was made by grinding GO nanopowder (200 mg) and ethyl cellulose (40 mg) in a mortar–pestle and slowly adding the solvent (-terpineol). Grind until you have a smooth, uniform paste [58]. The paste was made at room temperature and in the open air.

5.2.2.2 Doctor Blade Method

The FTO substrate is held flat, and scotch tape is used to define a region. The GO commercial paste is applied using a small spatula with the help of doctor blade method. A glass rod is used to spread the paste over the specified area. The rod is easily moved downward when pressed against the scotch tape. This causes the GO paste to fill the space between the scotch tape layers, resulting in the creation of a GO film in the region specified by the tape layers [34]. To ensure that nearly equal force/pressure is exerted along the length of the glass rod, it should be kept tightly and moved quickly. As a consequence, the films would be smooth and uniform. The resulting films will have an irregular texture and major differences in film thickness at various points if the pressure applied is not uniform. After a few tries, this coating phase can be mastered. The scotch tape is removed after the deposition to create a uniform film. The GO layered FTOs are placed on a hot plate at a temperature 100–120°C for 4–5 minutes followed by putting them inside the acetone poured around an inverted beaker for 30–45 seconds [59].

5.2.2.3 *Annealing*

Annealing is the process of heating for microstructural change, densification, and crystal growth of the material. Annealing of the film has been carried out in several steps of thermal heating. Annealing is the next step in which GO-coated FTO are placed in muffle furnace. The furnace's temperature is gradually raised to 500°C [60]. Thermal shock will harm the film if the temperature rises suddenly. Phase heating is essential for a crack free film. Once the temperature of 500°C is reached, it is sustained for 1 hour and then the furnace is switched off and left to cool down until it reaches room temperature [61].

5.2.3 Fabrication of Graphene Oxide-Based DSSCs

We have fabricated a DSSC device based on graphene oxide thin film. Figure 5.2 shows a flowchart of the steps undertaken from solution preparation to the final step of film

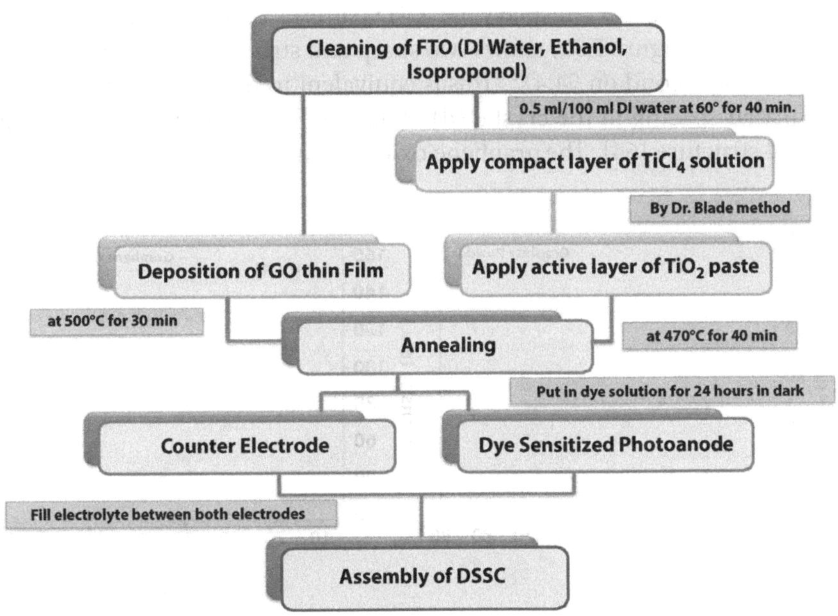

Figure 5.2 Flowchart of the process of DSSC fabrication.

deposition for fabrication of the device [59]. In this section, we explained every process step by step in detail.

5.2.4 Characterization Techniques

Crystalline properties were determined using a Bruker Advance X-ray diffraction (XRD) setup coupled to a Cu-K radiation source at a wavelength of 1.5406 over a 2Θ angle range of 20°–80° within the step size 0.05 at ambient temperature. Thermo Scientific's DXRxi Raman imaging microscope was used for the structural investigation, and a 532-nm excitation source was used [62]. In order to investigate the functional groups and chemical bonds of the electrode materials, an FTIR (Fourier Infrared) transmission spectrum was obtained using IR(ATR) (BRUKER TECH. PVT. LTD). To study the surface morphology a MIRA 3 TESCAN-field emission scanning electron microscope (FESEM) was used and EDX was used to study element identification [63]. A Verasol-Solar simulator was used to measure the current density voltage (J–V) curves of the DSSCs under AM 1.5G irradiation. The fill factor can be calculated using the J–V curves as follows: "FF= (Jmax * Vmax)/(Jsc*Voc), where Jmax and Vmax are the photocurrent density and photo voltage at the maximum power output, respectively, and Jsc and Voc are the short circuit photocurrent density and open circuit voltage, respectively. Pin in the intensity of the incident light (100 mW/cm^2), and the formula used to determine the energy efficiency(η) is η =(FF*Jsc*Voc)" [61].

5.3 Results and Discussion

5.3.1 X-Ray Diffraction Analysis

The crystal structure and purity of the graphene oxide as it was generated were described using XRD analysis. Figure 5.3 displays the XRD patterns of graphite and graphene oxide. The graphite pattern (Figure 5.3a) shows two sharp and strong peaks, with the first peak on 2θ = 26.38° and the second on 54.47°. This is equivalent to the diffraction line C (002) and has an interplanar spacing in the crystal (d) of 3.52 Å. The information exhibits graphite's average crystal structure [64]. The graphene oxide (Figure 5.3b) exhibits a wide diffraction

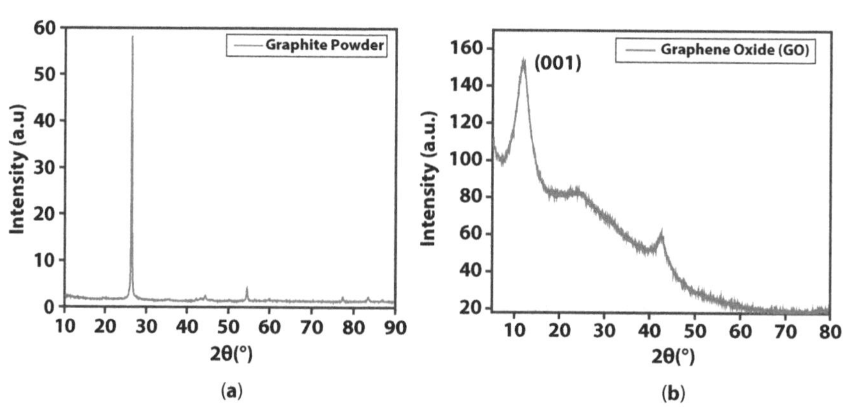

Figure 5.3 XRD pattern of (a) prepared graphite powder and (b) prepared GO powder.

peak that is located at 2θ = 11.79°, which corresponds to the (001) plane of crystal graphite structure with an interplanar spacing of 0.83nm. The spacing of GO after hydrothermal treatment again reduced to 3.70 Aa, which is still slightly higher than that of graphite [65].

The inclusion of oxygen functionalities and water molecules in the structure of the carbon layer increases the segregation of the graphene oxide field. These are typical graphene/graphite oxide vertical oxides [57].

5.3.2 Raman Spectroscopy Analysis

Raman spectroscopy is a common, non-destructive method for learning more about the structural characteristics of carbon-based materials. Raman spectra were used to characterize the synthesized GO. The Raman spectra of graphene oxide are presented in Figure 5.4 [66].

Graphene oxide shows a typical GO Raman spectrum with an intense tangential mode (G band) at 1604 cm^{-1} corresponding to the E2g phonons of sp2 C atoms. The existence of isolated double bonds resonating at higher frequencies may help explain this behavior. The other mode (band D) at 1355 cm corresponds to the breathing mode of the k-point phonon of symmetry A1g. The intensity ratio of these two bands indicates the quality of the product. The appearance of the D band indicates that the oxidation process affects the size of the planar sp2 domains [56].

5.3.3 Fourier Transform Infrared Spectroscopy (FTIR)

Fourier Transform Infrared Spectroscopy (also known as FTIR analysis or FTIR spectroscopy) is an analytical tool used to determine the chemical bonding of any substance [67]. Figure 5.5 shows an FTIR spectrum of graphene oxide.

The presence of hydroxyl groups in graphene oxide is indicated by broad peaks appearing at 3856 cm^{-1} and 3745 cm^{-1} in the high-frequency region belonging to the O–H bond stretching mode. The band observed at 2371 cm is related to the stretching vibrational band of the carbonyl group [65]. The resonance peaks associated with the stretching and bending

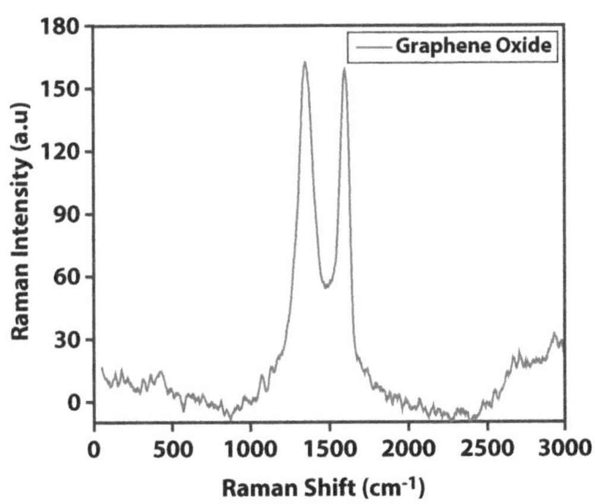

Figure 5.4 Raman spectrum of prepared GO powder.

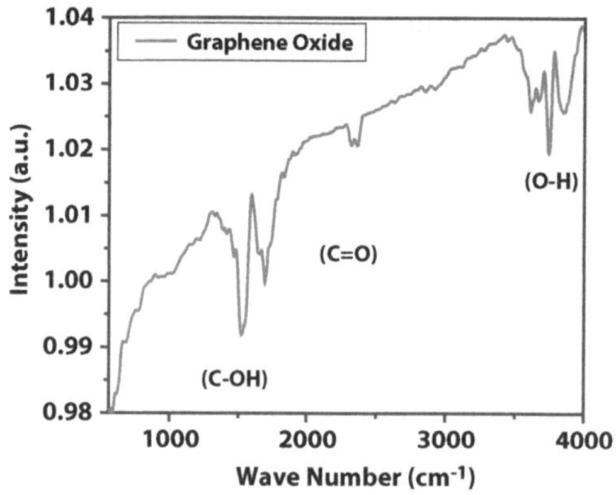

Figure 5.5 IR spectrum of prepared GO.

vibrations of the OH groups of water molecules adsorbed on graphene oxide are sharp peaks at 1615 cm^{-1}. A broad, sharp peak present at 1524 cm is associated with the C-OH bending vibrational band [64].

5.3.4 Field Emission Scanning Electron Microscopy (FESEM) and EDS Analysis

Field emission scanning electron microscopy (FESEM) is an effective technique for obtaining detailed information on morphological and microstructural properties [35]. The microstructure of graphene oxide is presented in Figure 5.6a. The generated aggregates are tightly bonded, and the graphene oxide exhibits a smaller flake size. The photos of graphene oxide show fewer stacks, indicating that the structure on the graphite has been exfoliated. The wrinkly structure of GO also results in folding. This situation arises from the GO's abundance of functional groups that include oxygen, which are mostly found at its boundaries [68]. The morphology of the prepared GO indicates that the grains are in a regular

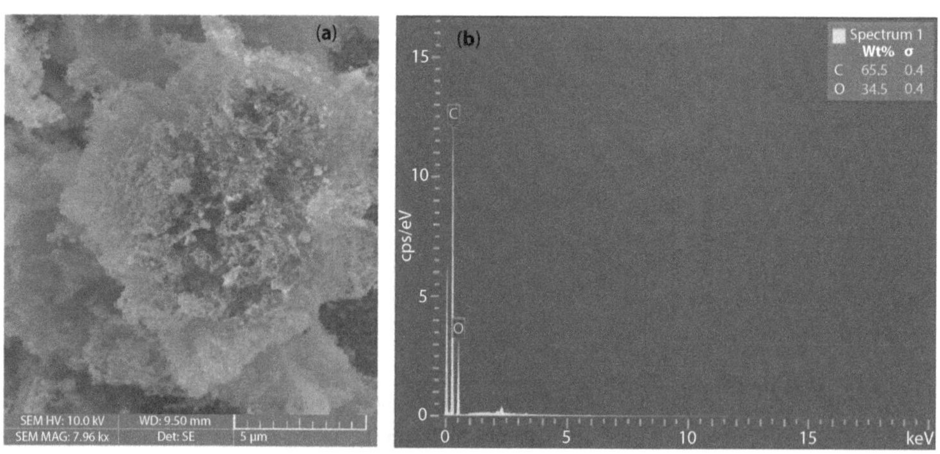

Figure 5.6 (a) FESEM image of prepared GO. (b) EDX spectra of prepared GO.

shape with a dense microstructure. It is also observed that the grains are agglomerated with almost uniform sizes [18].

The chemical homogeneity of the surface was further investigated using energy dispersive X-ray spectroscopy (EDX) as exhibited in Figure 5.6b. Elemental composition is studied using a setup connected to SEM, better known as energy dispersive X-ray analysis (EDAX or EDX) [35]. X-ray analysis is used to determine the elemental composition of any material with a sensitivity of 0.5 to 1 atomic percent. EDX spectra data of graphene oxide are measured at a voltage of 20 kV [69].

According to the graphic EDX spectrum of graphene oxide, the presence of both elements C and O at the sample surface suggests the uniform and homogeneous microstructure. The weight percentage of C and O agrees with their theoretical ratios [70]. The ratio of numbers of elements present in graphene oxide shows that the intensity of carbon atom detected by the EDX detector at 0.15 keV decreases and that at 0.20 keV oxygen atom increases [71].

5.3.5 Photovoltaic Performances Analysis

The photovoltaic parameters of the DSSCs fabricated with GO counter electrodes were measured under simulated solar irradiation of 100 mW cm^{-2} (AM 1.5G) [72]. The J–V curve of the DSSCs is represented in Figure 5.7a (dark condition) and B (under illumination) [73].

The photovoltaic parameters of DSSC, such as current density, voltage, and fill factor with a corresponding power conversion efficiency are measured under illumination conditions [74]. The fabricated DSSC represents an open-circuit voltage (Voc) of 0.70 V, a short current density (Jsc) of 11.10 mA/cm^2, and a fill factor (FF) of 57.91%, and overall photo conversion cell efficiency (η) is 4.49%. It exhibits higher open-circuit voltage, current density, and good duty cycle along with excellent power conversion efficiency, which is promising for DSSCs based on GO counter electrodes. GO counter electrode-based devices demonstrate excellent photovoltaic performance from the photovoltaic parameters [75]. It is an excellent alternative to expensive material-based DSSCs such as platinum-based DSSC with cost-effective factors. Moreover, GO material-based devices show a noticeable impact

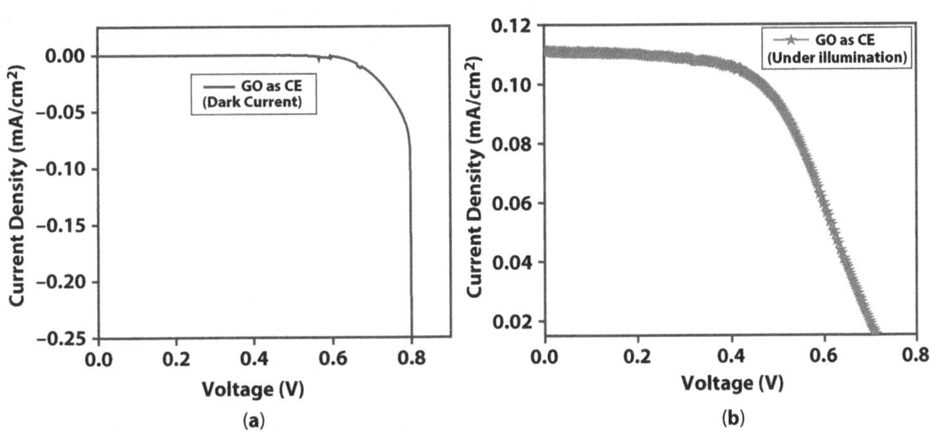

Figure 5.7 The J–V curve under the illumination of DSSC based on GO as CE (a) in the dark and (b) under illumination.

on the photovoltaic performance of the DSSC. Graphene-based GO smart materials exhibit promising impact factor in the field of optoelectronics devices such as solar cells [76].

5.4 Conclusions

Graphene oxide is successfully synthesized simply via the Hummers methods presented herein. The prepared graphene oxide XRD pattern shows a crystal structure and board and sharp peaks diffraction line C (002). A typical Raman spectra of GO shows intense two tangential modes such as G band and D band, demonstrating that the oxidation process affects the in-plane size of the sp2 domains. Hydroxyl group- and carboxyl group-based stretching bands are presented in the FTIR spectrum of GO. The morphology of the prepared GO indicates that the grains are in a regular shape with a dense microstructure. It is also observed that the grains are agglomerated with almost uniform sizes. The ratio of numbers of elements present in graphene oxide such as oxygen and carbon shows that the intensity of the carbon atom is detected by an EDX detector. We investigated the possibility of fabricating GO films as counter electrodes in optoelectronics, such as dye-sensitized solar cells. The typical DSSC power conversion efficiency of GO as CE achieves 4.49% with a current density (Jsc) of 11.10 mA/cm^2 and a fill factor (FF) of 57.91%, under 100 mV/cm^2 at AM 1.5 G. It is an excellent alternative to expensive material-based DSSCs such as platinum-based DSSC with cost-effective factors. Moreover, GO-based materials have a marked effect on the photovoltaic performance of DSSC. Graphene-based GO smart materials exhibit promising impact factor in the field of optoelectronics devices such as solar cells. This accomplishment opens the door for efficient research to increase the performance of this abundant and inexpensive material (GO) by carefully enhancing the structure and cell architecture to meet the growing demand for energy.

References

1. E. Coyle D. and R. Simmons A., *Understanding the Global Energy Crisis*. 2014.
2. V. Manieniyan, M. Thambidurai, and R. Selvakumar, "Study on Energy Crisis and the Future of Fossil," *Proceedings of SHEE*, no. October, pp. 7–12, 2009, doi: 10.13140/2.1.2234.3689.
3. EIA, "World Energy Demand and Economic Outlook- Chapter1," vol. 2016, no. May 2015, pp. 7–17, 2009, [Online]. Available: http://www.eia.doe.gov/oiaf/ieo/pdf/world.pdf.
4. R. G. Newell, D. Raimi, and G. Aldana, "Global Energy Outlook 2019: The Next Generation of Energy," *Resour. Future* , no. July, p. 46, 2019.
5. GEA, "Global Energy Assessment-Renewable Energy," pp. 761–900, 2012, [Online]. Available: http://www.iiasa.ac.at/web/home/research/Flagship-Projects/Global-Energy-Assessment/Home-GEA.en.html.
6. G. Climate and E. Project, "Global Climate & Energy Project An Assessment of Solar Energy Conversion Technologies and Research Opportunities," *Assessment*, pp. 1–20, 2006.
7. R. K. Musunuri, D. Sánchez, and R. Rodriguez, "Solar Thermal Energy Engineering," *Renew. Energy*, no. October, 2007.
8. S. Code, "Lecture Notes on Renewable Energy Sources," *Lecture Notes on Renewable Energy Sources*, pp. 1–178, [Online]. Available: http://www.vssut.ac.in/lecture_notes/lecture142891 0296.pdf.

9. J. Widén and J. Munkhammar, *Solar Radiation Theory*. 2019.
10. A. W. Copeland, O. D. Black, and A. B. Garrett, "The photovoltaic effect," *Chem. Rev.*, vol. 31, no. 1, pp. 177–226, 1942, doi: 10.1021/cr60098a004.
11. G. Teal and J. Little, "Principle and Applications of Solar Cells," 1877.
12. K. Ranabhat, L. Patrikeev, A. A. evna Revina, K. Andrianov, V. Lapshinsky, and E. Sofronova, "An introduction to solar cell technology," *J. Appl. Eng. Sci.*, vol. 14, no. 4, pp. 481–491, 2016, doi: 10.5937/jaes14-10879.
13. A. Mohammad Bagher, "Types of Solar Cells and Application," *Am. J. Opt. Photonics*, vol. 3, no. 5, p. 94, 2015, doi: 10.11648/j.ajop.20150305.17.
14. D. Wei, "Dye sensitized solar cells," *Int. J. Mol. Sci.*, vol. 11, no. 3, pp. 1103–1113, 2010, doi: 10.3390/ijms11031103.
15. I. Jinchu, C. O. Sreekala, and K. S. Sreelatha, "Dye sensitized solar cell using natural dyes as chromophores-review," *Mater. Sci. Forum*, vol. 771, pp. 39–51, 2014, doi: 10.4028/www.scientific.net/MSF.771.39.
16. J. Liu, Y. Li, S. Yong, S. Arumugam, and S. Beeby, "Flexible Printed Monolithic-Structured Solid- State Dye Sensitized Solar Cells on Woven Glass Fibre Textile for Wearable Energy Harvesting Applications," *Sci. Rep.*, vol. 9, no. 1, pp. 1–11, 2019, doi: 10.1038/s41598-018-37590-8.
17. K. Rokesh, A. Pandikumar, and K. Jothivenkatachalam, "Dye sensitized solar cell: A summary," *Mater. Sci. Forum*, vol. 771, pp. 1–24, 2014, doi: 10.4028/www.scientific.net/MSF.771.1.
18. H. Aliah *et al.*, "Dye Sensitized Solar Cells (DSSC) Performance Reviewed from the Composition of Titanium Dioxide (TiO2)/Zinc Oxide (ZnO)," *IOP Conference Series: Materials Science and Engineering*, vol. 288, no. 1, 2018, doi: 10.1088/1757-899X/288/1/012070.
19. U. Kingdom, "Efficiency enhancement in dye sensitized solar cells through step wise cosensitization of TiO2 electrode with N719 and metal free dye," vol. 54, no. October, pp. 656–664, 2016.
20. H. B. Abdullah and L. K. Wee, "Synthesis of a Nano-Silver Metal Ink for Use in Thick Conductive Film Fabrication Applied on a Semiconductor Package," vol. 9, no. 5, 2014, doi: 10.1371/journal.pone.0097484.
21. M. I. Khan, "A Study on the Optimization of Dye-Sensitized Solar Cells," no. January, 2013.
22. Q. Liu, N. Gao, D. Liu, J. Liu, and Y. Li, "Structure and photoelectrical properties of natural photoactive dyes for solar cells," *Appl. Sci. (Switzerland)*, vol. 8, no. 9, 2018, doi: 10.3390/app8091697.
23. J. Gong, J. Liang, and K. Sumathy, "Review on dye-sensitized solar cells (DSSCs): Fundamental concepts and novel materials," *Renew. Sustain. Energy Rev.*, vol. 16, no. 8, pp. 5848–5860, 2012, doi: 10.1016/j.rser.2012.04.044.
24. T. Setiawan, W. Y. Subekti, S. S. Nur'Adya, K. Ilmiah, and S. M. Ulfa, "Performance of Natural Dye and Counter Electrode from Robusta Coffee Beans Peel Waste for Fabrication of Dye-Sensitized Solar Cell (DSSC)," *IOP Conference Series: Materials Science and Engineering*, vol. 299, no. 1, 2018, doi: 10.1088/1757-899X/299/1/012073.
25. K. Sharma, V. Sharma, and S. S. Sharma, "Dye-Sensitized Solar Cells: Fundamentals and Current Status," *Nanoscale Res. Lett.*, vol. 13, 2018, doi: 10.1186/s11671-018-2760-6.
26. M. C. Ung, C. S. Sipaut, J. Dayou, K. S. Liow, J. Kulip, and R. F. Mansa, "Fruit based Dye Sensitized Solar Cells," *IOP Conference Series: Materials Science and Engineering*, vol. 217, no. 1, 2017, doi: 10.1088/1757-899X/217/1/012003.
27. R. Vittal and K. Ho, "Zinc oxide based dye-sensitized solar cells : A review," *Renew. Sustain. Energy Rev.*, vol. 70, no. November 2016, pp. 920–935, 2017, doi: 10.1016/j. rser.2016.11.273.
28. M. Wu *et al.*, "Economical Pt-free catalysts for counter electrodes of dye-sensitized solar cells," *J. Am. Chem. Soc.*, 2012, doi: 10.1021/ja209657v.

29. D. W. Zhang *et al.*, "Graphene-based counter electrode for dye-sensitized solar cells," *Carbon*, vol. 49, no. 15, pp. 5382–5388, 2011, doi: 10.1016/j.carbon.2011.08.005.
30. J. Ouyang, "Applications of carbon nanotubes and graphene for third-generation solar cells and fuel cells," *Nano Mater. Sci.*, vol. 1, no. 2, pp. 77–90, 2019, doi: 10.1016/ j.nanoms.2019.03.004.
31. P. Deb and J. C. Dhar, "Low Dark Current and High Responsivity UV Detector Based on TiO2 Nanowire/RGO Thin Film Heterostructure," *IEEE Trans. Electron. Devices*, vol. 66, no. 9, pp. 3874–3880, 2019, doi: 10.1109/TED.2019.2926797.
32. J. Nicasio-Collazo *et al.*, "Functionalized and reduced graphene oxide as hole transport layer and for use in ternary organic solar cell," *Opt. Mater.*, vol. 98, no. September, 2019, doi: 10.1016/j. optmat.2019.109434.
33. X. F. Zhang and X. Shao, "π-π binding ability of different carbon nano-materials with aromatic phthalocyanine molecules: Comparison between graphene, graphene oxide and carbon nano-tubes," *J. Photochem. Photobiol., A*, vol. 278, pp. 69–74, 2014, doi:10.1016/j. jphotochem.2014.01.001.
34. L. Wei *et al.*, "Enhanced performance of the dye-sensitized solar cells by the introduction of graphene oxide into the TiO2 photoanode," *Inorg. Chem. Front.*, vol. 5, no. 1, pp. 54–62, 2018, doi: 10.1039/c7qi00503b.
35. R. Siburian, H. Sihotang, S. Lumban Raja, M. Supeno, and C. Simanjuntak, "New route to synthesize of graphene nano sheets," *Orient. J. Chem.*, vol. 34, no. 1, pp. 182–187, 2018, doi: 10.13005/ojc/340120.
36. G. H. Guai *et al.*, "Graphene-Pt{minus 45 degree rule}ITO counter electrode to significantly reduce Pt loading and enhance charge transfer for high performance dye-sensitized solar cell," *Sol. Energy*, vol. 86, no. 7, pp. 2041–2048, 2012, doi: 10.1016/j.solener.2012.04.006.
37. A. G. Kannan, J. Zhao, S. G. Jo, Y. S. Kang, and D. W. Kim, "Nitrogen and sulfur co-doped graphene counter electrodes with synergistically enhanced performance for dye-sensitized solar cells," *J. Mater. Chem. A*, vol. 2, no. 31, pp. 12232–12239, 2014, doi: 10.1039/c4ta01927j.
38. Abid, P. Sehrawat, S. S. Islam, P. Mishra, and S. Ahmad, "Reduced graphene oxide (rGO) based wideband optical sensor and the role of Temperature, Defect States and Quantum Efficiency," *Sci. Rep.*, vol. 8, no. 1, pp. 1–13, 2018, doi: 10.1038/s41598-018-21686-2.
39. H. S. Kim *et al.*, "Reduced Graphene Oxide Improves Moisture and Thermal Stability of Perovskite Solar Cells," *Cell Rep. Phys. Sci.*, vol. 1, no. 5, 2020, doi: 10.1016/j.xcrp.2020.100053.
40. G. Zhu, L. Pan, T. Lu, T. Xu, and Z. Sun, "Electrophoretic deposition of reduced graphene-carbon nanotubes composite films as counter electrodes of dye-sensitized solar cells," *J. Mater. Chem.*, vol. 21, no. 38, pp. 14869–14875, 2011, doi: 10.1039/c1jm12433a.
41. H. Choi, H. Kim, S. Hwang, Y. Han, and M. Jeon, "Graphene counter electrodes for dye-sensitized solar cells prepared by electrophoretic deposition," *J. Mater. Chem.*, vol. 21, no. 21, pp. 7548–7551, 2011, doi: 10.1039/c1jm11145k.
42. G. Wang, W. Xing, and S. Zhuo, "The production of polyaniline/graphene hybrids for use as a counter electrode in dye-sensitized solar cells," *Electrochim. Acta*, vol. 66, pp. 151–157, 2012, doi: 10.1016/j.electacta.2012.01.088.
43. S. Jang, Y. Kim, D. Y. Kim, H. Kim, and S. M. Jo, "Am3005913_Si_001.Pdf."
44. Y. Xue *et al.*, "Nitrogen-doped graphene foams as metal-free counter electrodes in high-performance dye-sensitized solar cells," *Angew. Chem. Int. Ed.*, vol. 51, no. 48, pp. 12124–12127, 2012, doi: 10.1002/anie.201207277.
45. Z. Wei *et al.*, "Preparation of graphene-multi-walled carbon nanotube composite for quantitive determination of 2-hydroxy-3-methylanthraquinone in Hedyotis diffusa," *Int. J. Electrochem. Sci.*, vol. 12, no. 1, pp. 629–638, 2017, doi: 10.20964/2017.01.49.
46. J. Ma, C. Li, F. Yu, and J. Chen, "'Brick-like' N-doped graphene/carbon nanotube structure forming three-dimensional films as high performance metal-free counter electrodes

in dye-sensitized solar cells," *J. Power Sources*, vol. 273, pp. 1048–1055, 2015, doi: 10.1016/j.jpowsour.2014.10.003.

47. J. Dou, Y. Li, F. Xie, X. Ding, and M. Wei, "Metal–organic framework derived hierarchical porous anatase TiO2 as a photoanode for dye-sensitized solar cell," *Cryst. Growth Des.*, vol. 16, no. 1, pp. 121–125, 2016.
48. Y. Chen, K. Fu, S. Zhu, W. Luo, Y. Wang, Y. Li, E. Hitz *et al.* "Reduced graphene oxide films with ultrahigh conductivity as Li-ion battery current collectors." *Nano Lett.* 16, no. 6, 3616–3623, 2016.
49. J. I. Goldstein, D. E. Newbury, J. R. Michael, N. W. M. Ritchie, J. H. J. Scott, and D. C. Joy, *Scanning electron microscopy and X-ray microanalysis*. Springer, 2017.
50. V. D. Dao, L. L. Larina, J. K. Lee, K. D. Jung, B. T. Huy, and H. S. Choi, "Graphene-based RuO2 nanohybrid as a highly efficient catalyst for triiodide reduction in dye-sensitized solar cells," *Carbon*, vol. 81, no. 1, pp. 710–719, 2015, doi: 10.1016/j.carbon.2014.10.012.
51. T. Jiang *et al.*, "Economic synthesis of Co3S4 ultrathin nanosheet/reduced graphene oxide composites and their application as an efficient counter electrode for dye-sensitized solar cells," *Electrochim. Acta*, vol. 261, pp. 143–150, 2018, doi: 10.1016/j.electacta.2017.12.121.
52. S. Li *et al.*, "Vertically aligned carbon nanotubes grown on graphene paper as electrodes in lithium-ion batteries and dye-sensitized solar cells," *Adv. Energy Mater.*, vol. 1, no. 4, pp. 486–490, 2011, doi: 10.1002/aenm.201100001.
53. K. Kakiage, Y. Aoyama, T. Yano, K. Oya, J. I. Fujisawa, and M. Hanaya, "Highly-efficient dye-sensitized solar cells with collaborative sensitization by silyl-anchor and carboxy-anchor dyes," *Chem. Commun.*, vol. 51, no. 88, pp. 15894–15897, 2015, doi: 10.1039/c5cc06759f.
54. T. Wang, D. Huang, Z. Yang, S. Xu, G. He, X. Li, N. Hu, G. Yin, D. He, and L. Zhang. "A review on graphene-based gas/vapor sensors with unique properties and potential applications." *Nano Micro Lett.* 8, 95–119, 2016.
55. X. Yuan, X. Li, X. Zhang, Y. Li, and L. Liu, "MoS2 vertically grown on graphene with efficient electrocatalytic activity in Pt-free dye-sensitized solar cells," *J. Alloys Compd.*, vol. 731, pp. 685–692, 2018, doi: 10.1016/j.jallcom.2017.08.208.
56. F. T. Johra, J. W. Lee, and W. G. Jung, "Facile and safe graphene preparation on solution based platform," *J. Ind. Eng. Chem.*, vol. 20, no. 5, pp. 2883–2887, 2014, doi: 10.1016/j.jiec.2013.11.022.
57. D. K. Kumar, S. K. Swami, V. Dutta, B. Chen, N. Bennett, and H. M. Upadhyaya, "Scalable screen-printing manufacturing process for graphene oxide platinum free alternative counter electrodes in efficient dye sensitized solar cells," *FlatChem*, vol. 15, no. April, p. 100105, 2019, doi: 10.1016/j.flatc.2019.100105.
58. V. Yadav, S. Chaudhary, S. K. Gupta, and A. S. Verma, "Synthesis and characterization of TiO2 thin film electrode based dye sensitized solar cell," *East Eur. J. Phys.*, 2020, doi:10.26565/2312-4334-2020-3-16.
59. V. Yadav, S. Chaudhary, C. M. S. Negi, and S. K. Gupta, "Textile dyes as photo-sensitizer in the dye sensitized solar cells," *Opt. Mater.*, vol. 109, no. August, p. 110306, 2020, doi:10.1016/j.optmat.2020.110306.
60. A. Kumar, B. Khan, V. Yadav, A. Dixit, U. Kumar, and M. K. Singh, "Rietveld refinement, optical, dielectric and ac conductivity studies of Ba-doped SrSnO3," *J. Mater. Sci.: Mater. Electron.*, 2020, doi: 10.1007/s10854-020-04240-7.
61. V. Yadav, C. M. S. Negi, D. K. Kumar, and S. K. Gupta, "Fabrication of eco-friendly, low-cost dye sensitized solar cells using harda fruit-based natural dye," *Opt. Mater.*, vol. 122, no. PB, p. 111800, 2021, doi: 10.1016/j.optmat.2021.111800.
62. H. Cheema *et al.*, "Structural, optical and electrical properties of Mn-doped ZnFe2O4 synthesized using sol–gel method," *J. Mater. Sci.: Mater. Electron.*, 2021, doi:10.1007/s10854-021-06847-w.

63. S. Chaudhary, V. Yadav, C. M. Singh Negi, and S. K. Gupta, "Active layer thickness dependence of optoelectronic performance in CH3NH3PbI3 perovskite-based planar heterojunction photodiodes," *Opt. Mater.*, 2020, doi: 10.1016/j.optmat.2020.109960.
64. Z. Çiplak, N. Yildiz, and A. Cąlimli, "Investigation of graphene/Ag nanocomposites synthesis parameters for two different synthesis methods," *Fuller. Nanotub. Carbon Nanostructures*, vol. 23, no. 4, pp. 361–370, 2015, doi: 10.1080/1536383X.2014.894025.
65. D. Khalili, "Graphene oxide: a promising carbocatalyst for the regioselective thiocyanation of aromatic amines, phenols, anisols and enolizable ketones by hydrogen peroxide/KSCN in water Department of Chemistry, College of Sciences, Shiraz University, Shiraz 7," *New J. Chem.*, vol. 40, no. 3, pp. 2547–2553, 2016.
66. W. Vallejo, A. Rueda, C. Díaz-Uribe, C. Grande, and P. Quintana, "Photocatalytic activity of graphene oxide-TiO 2 thin films sensitized by natural dyes extracted from Bactris guineensis," *R. Soc. Open Sci.*, vol. 6, no. 3, 2019, doi: 10.1098/rsos.181824.
67. N. Purushothamreddy, R. K. Dileep, G. Veerappan, M. Kovendhan, and D. P. Joseph, "Prickly pear fruit extract as photosensitizer for dye-sensitized solar cell," *Spectrochim. Acta A Mol. Biomol. Spectrosc.*, vol. 228, p. 117686, 2020, doi: 10.1016/j. saa.2019.117686.
68. M. Kumar *et al.*, "Thermo-optical correlation for room temperature synthesis: cold-sintered lead halides," *J. Mater. Sci.: Mater. Electron.*, vol. 30, no. 6, pp. 6071–6081,2019, doi: 10.1007/s10854-019-00908-x.
69. E. Muchuweni, B. S. Martincigh, and V. O. Nyamori, "Recent advances in graphene-based materials for dye-sensitized solar cell fabrication," *RSC Adv.*, vol. 10, no. 72, pp. 44453–44469, 2020, doi: 10.1039/d0ra08851j.
70. I. Michalak, K. Marycz, K. Basińska, and K. Chojnacka, "Using SEM-EDX and ICP-OES to investigate the elemental composition of green macroalga Vaucheria sessilis," *Sci. World J.*, vol. 2014, 2014.
71. A. L. Palma *et al.*, "Reduced graphene oxide as efficient and stable hole transporting material in mesoscopic perovskite solar cells," *Nano Energy*, vol. 22, no. February, pp. 349–360, 2016, doi:10.1016/j.nanoen.2016.02.027.
72. Z. Yin *et al.*, "Graphene-Based materials for solar cell applications," *Adv. Energy Mater.*, vol. 4, no. 1, 2014, doi: 10.1002/aenm.201300574.
73. A. Devasia and S. K. Kurinec, " Teaching solar cell I-V characteristics using SPICE ," *Am. J. Phys.*, vol. 79, no. 12, pp. 1232–1239, 2011, doi: 10.1119/1.3636525.
74. M. J. Morgan, G. Jakovidis, and I. McLeod, "An experiment to measure the IV characteristics of a silicon solar cell," *Phys. Educ.*, vol. 29, no. 4, p. 252, 1994.
75. S. Das, P. Sudhagar, Y. S. Kang, and W. Choi, "Graphene synthesis and application for solar cells," *J. Mater. Res.*, vol. 29, no. 3, pp. 299–319, 2014, doi: 10.1557/jmr.2013.297.
76. J. Liu, Y. Xue, M. Zhang, and L. Dai, "Graphene-based materials for energy applications," vol. 37, no. December, pp. 1265–1272, 2012, doi: 10.1557/mrs.2012.179.

6

Synthesis and Characterization of Mechanical and Microstructural Properties of Fly-Ash-Reinforced Aluminum-Based Metal Matrix Composite

Rahul Bhatnagar[1]* and Varsha Yadav[2]

[1]*School of Engineering & Technology, Shri Venkateshwara University, Gajraula (Amroha), U.P., India*
[2]*School of Applied Sciences, Shri Venkateshwara University, Gajraula (Amroha), U.P., India*

Abstract
The increasing demand of composite materials leads researchers and scientists to developing alternate composites that are eco-friendly and contribute towards waste management. The utilization of industrial and agricultural waste produced by industrial processes has been the focus of waste reduction research for economic, environmental, and technical reasons. Bagasse fly ash is the waste product of sugar industries that causes air pollution. In this study, we take aluminum as the base metal with bagasse fly ash as reinforcement to make a metal matrix composite with different compositions of bagasse fly ash. Bagasse fly ash was used as reinforcement and varied from 6%, 10%, and 14% by weight in aluminum via the stir casting technique. The mechanical properties of the produced composites were determined. In this study, Tensile test, Izod test, Charpy test, and Hardness test were characterized. We found the best composition of the reinforcement in the metal matrix composite. In this study, microstructural characteristics of all the composites were determined by scanning electron microscopy. By using micrographs, we reported that the reinforcement is properly mixed in the base metal. It was concluded that bagasse ash can be used as reinforcement in aluminum composites and the produced composites could be used in the automobile industry for the production of engine blocks and pistons.

Keywords: Composite materials, aluminum, stir casting technique, fly ash, mechanical properties, microstructural, metal matrix composite

6.1 Introduction

Composite materials have been introduced into almost every industry in some form, which highlights the significance of composites as engineering materials (R. Sharma 2017). These composites were used to replace bronze and cast iron-based alloys at first, but due to their poor seizure resistance and wear, they were put through a lot of tests (Ramnath *et al.* 2014).

**Corresponding author*: rahulbhatnagarx@gmail.com

Upendra Kumar and Piyush Kumar Sonkar (eds.) Smart Materials for Science and Engineering, (105–120) © 2024 Scrivener Publishing LLC

Aluminum is frequently utilized in structural construction, mainly in the aerospace industries Because of its low weight (U. Umer *et al.* 2022), but there was always a problem with the low strength and melting point (Thandalam, Ramanathan, and Sundarrajan 2015). The introduction of reinforcing elements, such as Al_2O_3, SiO_2, and SiC particles, whiskers, and other compounds or elements as alloying elements was a low-cost solution to these issues (Dixon *et al.* 2018; Arivukkarasan *et al.* 2013). Several attempts have been taken to introduce hard ceramic such as Al_2O_3, B_4C, and SiC into an aluminum (Al)-based matrix; these ceramic and alloying element particles can be added to increase the specific elastic modulus of aluminum and enhance its thermal properties, among other things (Myriounis, Matikas, and Hasan 2012). Using numerous methods including powder metallurgy, stirring methods, etc. to prepare aluminum composite and alloys reinforced with reinforcing particles and other alloying elements will help to prepare a homogenous distribution of reinforcement in matrix (D.S. Chandra Mouli, R. Uma Maheswara Rao 2017). The term "composite" is most used to describe a material system made up of core elements (reinforcement materials) spread in a continuous phase (matrix), from which the material's important characteristics are derived. Typically, the physical and chemical properties of the matrix phase are used to categorize composite materials, such as polymer matrix, ceramic composites and metal matrix. Additionally, there are certain findings that point to the development of carbon– and intermetallic–matrix composites (Romanov, Chernyshov, and Romanov 2019). This study focuses on metal matrix composite (MMC), especially the aluminum matrix composites (AMCs). The building blocks of AMC, aluminum/aluminum alloys, create an infiltration network and are called the matrix phase (D.S.Chandra Mouli, R. Uma Maheswara Rao 2017). The other component, usually non-metallic and frequently ceramic like Al_2O_3 and SiC, is inserted in this matrix of aluminum/aluminum alloy and serves as reinforcement (Gobalakrishnan *et al.* 2021). By changing such kind of ingredients with their volume fraction, it is possible to alter the properties of AMCs. As potential matrix materials, aluminum alloys have been more widely used than other metals (V. Rama Koteswara Rao *et al.* 2016). The earliest industries to report their use were those in the aircraft, automotive, and kitchenware sectors (Lynch and Kershaw 2018). When it comes to reinforcing, silicon carbide (SiC) is the most often used material. In the current study, SiC (99% assay) and commercial-grade aluminum alloy 6061 (98% assay) were employed to create a composite (Ling and Gupta 2000). Adding a maximum amount fraction of whiskers or particles can improve the performance limits of MMCs (Pawar and Utpat 2014). One example is the production of aluminum MMCs by combining and condensing aluminum alloy powders and low-density, high-modulus, micron-sized carbides such as SiC (Meena *et al.* 2013). The MMC's specific modulus and wear resistance rise as a result of the addition of SiC particles, while at the same time, its ductility and fracture toughness decrease (Pawar, Wabale, and Utpat 2018). Thus, the physical and mechanical qualities are improved or degraded because of addition. A by-product of the sugar industry known as bagasse is frequently burned in factories to produce process power or steam, which powers process machinery (Dasgupta 2012). Since sugarcane bagasse has a high calorific value, ash is generated as waste when used as fuel, and there is no particular economic benefit (Ejiofor and Reddy 1997). These wastes, their ashes, and their gases must be recycled via a new procedure because they are harmful to the environment (Lakshmikanthan *et al.* 2022). When aluminum metal is reinforced with bagasse ash as a source of silica and alumina particles, the result is materials that exhibit a combination of the physical and mechanical properties of both metal matrix and silica (Kaczmar, Pietrzak, and Wlosiński 2000).

The key aim of this study is the utilization of abundant bagasse fly ash, an agricultural waste that is growing day by day worldwide with the introduction of new technologies in sugar cane production. Research has shown that sugar cane production has increased by about 2% annually since 2015 so waste is also increasing. To address this issue, it is necessary to use this by-product as a reinforcement in the construction of aluminum-based metal matrix composites to decrease the ecological effect of solid waste on the public and promote its economic utilization (Muley, Aravindan, & Singh 2015).

Fly ash is a combustion by-product of the bagasse that is found in very huge amounts in mainly the sugar mills, which is waste for the industry, making it suitable for use as a reinforcement in metal matrices (Fuzail Hashmi and Haq 2018). Because of their expensive manufacturing costs, produced glass microspheres now only have a few applications (Fuzail Hashmi and Haq 2018). Therefore, by adding fly ash to the matrix of polymers and metallic alloys, the material prices of composites can be greatly decreased (Zedadra *et al.* 2019).

Several attempts have been made to include fly ash in both polymer and metal matrices, but little information is available for composite design. Fly ash from the cenosphere is somewhat denser than hollow glass, although it is less dense than talc and calcium carbonate. Cenospheric gas will probably be significantly less expensive than hollow glass. In terms of price per volume, cenosphere can end up being one of the least expensive fillers (Zhou *et al.* 2020; Kumar, Maury, and Bahlawane 2016).

Hybrid materials are materials that contain at least three different materials (Chawla and Ganesh 2010). A "metallic matrix" (Cu, Mg, Al, Fe, etc.) with a distributed "ceramic (oxide, carbides) or metallic phase" make up metal matrix composites (Pb, Mo, W, etc.). While metallic reinforcement options include tungsten, beryllium, and others, ceramic reinforcement options include silicon nitride, silicon carbide, boron, boron carbide, boron nitride, and alumina (Johny James *et al.* 2014). MMCs are used in various applications, including the Space Shuttle, business jets, electronic substrates, bicycles, cars, golf clubs, and more. From a material standpoint, MMCs have an advantage over polymer matrix composites due to their strong abrasion and creep resistance qualities and ability to maintain strength and stiffness at high temperatures (Bhandare and Sonawane 2013). In comparison to polymer matrix composites, most MMCs are still in the early stages of development or fabrication (Ding, Biermann, and Hartmann 2002). The major drawbacks of MMCs are their high production costs, which have limited the actual applications they may be used for. Some of MMCs' physical properties are also beneficial, such as lack of significant hygroscopic properties, low electrical and thermal conductivity, and non-flammability. Since MMCs have been around for 30 years, several different MMCs have been researched (Sahraeinejad *et al.* 2015). Fly ash particles, which are cheap and have a low density and are found in huge numbers as a waste by-product in power plants, are employed in metal matrix composites. Fly ash is added to reinforced aluminum using the stir casting method, which lowers the price and density of the aluminum composition (Butola *et al.* 2020). When compared to other metals, metal composite methods have enhanced mechanical qualities such as low density, hardness, and good wear resistance (Bhatnagar, Trivedi, and Shankhdhar 2016). Kesavulu *et al.* performed a chemical investigation of aluminum clad and fly ash before and after they are mixed and formed as particulate metal matrix composites, and a comparison of the physical and mechanical characteristics of the MMC at various fly ash weight percentages.

In contrast, the MMC provides good mechanical and physical qualities at varying levels of 15% fly ash. This kind of MMC is extensively used in aerospace and lightweight vehicle applications. Due to their high strength, wear resistance, density, low weight, and coefficient of thermal expansion, aluminum alloys have seen a significant surge in use over the last several years, particularly in the automotive industry (A. Raju, and Kumar 2014). A lot of attention has been paid to aluminum metal matrix composites among the materials of tribological significance for both practical and theoretical reasons. Various automotive engine components are made using aluminum alloys and metal matrix composites using aluminum as a primary ingredient. The development of compound work pieces combines the beneficial qualities of various materials are reviewed by Vengatesh *et al.* Industrial and residential production' both use a lot of composite materials (Vengatesh 2014). Aluminum silicon carbide metal matrix composites are appealing to the automotive sectors. Ahemad *et al.* fabricated Aluminum LM25+SiC composites via casting process. Under dry sliding conditions, the wear and friction characteristics of Aluminum LM-25+SiC composites have been evaluated and compared to those seen in aluminum LM25 without any impurities. With a pin-on-disk machine and standard loads from 1 kg to 3 kg by increasing the weight by 1 kg per test, dry sliding wear tests have been conducted (Ahemad *et al.* 2013). Alloy pins are made in sizes of 8 mm in diameter and 30 mm in length. On the material mentioned above, testing is carried out by adjusting the speed, load, and sliding distance. It seems that the pin-on-disc test can provide a lot of valuable and instructive information about friction and wear. It was found that the wear rate in composite materials is lower than in base materials and that it varies linearly with normal load. Ravesh *et al.* fabricated and characterized aluminum-based metal matrix composite via stir casting technique. SiC, fly ash, and Aluminum 6061 (97.06% C.P.) have been selected as the matrix and reinforcing materials, respectively, and their mechanical properties like tensile strength, hardness, and toughness have been determined (Ravesh, S. and Garg 2012). The production of AMC by stir casting was studied by Upadhyaya *et al.*, and the casting processes used in the casting of an aluminum metal matrix composite was compared and it was observed that the stir casting method was the most suitable (Upadhyaya and Saxena 2021). Singh *et al.* fabricated HAMCs by stir casting route characteristics of their mechanical behavior. It has been observed that the presence of the hard ceramic phase gives composites their direct strengthening, whereas the thermal incongruity between the matrix alloy and reinforcing phase during solidification gives them their indirect strengthening. Based on the composition and combination of the reinforcement, the hybrid composites offer greater flexibility and dependability in the design of potential components (Singh and Chauhan 2016). The processes utilized to create these composites are continually changing, especially about eliminating uneven reinforcing distribution. Samal *et al.* developed a stir casting modification to create silicon carbide-reinforced aluminum–magnesium matrix composites. The created composite materials have substantially higher tensile strengths than the matrix because of the presence of SiC, demonstrating the value of the modified stir casting method (Samal, Panigrahi, and Sarangi 2013).

Boopathi *et al.* used stir casting technique to fabricate hybrid metal matrix composites and examined the mechanical properties such as co-efficient of friction and corrosion behavior of Al 2024 in the presence of fly ash and silicon carbide. At higher loads and farther sliding distances, it was found that the composites' wear rate and coefficient of friction were both reduced. An experimental evaluation of the volume percent corrosion properties of

SiC and fly ash mixtures on Al2024 was also performed. Waste reduction research has been concentrated on using agricultural and industrial waste generated by industrial operations for economic, environmental, and technological objectives (Boopathi *et al.* 2014). Sugar cane bagasse ash is a waste product that is already badly contaminating the environment; hence, there is an urgent need for techniques of garbage treatment. Bagasse ash mostly includes silica and aluminum ions. Srinivasan *et al.* studied the physically and chemically characterized bagasse ash replaced in various weight ratios of cement in concrete. At the ages of 7 and 28 days, results from fresh physical tests such as the slump cone test and compaction factor test, as well as tests on hardened concrete such as modulus of elasticity, split tensile strength, compressive strength, and flexural strength, were obtained. It was also found out that the strength of concrete with the percentage of bagasse ash replacement was increased (SRINIVASAN and Sathiya 2010). We created a metal matrix composite of aluminum in this study to create a new class of composites by incorporating bagasse fly ash into aluminum for use in the field of engineering, such as automotive manufacturers, to meet the requirements of turning to lighter and more efficient products, such as high-speed rotating shafts, high-speed machinery, robots, brake parts and automotive engine.

6.2 Materials and Methods

In this section, the methods and materials used to fabricate the composite are discussed.

6.2.1 Raw Materials

The matrix material used in the experiment is aluminum (LM4). The bagasse ash is used as reinforcement. The chemical composition of Aluminum LM4 (catalogue) and bagasse is mentioned in Tables 6.1 and 6.2, respectively.

6.2.2 Synthesis Process

6.2.2.1 Stir Casting Technique

Stir casting is the best manufacturing process for discontinuous metal matrix composites out of all the methods that are available. This method's benefits include economic effectiveness, ease of usage in large-scale production, and simplicity (Ghosh and Saha 2011).

Table 6.1 Chemical composition of aluminum LM4 from catalogue.

Content	Al	Cu	V	Mn	Zn	Si	Fe	Ti	Cr	Ni	Mg
%	99.09	0.022	0.011	0.122	0.003	0.362	0.248	0.009	0.007	0.003	0.106

Table 6.2 Chemical composition of bagasse ash.

Elements	Al_2O_3	Fe_2O_3	Sio_2	Na_2O	K_2O	MnO	BaO	P_2O_5	TiO_2	CaO
%	8.55	3.61	78.34	0.12	3.46	0.13	<0.16	1.07	0.50	2.15

In a stir casting method, mechanical stirring is used to disperse the reinforcing phases, which are typically in powder form, throughout the molten aluminum. Stir casting of metal matrix composites was invented in 1968 by S. Ray, who mixed molten aluminum alloys including ceramic powders to incorporate alumina particles into the aluminum melt. The following diagram depicts a typical stir casting procedure for an aluminum alloy matrix composite (A. K. Sharma *et al.* 2021). A very tough step in this procedure is stirring the molten metal mechanically in the furnace. The resulting molten alloy contains particles of ceramics that can subsequently be cast using a die, a permanent mold, or sand. The stir casting technique can be used to produce composite material with particles of reinforcement with vol. fractions up to 30%. To lessen porosity, improve the microstructure, and homogenize the distribution of the reinforcement, the cast composites are occasionally further extruded (U. Umer *et al.* 2022). The segregation of reinforcing particles during the stir casting process, which is brought on by the reinforcing particles rising to the surface or settling throughout the melting and casting processes, is a significant concern (Usama Umer *et al.* 2021). The final distribution of the particles in the solid is determined by the properties of the constituent materials as well as process variables, such as the degree to which the particles are wetted by the melt, the force of the mixing, the relative density, and the pace of solidification (Borgohain *et al.* 2013). The shape of the mechanical stirrer, the variables of the stirring process, the location of the stirrer inside the molten metal, the melting temperature, and the properties of added particles affect how the particles are distributed throughout the molten metal. A two-step mixing procedure is an interesting recent advancement in stir casting (Girish and Anandakrishnan 2021). Utilizing a suitable stirring mechanism, such as mechanical, ultrasonic, electromagnetic, or centrifugal force stirring, allows for the use of traditional metal processing techniques in stir casting. Stir casting's main advantage is that it can be used for mass manufacturing. Stir casting is the highly cost-effective technique of creating metal matrix composites out of all the well-known ones (Ks and Pinto 2018). For mass production, stir casting can be as inexpensive as one-third to one-tenth of other techniques. Because of this, stir casting is currently the most commonly used commercial method for producing aluminum-based composites. Figure 6.1 depicts the stir casting process (Inegbenebor *et al.* 2018).

The principal method of producing composites, known as stir casting, involves stirring the ingredient of reinforcement in the material into the molten metal. This entails adding ceramic particles to the melt and agitating it, after which the mixture must be allowed to harden. This can typically be processed using fairly standard equipment, and using a stirring mechanism allows for continuous and semi-continuous processing (Gillani, Khan, and Shah 2022). The importance of excellent particle wetting should be considered. When adding particles or, especially, fiber to a melt, difficulties may result from the increased viscosity. Even so, casting processes are permitted since the viscosity is so low. Particles can aggregate and settle in melts, which can lead to microstructure inhomogeneities (Wang *et al.* 2009). It can also be problematic when there is redistribution brought on by particle pushing caused by an increasing solidification front. If the solidification is quick, this decreases. The extended liquid–ceramic contact that occurs during stir casting frequently might result in significant interfacial reactivity (Aylor and Moran 1984). The production of Al4C3 and silicon will be common in Al-Sic, which has been researched in depth. These both weaken the composition's final quality and increase the slurry's viscosity, making further casting challenging (Cavaliere *et al.* 2004). If the melt is rich in Si, either through

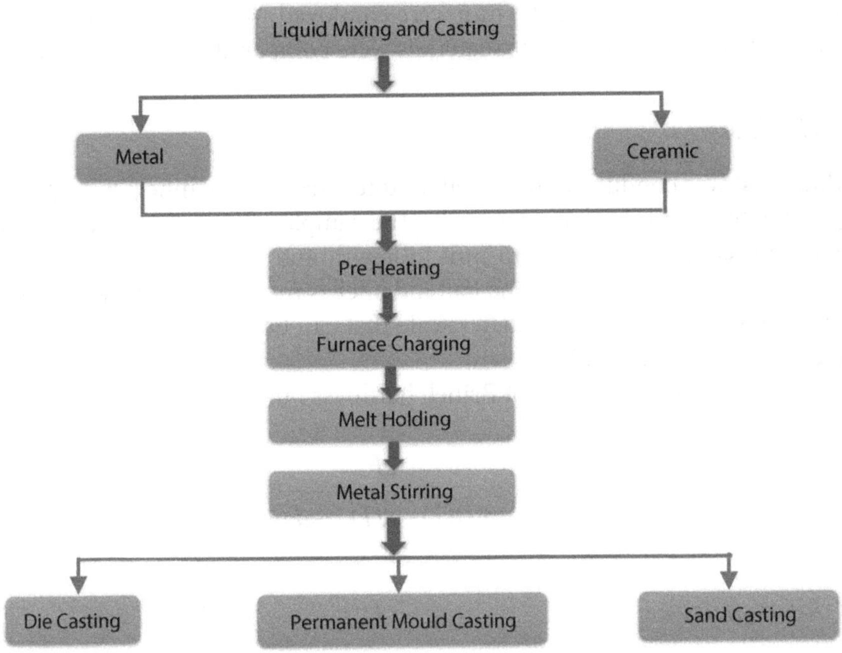

Figure 6.1 Flowchart of the stir casting method.

alloying beforehand or because of the reaction, the rate of reaction is lowered and may even become nil. It has been accomplished that casting of Al-Sic involving lengthy melt holding operations is suitable to traditional (high Si) casting alloys, but not to most wrought alloys, due to the reaction kinetics and Si levels required to eliminate it.

6.2.2.2 *Synthesis of Composite*

In a furnace, an alloy of aluminum was melted. For this, 1000 g of aluminum and the appropriate quantity of sugarcane bagasse ash were heated to 90 to 100°C. Melted aluminum was used up to 750°C, and the melt temperature was increased. The melt was then stirred using a mild steel stirrer. The stirring was done at a constant 200 rpm. The stirring continued for 5 to 7 minutes. When sugarcane bagasse ash was added, the melt temperature was kept at 700°C. The sand mold was filled with the melt with reinforced particles. At 600°C, the pouring temperature was held constant. After that, the melt was allowed to set up in the molds. The composites contained various amounts of fly ash (i.e. 6, 10, and 14 weight percentage) (Bhatnagar, Trivedi, and Shankhdhar 2016).

6.2.3 Testing of Composites

6.2.3.1 *Tensile Strength*

The ASTM B-557 M "Standard Test Methods of Tension Testing pure and Aluminum and their Composite" was used to evaluate the tensile performance of all the samples. The composite performed tensile testing on UTM. To measure ultimate tensile strength, standard specimens with a gauge length of 100 mm were employed. With commercially pure aluminum, the characteristics of composite materials were compared (Pawar and Utpat 2014).

6.2.3.2 *Hardness*

The Hardness test was conducted by Brinell hardness testing machine. On a lathe machine, the samples for the bending test were made. The samples were made in compliance with ASTM E10. The sample sizes were 15 mm in length and 10 mm in diameter. After that, the diameter of the indent made by the ball indenter is measured using a microscope. The microscope has an integrated micrometer scale that can be adjusted based on the position of the indent (Nayak *et al.* 2020).

The formula used for calculation of Brinell hardness number (BHN) is as follows:

$$\mathrm{BHN} = \frac{2F}{\pi D(D - \sqrt{D^2 - d^2})}$$

6.2.3.3 *Microstructure*

Scanning electron microscopy (SEM) is a beneficial tool to better understand the microstructural properties of materials (Krishna and Xavior 2014). Secondary electron (SE) mode was used to capture the photos. This investigation was carried out using a LEO scanning electron microscope outfitted with an Oxford data reference system energy dispersive X-ray (EDX) detector (Thandalam *et al.* 2015). To determine how reinforcement was distributed throughout the matrix, microstructural characterization experiments were performed. Scanning electron microscopes are used to do this. Prior to examination, the composite samples were polished metallographically. Characterization is carried out in etched circumstances. Using Keller's reagent, etching was performed (Yadav *et al.* 2020).

6.3 Results and Discussion

6.3.1 Tensile Strength Measurement

On a universal testing equipment, a tensile test was performed at room temperature. The samples were made in accordance with ASTM E8 standard (Padmavathi and Ramakrishnan 2014). The composites of various composition are prepared by adding sugarcane bagasse fly ash of different weight percentages.

The preparation of the first specimen involved adding 6 weight percentage fly ash, the second specimen involved adding 10 weight percentage fly ash, and the final specimen involved adding 14% fly ash by weight display in Figure 6.2. These compositions are compared with those of pure aluminum specimens mentioned in Table 6.3.

The tensile strength of the composite material was increased by increasing the weight fraction of reinforcement and the tensile strength is higher in hybrid composites with the 14% weight percentage fly ash composition compared to other compositions.

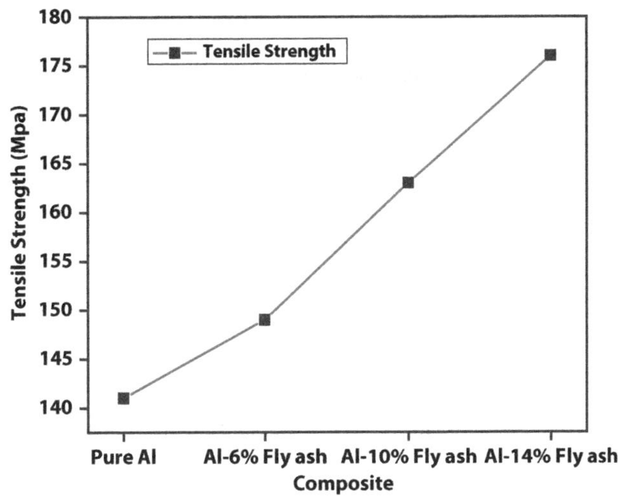

Figure 6.2 Comparison of tensile strength values.

Table 6.3 Tensile strength (MPa) results of pure aluminum and its composites.

		Hardness (BHN)
S. no.	Composite samples	Sample 1
1	Pure Al	46.31
2	Al-6% Fly ash	51.87
3	Al-10% Fly ash	62.41
4	Al-14% Fly ash	76.25

6.3.2 Hardness

The Brinell hardness test was performed for measuring the hardness of the reinforced metal composite and its compositions (Meena *et al.* 2013). The result of the hardness test is shown in Table 6.4 and the comparison of the average hardness at various concentrations of sugarcane bagasse fly ash are depicted in Figure 6.3.

Table 6.4 Hardness results of pure Al and its composites.

		Hardness (BHN)
S. no.	Composite samples	Sample 1
1	Pure Al	46.31
2	Al-6% Fly ash	51.87
3	Al-10% Fly ash	62.41
4	Al-14% Fly ash	76.25

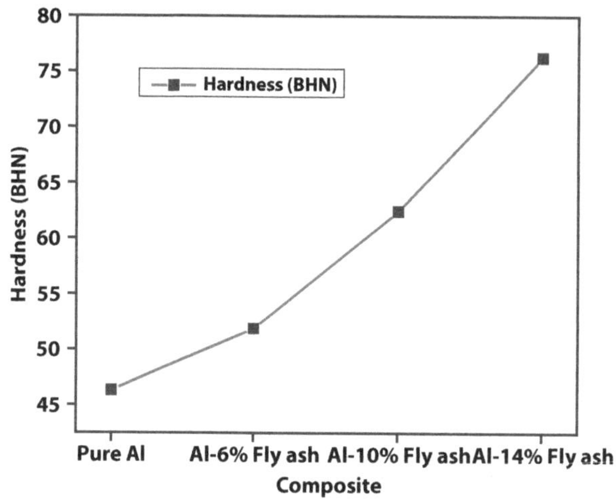

Figure 6.3 Comparison of hardness values.

Figure 6.3 shows that Brinell hardness number (BHN) are rapidly increased as with various fly ash ratio in the composite. As opposed to pure Al, Brinell hardness number (BHN) is maximal in Al-14 weight percentage fly ash composites and minimal in Al-6 weight percentage fly ash composites.

It is observed that the Brinell hardness number (BHN) at Al-14% fly ash composites is the suitable metal hardness as compared to other composites.

6.3.3 Microstructure

SEM is a beneficial tool to better understand the microstructural characteristics of materials (Krishna and Xavior 2014). The microstructural analysis of the 5% volume fraction of particle reinforced composites was performed using a SEM (Arivukkarasan *et al.* 2013). Figure 6.4 shows the SEM microstructure of pure Al and its composite of different Al-6% fly ash concentration ratio with fly ash.

After stir casting aluminum, a SEM micrograph is taken. Stir casting requires a melt temperature of 750°C. The stirring was kept going for 5 to 7 minutes at a 300-rpm impeller speed. The microstructure of pure Al, as shown in Figure 6.4a, is used as reference to compare the microstructure of MMC. The generated aggregates are tightly bonded, and the pure Al exhibits a larger grain size. The microstructure of pure Al with the addition of 6% fly ash, as shown in Figure 6.4b, clearly demonstrates the appearance of few amounts of fly ash over the pure Al gains accompanied by the pores. In Figure 6.4c, the microstructure of pure Al with the addition of 10% fly ash demonstrates the appearance of a higher amount of fly ash over the pure Al and no pores are observed. It seems that the microstructure of pure Al with the addition of 14% fly ash (Figure 6.4d) has simply overlaid the whole pure Al structure. Moreover, compared with pure Al and 6% and 10% ratio of fly ash, 14% fly ash with Al is more uniformly distributed. In addition, the density, porosity, and mechanical properties are found to be better for the 14% fly ash–Al composite. Thus, fly ash makes the morphology of aluminum denser, which enhances its mechanical strength.

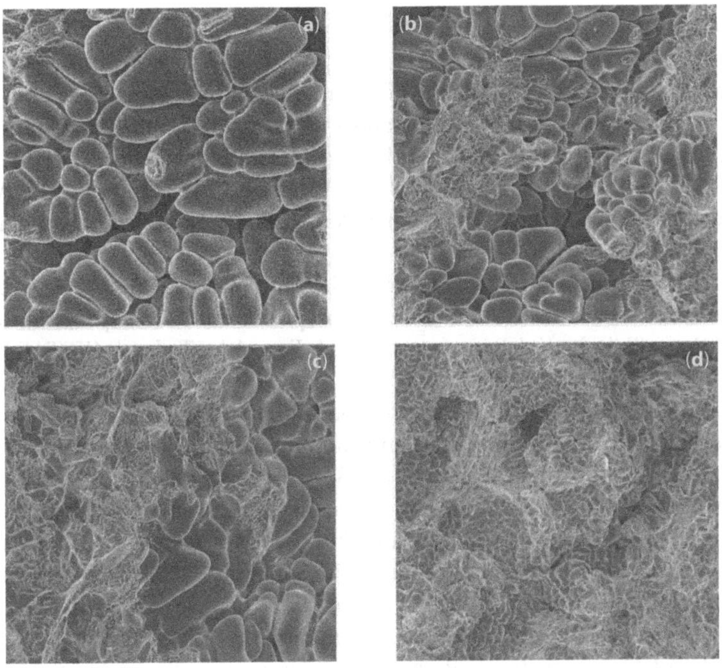

Figure 6.4 SEM images of (a) Pure Al, (b) Al with 6 weight percentage fly ash, (c) Al with 10 weight percentage fly ash, (d) Al with 14 weight percentage fly ash.

6.4 Conclusion

We have successfully fabricated the composite of aluminum matrix reinforced with bagasse fly ash via the stir casting technique presented herein. The stir casting technique was found to be suitable to synthesize aluminum matrix composites reinforced with various weight fractions of fly ash such as 6%, 10%, and 14%. Bagasse fly ash up to 14 weight percentage ratios was successfully added into aluminum to produce aluminum/fly ash composites. Microstructural and mechanical properties of prepared composites were analyzed and compared with pure aluminum. It has been shown that the weight fractions have a significant impact on the mechanical properties of composites, such as tensile strength and hardness. In a study on the tensile strength of aluminum matrix composites reinforced with fly ash composites of various weight fractions, the highest tensile strength recorded was 176 MPa at a concentration of 14% fly ash. Brinell hardness test was conducted using the Brinell hardness testing machine for measuring the hardness of the Al/fly ash metal composite. Its observed suitable metal hardness with Brinell hardness number (BHN) at Al–14% fly ash composites was compared to other weight fraction composites. Microstructure studies have shown that the matrix contains a homogeneous dispersion of reinforcing particles. As the amount of fly ash is increased and bagasse fly ash is added, the aluminum molecules become closer to one another, increasing the mechanical characteristics of the material. In the future, the findings of this study might be compared with those of various matrix and reinforcement combinations to develop materials that are cost-effective in terms of their uses.

References

A. Chennakesava Reddy. 2015. "Influence of Volume Fraction, Size, Cracking, Clustering of Particulates and Porosity on the Strength and Stiffness of 6063/Sicp Metal Matrix Composites." *Int. J. Res. Eng., Technol.* 04 (01): 434–42. https://doi. org/10.15623/ijret.2015.0401066.

A, Kesavulu, F.Anand Raju, and Dr. M.L.S.Deva Kumar. 2014. "Properties of Aluminium Fly Ash Metal Matrix Composite." *Int. J. Innov. Res. Technol. Sci. Eng.*, https://doi.org/10.15680/ijirset.2014.0311010.

Ahemad, J., Bhaskar, S.V., Kapgate, R.A., Sadaphal, N.D. and Kharde, S.B., 2013. Development of Al/Sic5%, Al/Sic10%, Al/Sic15%, "Metal Matrix Composite and Its Comparision with Aluminium Alloy–LM25 on Tribological Parameters." *Int. J. Eng. Adv. Technol. (IJETAE)*, 3(9), 545-549, 2013.

Arivukkarasan, S., V. Dhanalakshmi, A. Suresh babu, and M. Aruna. 2013. "Performance Study on Fatigue Behaviour in Aluminium Alloy and Alumina Silicate Particulate Composites." *J. Appl. Sci. Eng.*, 16 (2): 127–34. https://doi.org/10.6180/jase.2013.16.2.03.

Aylor, Denise M., and Patrick J. Moran. 1984. "Effect of Reinforcement on the Pitting Behavior of Aluminum-Base Metal Matrix Composites." *Proc. J. Electrochem. Soc.*, 84–9:584–95.

Bhandare, Rajeshkumar Gangaram, and Parshuram M. Sonawane. 2013. "Preparation of Aluminium Matrix Composite by Using Stir Casting Method." *Int. J. Eng. Adv. Technol. (IJEAT)*, 3 (2): 61–65.

Bhatnagar, Rahul, Vaibhav Trivedi, and Vivek Shankhdhar. 2016. "Preparation & Analysis of Mechanical and Microstructural Properties of Aluminium Based Metal Matrix Composite Reinforced with Bagasse Fly Ash," 15919–25. https://doi.org/10.15680/IJIRSET.2016.0509003.

Boopathi, M. Mahendra, K. P. Arulshri, N. Iyandurai, and P. Shanmughasundaram. 2014. "Dry Sliding Wear, Co-Efficient of Friction and Corrosion Behaviour of Aluminium Alloy 2024 Reinforced with Silicon Carbide and Fly Ash Hybrid Metal Matrix Composites." *Int. J. Mech. Mechatron. Eng.*, 14 (4): 44–53.

Borgohain, Chandan, Kaustubh Acharyya, Sidananda Sarma, Kula Kamal Senapati, K. C. Sarma, and Prodeep Phukan. 2013. "A New Aluminum-Based Metal Matrix Composite Reinforced with Cobalt Ferrite Magnetic Nanoparticle." *J. Mater. Sci.*, 48 (1): 162–71. https://doi. org/10.1007/s10853-012-6724-4.

Butola, Ravi, Susheem Kanwar, Lakshay Tyagi, Ranganath M. Singari, and Mohit Tyagi. 2020. "Optimizing the Machining Variables in CNC Turning of Aluminum Based Hybrid Metal Matrix Composites." *SN Appl. Sci.*, 2 (8): 1–9. https://doi.org/10.1007/s42452-020-3155-8.

Cavaliere, P., E. Cerri, P. Leo, L. Marzoli, and J. Dos Santos. 2004. "Erratum: Friction Stir Welding of Ceramic Particle Reinforced Aluminium Based Metal Matrix Composites (Applied Composite Materials (2004) 11: 4 (247-258))." *Appl. Compos. Mater.*, 11 (6): 399. https://doi.org/10.1023/B:ACMA.0000045336.48648.fa.

Chawla, N., and V. V. Ganesh. 2010. "Fatigue Crack Growth of SiC Particle Reinforced Metal Matrix Composites." *Int. J. Fatigue*, 32 (5): 856–63. https://doi.org/10.1016/j.ijfatigue.2009.08.005.

D.S.Chandra Mouli, R. Uma Maheswara Rao, A. Sarath Kumar. 2017. "A Review on Aluminium Based Metal Matrix Composites by Friction Stir Processing." *Int. J. Eng. Manuf. Sci. (IJEMS), Http://Www.Ripublication.Com* 07 (2): 203–24.

Dasgupta, Rupa. 2012. "Aluminium Alloy-Based Metal Matrix Composites: A Potential Material for Wear Resistant Applications." *ISRN Metall.*, 2012: 1–14. https://doi.org/10.5402/2012/594573.

Ding, H. Z., H. Biermann, and O. Hartmann. 2002. "A Low Cycle Fatigue Model of a Short- Fibre Reinforced 6061 Aluminium Alloy Metal Matrix Composite." *Compos. Sci. Technol.*, 62 (16): 2189–99. https://doi.org/10.1016/S0266-3538(02)00160-4.

Dixon, Mr J, Jim Joseph, Bavin Janis, Ghill Chrest, Abisal Zakaria, K Ilayaraja, and R Kavin. 2018. "Review on Various Aluminum Based Metal Matrix Composite" 6 (04): 1–2. www.ijert.org.

Ejiofor, J. U., and R. G. Reddy. 1997. "Developments in the Processing and Properties of Particulate Al-Si Composites." *Jom*, 49 (11): 31–37. https://doi.org/10.1007/s11837-997-0008-5.

Fuzail Hashmi, A., and Moin Ul Haq. 2018. "Indian Based Fly Ash & International Based Fly Ash: A Review Paper." *IOP Conference Series: Materials Science and Engineering*, 404 (1). https://doi.org/10.1088/1757-899X/404/1/012035.

Ghosh, Subrata Kumar, and Partha Saha. 2011. "Crack and Wear Behavior of SiC Particulate Reinforced Aluminium Based Metal Matrix Composite Fabricated by Direct Metal Laser Sintering Process." *Mater. Des.*, 32 (1): 139–45. https://doi.org/10.1016/j.matdes.2010.06.020.

Gillani, Fouzia, Muhammad Zubair Khan, and Owaisur Rahman Shah. 2022. "Sensitivity Analysis of Reinforced Aluminum Based Metal Matrix Composites." *Materials*, 15 (12). https://doi.org/10.3390/ma15124225.

Girish, G., and V. Anandakrishnan. 2021. "A Study on the Microstructure, Hardness, and Tribological Behavior of Aluminum-Based Metal–Matrix Composite Fabricated through Recursive Friction Stir Processing." *Proc. Inst. Mech. Eng. L.: Journal of Materials: Design and Applications*, 235 (3): 671–83. https://doi.org/10.1177/1464420720976686.

Gobalakrishnan, B., C. Rajaravi, Gobikrishnan Udhayakumar, and P. R. Lakshminarayanan. 2021. "Effect of Ceramic Particulate Addition on Aluminium Based *Ex-Situ* and *In-Situ* Formed Metal Matrix Composites." *Met. Mater. Int.*, 27 (9): 3695–3708. https://doi.org/10.1007/s12540-020-00868-6.

Inegbenebor, A. O., C. A. Bolu, P. O. Babalola, A. I. Inegbenebor, and O. S.I. Fayomi. 2018. "Aluminum Silicon Carbide Particulate Metal Matrix Composite Development Via Stir Casting Processing." *Silicon*, 10 (2): 343–47. https://doi.org/10.1007/s12633-016-9451-7.

Johny James, S., K. Venkatesan, P. Kuppan, and R. Ramanujam. 2014. "Hybrid Aluminium Metal Matrix Composite Reinforced with SiC and TiB2." *Procedia Eng.*, 97: 1018–26. https:// doi.org/10.1016/j.proeng.2014.12.379.

Kaczmar, J. W., K. Pietrzak, and W. Wlosiński. 2000. "Production and Application of Metal Matrix Composite Materials." *J. Mater. Process. Technol.*, 106 (1–3): 58–67. https://doi. org/10.1016/S0924-0136(00)00639-7.

Krishna, M. Vamsi, and Anthony M. Xavior. 2014. "An Investigation on the Mechanical Properties of Hybrid Metal Matrix Composites." *Procedia Eng.*, 97: 918–24. https://doi.org/10.1016/j.proeng.2014.12.367.

Ks, Lokesh, and Thomas Pinto. 2018. "Experimental Study on Effect of Metallic Oxides Mixture on Tribological Behaviour of Aluminum Based Metal Matrix Composites." *Int. Res. J. Eng. Tech.*, 1597–1603.

Kumar, Sunil, Francis Maury, and Naoufal Bahlawane. 2016. "Electrical Switching in Assembled VO_2 Disordered Metamaterial Coatings" 6.

Lakshmikanthan, Avinash, Santosh Angadi, Vinayak Malik, Kuldeep K. Saxena, Chandar Prakash, Saurav Dixit, and Kahtan A. Mohammed. 2022. "Mechanical and Tribological Properties of Aluminum-Based Metal-Matrix Composites." *Materials*, 15 (17). https://doi.org/10.3390/ma15176111.

Ling, P. S., and M. Gupta. 2000. "Recycling of Aluminium Based Metal Matrix Composite Using Disintegrated Melt Deposition Technique." *Mater. Sci. Technol.*, 16 (5): 568–74. https://doi.org/10.1179/026708300101508090.

Lynch, C. T., and J. P. Kershaw. 2018. "Metal Matrix Composites." *Metal Matrix Composites*, 43 (4):1–170. https://doi.org/10.1201/9781351074445.

Meena, K.L., Dr. A. Manna, Dr. S.S. Banwait, and Dr. Jaswanti. 2013. "An Analysis of Mechanical Properties of the Developed Al/SiC-MMC's." *Am. J. Mech. Eng.*, 1 (1): 14–19. https://doi.org/10.12691/ajme-1-1-3.

Muley, Aniruddha V., S. Aravindan, and I. P. Singh. 2015. "Nano and Hybrid Aluminum Based Metal Matrix Composites: An Overview." *Manuf. Rev.*, 2. https://doi.org/10.1051/ mfreview/2015018.

Myriounis, D. P., T. E. Matikas, and S. T. Hasan. 2012. "Fatigue Behaviour of SiC Particulate-Reinforced A359 Aluminium Matrix Composites." *Strain*, 48 (4): 333–41. https://doi.org/10.1111/j.1475-1305.2011.00827.x.

Nayak, Kanhu Charan, Parag Rajendra Deshmukh, Ankit Kumar Pandey, Premkumar Vemula, and Prashant P. Date. 2020. "Microstructural, Physical and Mechanical Characterization of Grinding Sludge Based Aluminium Metal Matrix Composite." *Mater. Sci. Eng., A* 773 (December 2019): 138895. https://doi.org/10.1016/j.msea.2019.138895.

Padmavathi, K. R., and R. Ramakrishnan. 2014. "Tribological Behaviour of Aluminium Hybrid Metal Matrix Composite." *Procedia Eng.*, 97: 660–67. https://doi.org/10.1016/j. proeng.2014.12.295.

Pawar, P. B., R. M. Wabale, and A. A. Utpat. 2018. "A Comprehensive Study of Aluminum Based Metal Matrix Composites: Challenges and Opportunities." *Mater. Today: Proc.*, 5 (11):23937–44. https://doi.org/10.1016/j.matpr.2018.10.186.

Particulate Metal Matrix Composite for Spur Gear." *Procedia Mater. Sci.*, 6 (Icmpc): 1150–56. https://doi.org/10.1016/j.mspro.2014.07.187.

Rama koteswara Rao, V., A. Balaji, D. S. Krishna, B. P. R. Bhavabhuthi, G. Sreevatsava, and K. Abhiram. 2016. "A Review on Properties of Aluminium Based Metal Matrix Composites via Stir Casting." *Int. J. Sci. Eng. Res.*, 7 (2): 742–49.

Ramnath, B. Vijaya, C. Elanchezhian, R. M. Annamalai, S. .Aravind, T. Sri Ananda Atreya, V. Vignesh, and C. Subramanian. 2014. "Aluminium Metal Matrix Composites - A Review." *Rev. Adv. Mater. Sci.*, 38 (1): 55–60.

Ravesh, S. and Garg, T. 2012. "Prepration & Analysis for Some Mechanical Property of Aluminium Based Metal Matrix Composite Reinforced with Sic & Fly ASh." *Int. J. Eng. Res., Appl.* 2 (6): 727-731.

Romanov, Ivan, Evgeny Chernyshov, and Alexey Romanov. 2019. "Assessment of the Possibility for Substituting Cast Iron in Vehicle Brake Disc with Aluminum-Based Metal-Matrix Composite Material Produced by Internal Oxidation." *Mater. Today: Proc.*, 19 (xxxx): 2125–28. https://doi.org/10.1016/j.matpr.2019.07.224.

Sahraeinejad, S., H. Izadi, M. Haghshenas, and A. P. Gerlich. 2015. "Fabrication of Metal Matrix Composites by Friction Stir Processing with Different Particles and Processing Parameters." *Mater. Sci. Eng. A*, 626: 505–13. https://doi.org/10.1016/j.msea.2014.12.077.

Samal, B P, S C Panigrahi, and B Sarangi. 2013. "Use of Modified Stir Casting Technique to Produce Metal Matrix Composites." *Int. J. Eng. Tech. Res., (IJETR)*, 1 (9): 63–65.

Sharma, Arun Kumar, Rakesh Bhandari, Chaitanya Sharma, Shrikrishna Dhakad, and Camelia Pinca-Bretotean. 2021. "A Study on Effects of Reinforcement Materials in Aluminum-Based Metal Matrix Composites." *Int. J. Eng. Trends Technol.*, 69 (9):24–28. https://doi.org/10.14445/22315381/IJETT-V69I9P203.

Sharma, Rohit, S. P. Jha, Khushboo Kakkar, Kushal Kamboj, and Pardeep Sharma. "A review of the aluminium metal matrix composite and its properties." *Int. Res. J. Eng. Tech.*, 4, no. 02 (2017): 2395–0056.

Singh, Jaswinder, and Amit Chauhan. 2016. "Characterization of Hybrid Aluminum Matrix Composites for Advanced Applications - A Review." *J. Mater. Res. Technol.*, 5 (2): 159–69. https://doi.org/10.1016/j.jmrt.2015.05.004.

SRINIVASAN, R., and K. Sathiya. 2010. "Experimental Study on Bagasse Ash in Concrete." *International Journal for Service Learning in Engineering, Humanitarian Engineering and Social Entrepreneurship*, 5 (2): 60–66. https://doi.org/10.24908/ijsle.v5i2.2992.

Tang, Shengyang, Raghunandan Ummethala, Challapalli Suryanarayana, Jürgen Eckert, Konda Gokuldoss Prashanth, and Zhi Wang. 2021. "Additive Manufacturing of Aluminum-Based

Metal Matrix Composites—A Review." *Adv. Eng. Mater.*, 23 (7): 1–17. https://doi.org/10.1002/adem.202100053.

Thandalam, Satish Kumar, Subramanian Ramanathan, and Shalini Sundarrajan. 2015. "Synthesis, Microstructural and Mechanical Properties of Ex Situ Zircon Particles (ZrSiO4) Reinforced Metal Matrix Composites (MMCs): A Review." *J. Mater. Res. Technol.*, 4 (3): 333–47. https://doi.org/10.1016/j.jmrt.2015.03.003.

Umer, U., M. K. Mohammed, M. H. Abidi, H. Alkhalefah, and H. A. Kishawy. 2022. "Machinability Analysis and Multi-Response Optimization Using NGSA-II Algorithm for Particle Reinforced Aluminum Based Metal Matrix Composites." *Adv. Prod. Eng. Manag.*, 17 (2): 205–18. https://doi.org/10.14743/apem2022.2.431.

Umer, Usama, Mustufa Haider Abidi, Jaber Abu Qudeiri, and Hisham Alkhalefah. 2021. "Evaulation of a Equivalent Homogeneous Material Model While Machining Aluminum Based Metal Matrix Composite." *Mater. Today: Proc.*, 44 (xxxx): 764–70. https://doi.org/10.1016/j.matpr.2020.10.679.

Upadhyay, Gaurav, and Kuldeep K. Saxena. 2021. "Role of Stir Casting in Development of Aluminium Metal Matrix Composite (AMC): An Overview." *IOP Conf. Ser.: Mater. Sci. Eng.*, https://doi.org/10.1088/1757-899x/1116/1/012022.

Velavan, K., and K. Palanikumar. 2015. "Effect of Silicon Carbide (SiC) on Stir Cast Aluminium Metal Matrix Hybrid Composites - A Review." *Appl. Mech. Mater.* 766–767: 293–300. https://doi.org/10.4028/www.scientific.net/amm.766-767.293.

Vengatesh. 2014. "Aluminium Alloy Metal Matrix Composite : Survey Paper." *Int. J. Eng. Res.Gen. Sci. Volume* 2 (6): 792–96.

Wang, Wei, Qing yu Shi, Peng Liu, Hong ke Li, and Ting Li. 2009. "A Novel Way to Produce Bulk SiCp Reinforced Aluminum Metal Matrix Composites by Friction Stir Processing." *J. Mater. Process Technol.* 209 (4): 2099–2103. https://doi.org/10.1016/j.jmatprotec.2008.05.001.

Yadav, Varsha, Swati Chaudhary, Saral Kumar Gupta, and Ajay Singh Verma. 2020. "Synthesis and Characterization of TiO2 Thin Film Electrode Based Dye Sensitized Solar Cell." *East Eur. J. Phys.*, https://doi.org/10.26565/2312-4334-2020-3-16.

Yanan Chen, Kun Fu, Shuze Zhu, Wei Luo, Yanbin Wang, Yiju Li, Emily Hitz *et al.* "Reduced graphene oxide films with ultrahigh conductivity as Li-ion battery current collectors." *Nano Lett.*, 16, no. 6 (2016): 3616-3623.

Zhou, Shengquan, Yongfei Zhang, Dawei Zhou, Weijian Wang, Dongwei Li, and Zhaibang Ke. 2020. "Experimental Study on Mechanical Properties of Fly Ash Stabilized with Cement." *Adv. Civ. Eng.*, 2020. https://doi.org/10.1155/2020/6410246.

7

Organic Smart Materials: Synthesis, Characterization, and Application

Shivaleela B.* **and S. M. Hanagodimath**

Department of Physics, Gulbarga University, Kalaburagi, India

Abstract

Organic smart materials are those that have carbon atoms. The presence of nitrogen, oxygen and other atoms adds variety to these carbon molecules. Organic molecules which are used in the present work have the property of fluorescence. Variety of techniques are used to synthesize fluorescent molecules. Based on molecular structure different types of fluorophores are synthesized from changing the positons of groups or adding groups like methyl, amino hydroxyl, carbonyl and inorganic elements. UV-Vis, Fluorescence spectra and Time correlated single photon counting (TCSPC) are used for characterization of the fluorescent molecules. Gaussian software can be used for computational or theoretical studies for spectroscopic analysis and molecular orbital study. Organic smart fluorophores have been used in the pharmaceutical industry as antibiotics, antiviral, antimicrobial and anticoagulant agents and as pH indicators in biological systems and medical sciences. Biological and chemical sensors, optical brightners, fluorescent probe colorants, photoinitiators, photo dimerization in polar, non-polar solvents and laser dyes, are some of their applications.

Keywords: Organic smart materials, fluorescent molecules, DFT

7.1 Introduction

Smart materials are a family of materials that are included in advanced materials. Based on their responses to environmental stimuli, including physical, chemical, and biological stimuli, these materials are categorized into different classes. These materials have the capacity to alter their properties or transform the absorbed energy. Smart materials are attracting a lot of attention due to their vast commercial applications as actuators or sensors. This study provides a thorough examination of several smart material kinds, their distinguishing characteristics and some related investigations that improved smart material attributes. Additionally, uses of smart materials are divided into groups based on the various application fields, such as waste reduction, medical implantation, and nanoengineered systems [1].

Molecules that modify fluorescence in response to chemical reaction or change in their immediate environment are known as smart fluorescent probes. Today, they are widely used

**Corresponding author*: shivaleelab17@gmail.com

Upendra Kumar and Piyush Kumar Sonkar (eds.) Smart Materials for Science and Engineering, (121–134) © 2024 Scrivener Publishing LLC

in the synthesis of new medications, cellular imaging, environmental research, and a number of medical applications. The ability to record fluorescence emission with great sensitivity, the amazing spatial and temporal resolution in cell imaging made possible by fluorescence microscopy, and the ease with which several fluorophores are currently available are all reasons for this. People who are not actively engaged in the scientific community can utilize fluorescent probes because there are examples for a wide range of applications. Excitation, emission wavelengths, and subcellular localization can all be impacted by changes in probe structure [2]. Fluorescence enables visual analysis detection, high-throughput analysis for screening applications, and the development of fiber optics for environmental and medicinal assessments.

Sensitive detection of cellular events is possible through smart organic fluorescence-based assays and imaging studies. Small-molecule fluorophores are frequently used in investigations as enzyme substrates, biological stains, ion indicators, and labels for macromolecules. Novel probes have been developed using synthetic organic chemistry terminology for a variety of biological purposes. By using enzyme catalysis or light, advanced imaging studies enable the chemical activation of fluorophores, enabling the visualization of a specific subset of fluorophores in comparison to a larger population of inactivated fluorophores [3]. Chemical methods may also be used in cellular settings, enabling the precise tagging of particular molecules and cellular areas. Here, we talk about new developments in smart fluorescent small-molecule chemistry and how they might be used in biological research.

Fluorescent materials are smart organic substances that alter their appearance and optical characteristics in reaction to UV-Vis light. In this study, recent developments and emerging ideas in the study of intelligent organic materials with acidochromic characteristics are discussed. In order to help researchers build and fine-tune new organic materials for desired applications, this chapter aims to provide an understanding of the design, characterization, mechanism, and applications of organic smart materials.

Gaussian 16 W software was used to perform the DFT and TDDFT computations. Basis set 6-31G (d) has been used to optimize the geometry at B3LYP level of theory. UV-Vis spectra were created from relevant output files. Origin software was used to plot the UV-Vis results' coordinates [4]. Through the use of ESP maps and HOMO–LUMO surfaces, the photophysical properties were investigated.

7.2 Organic Smart Materials

Organic smart materials are vast sources of naturally occurring carbon-based molecules, originally sourced from living things but today now include lab-created variants. The most significant field of fluorescence spectroscopy is fluorescence probes [5]. Fluorophores can only be organic fluorescent compounds. In recent years, fluorophores and fluorescent probes have been developed to help readers who want more information estimate the optical characteristics of luminous molecules. One day, it might be able to integrate computational and organic chemistry to create novel fluorescent compounds with improved characteristics. There are many thousands of fluorescence probes, and listing them all would be impractical. An overview of the numerous fluorophores, their spectrum characteristics, and their uses is provided in this chapter. Numerous organic fluorescent substances shown in Figure 7.1a, such as coumarins, indoles, hemicyanine dyes, hydroxycoumarin, acridione, fluorescein, flavones, quinazolines, acridines, phenazines, amino anthraquinones, and certain laser dyes, have been studied using various methodologies [6].

Since they offer important information on electronic structure of these transients, short-lived species dipole moments are of great relevance for direct measurement. Charges redistribute when a photon excites a molecule, altering its shape in excited state. Dipole moment of the ground and excited states may rise or fall depending on the charge distribution in the solute.

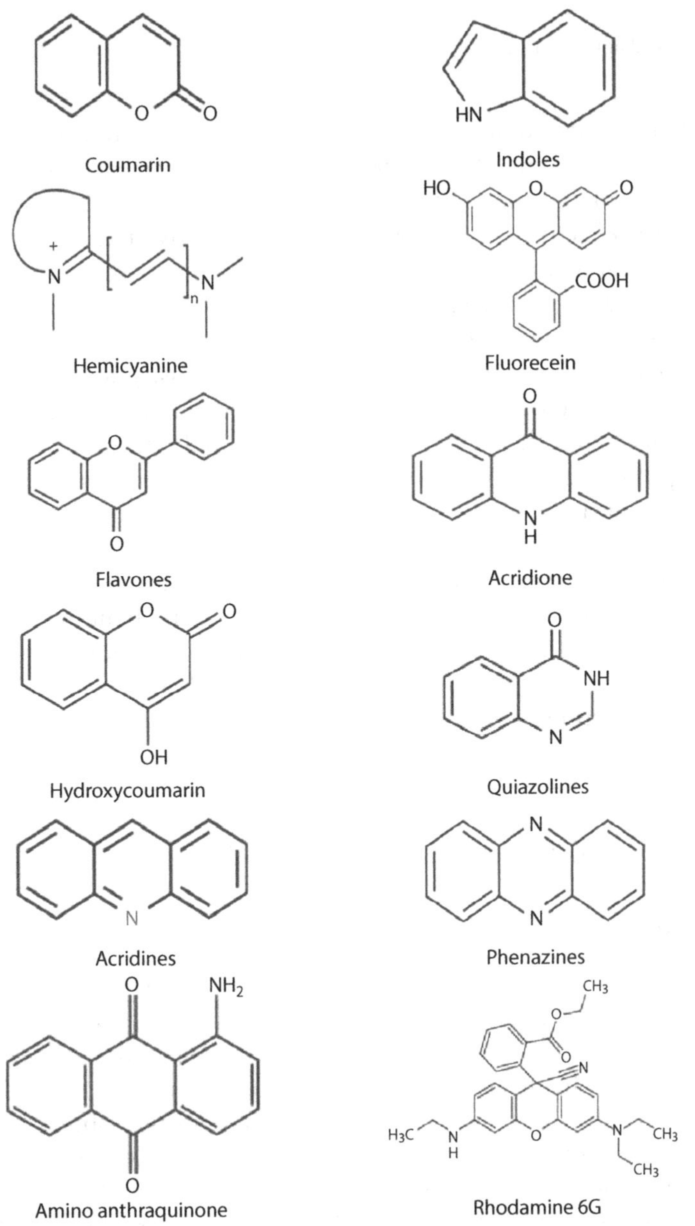

(a)

Figure 7.1a Molecular structures of different organic smart materials (organic fluorophores).

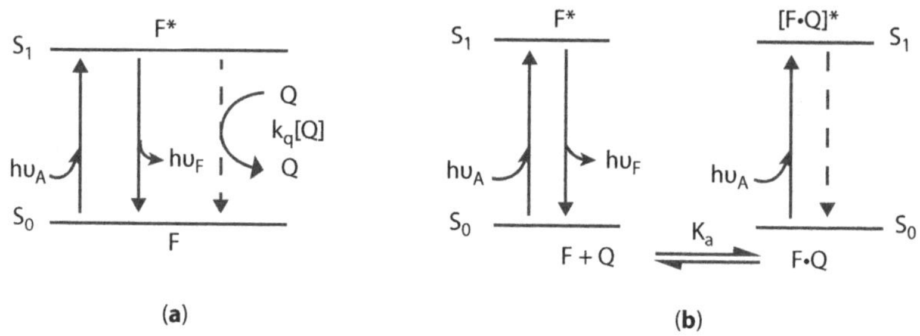

Figure 7.1b Jablonski diagram for quenching processes. (a) Dynamic quenching. (b) Static quenching.

Dipole moments of electronically excited species can be used to determine electrophilic and nucleophilic sites in some photochemical reactions, as well as to clarify the nature of excited states, course of any photophysical transformation, and design of non-linear optical materials [7, 8]. The effectiveness of quantum chemical derivations and the parameterization of semi-empirical methods are both determined by experiments on the excited state dipole moment. The growing interest in twisted internal charge transfer (TICT) processes has highlighted the significance of excited state properties.

Fluorescence quenching occurs when a fluorophore's intensity decreases and a quencher's concentration rises. Numerous molecular interactions, including as excited state processes, molecular rearrangements, the formation of complexes in the ground state, energy transfer, and collisional quenching, can result in the quenching of fluorescence. The two most typical types of quenching are dynamic (collisional) and static (contact) shown in Figure 7.1b. Both require that the quencher and the solute come into contact.

Computational tools like DFT (density functional theory) and TD-DFT (time-dependent density functional theory) have been shown to be helpful in obtaining information about various crucial molecular characteristics, including energy, dipole moment, absorption spectra, nonlinear optical properties, and charge transfer characteristics [9]. These findings are essential for predicting characteristics and prospective uses of compounds that have been synthesized.

7.3 Materials and Experimental Methods

7.3.1 Procedure to Record Spectra

Fluorescent molecules that have been used for the present work are 5-[(naphtho[2,1-b]furan-1-yl)methyl]-1,3,4-oxadiazole-2(3H)-thione (5NFMOT) and 4-[(5-amino-1,3,4-thiadiazol-2-ylthio) methyl]-7-methoxy-2H-chromen-2-one (4ATMMC) and phenyliodine (PID). S. D. Fine Chemicals Ltd. was selected to purchase all of the solvents. With a precision of 0.01 mg, the solutes were weighed using a semi-microbalance. An established solute concentration (10 M/L) was used to create the appropriate solutions. To reduce consequences of self-absorption, solute concentration was kept low. Absorption spectra were recorded using a PG Inst. Ltd., Model T-90+ UV-Visible spectrophotometer. Fluorescence spectra were recorded using an FP-8300 Fluorescence Spectrophotometer. The experiment was conducted at 302 K, which is considered room temperature [10].

The experimental setup and procedures for steady state and transient measurements are described in this section. The solutions for all of the fluorescent compounds were made by adjusting the quencher concentration 0.00, 0.02, 0.04, 0.06, 0.08, and 0.10 ML^{-1} in each solvent while maintaining the solute concentration [1×10^{-5} M/L] constant. The solutes were weighed using a semi microbalance with a precision of 0.01 mg. All of the solutions are uniform and clear. The solution was made fresh each time and stored in the dark. The fluorescence spectra of benzofuran derivative was recorded at absorption maxima in toluene. Using time-correlated single photon counting technique (TCSPC), fluorescence decay measurements were obtained in all solvents without the use of a quencher [11].

7.3.2 Computational Methods

Using computational techniques like TD-DFT, the UV-visible spectra of organic smart fluorescent molecules have been determined for the optimal geometry in various solvents. Using the Integral Equation Formalism Polarizable Continuum Model, solvent impact was examined. DFT was used at the B3LYP/6-31G basis set to map HOMO (highest-occupied molecular orbital), LUMO (lowest-unoccupied molecular orbital), and ESP (electrostatic potential) map [12].

7.4 Synthesis of Organic Smart Materials

The synthesis of dye scaffolds with suitable properties for biological studies is necessary for the quick and effective synthesis of fluorescent compounds. The structures of common fluorophores utilized in modern biological research are shown in Figure 7.1. Although colored chemicals were used to launch modern synthetic chemistry [13, 14], the development of dye chemistry has trailed much behind that of pharmaceutical and natural product chemistry. These ubiquitous core dyes are frequently created and modified using complex and often antiquated processes. Because fluorescent probes are useful in biological study, these substances need to be viewed as intriguing and significant targets.

7.4.1 Synthesis of Benzofuran Derivative 5NFMOT

The reaction of naphtha [2,1-b]furan-1-yl-acetic acid hydrazide 1 with potassium hydroxide and carbon disulfide in ethanol at reflux temperature resulted in the title compound 2 [1]. The preparation of compound 2 is shown in Figure 7.2a.

Preparation of Compound 2

Carbon disulfide (50 mmol) was progressively added to cold, stirring solution of naphtha [2,1-b]furan-1-yl-acetic acid hydrazide 1 (10 mmol) in pure ethanol (50 mL) containing potassium hydroxide (10 mmol). The reaction mixture was heated until hydrogen sulfide no longer evolved at reflux temperature on an oil bath. By distilling the excess ethanol under lower pressure, the excess was eliminated. The remaining material was then poured over crushed ice and filtered, and the filtrate was then neutralized with diluted hydrochloric acid. The mixture of ethanol and ethyl acetate in product 2 was filtered, washed with cold water, and recrystallized [1].

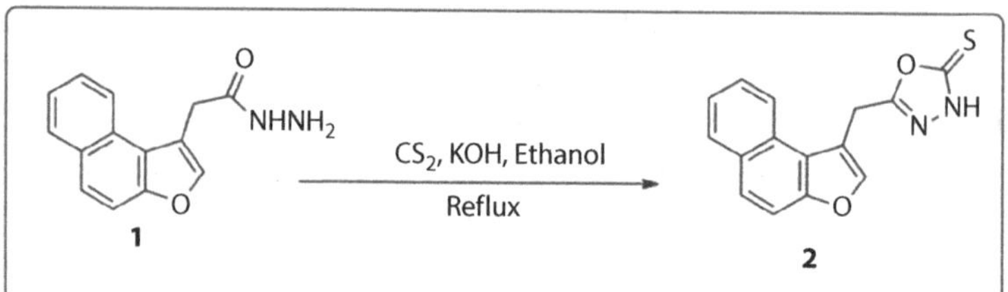

Scheme 1. Synthesis of benzofuran-oxadiazole derivative **2**

(a)

Figure 7.2a Synthesis of benzofuran derivative [13].

7.4.2 Synthesis of Coumarin Derivative

The reaction of 7,8-benzo-4-bromomethylcoumarin 1 with 5-amino-1,3,4-thiadiazole-2-thiol 2 at reflux temperature by dissolving sodium metal in absolute ethanol over a period of 2 hours resulted in the corresponding sulfide 3 in quantitative yield as shown in Figure 7.2b.

Preparation of Compound 3

Sodium metal (0.023 g, 0.001 M) was dissolved in absolute ethanol (25 mL) and treated with 5-amino-1,3,4-thiadiazole-2-thiol (0.133 g, 0.001 M) 2. To the resulting solution was added 7,8-benzo-4-bromomethylcoumarin (0.289 g, 0.001 M) 1 and the mixture was refluxed for 2 hours. The reaction mixture was concentrated and poured to the crushed ice. The turbid solution soon deposited a crystalline solid 3, which was filtered, washed with cold water, and crystallized from a mixture of ethanol and ethyl acetate 1.

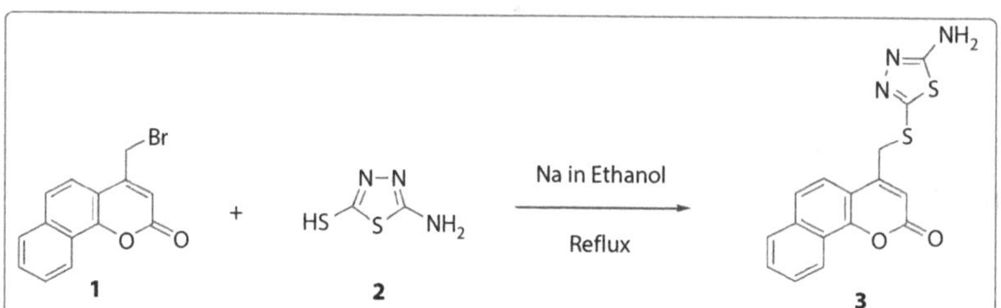

Scheme 1. Synthesis of **3**

(b)

Figure 7.2b Synthesis of coumarin derivative.

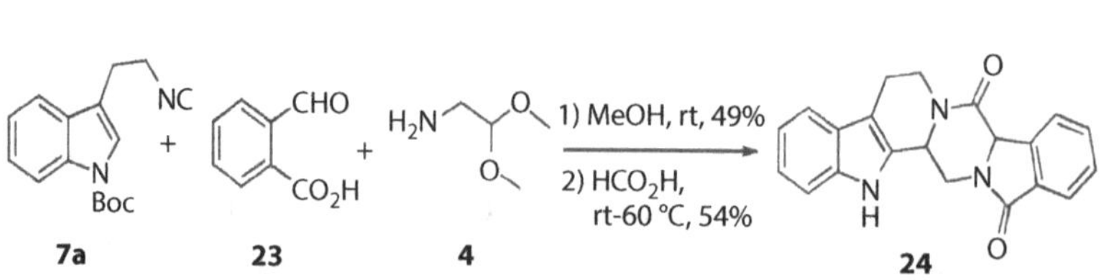

Figure 7.2c Synthesis of indole derivative [14].

7.4.3 Synthesis of Indole Derivative

Preparation of Compound 24

The synthesis of compound 24 is shown in Figure 7.2c and their detailed preparation was discussed in next subsection. At 0°C, isocyanide 7a (1.0 mmol) was added to a mixture of formylbenzoic acid 23 (1.2 mmol) and amine 4 (1.0 mmol) in MeOH (3.0 mL). For 24 hours, reaction mixture was agitated. The final product, Ugi (252 mg, 49%), was obtained by purifying the residue using column chromatography on silica gel while the solution was concentrated at reduced pressure. Cold HCO_2H (1.5 mL) was used to dissolve the aforementioned Ugi reaction product (100 mg, 0.20 mmol). The mixture was warmed to 60°C and swirled for 2 hours after being stirred at rt for 30 min. EtOAc was used to thin down the reaction mixture, then the saturating $NaHCO_3$ solution was carefully added to neutralize it. Organic layer was removed, brine washed, and dried [15].

7.5 Results and Discussion

In this chapter, the obtained experimental and theoretical results related to solvent effect on absorption and emission spectra of three organic smart fluorophores, viz., Coumarine (4ATMMC), benzofuran (5NFMOT), and indole (PID) derivative, are presented.

7.5.1 UV-Visible and Fluorescence Spectra

Solute–solvent interactions, which alter molecular absorption and emission spectra, can be used to interpret the structural changes that take place in molecules during physicochemical processes. 4ATMMC's emission and absorption spectra were studied in a variety of solvents with varied polarity. Figure 7.3a displays the absorption and emission spectra of these three compounds in common solvents. It is evident from absorption spectra that the absorption maxima move to shorter wavelengths. It relates to a blue or hypsochromic shift as the solvent polarity is increased. This suggests that the $n \rightarrow \pi^*$ transition is involved. This suggests that n orbitals are more stable than π^* orbitals due to hydrogen bonds or

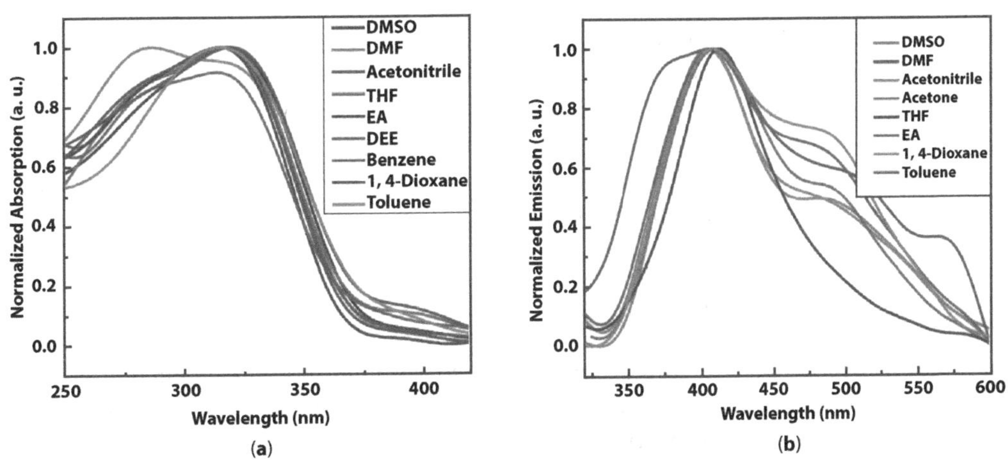

Figure 7.3 (a) UV-Visible and (b) fluorescence spectra of 4ATMMC in general solvents.

Table 7.1 Spectroscopic data of 4ATMMC in different solvents.

Solvents	λ_a	λ_f	$\overline{\nu_a}$	$\overline{\nu_f}$	$\overline{\nu_a} - \overline{\nu_f}$	$(\overline{\nu_a} + \overline{\nu_f})/2$
Water	339	398	29499	25126	4373	27312
DMSO	320	405	31250	24691	6559	27971
DMF	316	407	31645	24570	7075	28107
Acetonitrile	319	406	31348	24631	6717	27989
Acetone	334	410	29940	24390	5550	27165
THF	319	414	31348	24155	7193	27751
Ethyl acetate	317	407	31546	24570	6976	28058
DEE	317	407	31546	24570	6976	28058
Benzene	322	411	31056	24331	6725	27693
1,4-Dioxane	319	408	31348	24510	6838	27929
Toluene	316	375	31646	26667	4979	29156
Cyclohexane	258	312	38760	32051	6708	35405

polar solvents. The initial electrically excited state's vibrational structure is represented by the absorption peak [16, 17].

Fluorescence spectra reveal that the transition $\pi \rightarrow \pi^*$ correlates to a red or bathchromic shift when the solvent polarity increases, indicating that the excited state is more polar than the ground state (Figure 7.3b). Increased interaction of the solute with solvents in the excited states is shown by the fluorescence spectra shifting to a greater extent. Table 7.1 shows absorption maxima (λ_a) and wave number (ν_a), as well as fluorescence maxima (λ_f) and wave number (ν_f), which were used to determine Stokes shift and arithmetic mean in various solvents.

Fluorescence maxima for 4ATMMC is about 325 nm in general solvents. For 4ATMMC, fluorescence maxima is in the range 420–426 nm in alcohols as well as general solvents. As seen from these tables, fluorescence spectra are relatively more sensitive to solvent induced, solute-solvent interaction, and conformational changes compared to absorption spectra, whereas for 4ATMBC, larger Stokes shifts are in the range 4373–7193 cm^{-1}, and for general solvents, they are in the range 16018–10545 cm^{-1}.

7.5.2 Measurement of Fluorescence Lifetime

The fluorescence intensity I_0 and I were measured without and with quencher at varied quencher concentrations and at a fixed solute concentration of 1×10^{-5} M/L in toluene as a quencher. Table 7.2 lists the respective intensity levels. The fact that the fluorescence intensity drops as the quencher concentration increases can be seen in all of these tables, indicating

Table 7.2 Fluorescence intensity in toluene at varying quencher concentrations and at a fixed solute concentration.

Quencher: Bromobenzene
Solute concentration: 1×10^{-5} M/L
Excitation wavelength: 346 nm

[Q] M/L	I_0	I_0/I
0.00	1770	–
0.02	1743	1.01
0.04	1683	1.05
0.06	1602	1.10
0.08	1531	1.15
0.10	1367	1.29

$\text{Slope}\,(K_{SV}) = 2.599$. Intercept = 0.98.

that the quenching process has occurred. The fluorescence intensity graphs in Figure 7.4a confirm this. These graphs show that when the concentration of quencher increases, the intensity of the fluorophore decreases systematically in the solvents toluene for 5NFMOT [18]. Figure 7.4b shows the fluorescence decay curves of 5NFMOT at various quencher concentrations. Figure 7.4a shows that when quencher concentration rises, fluorescence lifetime drops significantly. Figure 7.4b demonstrates the fluorescence decay curves of three compounds with and without quencher concentrations. When we put quencher to the system, we can see that lifetime decreases [19].

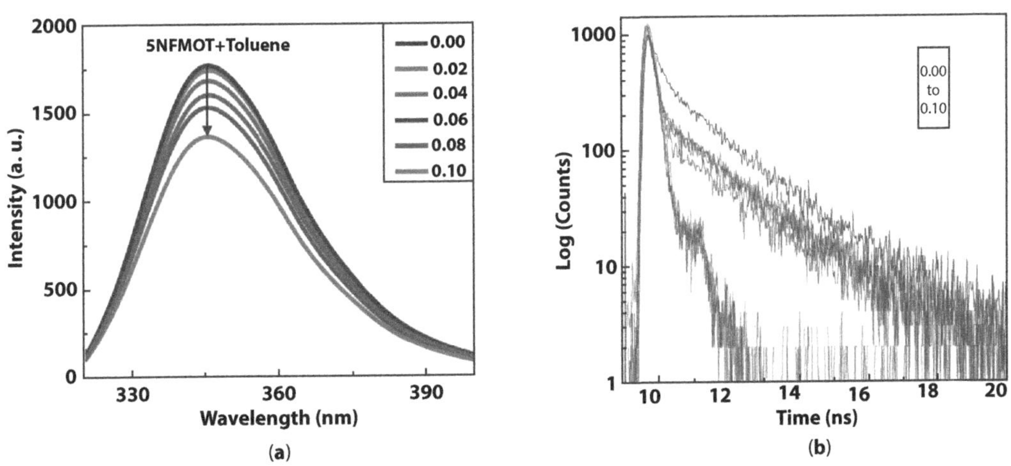

Figure 7.4 (a) Fluorescence quenching and (b) fluorescence decay of benzofuran derivative in toluene.

7.5.3 Quantum Chemical Calculations

Three different potential transitions are revealed by the UV-Vis absorption values of PID obtained from the TDDFT approach at the 6-311 (G)/ B3LYP basis set and presented in Table 7.3, with absorption maximum in acetonitrile displayed in Figure 7.5.

Only the first transition will be visible in the UV-Vis spectrum due to the oscillator strength for each transition, and the second and third transitions will have very low absorption coefficients. When the absorption coefficient is less, the excited state energies are larger; 3.54 Debye is the measured excited state dipole moment from TD-DFT, and 2.11 Debye is the measured ground state dipole moment from DFT. It is obvious to see that the excited state dipole moment is greater than its ground state counterpart [20].

Frontier molecular orbitals (FMOs) are HOMO and LUMO surfaces are shown in Figure 7.6. The ability to give an electron is represented by the HOMO, and the ability to accept an electron is represented by the LUMO.

Additionally, the FMOs have a significant impact on the optical, electric, quantum chemical, and UV-Vis spectra [21]. The electron transition from HOMO to LUMO is represented by the UV-visible absorption peaks of this molecule. In the gas phase, the HOMO–LUMO energy is -0.198 eV and -0.124 eV, while in acetonitrile, the values are -0.304 eV and -0.085 eV. The energy gap is respectively -0.074 eV and -0.219 eV.

The electronegativity and partial charges of a molecule are shown by the ESP map. The ESP map was used to research how potential was distributed in proportion to the different colors of bonded atoms in the molecule. Figure 7.7a and b displays the ESP map in

Table 7.3 Excited state geometry parameters from the TD-DFT method.

Parameters	I	II	III
Absorption maxima (nm)	424	627	695
Oscillator strength	2.9202	1.9760	1.7838
Excited state energy (eV)	0.1522	0.0026	0.4680

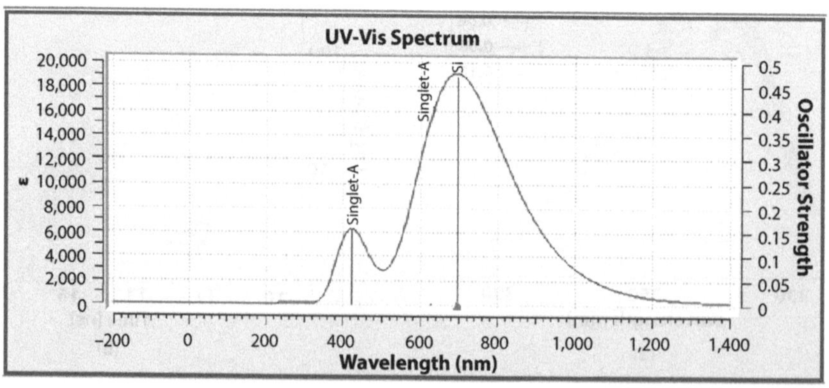

Figure 7.5 UV-Visible spectra of indole derivative.

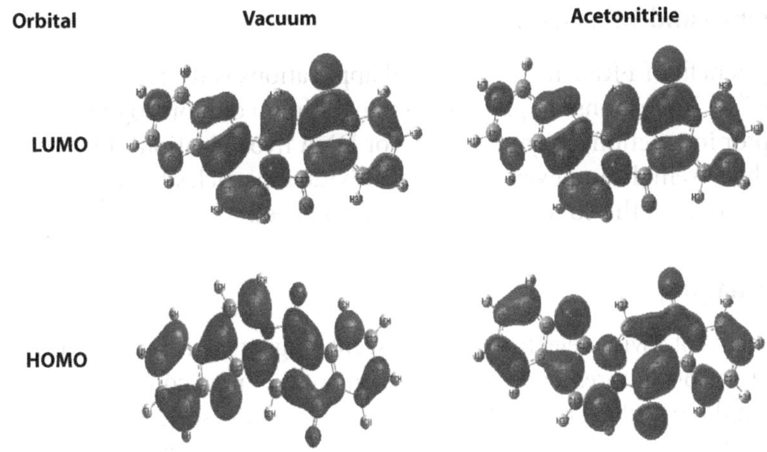

Figure 7.6 HOMO–LUMO surfaces of indole derivative in acetonitrile.

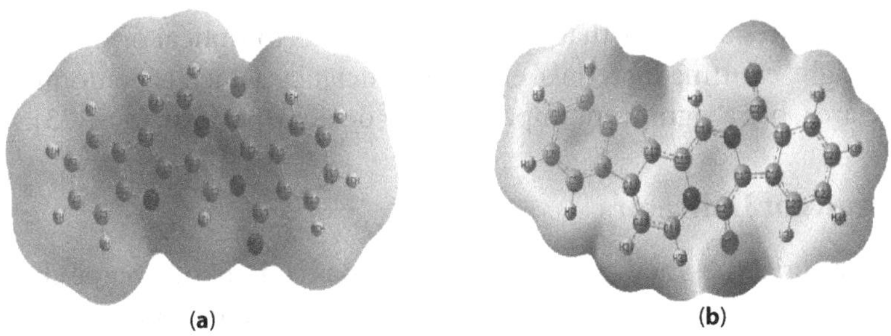

Figure 7.7 ESP map of indole derivative in acetonitrile.

both vacuum and acetonitrile. The ESP map is a yellowish glob in a vacuum. Due to the interaction of the molecule with acetonitrile, different colors are seen for the ESP map, indicating the extreme positive and negative sites ranging from the red to dark blue region. Positive sections are characterized by blue color nucleophilic reactivity, while negative parts are denoted by red color electrophilic reactivity [22]. The majority of negative areas are concentrated around the oxygen and nitrogen atoms. The majority of positive areas are concentrated on hydrogen atoms.

7.6 Applications

Huge amounts of research have been put into broadening the range of luminogen-based technology applications, drawn by the fascinating fluorescence property. This includes a variety of biological applications, chemosensing technology, and optoelectronic devices, among others. The inclusion of organic smart fluorescent materials has in turn opened up new directions for technical advancement in these fields. The most recent advancements in the field will be briefly covered in this section.

7.6.1 Chemo- and Biosensors

The primary benefit of FRET in bioanalytical applications is its capacity to switch fluorescence "on" or "off" in response to physicochemical stimuli or biorecognition events (such as ligand–receptor interaction, enzyme activity, or DNA hybridization). FRET techniques can be used on the ensemble and even down to individual particles because measured signals are not solely based on the concentration of fluorophores.

7.6.2 Bioimaging

Due to its benefits of a wide selection of fluorescent reagents, great temporal and spatial resolution, and low cost, fluorescence imaging is one of the most commonly used techniques in biological and biomedical research. The demand for fluorescent probes that can identify the intracellular environment, selectively illuminate specific organelles, and track biological activity is rising.

7.6.3 Optoelectronic Applications

Optoelectronic devices are tools that carry out electrical-to-optical or optical-to-electrical transduction. Light-emitting materials' emission efficiency, which is predominantly used in the solid state area, has a significant impact on the devices' performance [23].

7.6.4 Fluorescent Indicators

The pH affects the brightness and color of many luminous compounds. Some chemicals can be used as pH indicators because they are so sensitive to pH. Fluorescent or luminous indicators are what these are known as. Fluorescent indicators in acid–base indicators can be made from compounds that glow in ultraviolet light and change color or have their fluorescence quenched by changes in pH [24]. Such indicators have the advantage of being used in the titration of colored solutions, where the color change of conventional indications would be obscured.

7.6.5 Cosmetic Science

It is an area of study that is concerned with how people view aesthetic beauty. Researchers have looked into the makeup, traits, and functions of both skin and hair tissues in order to create cosmetics that meet (and maybe even exceed) customer expectations. The molecular architecture of the interiors of the skin and hair might be examined thanks to fluorescence. Using a combination of techniques, for instance, optical coherent tomography (OCT), second harmonic generation microscopy, and fluorescent molecular imaging (FMI), one can examine the deep internal components of the skin, such as the collagen fiber. This finding allowed for the collection of data on intricate biological and biochemical processes [25].

7.6.6 Laser Dyes

Generally, the absorption and emission spectra of laser dyes are complicated compounds with a variety of ring configurations. Due to their chemically similar structures, the laser dyes can be grouped into many groups. The coumarins, xanthenes, and pyrromethenes are typical examples. The molecule's structure and content have a significant impact on spectral emission. In addition to diagnostic operations, lithotripsy, and the activation of photosensitive drugs for photodynamic therapy, laser dyes are used in medicine for skin procedures like tattoo removal and port-wine stain removal. In the area of medical applications, dye lasers might be superior than other lasers [26]. Broadly tunable lasers have a substantial impact on a variety of fields in science and technology.

7.7 Conclusions

Organic smart materials used in this work have the property to fluoresce. Fluorescent molecules are created using a variety of methods. The fluorescent compounds are described using UV-Vis spectra, fluorescence spectra, and lifetime measurements. For computational or theoretical investigations for spectroscopic analysis and molecular orbital study, Gaussian software has been employed. In the pharmaceutical business, organic smart fluorophores have been utilized as pH indicators in biological systems and medical sciences, as well as antibacterial, antiviral, antimicrobial, and anticoagulant medicines. Some of its uses include laser dyes, optical brighteners, fluorescent probe colorants, photoinitiators, photodimerization in polar and non-polar solvents, and biological and chemical sensors.

References

1. "El-Cezeri » Makale » A review of smart materials: researches and applications." https://dergipark.org.tr/tr/pub/ecjse/issue/49096/562177 (accessed Dec. 26, 2022).
2. L. M. Wysocki and L. D. Lavis, "Advances in the chemistry of small molecule fluorescent probes," *Curr. Opin. Chem. Biol.*, vol. 15, no. 6, pp. 752–759, Dec. 2011, doi: 10.1016/j.cbpa.2011.10.013.
3. "Coumarin-Based Small-Molecule Fluorescent *Chemosensors | Chem. Rev.,*" https://pubs.acs.org/doi/abs/10.1021/acs.chemrev.9b00145 (accessed Dec. 26, 2022).
4. M. A. Robb, "New Chemistry with Gaussian 16 & GaussView 6," Jan. 1916, Available: http://spiral.imperial.ac.uk/handle/10044/1/57650
5. J. R. Lakowicz, "Fluorophores," in *Principles of fluorescence spectroscopy*, Springer, 1999, pp. 63–93.
6. B. G. Evale, S. M. Hanagodimath, I. A. Khan, and M. V. Kulkarni, Estimation of Dipole Moments of Some Biologically Active Coumarins by Solvatochromic Shift Method Based on Solvent Polarity Parameter, ETN, *Spectrochim. Acta, Part A. Mol. Biomol. Spectrosc.*, 73 (2009) 694–700.
7. C. Párkányi and J.-J. Aaron, "Dipole moments of aromatic heterocycles," in *Theor. Comput. Chem.*, vol. 5, Elsevier, 1998, pp. 233–258.
8. D. S. Chemla and J. Zyss, "Academic Press; New York: 1987," *Nonlinear Opt. Prop. Org. Mol. Cryst.*, 12Google Sch..
9. Y. G. Sıdır, C. Aslan, H. Berber, and İ. Sıdır, "The electronic structure, solvatochromism, and electric dipole moments of new Schiff base derivatives using absorbance and fluorescence spectra," *Struct. Chem.*, vol. 30, no. 3, pp. 835–851, Jun. 2019, doi: 10.1007/s11224-018-1228-8.
10. B. Siddlingeshwar and S. M. Hanagodimath, "Estimation of the ground and the first excited singlet-state dipole moments of 1,4-disubstituted anthraquinone dyes by the solvatochromic method," *Spectrochim. Acta. A. Mol. Biomol. Spectrosc.*, vol. 75, no. 4, pp. 1203–1210, Apr. 2010, doi: 10.1016/j.saa.2009.12.007.

11. S. Gounhalli, S. Basavaraj, and S. M. H. Hanagodimath, Comparative Study of NMR, IR, Raman Spectroscopy, HOMO-LUMO Surfaces of Benzofuran and Coumarin Derivative in Vacuum and Methanol, *Journal of Maharaja Sayajirao University of Baroda*, vol. 55, pp. 213–229, Dec. 2021.
12. S. Gounhalli, S. Basavaraj, and S. M. H. Hanagodimath, "Spectroscopic analysis of NMR, IR, Raman and UV-Visible, HOMO-LUMO, ESP and Mulliken charges of coumarin derivative by density functional theory," *Journal of Maharaja Sayajirao University of Baroda*, vol. 55, pp. 169–187, Dec. 2021.
13. Veerabhadrayya S. Negalurmath, Sathish Kumar Boda, ObelannavarKotresh, P.V. Anantha Lakshmi, MahanteshaBasanagouda, Benzofuran-oxadiazole hybrids: Design, synthesis, antitubercular activity and molecular docking studies, *Chem. Data Collect.*, 19 (2019) 100178. DOI: 10.1016/j.cdc.2019.100178.
14. M. Basanagouda, *Synthetic and biological studies on heterocycles, Doctoral dissertation*, Karnataka University, Dharwad 2009.
15. H. Liu and A. Dömling, "Efficient and Diverse Synthesis of Indole Derivatives," *J. Org. Chem.*, vol. 74, no. 17, pp. 6895–6898, Sep. 2009, doi: 10.1021/jo900986z.
16. B. Siddlingeshwar and S. M. Hanagodimath, "Estimation of first excited singlet-state dipole moments of aminoanthraquinones by solvatochromic method," *Spectrochim. Acta. A. Mol. Biomol. Spectrosc.*, vol. 72, no. 3, pp. 490–495, 2009.
17. D. S. Biradar and S. M. Hanagodimath, "Electronic excitation energy quenching of PPD by $ CCl_4 $ in different solvents," *J. Photoscience,* vol. 12, no. 1, pp. 51–54, 2005.
18. D. Nagaraja, R. M. Melavanki, and N. R. Patil, "Fluorescence quenching of 6BDTC by aniline in different solvents," *Can. J. Phys.*, vol. 91, no. 11, pp. 966–970, 2013.
19. S. V. Nishti, R. S. Kunabenchi, H. M. Suresh Kumar, S. M. Hanagodimath, and J. S. Biradar, "Electronic Excitation Energy Quenching of Indole Derivative By CCl 4 in Different Solvents," *Indian J. Phys.*, vol. 81, pp. 251–258, 2007.
20. H. Brahim, "DFT/TD-DFT investigation on the UV–vis absorption and phosphorescence spectra of platinum (II) and palladium (II) complexes with Schiff-base ligands," *J. Lumin.*, vol. 210, pp. 96–103, 2019.
21. M. Karabacak, S. Bilgili, and A. Atac, "Molecular structure, spectroscopic characterization, HOMO and LUMO analysis of 3, 3′-diaminobenzidine with DFT quantum chemical calculations," Spectrochim. *Acta. A. Mol. Biomol. Spectrosc.*, vol. 150, pp. 83–93, 2015.
22 .D. Rajaraman, G. Sundararajan, N. K. Loganath, and K. Krishnasamy, "Synthesis, molecular structure, DFT studies and antimicrobial activities of some novel 3-(1-(3,4-dimethoxyphenethyl)- 4,5-diphenyl-1H-imidazol-2-yl)-1H-indole derivatives and its molecular docking studies," *J. Mol. Struct.*, vol. 1127, pp. 597–610, Jan. 2017, doi: 10.1016/j.molstruc.2016.08.021.
23. Y. Hong, "Aggregation-induced emission—fluorophores and applications," *Methods Appl. Fluoresc.*, vol. 4, no. 2, p. 022003, 2016.
24. A. Bose, I. Thomas, G. Kavitha, and E. Abraham, "Fluorescence spectroscopy and its applications: A Review," *Int. J. Adv. Pharm. Anal.*, vol. 8, no. 1, pp. 1–8, 2018.
25. M. da Silva Baptista and E. L. Bastos, "Fluorescence in Pharmaceutics and Cosmetics," *Fluoresc. Ind.*, pp. 39–102, 2019.
26. G. S. Shankarling and K. J. Jarag, "Laser dyes," *Resonance,* vol. 15, no. 9, pp. 804–818, Sep. 2010, doi: 10.1007/s12045-010-0090-9.

8

Magnetostrictive Material-Based Smart Materials, Synthesis, Properties, and Applications

Pinki Singh[1]* and Sonam Perween[2]

[1]*Special Centre for Nanoscience, Jawaharlal Nehru University, New Delhi, India*
[2]*Department of Applied Sciences and Humanities, Jamia Millia Islamia University, New Delhi, India*

Abstract

Magnetostrictive materials are the passive class of smart materials, first discovered in Nickel by James Prescott Joule in the year 1842. He observed that a ferromagnetic material under the influence of an external magnetic field changes its size and regain their original dimension after removal of the field, also known as 'Joule effect'. After this discovery, Villari then discovered the reverse phenomenon in 1864, termed as the 'Villari Effect'. The phenomenon occurs due to migration of domains or re-orientation of magnetic domains under the influence of an applied magnetic field. Based on its origin, magnetostriction is categorized in two different types: spontaneous magnetostriction and induced magnetostriction. When the magnetostriction effect is very huge along with higher energy density and quicker response speeds, it is said to be giant magnetostriction.

All the ferromagnetic materials are believed to exhibit magnetostrictive property up to some extent. Out of general ferromagnetic materials, Co has been observed to show the maximum magnetostriction up to 50 μstrain which becomes more prominent if the FM specimen is in compressed state. Apart from conventional FM materials, a lot of research has been carried out to develop materials such as iron-based, rare-earth based, ferrite materials, alloys, polymer matrix composites and layered structure to enhance the magnetostrictive properties as well as to raise their Curie temperature (above which magnetostrictive property is lost).

The properties of materials are highly dependent on their synthesis techniques. Different methods are employed in order to obtain different shaped magnetostrictive materials. The common techniques such as citrate gel method, co-precipitation method, sol-gel hydrothermal route, electrochemical synthesis, auto-combustion method, magnetron sputtering, rapid quenching, rolling, electrodeposition, bonding method and directional solidification method are used according to the desired type of materials.

The reversible deformation has made them an excellent candidate for biomedical applications such as bone repair, sensors for articular movements, remote microactuators in cell operations. The magnetostrictive due to their superior mechanical properties in lower frequency band makes them suitable for deep sea measurements as SONAR system for submarine detection. They are useful as magnetostrictive transducers, actuator and various sensors in electronic applications.

This chapter presents the state of the art of magnetostriction and its research advances. The underlying principles and mechanism of magnetostriction are discussed. The detailed section about

**Corresponding author*: pinki9386@gmail.com

Upendra Kumar and Piyush Kumar Sonkar (eds.) Smart Materials for Science and Engineering, (135–150) © 2024 Scrivener Publishing LLC

preparation methods and measurement techniques for magnetostrictive properties is also presented. The chapter also discusses various examples of magnetostrictive materials with their possible applications.

Keywords: Magnetostriction, Joule effect, Villari effect, magnetic domains, Terfenol D

8.1 Introduction

Magnetostrictive materials are a class of smart materials consisting of ferromagnetic and ferrimagnetic materials that change their shape or size upon application of an external magnetic field. It was first observed in iron sample by James Prescott Joule in the year 1842. In this phenomenon, no change in the crystal structure is observed. It is a very different case from the changes that occur during phase transitions, which is driven by an applied magnetic field. These materials with bidirectional nature of energy exchange are capable of converting energy between elastic and magnetic states, and vice versa. The magnetostrictive materials, due to these properties of energy exchange, can be employed for sensing and actuation, and hence are often referred to as transducers [1]. The alloys based on the combination of transition metals (Fe, Ni, and Co) with few rare-earth elements are employed in several biomedical, industrial, and defense applications for sensor and actuator systems [2]. The sustainability of magnetostrictive properties is one of the most important characteristics of magnetostrictive materials. Their properties do not degrade over time since they are a completely inherent property of ferromagnetic materials, which makes them superior to some of the poled piezoelectric substances [3]. These materials also provide sufficient strains, energy densities, and forces along with coupling coefficients, which makes them suitable for a wider range of applications including the complex smart structures when compared to piezoelectric materials. Apart from many advantages, several modeling issues such as their bulkier size restrict their implementation in certain applications as compared to their electrostrictive and piezoelectric counterparts. These materials are also a bit costly to manufacture as they require complicated crystal growth setups to obtain directional solidification. The synthesis techniques including chemical reaction and physical methods to obtain magnetostrictive materials in several distinct forms such as sintered powder, crystalline thin films, and composite polymer matrix structures are employed to address the issues related to production cost. The non-linear effects and higher degree of hysteresis phenomena of these materials make them superior to electrostrictive materials.

The magnetostriction is defined as the magnetically induced strain, referred to as change in physical dimensions due to variation in magnetization (Figure 8.1). The value of magnetization depends on the direction of moments inside the magnetic domains of the material. This reorientation of magnetic moments can be induced inside the material by applying either of the magnetic fields, thermal energy, or directional stresses. Most of the ferromagnetic materials exhibit linear magnetostriction, $\Delta L/L$, which results due to the application of longitudinal magnetic field on a sample with length L. This linear magnetostriction is the most commonly used attribute of the magnetostrictive materials for their application as actuators. The compounds with rare-earth elements and some polymer matrices are found to show giant magnetostriction arising due to strong magneto-mechanical coupling.

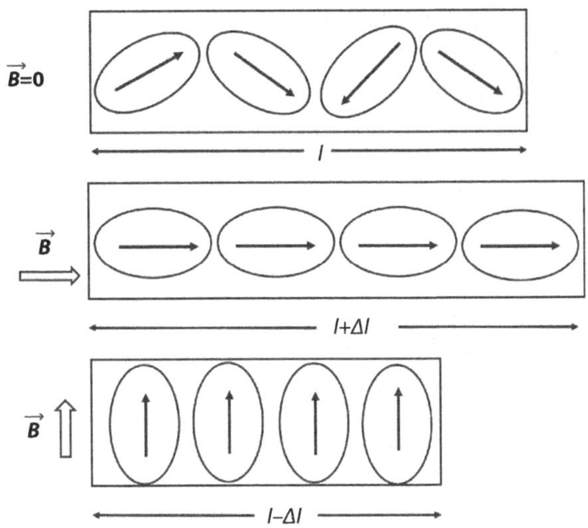

Figure 8.1 Magnetostriction effect.

This chapter provides an overview of smart materials based on magnetostrictive materials followed by different synthesis techniques, their properties, and a discussion on their applications.

8.2 Overview of Smart Materials Based on Magnetostrictive Materials

An extensive amount of research is being conducted to develop the magnetostrictive materials since their discovery in 1842. The ferromagnetic materials (Fe, Ni, Co, etc.) are primarily identified materials with saturated magnetostrictive strains 20–80 ppm. The materials such as cobalt, nickel, and their alloys were used in several applications including fog horns, hydrophones, oscillators, scanning sonar, and telephone receivers. The sonar transducers, at the time of World War II, essentially made use of nickel with magnetostrictive strains of the order of -50×10^{-6}. The major breakthroughs in the development of magnetostrictive materials were achieved in the early 1960s when rare-earth elements terbium- and dysprosium-based magnetostrictive materials were developed with saturated magnetostrictive strains several hundreds of times compared to those of conventional ferromagnetic materials (Rhyne, 1965). The magnetostrictive strains obtained in these materials were found to be higher than three orders of magnitude as compared to nickel ($\sim 10{,}000 \times 10^{-6}$), but at cryogenic temperatures, which was a major hindrance in gaining its technical maturity and led to the focus of research preference towards piezoelectric materials. Further, in 1972, the discovery of transducer materials, a combination of rare-earth elements (such as Tb, Sm, and Dy) with magnetic transition metals (Fe, Ni, and Co) with higher magnetostrictive strains at room temperature, was made by A.E. Clark (Cullen, 1977). The compounds R-Fe showed an increase in the Curie temperature with an increase in concentration of rare-earth elements whereas the compounds R-Ni and R-Co exhibited normal behavior of Curie temperature.

The compound $TbFe_2$ was found to show huge magnetostrictive strains up to 3,000 ppm at room temperature [4–6]. Since magnetostriction in these compounds originates from the strain arising due to magnetic anisotropy, higher values of magnetostrictive strains are attained at the cost of larger anisotropies. In order to achieve technical saturation, these compounds need to be applied with very large fields, which are practically impossible. This is a major technological limitation of these magnetostrictive compounds that restricts their applications.

Afterwards, the $TbFe_2$ system with typical composition $Tb_xDy_{1-x}Fe_{2-y}$ where terbium was partially substituted with dysprosium was developed, which produces giant magnetostrictive strains of the order 1000–2000 × 10^{-6} and improved anisotropic properties. The value of x and y lies between 0.27–0.35 and 0.05–0.1, respectively. The compound $Tb_{0.3}Dy_{0.7}Fe_{1.9-1.95}$ has been commercially available as Terfenol-D (**Ter**bium **Fe** **N**aval **O**rdinance **L**aboratory-**D**ysprosium) since 1980 (Savage, 1981). The magnetostrictive strain of Terfenol-D at room temperature is 1600 × 10^{-6} when subjected to a saturation field of 0.16 MA/m. At resonance, these values can increase to much higher values of up to 3600 ppm. Since the magnetic anisotropy of terbium and dysprosium is comparable in magnitude but directed opposite at room temperature, Terfenol-D is found to show the lowest anisotropy at room temperature. It is one of the most commonly used magnetostrictive material and finds applications in various fields such as ultra-large high-power ultrasonic devices, sonar systems, control systems with precision, and so on. The major limitation of Terfenol-D is larger losses due to eddy current at frequencies above a few kHz. In 1978, an amorphous metal obtained by rapid cooling of cobalt, nickel, and iron alloys altogether with one or more elements such as phosphorus, boron, and silicon were introduced, which were commercially termed Metglas [7]. These magnetostrictive materials are generally synthesized in thin-ribbon forms. These metallic glasses possess exceptionally high coupling coefficients that make them suitable to be used as sensors where mechanical energy is converted to voltage or electrical current.

Another giant magnetostrictive material, Fe-Ga alloy, without any rare-earth element, has been developed in the year 2000 by Clark [8]. The magnetostrictive strains were found to be higher for smaller Ga contents with best magnetostrictive properties exhibited for Fe-Ga alloys with 19% and 27.5% Ga atomic fractions. The inclusion of a third element in Fe-Ga alloys is reported to enhance the magnetostrictive properties of the material, which is the current research interest in this field. There are four main types of the third element added to Fe-Ga alloys: (i) 3d and 4d transition metals, (ii) Sn (slightly improves magnetostrictive strain) and Al (promotes malleability) elements, (iii) interstitial elements, and (iv) rare-earth elements.

8.3 Origin of Magnetostriction

There are two types of magnetic coupling inside the atoms, namely, spin–spin interactions and spin–orbit interactions. In the case of strongly magnetic materials, the spin–spin coupling within the magnetic domains that directs the adjacent spins to be aligned in parallel or anti-parallel direction are very strong. Since the exchange energy, in this case, is independent of the spin direction relative to the crystal lattice and depends only on the angle between neighboring spins, it is isotropic in nature [3].

The phenomenon of magnetostriction is mainly due to spin–orbit coupling, which is observed due to the interaction between the spin and orbital motion of individual electrons. The crystal anisotropy in a material arises due to spin–orbit coupling. When an external magnetic field is applied, it tries to align the electron spin as well as its orbit along the direction of the field. The electron orbit is coupled to crystal lattice strongly; thus, it resists the spin axis rotation. In order to overcome this spin–orbit coupling, energy is required, which can rotate the spins of magnetic domains in a direction other than the preferred one. The spin–orbit coupling of rare-earth metals is stronger when compared to those of most ferromagnetic materials by an order of magnitude. The rotation of spins and the orbital moments leads to considerable distortion, which results in magnetostriction. There are four distinct mechanisms causing magnetostriction in magnetic materials.

(i) Spontaneous magnetostriction

This type of magnetostrictive phenomenon is observed due to exchange forces. This force leads to the development of spontaneous magnetization in single crystals when temperature is cooled below Curie temperature. If the direction of the external magnetic field is perpendicular to spontaneous magnetization, the total magnetic moments of atoms tend to point in the direction of the applied magnetic field that causes magnetostrictive strain [3, 9].

(ii) Induced magnetostriction

The rotation of internal magnetic domains or movement of domain walls under the influence of an applied magnetic field is the basic underlying principle for magnetostriction phenomenon. This results in the redistribution of total energy to acquire the lowest energy state. The variation of magneto-elastic properties alters the distance between neighboring atoms leading to elongation or reduction of material size. The magnetization process of any material is the deciding factor for its magnetostrictive performance.

(iii) Shape effect

In the case of single-domain spherical sample, it is assumed that there are no spin–orbit or exchange interactions and only demagnetization energy exists. This demagnetization energy given by ($1/2NM_s^2V$) can be reduced by shrinking the sample volume, which is done via elongation along the direction of magnetization. The magnetostriction due to shape effects is much smaller than the other magnetostrictive effects [10].

(iv) Giant magnetostriction by rare-earth ions

The rare-earth metals, compounds, and alloys tend to exhibit giant magnetostrictive effect due to unfilled 4f electrons. The electrons of 4f orbit possess strong anisotropy for rare-earth ions. Upon spontaneous magnetization, the electrons in 4f layer acquire the lowest energy state in one or more directions causing larger lattice distortions. This is the major cause for giant magnetostriction in these systems [5].

8.4 Synthesis of Magnetostrictive Materials

The application of smart magnetostrictive materials has seen rapid surge since 1988 after the breakthrough of their manufacturing process. Since the microstructure has a direct consequence on magnetostrictive properties of materials, the preparation methods have a very significant role in obtaining materials with better properties. These materials are prepared using synthesis techniques like directional solidification, rapid quenching, rolling, magnetron sputtering, and bonding method. The choice of preparation method depends on the type of desired microstructure.

8.4.1 Directional Solidification Methods

This method is most commonly used to obtain Fe and rare-earth-based metallic magnetostrictive materials. The obtained materials using this technique are polycrystalline and have larger sizes, regular surface with uniform composition, and properties along the axis. In this method, the larger magnetostrictive materials are obtained via competitive grain growth in the preferred orientation. This method also produces columnar or single crystals that further improve the magnetostrictive properties of synthesized materials. This method consists of the Bridgman method, the Czochralski method, and the floating zone method depending on specific processes.

The Bridgman method is the simplest melt-based technique to obtain magnetostrictive rods with high performance. The materials with a larger diameter can be fabricated using this synthesis method [11]. In the Czochralski method, the crystal growth is directly observed, which allows the control of the crystal's diameter. In this method, the rate of crystal growth is faster and the materials obtained have higher optical homogeneity and smaller crystal dislocation density [12]. The floating zone method is used to synthesize small sized magnetostrictive materials. This method avoids the crucible contamination and reduces the loss of elements by burning [13]. The major drawback of the floating zone method is its high cost and difficulty to obtain fault-free crystals.

8.4.2 Rapid Quenching Method

The rapid quenching method is used to synthesize Fe-based and rare-earth-based materials. This technique is employed to obtain thin strips of magnetostrictive materials using rapid cooling and solidification. The vacuum arc furnace melting is used to obtain as-cast alloys from which samples are cut out to load in quartz tubes with holes of diameter 1 mm at the bottom. This sample alloy is heated to 1300°C using induction heating in order to melt the samples on a vacuum furnace with throwing strips from quartz tips along with an inert gas. The melt is sprayed by the nozzle to the copper roll surface, which is rotating at a high speed. It cools rapidly and solidifies as soon as it comes in contact with the roll surface. The melt is sprayed in forward direction by the effect of centrifugal force caused by roll rotation in order to form a strip. This extreme solidification process generates larger temperature and stress gradients that lead to formation of columnar crystals that possess a specific orientation of <100>. The magnetostrictive coefficient of materials obtained using this method is much higher than those obtained using the regular as-cast technique [14].

8.4.3 Rolling Method

The rolling method is most commonly used to obtain magnetostrictive materials, which are Fe-based alloys. In this method, the friction between roll and alloy drags the material alloys inside the roll gap, which is then compressed in order to form a sheet with increased length and reduced section [15]. The materials such as Fe-Al alloys, Fe-Ga alloys, etc. synthesized using this technique show good plasticity, better ductility, and preferred orientation, and have low production cost. The magnetostrictive sheet materials obtained via this technique show a lower eddy current loss in use. The major drawback of this method is its inability to produce stable form and optimization of the texture.

8.4.4 Magnetron Sputtering Method

The magnetron sputtering method is employed in order to obtain magnetostrictive films with strong adhesive properties and lower magnetic field saturation. In this method, argon is filled in a high vacuum chamber and applied with few hundreds of DC voltage. This produces a magnetically controlled glow discharge inside the vacuum coating chamber, thus ionizing the argon inside the chamber. The target surface is eroded by the use of high-energy ions inside the plasma chamber. The liberated atoms from the source travel all along the vacuum environment to deposit on the substrate and form a thin film [16].

8.4.5 Bonding Method

This method is used to prepare materials with certain complex shapes. This process requires mixing of alloys in powder form, followed by smelting in an electric arc furnace and grinding, using binders. The sample is then pressed, extruded, or injection molded to obtain composite materials of desired shape. The use of binders may alter the chemical composition and affect the magnetic properties and density of the material but also can improve the higher frequency characteristics and resistance of the magnetostrictive materials. This method has several advantages such as low cost, minimum loss of raw materials, and simple process [17].

In addition to the above-discussed preparation techniques, several other chemical synthesis methods, viz., co-precipitation, citrate, ceramic, and auto-combustion methods, are also used to obtain magnetostrictive materials. The end product formed using all these methods needs to be sintered at higher temperatures after making pellets. All the preparation techniques emphasize on the development of materials with larger magnetostrictive strains. The main focus of ongoing research on these materials is to explore the preparation techniques of magnetostrictive materials with larger dimensions and superior magnetostrictive properties.

8.5 Properties of Magnetostrictive Materials

The magnetostriction-based smart materials show advantageous features such as polarization, stress, and strain, which are the characteristics of magnetic anisotropies, hysteresis effects, and several multi-valued material properties.

8.5.1 Magnetic Anisotropy

The dependence of magnetic properties on the direction of measurement is referred to as magnetic anisotropy. There can be different types of anisotropy, viz., crystal, shape, stress, and exchange anisotropy. Among all these, only crystal anisotropy is the intrinsic property of the material whereas all others are induced externally. In most of the crystalline materials, the free rotation of magnetic moments in response to external fields is not possible since they tend to align themselves in the direction of easy crystallographic axis. The magnetic moments are subjected to an excess energy, known as magneto-crystalline or crystal anisotropy energy, to rotate their spins in any direction other than preferred crystallographic axis. There is a close relation between linear magnetostriction and crystal anisotropy energy, for example, as seen in rare-earth elements. The magnetostriction in the rare-earth elements is a consequence of direct relation between strain and magnetic anisotropy. In the case of absence of this dependence, no magnetostriction can be observed in the system. On application of an external magnetic field, the crystal lattice experiences deformations to achieve the minimum energy state of the system [1].

When a material with larger anisotropy is applied with a sinusoidal alternating magnetic field, a "jump" in magnetic moment is observed, as a result of which the obtained magnetization curve is not continuous. This jumping of magnetic moments and other associated nonlinearities can be reduced by alloying the elements with opposite anisotropies. Sometimes, the materials with the same constituents but different compositions (e.g. $Tb_{0.67}Dy_{0.33}$ and $Tb_{0.6}Dy_{0.4}$) are used to smooth the magnetization curve.

There are accurate models for relationship between crystal anisotropy and the magnetization process for simple crystal structures, i.e., cubic and hexagonal, but in case of models of complex structures, some simplifying assumptions are made to simplify the problem. For instance, in the modeling of complex structures of Terfenol-D, with growth in the form of dendritic twin sheets, aligned in the [112] direction, the operating regimes where crystal anisotropy is dominated by stress anisotropy are considered. These materials with a positive magnetostrictive coefficient have improved magnetoelasticity due to stress anisotropy caused by compression. This leads to an enhanced magnetostriction in these types of materials exploited for applications as transducers. For materials with negative magnetostrictive coefficient values such as nickel, the reverse effect is observed. In such type of system, tensile stresses generate enhanced magnetoelastic states.

8.5.2 Domain Processes and Magnetic Hysteresis

The modification in magnetic properties arising due to an external magnetic field can be reversible or irreversible. In case of reversible magnetization change, the energy is conserved and the material gains back its initial magnetic state soon after the removal of magnetic field whereas irreversible changes in magnetization are dissipative in nature. The reversible process occurs only for the small field increments whereas irreversible magnetization change occurs for large fields. The irreversible changes require the external restoring forces to return in its original state. The magnetization process in the system is due to the contribution of both reversible and irreversible processes, which are the consequence of two mechanisms: magnetic moment rotation and domain wall motion.

The ferromagnetic materials have domain walls as transition regions between two neighboring domains below their magnetic phase transition temperature (often referred as Curie temperature, T_c). At temperatures lower than T_c, the magnetic moments inside the domain are aligned parallel to each other, resulting in a spontaneous magnetization. In absence of an applied magnetic field, the direction of saturation magnetization is different for one domain than another and cancels out one another. Thus, the bulk magnetization of the material becomes zero on average. On application of a small external magnetic field, the domains with moments pointing in the same direction as the applied field tend to grow at the expense of others that causes motion of domain walls. When the strength of the applied field is increased, rotation of entire domains takes place in order to align along easy axis direction (Figure 8.2). This generates higher values of magnetization or strain changes, producing a *burst* region, even for small changes in field. The materials, in the final stage, behave as a single domain where the moments easily change their direction from easy crystallographic axis to the field direction and leads to the magnetization saturation. M.J. Dapino has reported the magnetization and strain curve for Terfenol-D as shown in Figure 8.3 [3]. These materials can be implemented for magnetic biasing, as discussed in a later section, designing them to operate in the burst region for best performance. The curves show nearly linear behavior in low field regions and hysteresis up to some extent. The domain wall motion along twin boundaries leads to irreversible impediment in the hysteresis for Terfenol-D. The properties of magnetostrictive materials can be designed and tuned by modeling the magnetic hysteresis and non-linear behavior [3].

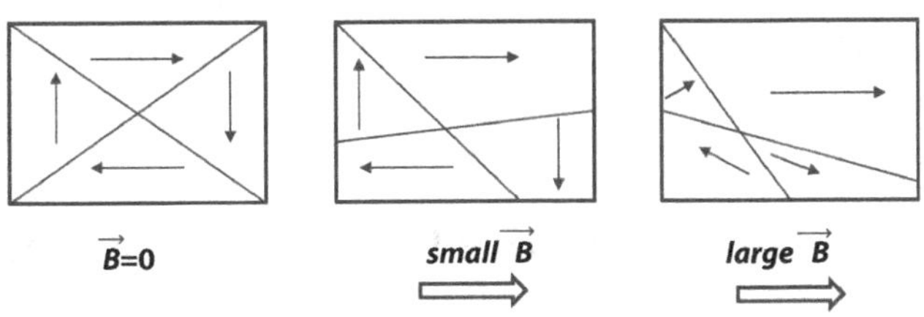

Figure 8.2 Domain process in ferromagnetic materials.

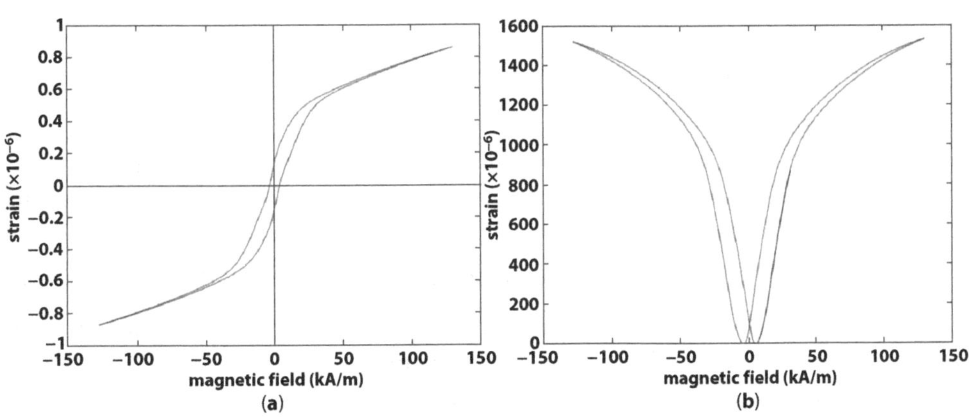

Figure 8.3 Experimental magnetic induction and total strain of Terfenol-D [3].

8.5.3 Multi-Valued Material Properties

The magnetic moments inside the magnetostrictive materials rotate to align themselves in the direction of applied field that leads to generation of strain. This process of alignment takes place via transduction of energy referred to as magneto-mechanical coupling. The coupling is bidirectional in nature and allows mechanisms of both actuation and sensing inside the materials. It is a complex function that depends on several other quantities such as stress, magnetic field, frequency and temperature collectively termed as "operating conditions" that vary at the time of device operation. For designing purposes, the efficiency of conversion between both mechanical and elastic states, termed linear coupling coefficient, should be nearly equal to unity. A small change in operating conditions strongly affects the material properties. The rotation of magnetic moments occur significantly inside a material only when the domain structure is present, which happens only in the ferromagnetic state at temperatures below T_c. Thus, the design of material should be chosen such that its range of operating temperature is well below the Curie temperature. In systems with smart structure, magnetostrictive materials experience large forces generated within the system that must be endured by them. The elastic modulus, E, quantifies the stiffness of the material, which should be higher in order to support larger forces. Also, the larger saturation magnetization and smaller magnetic anisotropy are the salient features of magnetostrictive materials [3].

Properties such as magnetostrictive strain, densities, magnetic field, Curie temperature, elastic modulus, and linear coupling coefficient are important magnetostrictive materials that need to be considered and optimized for their successful applications. An accurate assessment of material properties under different operating conditions is very important for an efficiently designed transducer.

8.6 Methods of Magnetostrictive Property Measurement

The magnetostriction inside the material can be measured either in a direct or in an indirect way, depending on whether strain is directly measured or derived from other strain-related parameters [18].

8.6.1 Direct Methods

Strain Gauge Method

In this method, the deformation induced by magnetostriction causes change in resistance of the strain sensor by making use of a strain gauge. This change in resistance is used to obtain the magnetostrictive strain (λ_s) of the material using the following relation: $\lambda_s = \frac{\Delta l}{l} = CK\frac{\Delta R}{R}$, where C, K, and R are the structural parameter of the strain resistor, the original value of the strain resistor, and magnification of the measurement system, respectively. The method has several advantages such as high sensitivity, performance stability, and cost-effectiveness. Since the strain changes are of the order of parts per million (ppm) in magnetostrictive materials, the measurement methods require high sensitivity [19, 20].

Capacitance Dilatometry Method
This method is known to have high sensitivity, which measures very small macroscopic changes inside the material [21]. In 1961, White and coworkers reported the development of a dilatometer with three-terminal capacitance for measuring magnetostrictive properties. These three terminal (3T) capacitance dilatometers have been developed nowadays to give a high-resolution measurement with an accuracy of 10^{-9} in static and pulsed magnetic fields with maximum values of 45 T and 60 T, respectively [22].

Optical Method
This is a simple traditional optical method to measure the magnetostriction with limited accuracy and used only for samples with larger dimensions. The measurement process involves generation of excited magnetic field by the use of solenoid and measurement of magnified micro-displacement using combined mechanical and optical levers [23]. The main advantage of the optical method is the separation between the probe and the sample, which leads to no contact between them.

8.6.2 Indirect Methods

The basic principle involved in the indirect measurement methods for magnetostriction is the Villari effect discovered in 1865 by E. Villari as the inverse Joule effect. It states that on application of a small stress, the magnetization curve for a magnetized iron rod varies with the stress. These methods do not involve any direct measurement for magnetostriction.

Strain Modulated Ferromagnetic Resonance Method
This method makes use of phase-sensitive detectors to measure a strain-modulated ferromagnetic resonance signal whose height is proportional to the strain modulation depth. The magnetostrictive strain can be obtained by comparing the signal height with the ferromagnetic resonance line. The method is suitable for materials in the form of thin sheets and has superiority of having a wider temperature range with a higher sensitivity [24].

Small-Angle Magnetization Rotation Method
It is an ideal method used to measure the magnetostrictive property of amorphous and filamentous materials. The axial direction of the sample material is subjected to a saturated static magnetic bias field followed by application of a transverse small alternating magnetic driving field produced by a solenoid. This leads to rotation of saturation magnetization of the material sample by a small angle lengthwise, which is converted to voltage induced in the sensing coil [25]. A small magnetostriction can be measured using this method.

8.7 Application of the Magnetostrictive Smart Materials

The magnetostrictive materials and alloys have gained the increased attention of science and engineering professionals due to their widespread use in numerous applications, including magnetostrictive transducers, sensors, actuators, and many more. These materials also find their suitability for implementation in biomedicine, for instance, used as a remote cell drive, physical sensors for pressure and temperature, and wireless implantable devices. Few of the important applications of magnetostrictive materials are discussed briefly in this section.

Magnetostrictive Transducers

Magnetostrictive transducers are used to convert the energy in magnetic field to mechanical energy and vice versa by utilizing the magnetostrictive property of the material. The magnetic field is generated by a wire coil wrapped around the magnetostrictive material. They are also known as non-contact linear transducers since they provide the information using magnetic sensing instead of measuring the resistance of contact point, which increases the life of magnetostrictive transducers. The major advantage of transducers made of magnetostrictive materials is that they are driven using low impedance amplifiers (i.e., require lower driving voltage), especially at below resonance frequencies. This property is useful in biomedical applications and can effectively simplify the amplifier design [1, 9].

The exact working of a transducer based on magnetostrictive material is explained in Figure 8.4. The figures show the effect of varying magnetic fields on the magnetostrictive strain. The mechanism is split in various magnetic and mechanical attributes with coupled effects on strain and magnetization. In absence of any external magnetic field, no change in length is observed (Figure 8.4a). When the strength of a magnetic field is increased from zero, the axial strain begins to develop, which increases with magnetic field strength and reaches up to a maximum value (Figures 8.4b and c). There will be no change even if we increase the strength of the magnetic field. Thus, a saturated maximum strain and magnetic induction are obtained at the maximum value of the magnetic field. The magnetic lines of force for applied magnetic field with domain position are represented in Figure 8.4d. The calculation of voltage generated in the conductor, which is placed perpendicular to the produced flux, provides information about the flux field value, which is proportional to the strain or force.

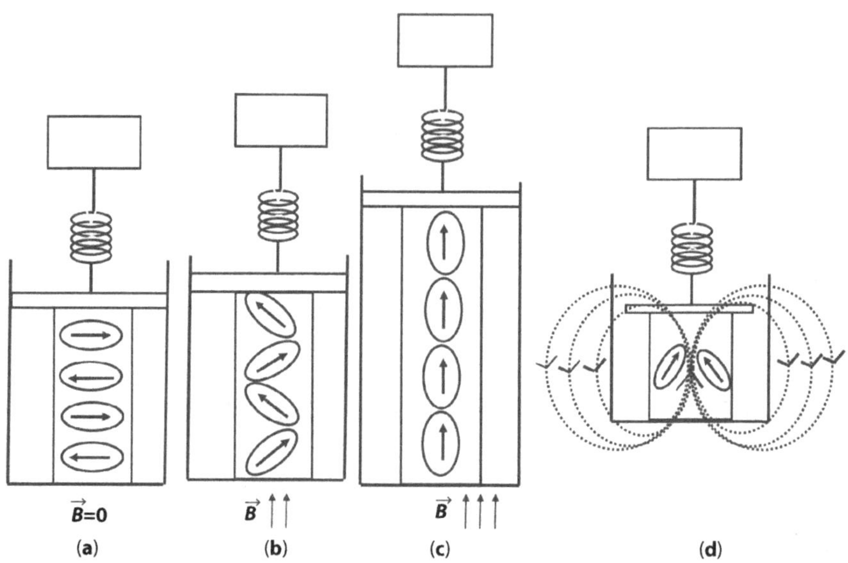

Figure 8.4 Working principle of magnetostrictive transducers (a) no external field, (b) applied external field, (c) external field causing maximum strain, and (d) magnetic forces of line with respect to domains.

When the magnetostrictive materials are used to convert magnetic energy into mechanical energy, they can generate force, which makes them useful as actuators. They can also be used for the detection of magnetic field leading to their implementation in sensor applications. The magnetostrictive materials can be used for the detection of force or motion when they are used to convert mechanical energy into magnetic energy. These transducers are used in several applications such as sonar transducers, rotational motors, linear motors, and hybrid magnetostrictive/piezoelectric smart material transducers.

Actuators

The actuators are a case of output transducers that produce mechanical response for an input signal. The applications of magnetostrictive materials as smart actuators, the most commonly used being Terfenol D, are due to their superior properties such as larger force, higher energy densities, faster response, and broader frequency bandwidth [3].

Sensor Applications

There are two more kinds of magnetostrictive effects known as Wiedemann and Matteuci effects. According to the Weidemann effect, on application of a longitudinal magnetic field to a current carrying ferromagnetic sample rod, a torsional strain is experienced by the rod due to helical anisotropy. The inverse Weidemann effect is called the Matteuci effect, which states that a torsional strain produces a change in axial magnetization of current carrying ferromagnetic sample rod. A magnetostrictive sensor works on the principle of the Weidmann effect and Villari effect. The sensor applications of magnetostrictive materials are attributed to its properties that generate the electrical responses when subjected to mechanical or magnetic excitations. A measurable change is produced in the magnetization anisotropy due to modification in the magnetoelastic state that arises due to magneto-mechanical coupling. A pick-up coil around the magnetostrictive material is used to sense the magnetization changes, which is further converted into electrical signals [3, 26]. Figure 8.5 shows the schematic of working mechanism of the magnetostriction based position sensor.

The e.m.f. induced across the pick-up coil is linked to magnetization in accordance with the Faraday–Lenz law of electromagnetic induction stated as $V = -NA\frac{dB}{dt}$, where N is the number of turns in the coil, A is the coil cross-sectional area, and B is magnetic induction to quantify the state of magnetization.

Based on the principles of fundamental operation, a large number of alternative sensor designs such as torque sensors, position and deformation sensors, force sensors, and magnetometers can be devised.

Remote Cell Actuation

The magnetostrictive materials have gained widespread interest in biomedicine for remote cell actuation [9]. A research to study the interaction between magnetic field- driven magnetostrictive Fe-Ga alloy and cells has been performed by Vargas-Estevez's group [27]. The alloy particles were allowed to enter the cells after their culture with macrophages. When an external static magnetic field is applied, a particle chain was formed inside the cells and on the cell membrane, which can be remotely controlled with a magnetic field vector. Also, Hart *et al.* reported surface strains larger than 900 ppm at a magnetic field of 170 kA/m for a

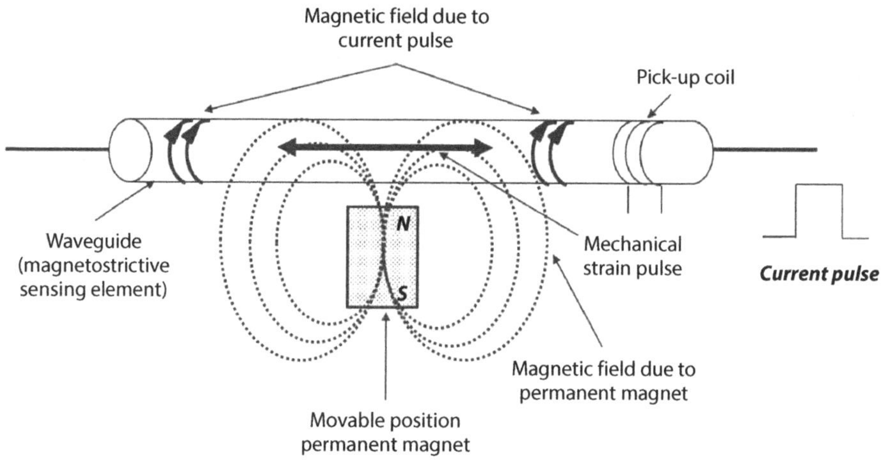

Figure 8.5 Working mechanism of the magnetostrictive position sensor.

magnetostrictive composite that consists of encapsulated Terfenol-D in epoxy resin. These composites were found to generate sufficient strain magnitude and accelerate the formation of bone tissues via remote regulation [28].

8.8 Conclusion

The magnetostrictive materials are a class of smart materials with the attributes of magneto-mechanical coupling that provides mechanism for bidirectional energy conversion between elastic and magnetic states. The widely used magnetostrictive materials like Terfenol-D, Fe-Ga alloys, and others with superior magnetostrictive properties such as crystal anisotropy, magnetic hysteresis, and nonlinear effects provide excellent combination of energy densities, strain, operating bandwidths, high forces, and coupling coefficients. All these properties have led their wider usage in actuators and sensor applications. The synthesis methods for magnetostrictive materials with larger strains and good machinability have been developed. The different preparation techniques allow the formation of magnetostrictive materials with different shapes and sizes. The accurate measurement of magnetostrictive property is also a major concern for their successful implication in various fields. The strain gauge method, a direct method, is the most commonly used method to measure magnetostriction. Few of the indirect methods of measurement have been developed to gather precise information about the magnetostrictive property. Magnetostrictive materials are widely being used in transducer, actuator, and sensor applications. Due to the biocompatibility of few magnetostrictive materials, they also find potential biomedical applications as remote cell actuators and wireless implantable devices, among others.

References

1. M. J. Dapino, 'On magnetostrictive materials and their use in smart material transducers', in *Proc. U.S. Korea Workshop on Smart Structural Systems*, Aug. 2002.

2. Y. Kikuchi, 'Magnetostrictive Materials and Applications', *IEEE Trans. Magn.*, vol. 4, no. 2, pp. 107–117, 1968.
3. M. J. Dapino, 'On magnetostrictive materials and their use in adaptive structures', *Struct. Eng. Mech.*, vol. 17, no. 4, pp. 0–000, 2004.
4. J. J. Rhyne, 'Magnetostriction of dysprosium, erbium and terbium single crystals'. [Online]. Available: https://lib.dr.iastate.edu/rtd
5. J. R. Cullen and A. E. Clark, 'Magnetostriction and structural distortion in rare-earth intermetallics', *Phys. Rev. B*, vol. 15, no. 9, pp. 4510–4515, 1977.
6. H. T. Savage and A. E. Clark, 'Magnetostriction and Characterization of Bridgman Grown Rare-Earth Iron Alloys', 1981.
7. D. C. Morris, 'Crystallisation of the METGLAS 2826 amorphous alloy', 1981.
8. A. E. Clark, J. B. Restorff, M. Wun-Fogle, T. A. Lograsso and D. L. Schlagel, 'Magnetostrictive Properties of Body-Centered Cubic Fe-Ga and Fe-Ga-Al Alloys', 2000.
9. C. Gao, Z. Zeng, S. Peng and C. Shuai, 'Magnetostrictive alloys: Promising materials for biomedical applications', *Bioact. Mater.*, vol. 8. KeAi Communications Co., pp. 177–195, Feb. 01, 2022. doi: 10.1016/j.bioactmat.2021.06.025.
10. H. Man-Gui and D. Long-Jiang, 'Mossbauer studies on the shape effect of Fe84 94Si 9.68Al5.38 particles on their microwave permeability', *Chin. Phys. B*, vol. 22, no. 8, p. 083303, Aug. 2013, doi: 10.1088/1674-1056/22/8/083303.
11. K. D. Choi, S. H. Kim, P. W. Jang, W. Y. Yoon and J. Y. Byun, 'Enveloping Fe–12%Al atomized powders in selectively-oxidized insulating films for soft magnetic composite (SMC) cores', *J. Alloys Compd.*, vol. 854, Feb. 2021, doi: 10.1016/j.jallcom.2020.157241.
12. S. Fujieda, S. Suzuki, A. Minato, T. Fukuda and T. Ueno, 'Growth of Fe-Ga alloy single crystals by the czochralski method and their application to vibration power generator', *IEEE Trans. Magn.*, vol. 50, no. 11, Nov. 2014, doi: 10.1109/TMAG.2014.2331139.
13. L. Kai and H. Wenrui, 'Effect of non-uniform magnetic field on crystal growth by floating-zone method in microgravity', *Sci. China Ser. A*, vol. 44, no. 8, pp. 1056–1063, 2001.
14. S. R. Kim, S. Y. Kang and S. H. Lim, 'The Magnetic and Magnetostrictive Properties of Melt-Spun Ribbons of B containing Terfenol-D Alloys', *J. Magn.*, vol. 2, no. 1, pp. 1–6, 1997.
15. W. Dong gang, S. Xue dao, W. Rui and X. Sheng, 'Mechanism of necking defect of 6082 aluminium alloy rolled by cross-wedge rolling method based on material thermal properties', *J. Cent. South Univ.*, vol. 27, no. 12, pp. 3721–3732, Dec. 2020, doi: 10.1007/s11771-020-4572-y.
16. S. Zhao *et al.*, 'The influence of nanoparticle size on the magnetostrictive properties of cluster-assembled Tb-Fe nanofilms', *Thin Solid Films*, vol. 518, no. 12, pp. 3190–3193, Apr. 2010, doi: 10.1016/j.tsf.2009.09.016.
17. H. Y. Liu *et al.*, 'Study on the preparation techniques and properties of bonded magnetostrictive materials', *J. Appl. Phys.*, vol. 91, no. 10, pp. 8213–8215, May 2002, doi: 10.1063/1.1447513.
18. N. B. Ekreem, A. G. Olabi, T. Prescott, A. Rafferty and M. S. J. Hashmi, 'An overview of magnetostriction, its use and methods to measure these properties', *J. Mater Process Technol.*, vol. 191, no. 1–3, pp. 96–101, Aug. 2007, doi: 10.1016/j.jmatprotec.2007.03.064.
19. Q. Xia, '*Testing principle of the strain gauge and its application in the practical engineering*', Shanxi Architect, vol. 28, 2008.
20. W. Wang *et al.*, 'Thermal expansion and magnetostriction measurements at cryogenic temperature using the strain gauge method', *Front. Chem.*, vol. 6, no. 72, Mar. 2018, doi: 10.3389/fchem.2018.00072.
21. G. K. White, 'Thermal expansion at low temperatures. II. Electronic component in metals', *Philos. Mag.*, vol. 6, no. 66. pp.;815–818, 1961. doi: 10.1080/14786436108238376.
22. A. Devishvili *et al.*, 'Measuring magnetostriction with neutrons', *J. Phys. Condens. Matter*, vol. 20, p. 104218, Mar. 2008, doi: 10.1088/0953-8984/20/10/104218.

23. H. Samata, Y. Nagata, T. Uchida and S. Abe, 'New optical technique for bulk magnetostriction measurement', *J. Magn. Mater*, vol. 212, pp. 355–360, 2000.
24. J. C. M. Henning and J. H. den Boef, 'Magnetostriction Measurement by Means of Strain Modulated Ferromagnetic Resonance (SMFMR)', *Appl. Phys.*, vol. 16, pp. 353–357, 1978.
25. A. Mitra and M. Vázquez, 'Measurement of the saturation magnetostriction constant of amorphous wire', *J. Appl. Phys.*, vol. 67, no. 9, pp. 4986–4988, 1990, doi: 10.1063/1.344698.
26. H. Kwun and K. A. Bartels, 'Magnetostrictive sensor technology and its applications", *Ultrasonics*, vol. 36, pp. 171–178, 1998.
27. C. Vargas-Estevez *et al.*, 'Study of Galfenol direct cytotoxicity and remote microactuation in cells', *Biomaterials*, vol. 139, pp. 67–74, Sep. 2017, doi: 10.1016/j.biomaterials.2017.05.049.
28. S. Hart, R. Bucio and M. Dapino, 'Magnetostrictive Actuation of a Bone Loading Composite for Accelerated Tissue Formation', *Smart Mater. Res.*, vol. 2012, pp. 1–7, Feb. 2012, doi: 10.1155/2012/258638.

9

Materials Development of Supercapacitors—Promising Device for Future Energy Storage Applications

Ram Chhavi Sharma

Department of Physics, Faculty of Science, SGT University, Gurugram, Haryana, India

Abstract

Advancement in energy storage technologies becomes essential for future applications that require certainty in energy storage devices in relation to their life cycle, power density, cyclability, stability, transient response (charging and discharging time), internal resistance, weight, and being environmentally friendly. The technology should be such that which provides high power density, long term stability, less charging time, large discharging time, and light weight to energy storage devices.

The well-established technologies for energy storage are conventional capacitors and battery. Low storage capacity of conventional capacitors and shorter life cycle, less power density, higher charging times, thermal instability, and environmental issues of batteries are not going to fulfill the energy storage needs for future applications. Thus, it becomes essential to use a device that bridges the gap between batteries and conventional capacitors and improve the power storage capacity and supply it at higher power outputs. The supercapacitors are promising devices in this respect, which combine the properties of batteries and conventional capacitors. It is vital to discover novel electrode materials that are inexpensive and environmentally sustainable and have a long life cycle in order to enable the large-scale production and application of supercapacitors. All solid-state supercapacitors will achieve amazing things in the future as a result of the development of solid-state electrolyte materials. Moreover, there is a need for flexible wearable supercapacitors for flexible electronic devices use.

This chapter critically analyzes the different energy storage devices with special reference to supercapacitors. A comparative analysis will be given with recent advancement in materials development, design, characteristics, and applications of energy storage devices.

***Keywords*:** Supercapacitors, energy storage, materials, stability, power density

9.1 Introduction

The development of more efficient technology that provides high power density, long-term stability, less charging time, large discharging time, and light weight to energy storage devices has attracted a lot of attention. Traditional capacitors have a lower energy density

Email: rcsharma@sgtuniversity.org

Upendra Kumar and Piyush Kumar Sonkar (eds.) Smart Materials for Science and Engineering, (151–172) © 2024 Scrivener Publishing LLC

and a higher power density, demonstrating a substantially quicker charge and discharge time. The electrical energy storage batteries offer an adequate amount of power for a variety of uses and requirements in daily life. Despite having a large quantity of stored energy in a comparably reduced volume and mass, batteries deliver power slowly. Their use is therefore limited in systems that require quick storage and lots of power. On the other hand, supercapacitors (SCs) have earned a lot of consideration as a substitute to or addition to batteries in systems of energy storage as a result of their greater power density, excellent lifetime, and rapid charging–discharging cycles [1–6]. Batteries and SCs as stand-alone technologies are unable to satisfy the business demands already created by the use of fossil fuels. Numerous attempts have been made to create hybrid SCs by fusing batteries and SCs with high energy density and high power density, respectively, to address these issues [7]. However, improving the power density of batteries to match SCs or raising the energy density of SCs to approach that of batteries comes at the expense of SCs losing power and batteries losing energy [8, 9]. Considerable development has taken place in supercapacitors over the past 10 years, and as a result, it has the potential to provide enhanced energy density together with high power density. The absence of dielectric material makes SCs superior to conventional capacitors [10]. Supercapacitors use thin electrolytic and electrode materials with high surface areas that could generate capacitances many orders of magnitude higher than those of conventional capacitors. SCs provide improved system protection and stability, as well as small weight, low heating, and adaptability [11]. This allows SCs to achieve larger energy densities while yet retaining the recognizable high power density of conventional capacitors. Figure 9.1 displays a Ragone plot comparing the energy and power densities of traditional capacitors, SCs, batteries, fuel cells, and hybrid SCs [12, 13].

Hybrid supercapacitors exhibit a large power density matching with batteries and fuel cells, but much smaller than ordinary capacitors. The regulation of a supercapacitor on an electrode's incredibly porous surface is dependent on the voltage created by the migration of ions from the electrolyte [14]. The higher power density of 10^3–10^4 W/kg that SCs can deliver is a result of this. They can meet a considerable need in electronic products and

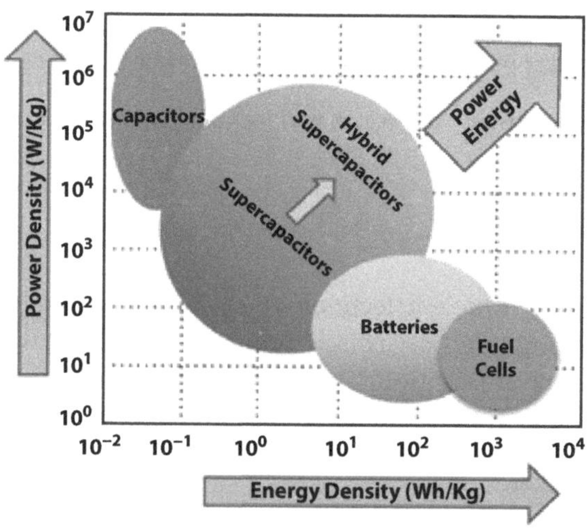

Figure 9.1 Ragone plot comparing the energy and power densities of several energy storage technologies.

electric vehicles because they have a far higher rate of energy provision than Li-ion batteries, while batteries are more expensive, have a shorter life cycle, and are less safe than SCs. Because of all these qualities, SCs are significantly better suited for renewable energy storage technologies. Consumer electronics, digital cameras, computers, security systems, inverters, communications, regenerative braking, etc. have all seen a significant increase in the use of SCs. The energy sources for electric and hybrid automobiles may also be extended into grid substations that may be used as the storage system [15, 16]. Table 9.1 compares and contrasts the technologies that have been chosen for electrical energy storage [17].

The present book chapter critically analyzes the different energy storage devices with special reference to supercapacitors. A comparative analysis will be given with recent advancement in materials development, design, characteristics and applications of energy storage devices.

After this introduction, in Section 9.2, the basic principle of operation of both conventional and supercapacitors is given. Section 9.3 provides the classification of the supercapacitors and their energy storage strategies. Development of advanced materials for SCs used in storage energy conversion is given in Section 9.4. An overview of the applications of SCs is given in Section 9.5. Section 9.6 concludes the present study. There is no perfect energy storage technology. Under deep, rapid, and high power discharge, the capacity of typical lead-acid batteries decreases. Low energy density and life spans contribute to environmental contamination. Applications for lithium-ion batteries include nickel-cobalt-manganese lithium batteries, lithium-ion phosphate batteries, and lithium titanate batteries. Ni-Co-Mn-Li batteries cannot be produced on a large scale due to a lack of cobalt resources, and bromine flow batteries have an issue with high pricing. The vanadium redox flow cell has a low energy and power density and a slow response time. The complex causing the electrode reaction causes the zinc bromine flow battery's rapid self-discharge rate. The metal sodium

Table 9.1 Comparison of main features of SCs, conventional capacitors, and batteries [17–19].

Characteristics	Supercapacitor	Conventional capacitor	Lead-acid battery	Lithium-ion battery
Charging duration (seconds)	1–30	10^{-3}–10^{-6}	3600–18000	1800–10800
Discharging duration (seconds)	1–30	10^{-3}–10^{-6}	>1080	Greater than 1080
Energy per unit mass (Wh kg^{-1})	1 to 10	Less than 0.1	10 to 100	150–200
Power density (W kg^{-1})	<10,000	<1000000	<1000	<2000
Life cycle	10^6	10^6	1000	5000
Discharge/charge efficiency	0.85–0.98	>0.95	0.70–0.85	0.99
Operation temperature	-40°C to 75°C	-40°C to 75°C	-5°C to 40°C	-30°C to 60°C

used to make the sodium sulfide battery has a safety risk as it operates in high-temperature conditions (300–350°C) [20].

The electrolyzed hydrogen or synthetic gas, which is pollution free, is used as a secondary energy carrier for the chemical energy storage technology. It is also feasible to synthesize methane gas using carbon dioxide. Although it can store more than 100 GHz, it has poor conversion efficiency (about 45%), a high investment cost, a large maintenance cost, and poor security.

Supercapacitors often exhibit a low range of energy density, making them suitable for incorporation into other energy storage methods. The rate of charge and discharge is an important statistic. It is expensive, requires little energy, and requires complicated upkeep [20].

9.2 Principle of Operation of Conventional Capacitors and Supercapacitor

Traditional capacitors are made by placing a dielectric among the two conducting electrodes. On application of potential difference across the electrodes, the accumulation of opposite charges takes place at electrodes, which are separated by the dielectrics. As a result, the capacitor generates an electric field and hence stores energy in its electrostatic field. This operation is depicted in Figure 9.2.

Capacitance (C) is the ratio of the amount of charge (Q) stored and the potential difference (V) applied:

$$C = \frac{Q}{V} \tag{9.1}$$

The capacitance of a conventional capacitor is given by

$$C = \frac{\varepsilon A}{D} \tag{9.2}$$

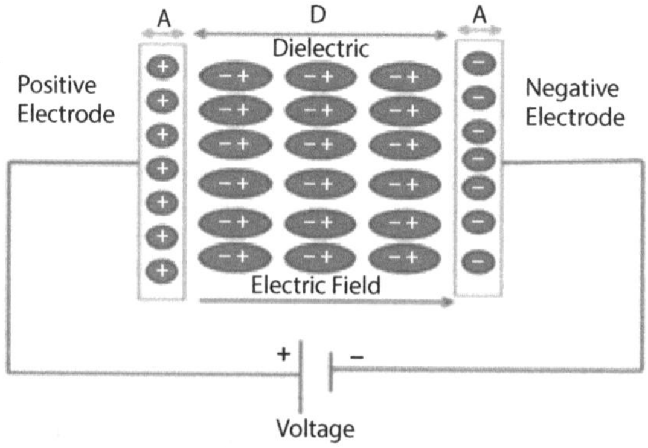

Figure 9.2 Principle of operation of conventional capacitor.

where $\varepsilon = \varepsilon_r\varepsilon_0$ is the permittivity of the dielectrics in between the two electrodes, ε_0 is the absolute permittivity of free space, ε_r is the dielectric constant of the dielectric material, A is the surface area of each electrode, and D is the distance between the two electrodes. The energy density, power density, and Coulomb efficiency are determined using the following relations [21]:

Energy per unit mass:

$$E(Wh/kg) = [(1/2)\ CV^2]/3.6 \tag{9.3}$$

Power per unit mass:

$$P(W/kg) = (E \times 3600)/t \tag{9.4}$$

Coulomb efficiency:

$$\eta = (t_d/t_c) \times 100\% \tag{9.5}$$

where t_c and t_d are the charging and discharging times of the capacitor, respectively.

To determine the power for a capacitor, potential difference is applied across a capacitor through a series resistance. It includes the resistance of electrodes, dielectrics, and current collectors. If we take equivalent series resistance as R, then the maximum power [3] is given by:

$$P_{max} = [V^2/4R] \tag{9.6}$$

This relationship illustrates how the equivalent series resistance might limit a capacitor's maximum power. Conventional capacitors have relatively high power densities but relatively low energy densities when compared to electrochemical batteries and fuel cells. In other words, although having the capacity to store more total energy, batteries have a lower power density than capacitors. A battery, however, is unable to transfer energy quickly. Contrarily, capacitors have a high power density despite having a relatively low energy density because the electrical energy they store can be swiftly discharged to produce a sizable amount of power. SCs are subject to the same fundamental laws as regular capacitors. The electrode spacing D is reduced by using thinner dielectrics, and surface area is increased by using porous electrode materials in order to increase the energy and capacitance of the supercapacitor (Equations 9.1 and 9.2). SCs are able to produce power densities that are comparable to regular capacitors while preserving the low equivalent series resistance feature. They also provide a greater power density, quicker charging times, longer cycle, and shelf life when compared to fuel cells and electrochemical batteries [2, 3].

9.3 Types of Supercapacitors

Electrochemical double-layer capacitors, pseudo-capacitors, and hybrid capacitors are the three primary types of supercapacitors. Each class is defined by a unique method of charge storage.

These methods include non-Faradaic, Faradaic, and a combination of both. A fundamental aspect of Faradic processes, such as oxidation–reduction reactions, is the transfer of charge between an electrode and an electrolyte. In contrast, non-Faradaic processes use physical mechanisms that do not involve the formation or rupture of chemical bonds to disperse charges on surfaces. The three types of supercapacitors are briefly described below.

9.3.1 Electrochemical Double-Layer Capacitors (EDLCs)

These supercapacitors are non-Faradaic type and are made of two carbon-based electrodes, an electrolyte, and a separator (Figure 9.3). The charge storing mechanism is electrostatic, and electrode and electrolyte do not exchange electrical charges. To store energy, they make use of an electrochemical double layer of charge. On application of voltage, the ions diffuse across the separator and accumulates into the pores of the oppositely charged electrodes. The recombination of ions is minimized by suitable design of electrodes. In this manner, at each electrode, a double layer of charge gets buildup, leading to higher energy densities than those of ordinary capacitors [1–3]. They can attain extremely high cycling stabilities since the charge storage is highly reversible. The performance characteristics may be adjusted by changing the nature of its electrolyte whose choice often depends on the intended application of the supercapacitor. The electrolyte may be either aqueous or organic. In comparison to organic electrolytes, aqueous electrolytes exhibit lesser breakdown voltages, reduced minimum pore size demands, and lower equivalent series resistance. To improve other

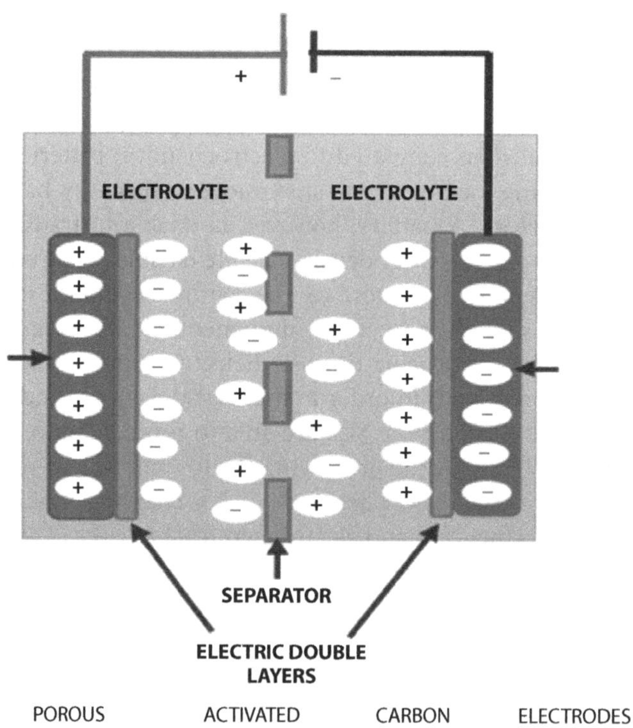

Figure 9.3 Electrochemical double layer capacitor (EDLC).

features such as increase in surface area of electrodes, reduced cost, and antiquated fabrication techniques, carbon materials such as carbon nanotubes, activated carbon, and carbon aerogels are preferred to the conducting polymers and metal oxides [1–3, 6].

Activated carbons are comparatively economical and have a larger surface area. They have a complicated porous structure made up of pores that range in size from less than 20 microns (micro-pores) to more than 500 microns (macro-pores). According to empirical data, the relationship between the size of pores and the concentration of ions determines the effective surface area that contributes to capacitance. It is thought that this disparity results from electrolyte ions that are too large to diffuse into some of the micro-pores, restricting them from assisting in storage of charge. Additionally, studies point to an empirical connection between the device's power density, energy density, and distribution of pore diameters. Greater power densities and higher energy densities are correlated with larger pore diameters. As a result, a significant field of research in EDLC design focuses on the distribution of pore size of electrodes [22, 23]. For a given ion size, the best pore size has been determined, and methods for controlling the distribution of pore sizes during fabrication have been improved.

Carbon aerogels are created by mesopores and conductive carbon nanoparticles, distributed throughout a continuous network. Since carbon aerogels have a continuous structure and can chemically bond to the current collector, additional adhesive binding agent is not required. As a binderless electrode, carbon aerogels have been shown to have a lower ESR than activated carbons [24]. Research on supercapacitors made of carbon aerogels is primarily focused on the decreased ESR that leads to enhanced power.

Another interesting EDLC electrode material is carbon nanotubes [25–27]. The typical structure of carbon nanotube-based electrodes generated from this material is an entangled mat having a mesopore network that is open and accessible. The mesopores of carbon nanotubes are linked, in contrast to other electrodes based on carbon, enabling a continuous charge distribution that practically utilizes the entire surface area. Activated carbon electrodes have a larger surface area, but carbon nanotube electrodes employ their surface area more effectively to produce capacitance resembling those of SCs based on activated carbon. Due to the electrolyte ions' easier migration into the network of mesopores, they exhibit a reduced ESR as compared to activated carbon. To further reduce the ESR, several manufacturing techniques have been devised. Particularly, carbon nanotubes can be produced directly on current collectors, heated, or cast into thin films made of colloidal suspension [28, 29]. The tangled mat structure's efficiency enables energy densities that are on par with those of other materials based on carbon, and the lower ESR enables larger power densities.

9.3.2 Pseudocapacitors

The process of storing charge in these supercapacitors is Faradaic. The transfer of charge takes place by three processes. These are intercalation, oxidation–reduction reactions, and electrosorption. These mechanisms might make it possible for pseudocapacitors to potentially surpass EDLCs' capacities and energy densities [30]. Metal oxides and conducting materials are used as electrode materials to store charge in pseudocapacitors. The configuration of a pseudocapacitor is depicted in Figure 9.4.

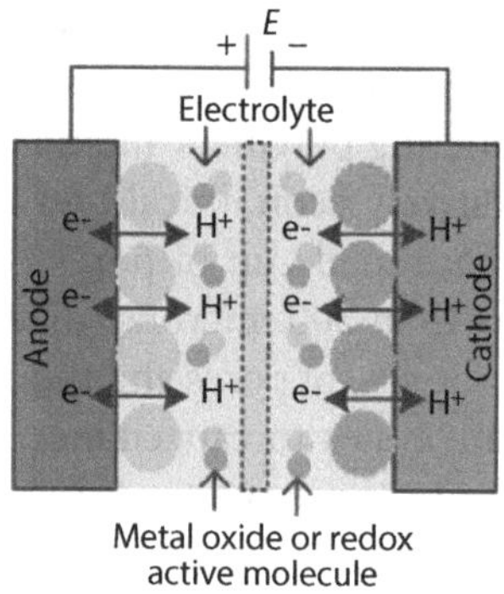

Figure 9.4 Configuration of a pseudocapacitor.

As compared to carbon-based electrodes, conducting polymers are cheaper and have a comparatively high capacitance and low ESR [7]. The maximal power densities and potential energy are found in the configuration, having one positively charged and one negatively charged conducting polymer electrode. Despite this, these pseudocapacitors have not yet reached their maximum capabilities due to the emergence of conducting polymer as a new material that is n-doped [31]. The decreased cycling stability of these pseudocapacitors is also thought to be limited by the mechanical stress that reduction–oxidation reactions place on conducting polymers [32]. Due to their high conductivity, metal oxides have also been investigated as a viable material to make electrodes of pseudocapacitors. In the group of metal oxides, ruthenium oxide has a larger capacitance and, thus, the majority of pertinent research focuses on it. Protons are intercalated into the amorphous structure of ruthenium oxide to increase and decrease its capacitance. The capacitance in its hydrous phase is greater than that of materials made of carbon and conducting polymers [33]. Additionally, hydrous ruthenium oxide is superior to other electrode materials due to lesser ESR. In comparison to comparable EDLCs and conducting polymer pseudocapacitors, ruthenium oxide pseudocapacitors may attain better power and energy densities. Ruthenium oxide has a lot of potential, but its high price has prevented it from reaching its full potential. Development of composite materials and production techniques in order to lower its cost without any compromising its performance is thus a key field of research [34].

9.3.3 Hybrid Capacitors

For improved performance characteristics, hybrid capacitors work to maximize the relative merits of pseudocapacitors and EDLCs, while minimizing their relative disadvantages. Because they use both non-Faradaic and Faradaic processes to hold charge, they have power and energy densities better than EDLCs without compromising cost-effectiveness and cycling stability. The configuration of electrodes (Figure 9.5) is the key parameter to differentiate these capacitors. The three distinct types of electrode configurations are composite, asymmetric, and battery type.

Composite electrodes merge materials based on carbon with either metal oxide or conducting polymer components to blend physicochemical charge storage processes in an electrode. The carbon-based materials help create a double layer of charge and also act as

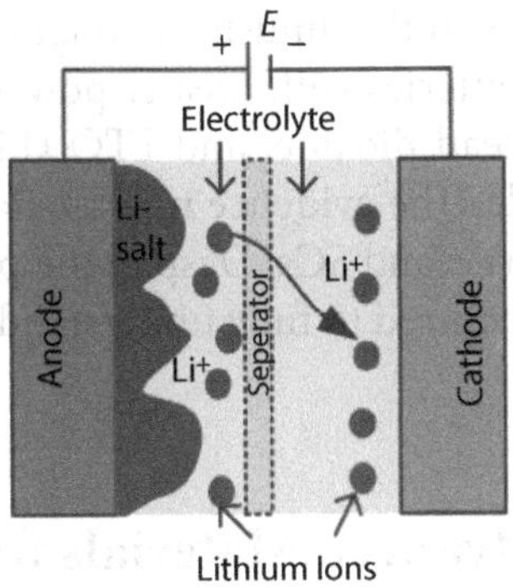

Figure 9.5 Configuration of a hybrid supercapacitor.

a foundation with a large surface area to improve the interaction between the electrolyte and the coated pseudocapacitive materials. Through Faradaic processes, the capacitance of the composite electrodes is further enhanced by the pseudocapacitive materials. Successful composite electrodes made of the conducting polymer polypyrrole and carbon nanotubes have been developed. This electrode can obtain better capacitances than an electrode made only of a polypyrrole polymer or a carbon nanotube. It is due to the convenience of entangled mat-like structures, which enables a three-dimensional charge distribution and a homogeneous coating of polypyrrole. It has also been demonstrated that the entangled mat's structural integrity reduces the mechanical stress resulting from the introduction and withdrawal of ions from the implanted polypyrrole. As a result, these composites, as opposed to conducting polymers, attain cycling stability on par with EDLCs [35].

Asymmetric hybrids integrate non-Faradaic and Faradaic processes by coupling electrodes of an EDLC and a pseudocapacitor. Notably, the connection of a conducting polymer and activated carbon as positive and negative electrodes has drawn a lot of interest. Conducting polymer electrodes frequently have greater capacitances and lower resistances when compared to activated carbon electrodes, but they also have lower peak voltages and less stable cycling. These electrodes are combined to create asymmetric hybrid capacitors, which perform better in terms of energy and power density than equivalent EDLCs while reducing the impact of this compromise. Additionally, compared to comparable pseudocapacitors, they offer superior cycling stability [36].

The most promising methods for enhancing the energy density of SCs are those that increase the operating potential window by creating asymmetric SCs with various electrodes. The anode and cathode materials are produced separately for the creation of asymmetric SCs. Moreover, each electrode can run at a distinct operating voltage window during an electrochemical demonstration. Additionally, by running electrodes at various potential windows, the overall operating potential and thus the energy density of asymmetric SCs are improved. Asymmetric SCs are observed to perform electrochemically better than carbon-based SCs. The potential candidates for asymmetric SCs are π-CPs. Several research has been published to date that successfully combine metal oxides, reduced graphene oxides, and CNTs as anodes and cathodes in asymmetric SCs with π-CPs [37–41].

Battery-type hybrids are unique because they combine a supercapacitor electrode with a battery electrode. The recharge time, power, and life cycle of SCs are combined with the energy characteristics of batteries in this unusual configuration. It answers to the demand for SCs with more energy and batteries with greater power. The principal materials investigated were nickel hydroxide, lead dioxide, and LTO ($Li_4Ti_5O_{12}$), with activated carbon acting as the secondary electrode. The evidence now available indicates that these hybrids can close the gap between batteries and SCs. Despite the positive outcomes, it is generally agreed that more study will be required to fully understand the possibilities of battery-type hybrids [41, 42].

9.4 Development of Advanced Materials for Supercapacitors

Supercapacitors were used for typical tasks including internal battery backup and memory preservation. The application field has, however, greatly expanded recently to include hybrid carriers, smartphones, and energy gathering. Supercapacitors are encouraged to be produced and placed in direct rivalry with rechargeable batteries by the newest technologies on the horizon. The development of improved electrode materials, such as spinel ferrites, perovskite oxides, transition metal sulfides, carbon materials, and conducting polymer materials, is thus the focus of recent research on supercapacitors. The materials based on spinel ferrite and perovskite oxides exhibit remarkable discharge capabilities of 1000 mA h g^{-1}, which are, on average, 2.5 times more than those attained using graphite anodes [43]. In magnetic oxides, the crystal structure is destroyed into various mineral particles during the initial discharge cycle, and then the Li_2O form is produced. As they work, mineral particles encourage the electrochemical reaction that provides the pathway for the conversion reaction mechanism and uses the creation and destruction of Li_2O. The shape of the magnetic oxides' numerous crystals depends on the method used to make them and the temperature at which they were annealed. Additionally, the crystals' form affects their particular capacitance and greater cycling stability. The symmetry of the original structure can be altered by substituting numerous cations into the A- or B-sites, which will affect the structure's physical and chemical characteristics [44]. For supercapacitors and hybrid supercapacitors, the use of magnetic oxides as anodes provides an advantage [45]. Therefore, their high energy densities are made possible by the enormous amount of oxygen vacancies and outstanding conductivity. Likewise, the perovskites retain charge by oxygen insertion, and their superior diffusion channels together with crystal domain borders help to increase the dispersion rate [46]. However, due to their good electrochemical properties, transition metal sulfides are intriguing materials for use in energy storage. The inclusion of sulfur atoms rather than oxygen atoms explains why transition metal sulfides exhibit substantially better electrochemical properties than transition metal oxides.

Therefore, electron transport in metal sulfide is more frequent than that in the metal oxide, because sulfur has lesser electronegativity in comparison to oxygen. Sulfur thus provides more flexibility for the synthesis and production of nanomaterials when oxygen is substituted. However, metal oxides and transition metal sulfides need improvement in conductivity and cycling stability for supercapacitors and so they were combined with conducting polymers [47]. Conducting polymer hydrogels have been widely used for the making of supercapacitors because of their many exceptional and promising features, such as strong electrochemical activities, improved electrical conductivity, distinctive solid–liquid interface, high stretchability, and distinctive elastic resilience [48, 49]. Additionally, due to its exceptional properties, such as high conductivity, strong tensile strength, huge specific area, porosity, and electrochemical activity, graphene has drawn a lot of study interest. The outcome demonstrated that the composites with the peak capacitance activity and cycle stability were made of conducting polymers or materials and magnetic oxides or transition metal sulfides. The oxygen and S active sites in these composites, which encouraged electrolyte uptake while cycling and permitted more active sites, were responsible for the improved characteristics of these materials [50]. PANI/CNTs/graphene, MnO_2-doped PANI/CNTs, and MnO_2-doped PANI composites have recently been tested for their electrochemical performance for asymmetric hybrid SCs and used to produce electrode materials by *in situ* redox deposition. Out of the composites listed above, it was determined that the PANI/CNTs/graphene composite displayed the best electrochemical performance and produced the highest specific capacitance. It has been possible to attain a specific capacitance of 1360 F g^{-1} at a scan rate of 5 mV s^{-1} and a cyclic stability of 85% through 5000 galvanostatic charge–discharge cycles. It is interesting to note that the obtained specific capacitance is 50% and 30% greater than that of MnO_2-doped PANI composite and PANI/CNTs, respectively [51]. The best potential materials for the next wave of energy storage applications are therefore mixed-magnetic oxides or transition metal sulfides and their composites. The electrochemical properties of some advanced materials for development of supercapacitors are given in Table 9.2.

Table 9.2 Electrochemical properties of some advanced materials along with references.

Material	**Specific capacitance (F g^{-1})**	**Energy density (Wh kg^{-1})**	**Power density (W kg^{-1})**	**Cyclic stability**	**References**
$MnFe_2O_4$	25.21	12.60	0.74	No data	[52]
$MnFe_2O_4$/rGO	271	15.9	324.5	104% retention after 5000 cycles	[53]
Polyaniline/acetylene black/$CuFe_2O_4$	732.35	26.757	3165.25	78% retention after 5000 cycles	[54]
$Ni1-xMgxFe_2O_4$	259.89	11.96	1439	88.79% retention after 1000 cycles	[55]
$Co0.5Ni0.5Fe_2O_4$	630.00	7.43	60.45	82% retention after 1500 cycles	[56]

(*Continued*)

Table 9.2 Electrochemical properties of some advanced materials along with references. (*Continued*)

Material	Specific capacitance ($F\ g^{-1}$)	Energy density ($Wh\ kg^{-1}$)	Power density ($W\ kg^{-1}$)	Cyclic stability	References
$MnFe_2O_4$-$ZnFe_2O_4$/graphene	263.00 $mAhg^{-1}$	75.65	6629.53	96.89% retention after 5000 cycles	[57]
$CoFe_2O_4$/MWCNTs	390.00	26.7	319	93% retention after 2000 cycles	[58]
rGO–$NiFe_2O_4$	210.9	23.7	225	94.2% retention after 5000 cycles	[59]
$CuFe_2O_4$/rGO	797.00	11	543	92% retention after 2000 cycles	[60]
Carbon fiber cloth/$CoFe_2O_4$	237.8	84.6	1334	-	[61]
$ZnMoO_4$	704.8	22.45	800.06	93.6% retention after 2000 cycles	[62]
Ag/$NiMoO_4$	3342.7	48.5	212.5x10^3	84.4% retention after 5000 cycles	[63]
Zn–Ni–Co oxide@ $NiMoO_4$	87.5mAh g^{-1}	70	5115.1	91% retention after 10000 cycles	[64]
MoS_2/$NiCo_2O_4$	51.7	18.4	1200.2x10^3	98.2% retention after 8000 cycles	[65]

(*Continued*)

Table 9.2 Electrochemical properties of some advanced materials along with references. (*Continued*)

Material	Specific capacitance ($F\ g^{-1}$)	Energy density ($Wh\ kg^{-1}$)	Power density ($W\ kg^{-1}$)	Cyclic stability	References
α-$ZnMoO_4$	234.75	20.808	199.44	82% retention after 1600 cycles	[66]
$MnCo_2O_4$@ $NiMoO_4$	1244	42	852.3	93% retention after 8000 cycles	[67]
$FeCo_2O_4$	960.0	34.5	6391.7	94% retention after 10000 cycles	[68]
$MnCo_2O_4$@ graphene	406.50	20.32	300000	95% retention after 5000 cycles	[69]
$LaMnO_3$	520	52,5	1000	117% retention after 7500 cycles	[70]
Reduced graphene oxide/$LaAlO_3$	283	57	569	74% retention after 7500 cycles	[71]
$SrTiO_3$	592.0	27.8	1921.0	99% retention after 5000 cycles	[72]
Ag@La2NiO_4+δ	466.4 $C\ g^{-1}$	44.7	800	93% retention after 10000 cycles	[73]
$LaMnO_3$@ $NiCo_2O_4$	811 $C\ g^{-1}$	36.6	800	96% retention after 2000 cycles	[74]

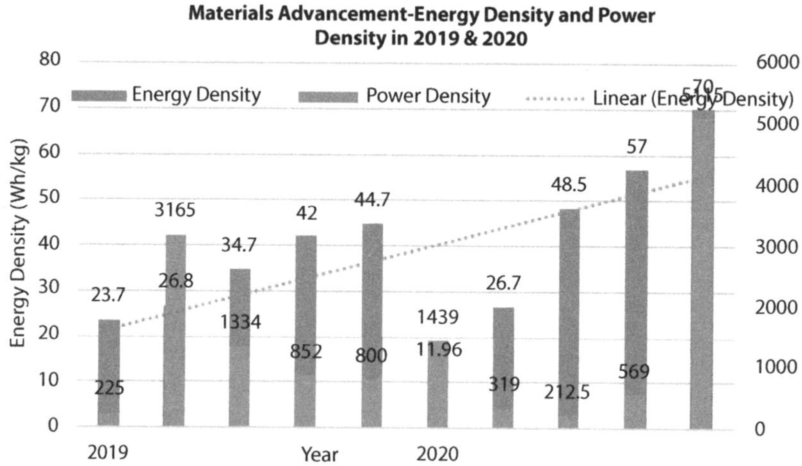

Figure 9.6 Trends in improvement in energy density of supercapacitors with the advancement in electrodes material in years 2019 and 2020.

The trends in improvement in energy density with the advancement in electrode materials in 2019 and 2020 are shown in Figure 9.6.

9.5 Applications of Supercapacitors

Applications requiring a reasonably high power density and prolonged life cycle can benefit greatly from the SC. SCs are utilized in a variety of industrial processes, mobile base stations, electrical gadgets, and wind turbines [75]. Due to their advantages over lead-acid batteries, they have also begun to be employed in UPS systems, electric cars, and a variety of power electronics applications. Recently, SCs have been used for voltage stability in hybrid and renewable energy storage systems to manage the source and grid as a device for energy storage [76]. In applications where the load is erratic, SCs help stabilize the power supply [77]. Flashes and portable speakers are both powered by SCs, which can be charged quickly [78]. All transportation systems face a major challenge in reducing energy use and CO_2 emissions, and brake energy recovery can do just that. SCs meet the needs for components that can quickly store and transfer energy in a variety of applications in all types of vehicles. Consumer electronics, power supply, tools, street lighting, voltage stabilization, microgrid, medicinal applications, renewable energy storage, energy recovery, and military and automotive applications are a few examples of SC uses. Electric machines, other power systems, and many variable loads all contribute to power and harmonic oscillations on the grid. To get around all such issues, SCs can be employed as a terminal between the grid and the load [79]. One SC-equipped wireless screwdriver charges completely in 90 seconds and can keep 85% of its stored energy even after three months of inactivity [80]. SCs also offer backup power in wind turbine pitch mechanism for actuators. Systems for generating energy from the sun and wind fluctuate due to the weather. By serving as dampeners, SCs can stabilize some voltage oscillations for power lines. SCs can be utilized in microgrid storage, which is typically energized by renewable energy sources that may not instantly fulfill the requirement to supply power when the need arises and the production momentarily falls. For temporary energy storage, SCs have a good energy collecting mode. Additionally, hybrid SCs are used in battery-powered communication, navigation, and sensor systems. A proper installation of hybrid SCs can also be used to operate other systems like radar, torpedoes, GPS, and electromagnetic pulse weapons [81]. SCs also meet the criteria for a few of the transportation applications list. Several hybrid buses, according to the Maxwell Technologies manufacturer corporation, use SC-based devices for increasing acceleration. In the starter systems of diesel railroad locomotives with hybrid gear boxes, batteries can be substituted by SCs. SCs are used to move stack containers using mobile hybrid gantry cranes [82] and to store the braking energy of buffer power peaks.

In order to reduce braking energy consumption and replace overhead wires in Mannheim, a prototype light-rail car with an SC unit installed on the roof was built in 2003. In 2012, SCs were used to power the Geneva Public Transport tram and the Paris T3 tram line to recover energy during braking [83]. The so-called "Ultracap Bus," which was put through its paces in Nuremberg, Germany, in 2001, was the first hybrid bus in Europe featuring SCs. Then, in 2002, a fleet of electric buses was put to the test in Luzern, Switzerland. With a high-speed power charger, the SCs may be recharged in 3 to 4 minutes after each cycle of transportation. In 2005, tests were conducted in Shanghai on a brand-new SC-powered electric bus known as the "Capabus," which travels without power lines and charges completely at the final destination. A hybrid drive train using SCs was utilized in a Toyota hybrid racing car and was continuously improved. Hybrid electric vehicles (HEVs) have attracted more study attention. SCs are excellent for use in electric vehicles due to their ability to charge batteries significantly more quickly, as well as their longer life span, steady electrical characteristics, and wide temperature range. However, SCs are inadequate as a stand-alone energy source for long-distance transportation due to their lower specific energy density. Almost all HEV automakers have the passive HESS-created prototypes as of 2013 that use SCs rather than batteries to increase power train efficiency and braking energy storage. Using power electronics converters, the present HEVs technology has created new topologies to boost EV efficiency, enhance the environmental sustainability, and cut costs [84, 85]. In a HEV's electric drive system, the maximum prolonged battery lifetime is still provided by SCs along with fuel cell arrangement [86]. Another study looked at light-duty vehicle energy systems and outlined the key features of EV and HEVs centered on their drivetrain designs. For these power trains, many energy-efficient management techniques and topologies have been studied [87, 88]. For PV–battery–SC systems, SCs have been compared to various energy storage devices, and demonstrated the advantages of SCs over other technologies. Additionally, the hybrid energy storage system (HESS), a combination of SCs, PV–battery–fuel cell, and having active, semi-active, and passive variants are recommended as alternatives. A passive HESS does not use a DC–DC converter and is made up of several energy storage equipment that are connected precisely to the DC bus. A semi-active HESS makes use of a DC–DC converter to connect its one side to a DC bus. An active HESS is one that has converters connecting both sides of the energy storage device. In a case study, to determine the pricing function over the course of a day, it was established that SC semi-active HESSs enhance the battery life by 33% over the passive HESSs. In a further investigation in 2017, in four different days, temperature and solar irradiance data were tested for a solar farm model in MATLAB/Simulink to calculate the cost of energy storage annually. The investigation concluded that the battery along with SC HESS cost was 25% less per year [89].

In a different investigation, a passive HESS was suggested for a stand-alone solar and wind energy, and the functioning was examined both theoretically and empirically. Over the past 10 years, a number of research have provided in-depth analyses of control mechanisms of HESS for enhancing power quality in micro-grids [90]. Due to the increasing use of renewable energy sources and the interruption of the electricity generated, the primary electrical grid is having problems with stability, dependability, and power quality. HESSs are utilized to lessen the impact of these variations because the microgrid is particularly susceptible to changes in demand or production due to its fragile electrical infrastructure. A battery-supercapacitor HESS for a stand-alone PV power system was reviewed in a paper [91]. A stand-alone PV system with a HESS is constructed and its performance is evaluated for a rural Indian location in another study [91, 92].

9.6 Conclusion

Supercapacitors do not appear likely to displace batteries as the primary power storage option, according to the evaluation of the literature indicated above. This is mainly because supercapacitor systems, as they are currently envisioned, have lower energy density than batteries. Supercapacitors, however, may be employed in circumstances where batteries are less suitable due to their high flexibility. Supercapacitors, due to a few inborn characteristics, are also ideal for specialized roles and applications that increase the benefits of batteries. Particularly for applications requiring a balance of fast charging, superior cycle stability, longer shelf life, and high power, supercapacitors hold a lot of potential. Supercapacitors may therefore prove to be the best power source for a range of application-specific requirements. Particularly, there has been a lot of interest in creating supercapacitors for backup and emergency power sources, pulse power applications, and hybrid electric vehicle power systems. Supercapacitors have advantages in these specialized fields, but up until now, production and application have been restricted. This lack of commercial penetration could be associated to an array of factors, which includes high cost, problematic packing, and self-discharge. However, due to advancement in research in electrode material development, at least some of these problems may be resolved. Due to all of these factors, supercapacitors may develop into a practical, widely accessible power source for an expanding range of applications as R&D efforts' outcomes continue to advance. It is believed that this poll will help to further energize the R&D necessary to achieve this result and act as a springboard for creating future applications.

References

1. Conway, B. E.; Electrochemical Supercapacitors: Scientific Fundamentals and Technological Applications (New York, Kluwer-Plenum, 1999).
2. Kotz, R.; Carlen, M.; Principles and applications of electrochemical capacitors, *Electrochim. Acta*, 2000, 45(15-16), 2483-2498.
3. Chu, A.; Braatz, P.; Comparison of commercial supercapacitors and high power lithium-ion batteries for power-assist applications in hybrid electric vehicles I. Initial characterization. *J. Power Sources*, 2002, 112(1), 236-246.

4. Frackowiak, E.; Beguin, F.; Carbon materials for the electrochemical storage of energy in capacitors. *Carbon,* 2001, 39(6), 937-950.
5. Arbizzani, C.; Mastragostino, M.; Soavi, F; New trends in electrochemical supercapacitors. *J. Power Sources,* 2001, 100(1-2), 164-170.
6. Qu, D. Y.; Shi, H.; Studies of activated carbons used in double-layer capacitors. *J. Power Sources,* 1998, 74(1), 99-107.
7. Aricò, A. S.; Bruce, P.; Scrosati, B.; Tarascon, J.M.; Schalkwijk, W.; Nanostructured materials for advanced energy conversion and storage devices. *Nature Mater.,* 2005, (4), 366-377.
8. Gamby, J.; Taberna, P. L; Simon, P.; Fauvarque, J. F.; Chesneau, M.; Studies and characterisations of various activated carbons used for carbon/carbon supercapacitors. *J. Power Sources,* 2001, 101(1): 109-116.
9. Shi, H.; Activated carbons and double layer capacitance. *Electrochim. Acta,* 1996, 41(10): 1633-1639.
10. Wang, G.; Zhang, L.; Zhang, J. A review of electrode materials for electrochemical supercapacitors. *Chem. Soc. Rev.,* 2012, 41, 797–828.
11. Forse, A.C.; Merlet, C.L.; Griffin, J.M.; Grey, C.P.; New perspectives on the charging mechanisms of supercapacitors. *J. Am. Chem. Soc.,* 2016, 138, 5731–5744.
12. Forouzandeh, P.; Kumaravel, V.; Pillai, S.C.; Electrode Materials for Supercapacitors: A Review of Recent Advances. *Catalysts,* 2020, 10, 969.
13. Humayun, M.; Ullah, H.; Shu, L.; Ao, X.; Tahir, A.A.; Wang, C.; Luo, W.; Plasmon Assisted Highly Efficient Visible Light Catalytic CO2 Reduction Over the Noble Metal Decorated Sr-Incorporated g-C3N4. *Nano-Micro Lett.,* 2021, 13, 1–18.
14. Shetti, N.P.; Dias, S.; Reddy, K.R.; Nanostructured organic and inorganic materials for Li-ion batteries: A review. *Mater. Sci. Semicond. Process,* 2019, 104, 104684.
15. Lin, Z.; Goikolea, E.; Balducci, A.; Naoi, K.; Taberna, P. L.; Salanne, M.; Yushin, G.; Simon, P.; Materials for supercapacitors: When Li-ion battery power is not enough. *Mater. Today,* 2018, 21, 419–436.
16. Zhu, J.; Wei, S.; Gu, H.; Rapole, S.B.;Wang, Q.; Luo, Z.; Haldolaarachchige, N.; Young, D.P.; Guo, Z.; One-pot synthesis of magnetic graphene nanocomposites decorated with core@ double-shell nanoparticles for fast chromium removal. *Environ. Sci. Technol.,* 2012, 46, 977–985.
17. Rafik, F.; Gualous, H.; Gallay, R.; Crausaz, A.; Berthon, A.; Frequency, thermal and voltage supercapacitor characterization and modelling, *J. Power Sources,* 2007, 165, 928 - 934.
18. Khalid, M.; A review on the selected applications of battery-supercapacitor hybrid energy storage systems for microgrids. *Energies,* 2019, 12, 4559.
19. Sahin, M.E.; Blaabjerg, F.; A hybrid PV battery/supercapacitor system and a basic active power control proposal in MATLAB/ Simulink. *Electronics,* 2020, 9, 129.
20. Arani, A.K; Gharehpetian, G.B.; Abedi, M.; Review on energy storage systems control methods in microgrids. *Int. J. Electr. Power Energy Syst.,* 2019, 107, 745-757.
21. Muhammad, Y. 1,; Muhammad, A. K..K; Muhammad, H.; Muhammad, U; Syed, S.S; Shaista, B; Bakhtier, S.H; Shah, M.A; Abbas, K; Nasrullah, S; Asif, A.T; Habib, U; A Review of Supercapacitors: Materials Design, Modification, and Applications, *Energies,* 2021, 14, 7779.
22. Qu, D. Y.; Shi, H.; Studies of activated carbons used in double-layer capacitors. *J. Power Sources,* 1998, 74, 99-107.
23. Gamby, J.; Taberna, P.L.; Simon, P; Fauvarque, J. F; Chesneau, M.; Studies and characterisations of various activated carbons used for carbon/carbon supercapacitors. *J. Power Sources,* 2001, 101(1), 109-116.
24. Wang, J.; Zhang, S.Q.; Guo, Y.Z; Shen,J; Attia, S.M; Zhou,B;Zheng, G; Gui, Y; Morphological effects on the electrical and electrochemical properties of carbon aerogels. *J. Electrochem. Soc.,* 2001, 148(6): D75-D77.

25. Frackowiak, E.; Beguin, F.; Electrochemical storage of energy in carbon nanotubes and nanostructured carbons. *Carbon,* 2002, 40(10), 1775-1787.
26. Pico, F.; Rojo, J.M.; Sanjuán, M.L.; Ansón C.A; Alejandro, A.C; Benito, A M.; Callejas, M. A.; Maser, W. K;.; Martínez F ; Teresa, M; Single-walled carbon nanotubes as electrodes in supercapacitors. *J. Electrochem. Soc.,* 2004, 151(6), A831-A837.
27. Du, C. S.; Yeh, J.; Panet, N; High power density supercapacitors using locally aligned carbon nanotube electrodes. *Nanotechnology,* 2005, 16(4), 350-353.
28. An, K.; Kim, W.Y; Park, Y.S; Moon, J; Bae, D; Lim, S; Lee, Y.S; Lee, Y.H; Electrochemical properties of high-power supercapacitors using single-walled carbon nanotube electrodes. *Adv. Funct. Mater.,* 2001, 11(5), 387-392.
29. Yoon, B. J.; Jeong, S.H.; Lee, K.H; Seok, K.H; Gyung, P.C; Hun Han, J; Electrical properties of electrical double layer capacitors with integrated carbon nanotube electrodes. *Chem. Phys. Lett.,* 2004, 388(1-3), 170-174.
30. Conway, B. E.; Birss, V.; Wojtowicz, J; The role and utilization of pseudocapacitance for energy storage by supercapacitors. *J. Power Sources,* 1997, 66(1-2), 1-14.
31. Mastragostino, M.; Arbizzani, C.; Soavi, F; Polymer-based supercapacitors. *J. Power Sources,* 2001, 97-8, 812-815.
32. Frackowiak, E., Khomenko, V.; Jurewicz,, K; Lota, K; Béguin, F; Supercapacitors based on conducting polymers/nanotubes composites. *J. Power Sources,* 2006.
33. Zheng, J. P.; Cygan, P.J.; Jow, T.R; Hydrous Ruthenium Oxide as an Electrode Material for Electrochemical Capacitors. *J. Electrochem. Soc.,* 1995, 142(8), 2699-2703.
34. Kim, I. H.; Kim, K.B.; Ruthenium oxide thin film electrodes for supercapacitors. *Electrochem. Solid-State Lett.,* 2001, 4(5), A62-A64.
35. Jurewicz, K.; Delpeux, S.; Bertagna, V; Eguin, F.B; Frackowiak, E; Supercapacitors from nanotubes/polypyrrole composites." *Chem. Phys. Lett.,* 2001, 347(1-3), 36-40.
36. Laforgue, A.; Simon, P.; Fauvarque, J.F; Mastragostino, M; Soavi, F; Sarrau, J.F; Lailler, P; Conte, M; Rossi, E; Saguatti, S; Activated carbon/conducting polymer hybrid supercapacitors. *J. Electrochem. Soc.,* 2003, 150(5), A645-A651.
37. Chen, X.; Tung, W.Y.; Yang, K.; Chen, Y.M.; Liu, K.; Cheng, C.F.; Zhu, Y.; Electropolymerized porous conjugated polymer film as an electrode for a stable, high energy density thin film asymmetric supercapacitor. *ACS Appl. Polymer Mater.,* 2019, 1, 1634-1640.
38. Liu, T.; Zhao, H.; Li, J.; Zhang, W.; Wang, L.; POSS-tetraaniline based giant molecule: synthesis, self-assembly, and active corrosion protection of epaxy-based organic coatings. *Corros. Sci.,* 2020, 168, 108555.
39. Jarosz, K.; Stolarczyk, A.; Glosz, K.; Recent advances in the electrochemical synthesis of copolymers bearing Pi-conjugated systems and methods for the identification of their structure. *Curr. Org. Chem.,* 2020, 24, 339-353.
40. Wang, K.; Huang, L.; Eedugurala, N.; Zhang, S.; Sabuj, M.A.; Rai, N.; Gu, X.; Azoulay, J.D.; Ng, T.N.; Wide potential window supercapacitors using open cell donor-acceptor conjugated polymers with stable N-doped states. *Adv. Energy. Mater.,* 2019, 9, 1902806.
41. Ling, Y.; Chen, H.; Zhou, J.; Tao, K.; Zhao, S.; Yu, X.; Han, L.; Metal-organosulfide coordination polymer nano-sheet array as a battery type electrode for an asymmetric supercapacitors. 2020, 59, 7360-7369.
42. Du Pasquier, A.; Plitz, I.; Menocal, S; Amatucci, G; A comparative study of Li-ion battery, super-capacitor and nonaqueous asymmetric hybrid devices for automotive applications. *J. Power Sources,* 2003, 115(1): 171-178.
43. Pell, W. G.; Conway, B.E.; Peculiarities and requirements of asymmetric capacitor devices based on combination of capacitor and battery-type electrodes. *J. Power Sources,* 2004, 136(2): 334-345.

44. Yin, Y.; Zhang, B; Zhang, X; Xu, J; Yang, S; Nano MgFe2O4 synthesized by sol–gel auto-combustion method as anode materials for lithium ion batteries. *J. Sol.-Gel. Sci. Technol.*, 2013, 66, 540–543.
45. Zhang, G,; Liu, G; Wang, L; Irvine, J.T.S; Inorganic perovskite photo catalysts for solar energy utilization. *Chem. Soc. Rev.*, 2016c, 45, 5951–5984.
46. Liu, Y.; Wang, Z.; Veder, J.P.M.; Xu, Z.; Zhong, Y.; Zhou, W.; Tade, M.O.; Wang, S.; Shao, Z.; Highly defective layered double perovskite oxide for efficient energy storage via reversible pseudocapacitive oxygen- anion intercalation. *Adv. Energy Mater.*, 2018c, 8, 1702604.
47. Nan, H.S.; Hu, X; Tian, H; Recent advances in perovskite oxides for anionintercalation supercapacitor: a review. *Mater. Sci. Semicond. Process,* 2019, 94, 35–50.
48. Qiao, H.; Li, R.; Yu, Y. ; Xia, Z.; Wang, L.; Wei, Q.; Chen, K.; Qiao, Q.; Fabrication of PANI-coated ZnFe2O3 nanofibers with enhanced electrochemical performance for energy storage. *Electrochim. Acta,* 2018, 273, 282-288.
49. Wang, Y.; Ding, Y.; Guo, X.; Yu, G.; Conductive polymers for stretchable supercapacitors. *Nano Res.*, 2019c, 12, 1978–1987.
50. Xu Z.; Zhang, Z.; Yin, H.; Hou, S.; Lin, H.; Zhou, J.; Zhuo, S.; Investigation on the role of different conductive polymers in supercapacitors based on a zinc sulfide/reduced graphene oxide/conductive polymer ternary composite electrode. *RSC Adv.*, 2020, 10, 3122-3129.
51. Kaushal, I.; Sharma, A.K.; Saharan, P.; Sadasivuni, K.K.; Duhan, S.; Superior architecture and electrochemical performance of MnO2 doped PANI/CNT graphene fastened composite. *J. Porous Mater.*, 2019, 26, 1287-1296.
52. Singh, G.; Chandra, S.; Electrochemical performance of $MnFe_2O_4$ nano-ferrites synthesized using thermal decomposition method. *Int. J. Hydrog. Energy,* 2018, 43, 4058–4066.
53. Tabrizi, A.G.; Arsalani,N.; Mohammadi, A.; Namazi, H.; Ghadimi, L.S.; Ahadzadeh, I.; Facile synthesis of a $MnFe_2O_4$/rGO nanocom-posite for an ultra-stable symmetric supercapacitor. *New J. Chem.*, 2017,41, 4974–4984.
54. Das, T.; Verma, B.; Synthesis of polymer composite based on polyaniline-acetylene black-copper ferrite for supercapacitor electrodes. *Polymer,* 2019, 168, 61–69.
55. Wongpratat, U.;Tipsawat, P.; Khaionrit, J.; Swatsitang, E.; Maensiri,S.; Effects of nickel and magnesium on elec-trochemical performances of partial substitution in spinel ferrite. *J. Alloys Compd.*, 2020, 831, 154718.
56. Sharifi, S.; Rahimi, K.; Yazdani, A.; Effect of Co^{2+} content on supercapacitance properties of hydrothermally synthesized $Ni_{1-x}Co_xFe_2O_4$ nano-particles. *Mater. Sci. Semicond. Process,* 2020, 108, 104902.
57. Gopi, C.V.V.M.; Vinodh, R.; Sambasivam, S.; Obaidet, I.; Singh, S.; Kim, H.J.; Co9S8–Ni3S2/$CuMn_2O_4$–$NiMn_2O_4$ and $MnFe_2O_4$–$ZnFe_2O_4$/graphene as binder-free cathode and anode materials for high energy density supercapacitors. *Chem. Eng. J.*, 2020, 381:122640.
58. Acharya, J.; Raj, B.G.S.; Ko, T.H.; Khil, M.S.; Kim, H.Y.; Kim, B.S.; Facile one pot sonochemical synthesis of $CoFe_2O_4$/MWCNTs hybrids with well-dispersed MWC-NTs for asymmetric hybrid supercapacitor applications. *Int. J. Hydrog. Energy,* 2020, 45, 3073–3085.
59. Cai, Y.Z.; Cao, W.; Zhang, Y..L; He, P.; Shu,J.; Cao,M.; Tailoring rGO-$NiFe_2O_4$ hybrids to tune transport of electrons and ions for supercapacitor electrodes. *J. Alloys Compd.*, 2019, 811:152011.
60. Chandel, M.;Moitra, D.; Makkar,P.; Sinha, H.; Hora, H.S.; Ghosh, N.N.; Synthesis of multifunctional $CuFe_2O_4$-reduced graphene oxide nanocomposite: an efficient magnetically separable catalyst as well as high performance supercapacitor and first-principles calculations of its electronic structures. *RSC Adv.*, 2018, 8, 27725–27739.

61. Song, K.; Wang, X.; Wang, J.; Zhang,B.; Yang, R.; Hierarchical structure of $CoFe_2O_4$ core–shell microsphere coating on carbon fiber cloth for high-performance asymmetric flexible supercapacitor applications. *Ionics,* 2019, 25, 4905–4914.
62. Gao, Y.P.; Huong, K.J.; Zhang, X.; Song, S.S.; Wu, X.;. High-performance symmetric supercapacitor based on flower-like zinc molybdate. *J. Alloys Compd.,* 2018a, 731, 1151–1158.
63. Zhang, X.; Li, Z.; Yu,Z.; Wei, L.; Guo, X.; Mesoporous $NiMoO_4$ microspheres decorated by Ag quantum dots as cathode material for asymmetric super-capacitors: enhanced interfacial conductivity and capacitive stor-age. *Appl. Surf. Sci.,* 2020a, 505.
64. Bandyopadhyay, P.; Saeed, G.; Kim, N..H.; Lee, J.H.; Zinc–nickel–cobalt oxide@NiMoO4 core–shell nanowire/nanosheet arrays for solid state asymmetric supercapacitors. *Chem. Eng. J.,* 2020, 384, 123357.
65. Wen, S.; Liu, Y.; Zhu,F.; Shao,R.; Xu,W.; Hierarchical MoS_2 nanowires/$NiCo_2O_4$ nanosheets supported on Ni foam for high-performance asymmetric super-capacitors. *Appl. Surf. Sci.,* 2018, 428, 616–622.
66. Reddy, B.J.; Vickraman, P.; Justin, A.S.; A facile synthesis of novel α-$ZnMoO_4$ micro-spheres as electrode material for supercapacitor applications. *Bull. Mater. Sci.,* 2019, 42, 52.
67. Mehrez, J.A.A; Owusu, K.A.: Chen, Q.; Li, L.; Hamwi, K.; Wen Luo, W.; Mai, L.; Hierarchical $MnCo_2O_4$@$NiMoO_4$ as free-standing core–shell nanowire arrays with synergistic effect for enhanced supercapacitor performance. *Inorg. Chem. Front.,* 2019, 6, 857–865.
68. Lalwani, S.; Munjal, M.; Singh, G.; Sharma, R. K.: Layered nanoblades of iron cobaltite for high performance asymmetric supercapacitors. *Appl. Surf. Sci.,* 2019, 476,1025–1034.
69. Saren, P.; De Adhikari, A.; Khan, S.; Nayak, G. C.; Self-assembled GNS wrapped flower-like MnCo2O4 nanostructures for supercapacitor application. *J. Solid State Chem.,* 2019, 271, 282–291.
70. Shafi. P.M.; Ganesh, V.; Bose, A.C.; Enhanced electrochemical performances of agglomeration-free $LaMnO_3$ perovskite nanoparticles and achieving high energy and power densities with symmetric supercapacitor design. *Chem. Eng. J.,* 2018, 338, 147–156.
71. Vinuth Raj, T.N.; Hoskeri, P.A.; Murlidhara, H.B.; Manjunatha C.R.; Yogesh, K.K.; Facile synthesis of perovskite lantha- num aluminate and its green reduced graphene oxide composite for high performance supercapacitors. *J. Electroanal. Chem.,* 2020, 858, 113830.
72. Tomar, A.K.; Singh, G.; Sharma, R.K.; Charge storage characteristics of mesoporous strontium titanate perovskite aqueous as well as flexible solid-state supercapacitor cell. *J. Power Sources,* 2019, 426, 223–232.
73. Wei, M.; Che, W.; Li, H.; Wang, Z.; Yan, F.; Liu, Y.; Liu, J.; Ruddlesden–Popper type $La_2NiO_{4+\delta}$ oxide coated by Ag nanoparticles as an outstanding anion intercalation cathode for hybrid supercapacitors. *Appl. Surf. Sci.,* 2019, 484, 551–559.
74. Tian, H.; Lang, X.; Nan, H.; An, P.; Zhang, W.; Hu, X.; Zhang, J.; Nanosheet-assembled $LaMnO_3$@$NiCo_2O_4$ nanoarchitecture growth on Ni foam for high power density supercapacitors. *Electrochim. Acta,* 2019, 318, 651–659.
75. Libich, J.; Máca, J.; Vondrák, J.; Čech, O.; Sedlaříková, M.; Supercapacitors: Properties and applications. *J. Energy Storage,* 2018, 17, 224–227.
76. Rawat, M.S.; Vadhera, S..; A comprehensive review on impact of wind and solar photovoltaic energy sources on voltage stability of power grid. *J. Eng. Res.,* 2019, 7, 178–202.
77. Joshi, P.S.; Sutrave, D.S.; Supercapacitor: Basics and Overview. *J. Inf. Comput. Sci.,* 2019, 9, 609–625.
78. Berrueta, A.; Ursúa, A.; San Martín, I.; Eftekhari, A.; Sanchis, P.; Supercapacitors: Electrical characteristics, modeling, applications, and future trends. *IEEE Access,* 2019, 7, 50869–50896.
79. Farhadi, M.; Mohammed, O.; Adaptive Energy Management in Redundant Hybrid DC Microgrid for Pulse Load Mitigation. *IEEE Trans. Smart Grid.,* 2015, 6, 54–62.

80. Liszewski, A.; Coleman,; Flash Cell Cordless Screwdriver Recharges in Just 90 Seconds. 2007. Available online: https://ohgizmo. com/coleman-flashcell-cordless-screwdriver-recharges-in-just-90-seconds (accessed on 30 October 2021).
81. Tecate Group. Military Applications. Available online: https://www.tecategroup.com/markets/?market =Military-Aerospace.
82. Miller, J.R.; Burke, A.F.; Electrochemical Capacitors: Challenges and Opportunities for Real-World Applications. *ECS Spring,* 2008, 17, 53–57.
83. Rhein-Neckar Verkehr Orders More Supercapacitor Trams. Railway Gazette, 5 April 2011. Available online: https://www. railwaygazette.com/rhein-neckar-verkehr-orders-more-super-capacitor-trams/35818.article
84. Namirian, Z.; Comprehensive overview of hybrid vehicle drivetrains. *Int. Res. J. Eng. Technol.,* 2020, 7, 1–10.
85. Balali, Y.; Stegen, S.; Review of energy storage systems for vehicles based on technology, environmental impacts, and costs. *Renew. Sustain. Energy Rev.,* 2021, 135, 110185.
86. Kumar, D.; Nema, R.K.; Gupta, S. A.; Comparative review on power conversion topologies and energy storage system for electric vehicles. *Int. J. Energy Res.,* 2020, 44, 7863–7885.
87. Tran, D.D.; Vafaeipour, M.; El Baghdadi, M.; Barrero, R.; Van Mierlo, J.; Hegazy, O.; Thorough state-of-the-art analysis of electric and hybrid vehicle powertrains: Topologies and integrated energy management strategies. *Renew. Sustain. Energy Rev.,* 2020, 119, 109596.
88. Lü, X.; Qu, Y.; Wang, Y.; Qin, C.; Liu, G.; A comprehensive review on hybrid power system for PEMFC-HEV: Issues and strategies. *Energy Convers. Manag.,* 2018, 171, 1273–1291.
89. Roy, P.K.S.; Design of a Cost-Effective Battery-Supercapacitor Hybrid Energy Storage System for Hourly Dispatching Solar PV Power. Ph.D. Thesis, Western Carolina University, Cullowhee, NC, USA, 2018.
90. Khalid, M.; A review on the selected applications of battery-supercapacitor hybrid energy storage systems for microgrids. *Energies,* 2019, 12, 4559.
91. Jing, W.; Lai, C.H.; Wong, W.S.H.; Wong, M.L.D. A comprehensive study of battery-supercapacitor hybrid energy storage system for stand-alone PV power system in rural electrification. *Appl. Energy,* 2018, 224, 340–356.
92. Javed, K.; Ashfaq, H.; Singh, R.; Hussain, S.M.; Ustun, T.S. Design and performance analysis of a stand-alone PV system with hybrid energy storage for rural India. *Electronics,* 2019, 8, 952.

10

Smart Solid Electrolyte Materials in Energy Storage Devices: Batteries

Pawan Kumar[1]*, **Shalu Rani**[2†] **and Sanjay Kumar**[3‡]

[1]*Electrical Engineering, Indian Institute of Technology Delhi, New Delhi, India*
[2]*University of Glasgow, Scotland, United Kingdom*
[3]*School of Engineering, University of Edinburgh, Scotland, United Kingdom*

Abstract
In the past few decades, Li-ion-based conducting solid electrolytes have gained much attention from researchers around the globe to design, develop, and implement high-energy density Li-ion batteries, electrochemical sensors, and supercapacitors that are extensively used in the versatile areas of energy storage and transportation. For this purpose, stable, highly ion-conductive, and lightweight solid electrolyte materials as an ionic transportation medium in the energy storage devices play a very important role to fabricate the lightweight, portable, and flexible electronic gadgets that further make daily human life easy. Therefore, by considering the aforementioned items, we have comprehensively outlined the different types of solid electrolyte materials such as solid polymer electrolytes, non-oxide solid electrolytes that are based on polymer and glass, oxide-type solid-state electrolytes (inorganic solid electrolytes), and quasi-solid-state electrolytes that are widely used in the current battery technologies and have also discussed the fundamental advantages and disadvantages of these electrolytes as per the current demand of energy storage devices. The book chapter also covers all fundamental aspects of the solid electrolyte materials as per the current scenario and also lines up the future of the battery technology by utilizing these electrolyte materials.

Keywords: Li^+ ion, solid-state electrolytes, energy storage devices, batteries

10.1 Introduction

In the last few decades, scientists have paid much attention to lithium ion conducting solid-state electrolytes for potential application in electrochemical devices, such as high-energy-density Li-ion batteries, electrochemical sensors, and supercapacitors. A battery is a device that transforms chemical energy into electricity.

**Corresponding author*: kumarpawan.ece@gmail.com
†*Corresponding author*: shalu29singh@gmail.com
‡*Corresponding author*: sanjaysihag91@gmail.com

Upendra Kumar and Piyush Kumar Sonkar (eds.) Smart Materials for Science and Engineering, (173–190) © 2024 Scrivener Publishing LLC

Batteries have many applications in our daily life; they are used in industrial applications, transportation, medical industry, and power storage. The lithium (Li) metal-based battery has become a dominant battery for consumers in the electronics industry. It has two types: (i) Li battery: this type is not rechargeable, which means it is a primary battery. In this battery system, the anode is made of pure metallic lithium. (ii) Li^+ ion battery: this is a rechargeable type of battery, which means it is a secondary battery. In this battery system, the anode is made of compound lithium.

Out of these, the Li^+ ion battery is the most prominent rechargeable battery for consumers. It is different from previously popular rechargeable battery chemistries (e.g., Ni metal hybrid, Ni cadmium, and lead acid) in a number of ways. For example, Ag^+ and Cu^+ ion-based solid electrolytes have good conductivity (e.g., $RbAg_4I_5$, 0.21 S/cm at 298 K and $Rb_4Cu_{15}I_7Cl_{13}$, 0.34 S/cm at 298 K) as compared to Li^+ ion-based solid electrolytes. However, they are not suitable to use in practical applications, due to their low decomposition potential (0.5–0.7 V) and they can easily absorb moisture from atmosphere. In addition, the energy density of Ag and Cu is also low, because they have high mass number as compared to Li. Ni-Cd and lead acid also have low energy density, and Li^+ ions have 10 times more energy density than Ni-Cd and lead acid storage batteries. In addition, toxic elements are present, like cadmium and lead acid. These toxic elements create a serious problem for the environment. The electromotive force (EMF) of these batteries is 2 and 1.3 V. The environmental problem of Ni-Cd and lead acid is overcome by the Ni-hydrogen battery, which is rapidly entering the market. However, its theoretical energy density value is one-fifth of the lithium-ion battery and the EMF is 1.4 V. Lithium is the lightest of all the metals; therefore, Li-ion battery has the lowest weight as compared to other batteries. The preceding conversation is duly documented and illustrated in both Table 10.1 and Figure 10.1 for comprehensive reference and visual representation.

Table 10.1 Comparison of different battery types [1].

	Ni-Cd	NiMH	Lead acid	Li-ion
Gravimetric energy density (Wh/kg)	200	60–120	180	2000
Cycle life	1500	300 to 500	200 to 300	500 to 1000
Charging time	1 h	2–4 h	8–16 h	2–4 h
Self-discharge/Month (room temp.)	20%	30%	5%	10%
Cell voltage (V)	1.25	1.25	2.1	4.1
Operating temperature (discharge only)	-40 to 60°C	-20 to 60°C	-20 to 60°C	-20 to 60°C

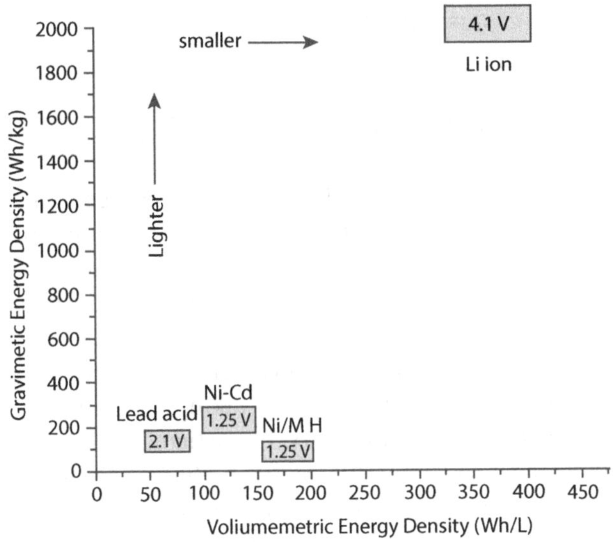

Figure 10.1 Energy densities of different types of batteries with EMF [6].

10.2 Fundamental Aspects, Different Types of Electrolytes, and the Role of the Electrolyte in Battery Technology

The electrolyte is a separator between anode and cathode. Electrolytes provide a medium to move the Li^+ ions between cathode and anode, and vice versa, as shown in Figure 10.2. On the charging and discharging mechanism, the mobility of the Li ion, the required activation energy for moving the Li ion from one grain to another, and all other properties of the Li ion depend on the electrolyte.

Electrolytes are of two types:

a) Solid electrolyte (non-oxide and oxide based)
b) Aqueous electrolyte (organic solvent)

Key parameters to evaluate the performance of solid electrolyte are as follows:

a) **Ionic Conductivity:** Lithium ionic conductivity (σ) is mentioned here as one of the most significant metrics to assess a solid-state electrolyte. In general, the Arrhenius or/and Vogel-Tammann-Fulcher (VTF) behaviors may be used to simulate ionic conductivity. The former is typically used to explain both the ionic transport in crystalline polymers and the ionic transport in inorganic and ceramic solid-state electrolyte materials, which is further related to the lattice defect mechanism and will be discussed in-depth later. The conductivity can be described by the sum of all the moving charged species (i) and the following equation can illustrate the Arrhenius behavior of ionic conductivity (Equation 10.1) [2].

$$\sigma = \sum q_i c_i u_i = \sigma_o / T \exp(-E_a/KT) \tag{10.1}$$

$$E_a = (\text{slope of Arrehenius plot} \times K \times 1000 \times 2.303\text{Log}10)/1.6 \times 10^{-19}\ \text{eV} \tag{10.2}$$

where q_i, c_i, and m_i are the number of charge, charge concentration, and changed species mobility, respectively. σ_o is the pre-exponential factor that is associated with the charge carriers' concentration, E_a denotes the activation energy of Li ion, and k and T denote the Boltzmann constant and temperature, respectively.

b) **Li^+ ion Transference Number:** Li^+ transference number is a crucial quantity to consider when assessing the activities taking place in solid-state electrolytes together with ionic conductivity. In particular, a high Li^+ transference number electrolyte can demonstrate quick charge–discharge performance even with a poor ionic conductivity, suppression of lithium dendrites, and even lengthy cycling with the Li^+ metal anode [3, 4].

c) **Mechanical Properties:** Strength, toughness, brittleness, hardness, and resilience are examples of mechanical qualities that have an impact on a material's capacity to be molded into a suitable shape as well as its mechanical strength. The mechanical strength of solid-state electrolytes used in lithium batteries, or their capacity to bear the strain of physical forces, is crucial for lengthy cycling, high energy density, and high voltage [5].

d) **Electrochemical Stability:** Designing electrolytes for lithium batteries should take good electrochemical stability into account. In particular, electrochemical impedance spectroscopy is used to assess various parameters, such as interface resistance between electrode/solid electrolyte, formation of solid electrolyte interface, and the electrochemical behavior of materials during battery cycle tests (EIS) [6].

e) **Battery Test:** In addition to analyzing the performance of solid-state electrolytes, the performance of the final battery product is typically recognized as one of the most important evaluation parameters, and the viability and

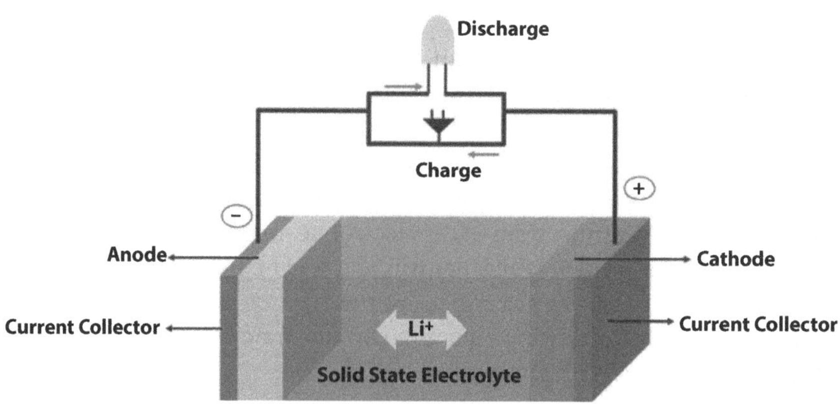

Figure 10.2 Schematic of lithium-ion battery with solid-state electrolyte.

stability of their functionality as a component in an assembled Li-ion battery are crucial. In order to evaluate the cycling behavior at various temperatures, current densities, and discharge rates, the anode/electrolyte/cathode, and symmetrical Li/electrolyte/Li cells are often built and cycled in a suitable voltage range. This method allows one to calculate the specific capacity ($mA.h.g^{-1}$) and charging/discharging efficiency based on the voltage profile of the ongoing Li plating/stripping cycle over time [6].

Performance requirements for solid-state electrolyte

- **High Ionic Conductivity:** Ideally, the ionic conductivity of a solid-state electrolyte is more than 10^{-3} S/cm at room temperature and even increases to match that of conventional liquid electrolytes that have ionic conductivity in the range of 10^{-2} to 10^{-3} S/cm) [7]. The low ionic conductivity of all-solid-state polymer electrolytes for composite solid-state electrolytes is one of the main obstacles to their practical use. In contrast to organics, it is significantly better (e.g., the conductivity of PEO is about 10^{-4} S/cm at 65–78°C) [8]. Even certain ceramic electrolytes have attained ionic conductivities in the neighborhood of 10^{-2} S/cm, as they are typically having dense pellets with high interfacial resistance and are therefore not suitable as a stand-alone choice for solid-state electrolytes.
- **Appreciable Li^+ ion Transference Number:** A higher Li^+ ion transference number (t_{Li+}) can decrease the polarization concentration of the electrolyte, which further increases the power density of the batteries [9]. Since composite solid-state electrolytes (CSSEs) may contain polymer components with a lower t_{Li+}, the usual inorganic solid electrolyte has a t_{Li+}, which is near to one and obviously superior to that of CSSEs. Therefore, prior to their structural outlet and material selection, several techniques for increasing the t_{Li+} of CSSEs must be established. By considering anions to the polymer backbone, for instance, it is possible to enhance the t_{Li+} considerably while restricting anions' mobility in the polymer components of CSSEs [10, 11]. To achieve this, anion receptors that can selectively interact with anions are introduced.
- **Good Electrochemical Stability:** The electrolyte needs to be connected to both the cathode and the anode in order to guarantee the stability and reversibility of the oxidation reaction and the reduction process. The max reduction potential should be lower than the max oxidation potential, which should be greater than the max embedding potential of Li ions in the cathode [9]. Therefore, the electrochemical window, i.e., the difference between the potentials of the oxidation and the reduction should be sufficiently large enough to ensure the battery's reversibility and long-term life cycle. For an optimal solid-state electrolyte (SSE), a large electrochemical stable window (0.0–5.0 V) is preferred in order to permit the greatest voltage output of an all-solid-state lithium-ion battery (ASSLB) [12].
- **Excellent Chemical and Thermal Stability:** The primary and essential requirement for a solid-state electrolyte (SSE) is the chemical stability with other associate components, which is slightly more challenging for composite

solid-state electrolytes (CSSEs) since they consist of two or more different parts, in order to ensure the safety, charge retention, and cyclic performance of the battery. The side reactions should be eliminated when the solid-state electrolyte is in contact with the two electrodes. The thermal stability of CSSE materials is crucial for fundamental research and even more advanced applications of ASSLBs since, on the other hand, it is highly encouraging if the Li-ion battery can run over a larger temperature range. This can guarantee the battery's safety even under extreme conditions like overcharging, short circuiting, and thermal damage [9]. Before designing CSSEs, it is essentially very important to take into account that the chemical and thermal stabilities of the composite materials along with the interface between SSE and electrodes.

- **Preferable Mechanical Strength and Flexibility:** Solid electrolytes must have good mechanical strength and flexibility in order to be produced on a big scale and used in lithium batteries going forward. The common benefit of CSSEs employed in ASSLBs is that they integrate some advantageous characteristics of their separate components. For instance, ceramic/polymer CSSEs, which combine the excellent dimensional stability of ceramics and the elasticity of polymers, have preferred mechanical strength and flexibility. Many studies have focused on plasticizing the polymer phase in polymer electrolytes and CSSEs that contain polymer electrolytes in order to reduce the temperature needed to attain desired ionic conductivity at ambient temperature. However, particularly for organic plasticizers, this occasionally comes at the expense of mechanical strength.

For the application of Li^+ ion battery, some organic solvents (e.g., $LiClO_4$ dissolved in propylene carbonate) as electrolytes are available in the market. These electrolytes have good conductivity (10^{-2}–10^{-1} S/cm) as compared to solid electrolytes, but at the time of application, these electrolytes face some problems: (1) leakage of organic electrolyte, (2) freezing of electrolyte at lower temperature, and (3) burning at higher temperature and other side reactions can take place, because in solution-type electrolytes, ions are free to move. In solution electrolytes, in the application of Li-ion battery, individual space in between cells is required, but in solid electrolytes, the individual space in cells is not required, so it is a compact package [1]. To overcome the above leakage problems in an organic solvent, an organic polymer solid electrolyte is used in place of an aqueous electrolyte. An organic polymer (e.g., ethylene oxide copolymers consist of $LiN(SO_2CF_3)_2$) has good conductivity i.e., 10^{-5}–10^{-4} S/cm at room temperature. In 1993, Angell *et al.* have reported rubbery Li^+ electrolytes (e.g., poly-propylene oxide mixed with $LiClO_4$-$LiClO_3$-based salt), which offer good ionic conductivity comparatively higher than 10^{-4} S/cm even at 298 K. However, these organic polymers could not use below glass transition temperature because its ionic conductivity is reduced [14].

Therefore, to overcome all of the above problems, these electrolytes are replaced by inorganic Li^+ ion conducting solid material, which has the following advantages: [13] no leakage problem; [14] a wide range of operating temperature; hence, the inorganic solid electrolytes do not freeze and burn in an applied situation; [15] ions are not moving anywhere inside the solid electrolyte (in solution-type electrolytes, ions are free to move anywhere inside the electrolyte), so no side reactions occur and only one kind of carrier ion migrates; and [16] lower memory effect, which means they have a long lifetime because of little self-discharge.

The ionic conductivity is high, i.e., 1.2×10^{-3} S/cm, at room temperature (RT) due to its two-dimensional layered structure; hence, the ionic conduction is very limited via anisotropic two-dimensional planes. A significant number of efforts and time are required to develop a single crystal as compared to the polycrystalline. Therefore, polycrystalline materials are thoroughly investigated, in order to obtain an isotropic Li^+ ion conduction and maximum conductivity is achieved for polycrystalline Li_3N, which is 4×10^{-4} S/cm. However, they have several disadvantages such as theoretically lower decomposition potential of 0.445 V and highest electromotive force (EMF) output of battery, which is below the decomposition potential. After that, Li_3N-LiI, a cubic structure solid electrolyte, has been reported, and it has a higher decomposition potential but has a conductivity of 1×10^{-6} S/cm at room temperature, which is three orders of magnitude lower as compared to that of the Li_3N single crystal [15, 16].

Glass-based electrolytes (Li_2S) also have a good ionic conductivity in the order of 10^{-4}–10^{-3} S/cm at room temperature. However, still, these types of electrolytes face critical issues as these are deliquescent in the air-dominated ambient. In addition, electrolytes of Li_2S-SiS_2-Li_3PO_4 generate poisonous H2S wherein SiS_2 in the glass decomposes to SiO_2 while glassy electrolyte absorbs water. For the fabrication of these types of cells, a dry environment is required [17–19]. To overcome the above problems, Li^+ conducting oxides are other solid-state electrolytes for battery applications, wherein the oxides can be handled and prevented easily even in humid air. Generally, ions cannot move with lower activation energy and higher ionic conductivity; however, some solid ions can be easily migrated with lower activation energy and high ionic conductivity and these types of solid materials are called solid-state electrolytes. In the mid-1960, β-alumina ($Na_2O.11Al_2O_3$) has been prepared as a fast Na^+ ion conductor, which has a two-dimensional layered structure wherein Na^+ ions conduct between the layers, and due to the two-dimensional layered structure, the ionic migration is limited between the layers because of the anisotropic conduction mechanism [20, 21]. In 1976, to obtain an isotropic mechanism, Goodenough and Hong have designed the 3-dimensional network structure with a suitable tunnel size to move Na^+ ion, and the material is a Na^+ super ionic conductor ($Na_{1+x}Zr_2Si_xP_{3-x}O_{12}$) and is called NASICON. This Na^+ ion conductor has a good conductivity of around 10^{-3}–10^{-1} S/cm, which is comparable to aqueous electrolyte [22, 23]. However, Na^+ ions also have low energy density as compared to Li^+.

As aforementioned, various types of solid electrolytes, such as Li^+ ion-based NASICON electrolytes, have desirable materials properties, including high Li^+ ion conductivity ($>10^{-3}$ S/cm at room temperature) and a widely acceptable electrochemical stability. Li^+ ion has higher energy density and decomposition potential (3.5 V), has less memory effect, is less toxic, and has a high lifetime. In addition, NASICON is stable in water, so a variety of cathodes, even aqueous cathodes, can be used [1].

$LiA_2(PO_4)_3$, where A(tetravalent) = Ge, Ti, HF, and Zr, also formed a NASICON-type structure [24–28]. Here, Figure 10.3 shows the NASICON-type structure whereas 2 polyhedra, AO_6 octahedra and PO_4 tetrahedra, have been linked and associated by their corners to form the $[A_2(PO_4)_3]^-$ rigid skeleton and the Li^+ ions move through the tunnel in the 3D structure [29]. In this structure, M1 and M2 are two different Li-ion sites. For $LiA_2(PO_4)_3$, the M_1 site is fully connected but the M_2 site is completely vacated. Thus, the M_2 site is

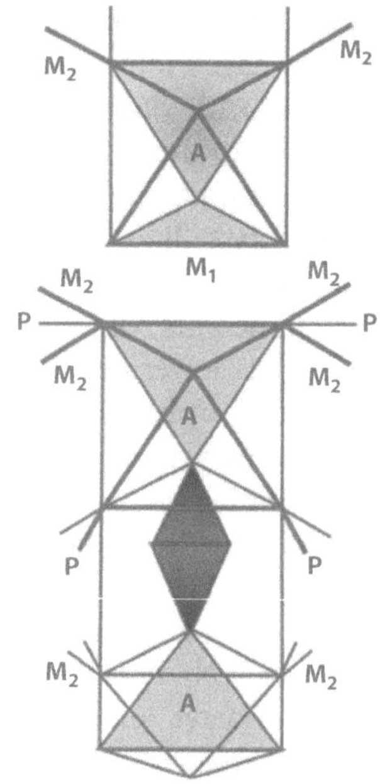

Figure 10.3 Structure of the $LiA_2(PO_4)_3$ electrolyte.

filled with the help of trivalent material (B = Al^{3+}, Y^{3+}), because a tetravalent (A = Ti^{4+}, Ge^{4+}, HF^{4+}, Zr^{4+}) ion was partially substituted by the trivalent (B = Al^{3+}, Y^{3+}) ion. Hence, due to this, additional Li^+ ions occupied the M_2 site with an increase in x for the material $Li_{1+x}B_xA_{2-x}(PO_4)_3$ [25, 28].

Li^+ ion batteries use a variety of solid electrolyte types, and each type has several advantages and disadvantages. Table 10.2 provides a comprehensive summary for the various solid-state electrolytes' properties.

The benefits and drawbacks of the aformentioned solid electrolytes outlined in Table 10.2 are as follows:

- Higher ionic conductivity and better chemical and electrochemical stability are characteristics of LAGP of the NASICON type. The price of GeO_2, one of the LAGP precursors, however, can dramatically incease the cost of LAGP-based ASSBs. It is essential to create new NASICON that is reasonably priced.
- The perovskite-type of LSTZ/LSTH is promising to use in ASSBs. The mateials preparation conditions are extremely harsh, and the ionic conductivity of the material is not yet high enough.
- The Al-stabilized LLZO garnet type is stable with lithium and has a strong ionic conductivity. However, it loses its stability when in contact with atmospheric CO_2 or moisture. It is crucial to modify the materials in order to increase LLZO stability.
- Solid electrolytes of the LISICON type are chemically and electrochemically stable, but they offer lower ionic conductivities as compared to the solid-state electrolytes of other oxides.

Table 10.2 Comprehensive list of the solid-state electrolytes with their characteristics.

Electrolyte	Prototype	Composition	Ionic conductivity (S/cm)	Electrochemical stability vs. Li/Li^+
NASICON	$LiTi_2(PO_4)_3$	$LiTi_2(PO_4)_3$-$0.2Li_3BO_3$ [29]	3×10^{-4} (25°C)	>2.5 V
		$2[Li_{1.4}Ti_2Si_{0.4}P_{2.6}O_{12}]$-$AlPO_4$ [30]	1.5×10^{-3} (RT)	
		$100[Li_{1.5}Cr_{0.5}Ti_{1.5}(PO_4)_3]$-$5SiO_2$ [31]	2.14×10^{-2} (25°C)	
	$LiGe_2(PO_4)_3$	$Li_{1.5}Al_{0.4}Cr_{0.1}Ge_{1.5}(PO_4)_3$ [32]	6.65×10^{-3} (26°C)	0.44–7 V
Perovskite	$Li_{3x}La_{2/3-x}TiO_3$	$Li_{0.34(1)}La_{0.51(1)}TiO_{2.94(2)}$ [33]	2×10^{-5} (RT)	>1.8 V
		$LiSr_{1.65}Zr_{1.3}Ta_{1.7}O_9$ [34]	1.3×10^{-5} (30°C)	
Garnet	$Li_7La_3Zr_2O_{12}$	$Li_7La_3Zr_2O_{12}$ [35]	3×10^{-4} (25°C)	0–9 V
		$Li_{6.55}Ga_{0.15}La_3Zr_2O_{12}$ [36]	1.3×10^{-3} (24°C)	0–5 V
LISICON	γ-Li_3PO_4	$Li_{10.42}Si_{1.5}P_{1.5}Cl_{0.08}O_{11.92}$ [37]	1.03×10^{-5} (27°C)	0–9 V
		$Li_{14}Zn(GeO_4)_4$ [38]	1.3×10^{-6} (33°C)	
Sulfide	Li_4GeS_4	$Li_{4.25+\delta}(Ge_{0.75-\delta'}Ga_{0.25})S_4$ [39]	6.5×10^{-5} (25°C)	0–5 V
		$Li_{3.25}Ge_{0.25}P_{0.75}S_4$ [40]	2.17×10^{-3} (25°C)	0–5 V
	$Li_{10}GeP_2S_{12}$	$Li_{10}GeP_2S_{12}$ [41]	1.2×10^{-2} (27°C)	
		$Li_{10}GeP_2S_{11.7}O_{0.3}$ [42]	1.03×10^{-2} (25°C)	
	Li_2S-P_2S_5	$75Li_2S$-$25P_2S_5$ glass [43]	2×10^{-4} (25°C)	
		$80Li_2S$-$20P_2S_5$ glass [43]	2×10^{-4} (RT)	
Argyrodite	Li_6PS_5X (X=Cl, Br, I)	Li_6PS_5Cl [44]	1.33×10^{-3} (25°C)	0–7 V
		Li_6PS_5Br [45]	6.8×10^{-3} (RT)	
		Li_6PS_5I [46]	3.6×10^{-7} (RT)	
Anti-perovskite	Li_3OX (X=Cl, Br, I)	$Li_3OCl_{0.5}Br_{0.5}$ [47]	1.94×10^{-3} (RT)	
		Li_3OCl [47]	8.5×10^{-4} (RT)	

- At normal temperature, solid-state electrolytes of the sulfide type offer highest ionic conductivities, which are measured in units of 10^{-2} S/cm. However, due to an interaction with moisture, they are not stable in the surrounding atmosphere. Because of the insufficiently developed deposition technologies for sulfide electrolytes, they must therefore be utilized in a controlled environment for bulk batteries but not for thin-film ones.
- Higher ionic conductivities can be found in solid-state electrolytes of the argyrodite type. They are not stable in the surrounding atmosphere similar to the solid electrolytes of the sulfide type. It is crucial to properly encapsulate cells by utilizing argyrodite-type solid electrolytes; otherwise, safety concerns could arise.
- Solid electrolytes based on the anti-perovskite have strong ionic conductivities and can be utilized in both bulk and thin-film ASSBs. These are operated in an inert environment due to their high hygroscopicity.

10.3 Conductivity Enhancement Approach in Solid Electrolyte Materials

Table 10.3 presents a comprehensive overview of the performance parameters associated with the lithium-ion (Li-ion) based electrolyte. The performance parameters outlined in Table 10.3 serve as crucial benchmarks for assessing the porosity, conductivity and activation energy of the solid electrolyte. For $LiA_2(PO_4)_3$ at x = 0, a highly densified material cannot be achieved by calcination, even at higher processing temperature; hence, the conductivity of the electrolyte is low. However, due to the high density or porosity at the grain boundary, high Li^+ ion conducting materials were obtained by A^{4+} site by substitution of B^{3+}. The lithium inserts into M_2 sites by B^{3+}, and it does not influence the Li^+ ion mobility in the bulk. If X is increasing in the material $Li_{1+x}B_xA_{2-x}(PO_4)_3$, the required activation energy for Li^+ ion migration from one grain to another is reduced [24, 25, 27]. At room temperature, the ionic conductivity of the NASICON-type polycrystalline electrolytes is mainly managed by the grain boundary component of the used materials. The overall conductivity of the polycrystalline electrolyte is the reciprocal of the addition of the bulk resistivity (ρ_b) and grain boundary resistivity (ρ_g), which can be determined with the help of complex impedance analysis and Cole–Cole plot individually [1, 48].

Table 10.3 Performance parameters of the Li+ ion-based electrolyte.

Sample	Porosity (%)	Conductivity (S/cm)	Activation energy (KJ/mol)		
			Bulk grain boundary		
$LiTi_2(PO_4)_3$	34.0	2.0×10^{-6}	29	46	Ref. [1]
$Li_{1.2}Al_{0.2}Ti_{1.8}(PO_4)_3$	5.6	6.6×10^{-4}	28	34	
$Li_{1.3}Al_{0.3}Ti_{1.7}(PO_4)_3$	4.1	7.0×10^{-4}	28	35	
$Li_{1.4}Al_{0.4}Ti_{1.6}(PO_4)_3$	1.4	5.0×10^{-4}	28	35	
$Li_{1.5}Al_{0.5}Ti_{1.5}(PO_4)_3$	2.0	2.5×10^{-4}	29	34	

Another effective way to obtain low porosity or highly dense solid electrolyte material is to add lithium compounds like Li_2O and Li_3PO_4 into $LiA_2(PO_4)_3$, whereas A = Ge, Ti, Zr, and HF. Hence, the activation energy required for Li-ion migration from one grain to other grain boundary is reduced after the addition of Li compound, for example, $LiA_2(PO_4)_3$ + xLi_2O, but in the bulk, the component was not changed. The activation energy is reduced in the lithium compound at the grain boundary and this similarly happened in the B^{3+}-substituted $Li_{1+x}B_xA_{2-x}(PO_4)_3$ in the solid electrolyte [27, 49, 50].

The other factors affecting the conductivity of NASICON-type solid electrolytes are tunnel size and cell volume. The grain and grain boundary for Li^+ migration are directly proportional to the change in the lattice size, because of tetravalent cation A^{4+} and trivalent cation B^{3+} substitution in the $Li_{1+x}B_xA_{2-x}(PO_4)_3$. The sample tunnel size can be easily adjusted by changing the tetravalent cation A^{4+}, for example, A^{4+} = Ge, HF, Ti, Zr in $LiA_2(PO_4)_3$. The activation energy of the material is changed with the cell volume, which can become smaller or larger. A smaller tunnel size of solid electrolyte makes the carriers' Li^+ ion migration more difficult and a larger tunnel size of electrolyte leads the mobile Li^+ ions to approach the oxygen in polyhedra from the center of the Li^+ ion site because of a high Li^+-O bonding energy. For example, $LiGe_2(PO_4)_3$ has a lower cell volume and $LiHF_2(PO_4)_3$ has a higher cell volume as compared to $LiTi_2(PO_4)_3$, and due to this, they require a high activation energy [1, 51]. Therefore, the $LiTi_2 (PO_4)_3$ structure is the most suitable and preferable lattice or tunnel size for Li^+ ion migration. Note that the Ti^{4+} ion in the $LiTi_2(PO_4)_3$ solid electrolytes is easily reduced to Ti^{3+} by the Li metal used as an anode [50]. Hence, this problem can be overcome by replacing tetravalent cation (Ti^{4+}) with another tetravalent cation (i.e., Zr).

Due to the stable oxidation, the M-site, an isomorphous $LiZr_2(PO_4)_3$ (LZP), is anticipated to be electrochemically stable against Li metal. Zr^{4+} shows lower ionic conductivity at ambient temperature, which is a drawback with LZP. $LiZr_2(PO_4)_3$ is well known for having a highly complex polymorphism; the low-temperature synthesized "β" phase is reported to go through an orthorhombic-to-monoclinic transition at 300°C, whereas the high process temperature synthesized "α" phase is represented to go through a rhombohedral-to-triclinic structural transition at 60°C [52]. The rhombohedral phase's ionic conductivity at ambient room temperature is $10^{-5}\ \Omega^{-1}\ cm^{-1}$ while the triclinic phase's is approximately $10^{-8}\ \Omega^{-1}cm^{-1}$ [53]. The essential requirements of the solid electrolytes in rechargeable batteries for an energy storage material is high ionic conductivity. The ionic conductivity of the electrolyte thin sheet is anticipated to be at least $10^{-5}\ \Omega^{-1}\ cm^{-1}$, even for Li microbatteries, wherein electric current densities are rather modest [54]. As a result, the value of ionic conductivity in LZPs at room temperature is lower than the ideal for solid electrolytes in actual applications. Aliovalent doping may significantly enhance the ionic conductivity in the NASICON-type materials [55]. The size of bottlenecks in the Li^+ ion transport channels might be adjusted by introducing elements with varying ionic sizes and charges to the crystal lattice of LZP, which increases the ionic conductivity. The rhombohedral structure of LZP has stabilized at RT as discussed by Mariappan *et al.* [56] and Li *et al.* [57] by replacing Zr^{4+} with Y^{3+}. $Li_{1.15}Y_{0.15}Zr_{1.85}(PO_4)_3$ is densified by spark plasma sintering, shown as a significantly enhanced overall ionic conductivity of 0.7×10^{-4} S.cm^{-1} at 25°C when the authors evaluated the electrical characteristics of the compounds $Li_{1+x}Y_xZr_{2-x}(PO_4)_3$ where $0.1 \leq x \geq 0.2$ and $0.05 \leq x \geq 0.2$, respectively. As stated by the authors, the difference in the ionic radii of Zr^{4+} (0.72) and Y^{3+} (0.9) makes Y^{3+} a more viable replacement ion for Zr^{4+} in LZP, increasing the lattice parameters and cell volume.

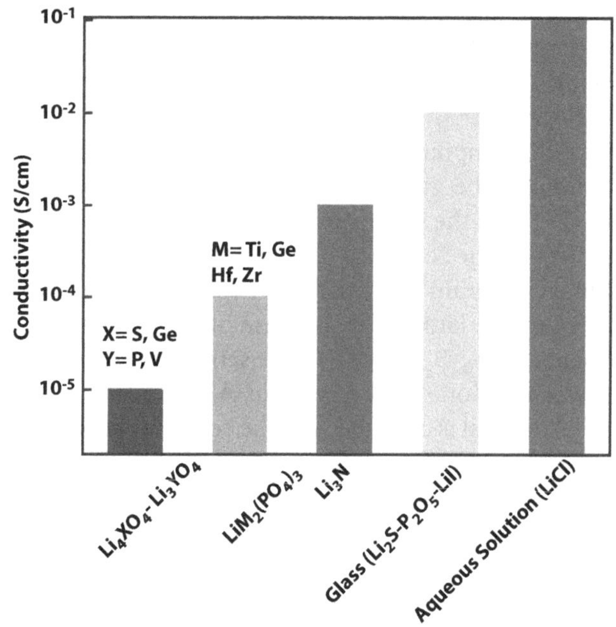

Figure 10.4 The ionic conductivity at room temperature for various electrolytes of Li.

The effects of Al^{3+} substitution with Zr^{4+} in $LiZr_2(PO_4)_3$ in its structure are determined, and afterward Li-ion conduction is investigated. The ionic conductivities of 4.4×10^{-5} S/cm for the sample $Li_{2.2}AlZr(PO_4)_3$ at 400°C, 3.1×10^{-6} S/cm for $Li_{1.0.275}Al_{0.275}Zr_{1.725}(PO_4)_3$ at RT, and 1.83×10^{-6} S/cm for the sample $Li_{1.25}Zr_{1.75}Al_{0.25}(PO_4)_3$ at RT have been reported by Savitha *et al.* [58], Zhang *et al.* [59], and Pareek *et al.* [60], respectively, by studying the systems $Li_xAlZr(PO_4)_3$ (x = 1.8, 2.0, 2.2), $Li_{1+x}Al_xZr_{2-x}(PO_4)_3$ (x = 0.2, 0.225, 0.25, 0.275, 0.3, 0.5), and $Li_{1+x}Al_xZr_{2-x}(PO_4)_3$ (x = 0, 0.1, 0.2, 025, 0.3, 0.4 and 0.5).

$Li_{1.2}Ca_{0.1}Zr_{1.9}(PO_4)_3$ was used by Cassel *et al.* [61] to create a stable rhombohedral phase, which demonstrated enhanced ionic conductivity at RT of 1.2×10^{-7} S/cm and 7.2×10^{-7} S/cm, respectively, for the materials synthesized by solid-state reaction and sol-gel synthesis. In their study on Sr substitution, Kumar *et al.* [62] have found that the molecule $Li_{1.2}Sr_{0.1}Zr_{1.9}(PO_4)_3$ has shown the highest ionic conductivity, which is 3.4×10^{-5} S/cm. Figure 10.4 provides valuable insights for researchers by visually presenting the range of ionic conductivity exhibited by various electrolytes for lithium-ion batteries. This graphical representation serves as a helpful tool for researchers to observe and compare the performance of different electrolytes in terms of their ability to conduct ions.

10.4 Synthesis Approaches for Solid Electrolytes

There are two methods for the synthesis of electrolyte:

1) Solid-state reaction process
2) Pechini polymerization process

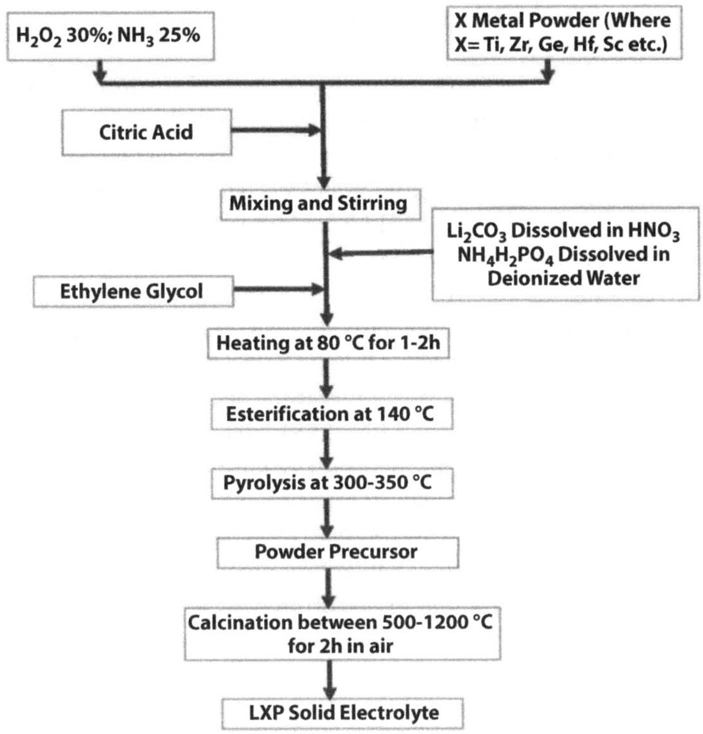

Figure 10.5 Flowchart of synthesis of LXP solid electrolyte by a Pechini-type polymerization method.

NASICON-type solid-state electrolytes are synthesized by a conventional solid-state reaction process wherein all the precursors are mixed in powder form and ground up to 50–60 minutes, followed by sintering. This process requires a high temperature (1200°C) for a longer time period (24 h), which further leads to the Li-ion loss in electrolyte materials. The crystal size is in the order of µm, and due to this, the electrical performance of the sample is affected. Therefore, to overcome the problems of solid-state reaction, the Pechini polymerization method is utilized. It is a solution and combustion method that requires comparatively lower synthesis and treatment temperatures, and the pure phase of the sample with crystal size in the order of nm is obtained. Figure 10.5 presents a detailed flowchart illustrating the Pechini-type polymerization process. This visual representation serves as a valuable guide for understanding the sequential steps involved in the Pechini synthesis method. The Pechini-type process is utilized to synthesize the homogenous bulk multi-component metal oxide, and the Pechini process is a process involving metal complex formation and *in situ* polymerization of organics. In this process, α-hydroxycarboxylic acid such as citric acid (CA) is used to create stable metal complexes and their polyesterification along with a polyhydroxy alcohol, such as ethylene glycol (EG), forms a polymeric resin. After that, calcinations of polymeric resin are performed to synthesize pure phase multi-component metal oxides [18]. The comprehensive advantage of the Pechini polymerization process is that the porosity or density at the grain boundary is controlled by the viscosity and the molecular weight of the polymer can be manipulated by the CA/EG molar ratio with synthesis temperature [18].

10.5 Conclusion and Future Perspective

Batteries have many applications in our daily life; it is used for industrial, transportation, medical industry, and power storage purposes. There are different types of batteries available in the market, but out of these, the Li-ion battery is the most beneficial because Li^+ ion has higher energy density than other commercialized Ni-Cd or lead acid batteries. For the application of batteries, an electrolyte should have good electrical properties.

Herein, different types of electrolytes were discussed for the Li^+ ion battery system. The ionic conductivity of aqueous electrolytes is higher than other electrolytes, but it is facing problems like leakage, freezing at low temperatures, and burning at high temperatures. Therefore, to overcome these problems, polymer-based electrolytes came into the market but they cannot be used below glass temperature because its conductivity is affected. Hence, solid electrolytes i.e., non-oxide (Li_3N, Li_2S)-based, were utilized to develop a stable battery technology. For the fabrication of non-oxide-based electrolytes, a dry environment is needed because it is more moisture sensitive. Oxide-based lithium-type (Li_3PO_4, $LiA_2P_3O_{12}$) solid electrolytes resolved this problem. Nowadays, lithium-ion batteries (LIBs) are highly optimized, well operable from several months to several years with some unexpected functions. This is a remarkable breakthrough wherein several materials systems are operated outside their thermodynamic stability windows. The Li-based anodes have potentials close to Li metal (~0.08 V vs Li/Li+) on charging, whereas no electrolyte materials are stable. Moreover, the passivation layer helps the battery to survive, and the solid–electrolyte interphase (SEI) prevents further degradation in electrolyte materials. On the other hand, cathodes, wherein aluminum current collector corrosion is responsible for the decomposition of the salts of the electrolyte, produce a stable passivation layer. Therefore, cathode electrode materials have been optimized to minimize oxygen loss at higher working temperatures, which further enhanced the "thermal runaway" problem and withstood the mechanical stresses under the repeated volume changes, which is further associated with Li-ion removal and insertion. The current commercial systems have materials that are operated with energy densities increasingly closer to their standard fundamental limits, which further leads to the Li-ion removal from the cathode and, as a result, to irreversible structural changes or oxygen losses at the anode side wherein no vacancies in the lattice structure remain to occupy more Li ions. The separators and current collectors should be much thinner, which enhances the function of batteries with higher voltages via surface coatings, electrolyte additives, and morphology optimization.

References

1. Adachi, Gin-ya; Imanaka, N.; Aono, H.; Fast Li+ Conducting Ceramic Electrolytes. *Adv. Mater.*, 1996, 2, 8.
2. Goodenough, J. B.; Oxide-ion Electrolytes. *Annu. Rev. Mater. Res.*, 2003, 33, 91–128.
3. Diederichsen, K. M.; McShane, E. J.; McCloskey, B. D.; Promising Routes to a High Li^+ Transference Number Electrolyte for Lithium-Ion Batteries. *ACS Energy Lett.*, 2017, 2, 2563–2575.
4. Shigenobu, K.; Dokko, K.; Watanabe, M.; Ueno, K.; Solvent effects on Li ion transference number and dynamic ion correlations in glyme- and sulfolane-based molten Li salt solvates. *Phys. Chem. Chem. Phys.*, 2020, 22, 15214–15221.

5. Mackanic, D. G.; Yan, X.; Zhang, Q.; Matsuhisa, N.; Yu, Z.; Jiang, Y.; Manika, T., Lopez, J.; Yan, H.; Liu, K.; Chen, X.; Cui, Y.; Bao, Z.; Decoupling of mechanical properties and ionic conductivity in supramolecular lithium-ion conductors. *Nat. Commun.*, 2019, 10, 1–11.
6. Quartarone, E.; Mustarelli, P.; Electrolytes for solid-state lithium rechargeable batteries: recent advances and perspectives. *Chem. Soc. Rev.*, 2011, 40, 2525–2540.
7. Scrosati, B.; Croce, F.; Panero, S.; Progress in lithium polymer battery R&D. *J. Power Sources*, 2001,100, 93–100.
8. Manthiram, A.; Yu, X.; Wang, S.; Lithium battery chemistries enabled by solid-state electrolytes. *Nat. Rev. Mater.*, 2017, 2, 1–16.
9. Long, L.; Wang, S.; Xiao, M.; Meng, Y.; Polymer electrolytes for lithium polymer batteries. *J. Mater. Chem. A*, 2016, 4, 10038–10039.
10. Cowie, J. M. G.; Spence, G. H.; Novel single ion, comb-branched polymer electrolytes. *Solid State Ionics, 1999*,123, 233–242.
11. Lee, H. S.; Ma, Z. F.; Yang, X. Q.; Sun, X.; Mcbreen, J.; Synthesis of a series of fluorinated boronate compounds and their use as additives in lithium battery electrolytes. *J. Electrochem. Soc.*, 2004, 151, A1429–A1435.
12. Han, F.; Zhu, Y.; He, X.; Mo, Y.; Wang, C.; Electrochemical Stability of $Li_{10}GeP_2S_{12}$ and $Li_7La_3Zr_2O_{12}$ Solid Electrolytes. *Adv. Energy Mater.*, 2016, 6, 1–9.
13. C. A. Angell, C. Liu, E. Sanchez, "Rubbery Solid Electrolytes with dominant cationic transport and high ambient conductivity" *Nature*, 1993,362,137-139.
14. B. A. Boukamp, R. A. Huggins, "Fast ionic conductivity in lithium nitride" *Mater. Res. Bull.*, 1978, 13, 23.
15. J. R. Rea, D. L. Foster, "High ionic conductivity in densified polycrystalline lithium nitride" *Mater. Res. Bull.*, 1979, 14, 841.
16. J. H. Kennedy, Y. Yang, "A Highly Conductive Li+ - Glass System: (1-x) ($0.4SiS_2$ - $0.6Li_2$ S)- xLiI" *J. Electrochem. Soc.*, 1986, 133, 2437-2438.
17. N. Aotani, K. Iwamoto, K. Takada, S. Kondo, "Synthesis and electrochemical properties of lithium-ion conductive glass , $Li_3PO_4Li_2SSIS_2$"*Solid State Ionic*, 1994, 68, 35-39.
18. K. Takada, N. Aotani, K. Iwamoto, S. Kondo, "Electrochemical behavior of LixMO2 (M= Ni, Co) all solid-state cells using a glass electrolyte" *Solid State Ionics*, 1995, 79, 284-287.
19. N. Weber, J.T. Kummer, "Sodium-Sulfur Secondary battery", *Proc. Annu. Power Source Conf.*, 1967, 21,37.
20. M. S. Whittingham, R. A. Huggins, "Measurement of sodium ion transport in beta alumina using reversible solid electrodes" *J. Chem. Phys.*, 1971, 54, 414.
21. J. B. Goodenough, H. Y. P. Hong, J. A. Kafalas, "Fast Na^+ ion transport in skeleton structures" *Mater. Res. Bull.*, 1976, 11, 203-220.
22. H.Y.P Hong, "Crystal structures and crystal chemistry in the system $Na_{1+x}Zr_2Si_xP_{3-x}O_{12}$" *Mater. Res. Bull.*, 1976, 11, 173-182.
23. Shi-chun Li, Jian-yiCai, Zu-xaing Lin," Phase relationships and electrical conductivity of $Li_{1+x}Ge_{2-x}Al_xP_3O_{12}$ and $Li_{1+x}Ge_{2-x}Cr_xP_3O_{12}$ systems" *Solid State Ionics*, 1988, 28-30, 1265-1270.
24. H. Aono, E. Sugimoto, Y. Sadaoka, N. Imanaka, G. Adachi, "Electrical Properties of Sintered Lithium Titanium Phosphate Ceramics" *Chem. Lett.*, 1990, 19, 1825-1828.
25. H. Aono, E. Sugimoto, Y. Sadaoka, N. Imanaka, G. Adachi, "Electrical Properties and Crystal Structure of Solid Electrolyte based on Lithium Hafnium Phosphate $LiHf_2(PO_4)_3$" *Chem. Lett.*, 1993, 62, 309-316.
26. B. E. Talyor, A. D. English, T. Berzins, "New Solid Ionic Conductor" *Mater. Res. Bull.* 1977, 12, 171-181.
27. H. Aono, E. Sugimoto, Y. Sadaoka, N. Imanaka, G. Adachi, "The Electrical Properties of Ceramic Electrolytes for $LiM_xTi_{2-x}(PO_4)_3$+yLi_2O, M= Ge, Sn, Hf, and Zr systems" *J. Electrochem. Soc.* 1993, 140, 1827-1833.

28. L. Hagman, P. Kierkegaard, "The Crystal Structure of $NaMe_2^{IV}(PO_4)_3$; Me^{IV} = Ge, Ti, Zr" *Acta Chem. Scand.*, 1968, 22, 1822-1832.
29. Aono, H.; Sugimoto, E.; Sadaoka, Y.; Imanaka, N.; Adachi, G.; Electrical property and sinterability of $LiTi_2(PO_4)_3$ mixed with lithium salt (Li_3PO_4 or Li_3BO_3). *Solid State Ionics,* 1991, 47, 3–4, 257-264.
30. Fu, J.; Fast Li+ ion conduction in Li_2O-Al_2O_3-TiO_2-SiO_2-P_2O_5 glass-ceramics. *J. Am. Ceram. Soc.,* 1997, 80, 7, 1901-1903.
31. Goharian, P.; Yekta, B. E.; Aghaei, A. R.; Banijamali, S.; Lithium ion-conducting glass-ceramics in the system Li_2O–TiO_2–P_2O_5–Cr_2O_3–SiO_2. *J. Non-Cryst. Solids,* 2015, 409, 120-125.
32. Xu, X.; Wen, Z.; Wu, X.; Yang, X.; Gu, Z.; Lithium Ion-Conducting Glass–Ceramics of $Li_{1.5}A_{l0.5}Ge_{1.5}(PO_4)_{3-x}Li_2O$ (x=0.0–0.20) with Good Electrical and Electrochemical Properties. *J. Am. Ceram. Soc.,* 2007, 90, 9, 2802-2806.
33. Inaguma, K.; Liquan, C.; Itoh, M.; Nakamura, T.; Uchida, T.; Ikuta, H.; Wakihara, M.; High ionic conductivity in lithium lanthanum titanate. *Solid State Commun.,* 1993, 86, 689-693.
34. Thangadurai, V.; Shukla, A. K.; Gopalakrishnan, J.; $LiSr_{1.65\Delta0.35}B_{1.3}B'_{1.7}O_9$ (B= Ti, Zr; B′ =Nb, Ta): New Lithium-Ion Conductors Based on the Perovskite Structure. *Chem. Mater.,* 1999, 11, 835-839.
35. Murugan, R.; Thangadurai, V.; Weppner, W.; Fast Lithium-Ion Conduction in Garnet-Type $Li_7La_3Zr_2O_{12}$. *Angew. Chem. Int. Ed.,* 2007, 46, 7778–7781.
36. Lopez, C. B.; Manalastas, W.; Jr.; Amo, J. M. L. D.; Aguadero, A.; Aguesse, F.; Kilner; Atmosphere Controlled Processing of Ga-Substituted Garnets for High Li-Ion Conductivity. *Ceram. Chem. Mater.,* 2014, 26, 12, 3610–3617.
37. Song, S.; Lu, J.; Zheng, F.; Duong, H. M.; Lu, L.; A facile strategy to achieve high conduction and excellent chemical stability of lithium solid electrolytes. *RSC Adv.,* 2015, 5, 6588-6594.
38. Mazumdar, D.; Bose, D. N.; Mukherjee, M. L.; Transport and dielectric properties of lisicon. *Solid State Ionics,* 1984, 14, 2, 143-147.
39. Kanno, R.; Hata, T.; Kawamoto, Y.; Irie, M.; Synthesis of a new lithium ionic conductor, thio-LISICON–lithium germanium sulfide system. *Solid State Ionics,* 2000, 130, 1–2, 97-104.
40. Kanno, R.; Murayama, M.; Lithium Ionic Conductor Thio-LISICON: The Li_2S GeS_2 P_2S_5 System. *J. Electrochem. Soc.,* 2001,148, A742.
41. Kamaya, N.; Homma, K.; Yamakawa, Y. *et al.*; A lithium superionic conductor. *Nature Mater.,* 2011, 10, 682–686.
42. Sun, Y.; Suzuki, K.; Hara, K.; Hori, S.; Yano, T.; Hara, M.; Hirayama, M.; Kanno, R.; Oxygen substitution effects in $Li_{10}GeP_2S_{12}$ solid electrolyte. *J. Power Sources,* 2016, 324, 798-803.
43. Hayashi, A.; Noi, K.; Sakuda, A. *et al.*; Superionic glass-ceramic electrolytes for room-temperature rechargeable sodium batteries. *Nat. Commun.,* 2012, 3, 856.
44. Boulineau, S.; Courty, M.; Tarascon, J.; Viallet, V.; Mechanochemical synthesis of Li-argyrodite Li_6PS_5X (X=Cl, Br, I) as sulfur-based solid electrolytes for all solid-state batteries' application, *Solid State Ionics,* 2012, 221, 1-5.
45. Rao, R. P.; Adams, S.; Studies of lithium argyrodite solid electrolytes for all-solid-state batteries. *Phys. Status Solidi A,* 2011, 208: 1804-1807.
46. Pecher, O.; Kong, S.-T.; Goebel, T.; Nickel, V.; Weichert, K.; Reiner, C.; Deiseroth, H.-J.; Maier, J.; Haarmann, F.; Zahn, D.; Atomistic Characterisation of Li^+ Mobility and Conductivity in $Li_{7-x}PS_{6-x}I_x$ Argyrodites from Molecular Dynamics Simulations, Solid-State NMR, and Impedance Spectroscopy. *Chem. - Eur. J.,* 2010, 16: 8347-8354.
47. Zhao, Y.; Daemen, L. L.; Superionic Conductivity in Lithium-Rich Anti-Perovskites. *J. Am. Chem. Soc.,* 2012, 134, 36, 15042–15047.
48. Mariappan, C. R.; Galven, C.; Crosnier-lopez, M. P.; Berre, F. L.; Bohnke, O.; Synthesis of Nanostructured $LiTi_2(PO_4)_3$ Powder by Pechini-type Polymerization Complex Method. *J. App. State Chem.,* 2006, 179, 450-456.

49. Aono, H.; Sugimoto, E.; Sadaoka, Y.; Imanaka, N.; Adachi, G.; Electrical Property and Sinterability of LiTi2(PO4)3 mixed with Lithium Salt (Li_3PO_4 or Li_3BO_3). *Solid State Ionic,* 1991, 47, 257-264.
50. Aono, H.; Sugimoto, E.; Sadaoka, Y.; Imanaka, N.; Adachi, G.; DC Conductivity of $Li_{1.3}Al_{0.3}Ti_{1.7}(PO_4)_3$ Ceramic with Li Electrodes. *Chem. Lett.,* 1991, 20, 1567-1570.
51. Aono, H.; Imanaka, N.; Adachi, G.; High Li+ Conducting Ceramics. *Acc. Chem. Res.,* 1994, 27, 265.
52. Catti, M.; Comotti, A.; Blas, S. D.; High-Temperature Lithium Mobility in α-$LiZr_2(PO_4)_3$NASICON by Neutron Diffraction. *Chem. Mater.,* 2003, 15, 1628–1632.
53. Zheng, F.; Kotobuki, M.; Song, S.; Lai, M. O.; Lu, L.; Review on solid electrolytes for all-solid- state lithium-ion batteries. *J. Power Sources,* 2018, 389, 198–213.
54. El-Shinawi, H.; Regoutz, A.; Payne, D. J.; Cussen, E. J.; Corr, S. A.; NASICON $LiM_2(PO_4)_3$ electrolyte (M = Zr) and electrode (M = Ti) materials for all solid-state Li-ion batteries with high total conductivity and low interfacial resistance. *J. Mater. Chem. A,* 2018, 6, 5296–5303.
55. Smith, S.; Thompson, T.; Sakamoto, J.; Allen, J. L.; Baker, D. R.; Wolfenstine, J.; Electrical, mechanical and chemical behavior of $Li_{1.2}Zr_{1.9}Sr_{0.1}(PO_4)_3$. *Solid State Ionics,* 2017, 300, 38–45.
56. Mariappan, C. R.; Kumar, P.; Kumar, A.; Indris, S.; Ehrenberg, H.; Prakash, G. V.; Jose, R.; Ionic conduction and dielectric properties of yttrium doped $LiZr_2(PO_4)_3$ obtained by a Pechini-type polymerizable complex route. *Ceram. Int.,* 2018, 44, 15509–15516.
57. Li, Y.; Liu, M.; Liu, K.; Wang, C. A.; High Li^+ conduction in NASICON-type $Li_{1+x}Y_xZr_{2-x}(PO_4)_3$ at room temperature. *J. Power Sources*, 2013, 240, 50–53.
58. Savitha, T.; Selvasekarapandian, S.; Ramya, C. S.; Structural and electrical conductivity studies of $Li_xAlZr[PO_4]_3$ (x= 1.8, 2.0, 2.2), solid electrolyte for lithium-rechargeable batteries. *J. Solid State Electrochem.,* 2008, 12, 857–860.
59. Zhang, Y.; Chen, K.; Shen, Y.; Yuanhua, L.; Nan, C. W.; Enhanced lithium-ion conductivity in a $LiZr_2(PO_4)_3$ solid electrolyte by Al doping. *Ceram. Int.,* 2017, 43, S598-S602.
60. Pareek, T.; Singh, B.; Dwivedi, S.; Yadav, AK.; Sen, S.; Kumar, P.; Kumar, S.; Ionic conduction and vibrational characteristics of Al^{3+} modified monoclinic $LiZr_2(PO_4)_3$. *Electrochim. Acta,* 2018, 263, 533-543.
61. Cassel, A.; Fleutot, B.; Courty, M.; Viallet, V.; Morcrette, M.; Sol-gel synthesis and electrochemical properties extracted by phase inflection detection method of NASICON-type solid electrolytes $LiZr_2(PO_4)_3$ and $Li_{1.2}Zr_{1.9}Ca_{0.1}(PO_4)_3$. *Solid State Ionics,* 2017, 309, 63–70.
62. Kumar, S.; Balaya, P.; Improved ionic conductivity in NASICON-type Sr^{2+} doped $LiZr_2(PO_4)_3$. *Solid State Ionics,* 2016, 296, 1–6.

11

Smart Materials in Energy Storage Devices: Solar Cells

Indu Sharma*, Neha Bisht, Parag R. Patil, Pravin S. Pawar, Rahul Kumar Yadav, Yong Tae Kim and Jaeyeong Heo†

Department of Materials Science and Engineering, and Optoelectronics Convergence Research Center, Chonnam National University, Gwangju, Republic of Korea

Abstract

The development of civilization in international communities causes many critical environmental and economic issues in the generation of clean energy. New strategies are required to address or resolve these issues to reach the renewable energy goals of 2030. Among those, solar cell technology is one strategy for resolving these issues. Recently, researchers introduced various smart materials to the solar cell industry. The fabrication cost can be significantly reduced by using these smart materials. In this chapter, we will be giving a brief survey of solar cell technology emphasizing mainly the material aspect. Beginning with the importance or status of solar cell technology in the energy sector, the types of solar cells and prospective smart materials used will be presented. The smart materials can be used as an absorber layer, a buffer layer, or electrode materials in the thin-film solar cell. Further, future trends and possibilities for tackling the challenges in the improvement of smart materials to make the solar cell industry very promising have been discussed in detail. Finally, we summarized key issues, cell concepts, and the state of the art of recently developed high-efficiency solar cells.

Keywords: Photovoltaics, solar cells, power conversion efficiency, smart materials, nanomaterials

11.1 Introduction

Smart materials are a class of materials that act smartly with the environment, and enable the generation of clean energy for powering an ever-developing civilization in international communities. Solar cells, i.e., also known as photovoltaic cells, are one such type of energy storage device that converts solar radiation energy into electrical energy via the photovoltaic effect. Solar energy, which is the best renewable energy source, has the capability of generating ~2×10^{13} W of electricity as sunlight covers 0.16% of Earth's surface with cells of 10% efficiency. Hence, this amount of electricity is more than enough for the overall energy demand of humans [1]. Hence, in the last few decades, photovoltaic (PV) research has been

**Corresponding author*: indusharma.86@gmail.com
†*Corresponding author*: jheo@jnu.ac.kr

Upendra Kumar and Piyush Kumar Sonkar (eds.) Smart Materials for Science and Engineering, (191–220) © 2024 Scrivener Publishing LLC

conducted for finding highly efficient and cost-effective solar cells. PV technology has significant benefits over other non-renewable energy source technologies:

- Solar cell technology is cost-effective owing to low maintenance and no fuel required for operation.
- Solar cell technologies are compact and highly modular. Hence, they are easy to handle and can be used anywhere.
- Solar cell technology is a reliable and environmentally safe technology for producing energy as long as the sun radiates heat.

The technical or economic feasibility, power conversion efficiency (PCE), durability, and production cost are the primary criteria to measure the performance of solar cells. For highly efficient solar cell fabrication, a solar cell needs absorption of a certain energy range corresponding to the solar radiation spectrum, i.e., 100 nm to 1 mm. However, most of the irradiation occurs within the range of 250 nm to 2500 nm as indicated in Figure 11.1, having maxima in the visible range of 400–700 nm for zero Air Mass (AM). Here, AM is the path length taken by the light through the atmosphere perpendicular to the shortest feasible path length. The value of AM = 1/ cos θ signifies the light power reduction while crossing the atmosphere and is captivated by the external particles like air and dust. Here, θ signifies the angle from the vertical, also known as the zenith angle. For highly efficient solar cells, maximum absorption should occur within the visible region of the solar radiation spectrum.

The motivation of the present chapter is to provide an insight into the fundamental working principles of different types of photovoltaic cells, their mechanisms, and smart materials used for the production of photovoltaic cells. Smart materials used in the photovoltaic cell have a high impact on the functioning of the photovoltaic cell as it affects the ability to absorb radiation in the photovoltaic cells. The performance of photovoltaic cells depends on various parameters such as short-circuit current density (J_{SC}), open-circuit

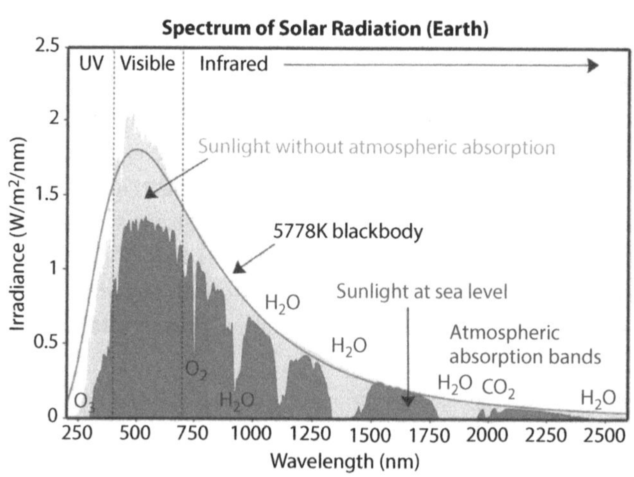

Figure 11.1 Solar radiation spectrum [2].

voltage (V_{OC}), fill factor (FF), and power conversion efficiency (PCE). Most commonly, PCE is also known as the efficiency of photovoltaic cells and is the ratio of maximum power generated by the photovoltaic cell to the incident solar radiant energy.

An overview of solar technologies is displayed in Figure 11.2. The photovoltaic effect was discovered by A. E. Becquerel in 1839 and has been known among scientists since then [4]. In 1946, R. Ohl invented silicon-based solar cells; hence, a journey of photovoltaic technology started. A selenium cell was manufactured in 1876, and came to market as a true photovoltaic cell with less than 1% efficiency. The first report on photoelectric effect was published by A. Einstein in 1904 [5]. Copper and copper oxide solar cells with an efficiency of less than 1% have been developed in 1927. Additionally, silicon solar cells have been developed by R. Ohl in 1941 [4]. Bell Laboratories introduced silicon solar cells with an enhanced efficiency of 4% in 1954, which is a benchmark experience for the photovoltaic industry [6]. Photovoltaic technology accomplished novel achievement of usage of solar cells in space in 1958 [5], and thus, the journey of enhancement in the PCE of solar cells begins. Figure 11.3 demonstrates state of the art for solar cell efficiency since 1975 reported by the National Renewable Energy

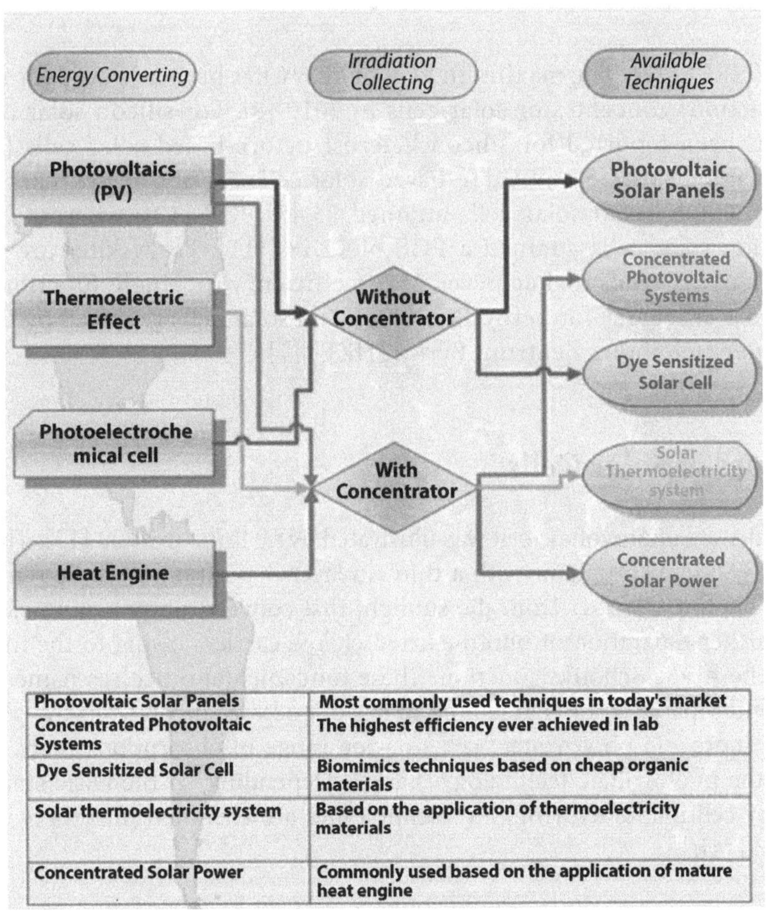

Figure 11.2 Overview of solar technologies [3].

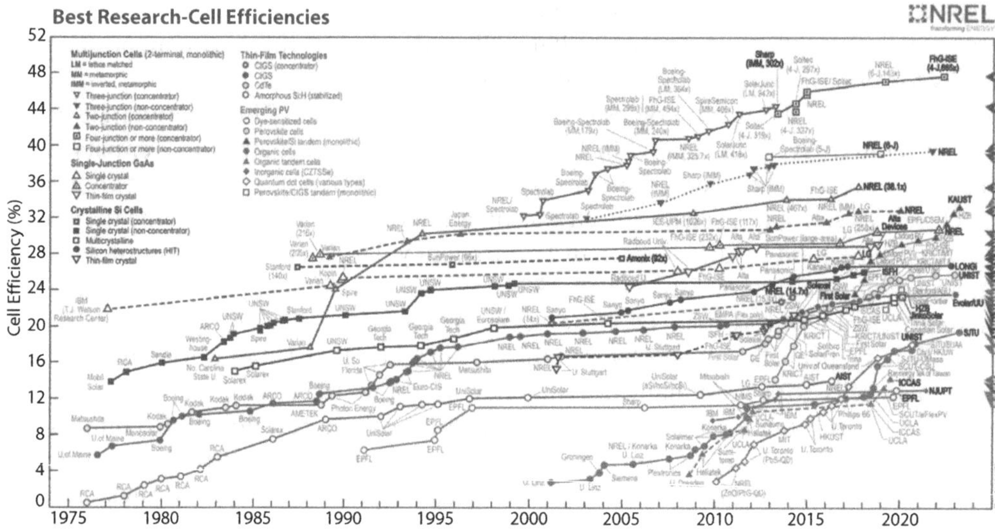

Figure 11.3 Best reported timeline of PCE of solar cells [7].

Laboratory (NREL) [7]. The maximum PCE of 47.6% has been achieved for multi-junction (six-junction) concentrator solar cells in 2019 [8]. For silicon solar cells, a PCE of 27.6% has been reported for silicon heterostructure-based solar cells [9]. Among thin-film solar cells, CIGS and CdTe-based solar cells are above the market in terms of efficiency. CIGS-based solar cells attained 23.4% PCE [10], whereas CdTe-based single-junction solar cells attained a PCE of 22.1% [11]. Perovskites are an upcoming new class of materials that achieved 25.7% efficiency for single-junction solar cells [12] and 32.5% efficiency for perovskite/Si tandem solar cells reported by the German research center Helmholtz Zentrum Berlin (HZB) [7].

11.2 Types of Solar Cells

The first solid-state photovoltaic cell was illustrated by C. Fritts in 1883 [13]. This solar cell consisted of a Se semiconductor with a thin Au layer deposition on it. Se semiconductor acted as the photon absorber from the sunlight that converts photons into electron–hole pairs, and further separation of photo-excited charge carriers owing to the internal electric field in the Au/Se Schottky junction. These fundamental processes named solar light absorption and separation of charge are the basic processes involved in the photoelectric effect in the inorganic photovoltaic cell. A wide range of absorption smart materials is available in the photovoltaic technology market. Depending on the basic smart material used for solar cell manufacturing, PV technologies are classified into three generations (Figure 11.4) [14].

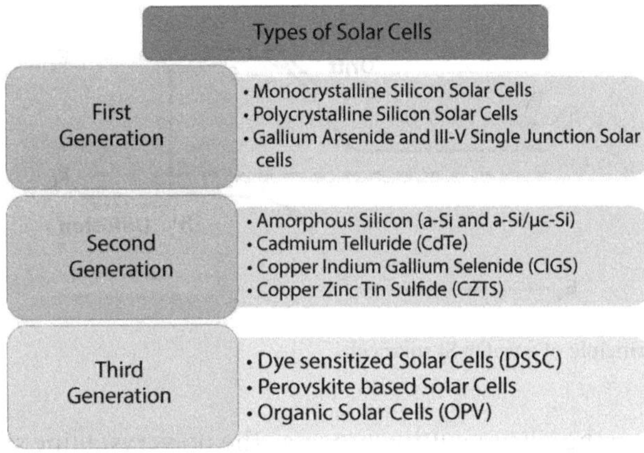

Figure 11.4 Classification of photovoltaic cells.

11.2.1 First-Generation Solar Cells

11.2.1.1 Crystalline Silicon Solar Cells

Silicon is an indirect bandgap semiconducting material exhibiting an optimum energy bandgap of 1.1 eV with a lower absorption coefficient of 100 cm^{-1}. Depending on the crystallinity of Si wafers, it is classified into three types: monocrystalline (mono c-Si) (or single crystalline) and polycrystalline (poly c-Si) silicon solar cells. Silicon solar cells make up approximately 93% of the market and are evidenced as highly efficient long-lasting solar cells. Modern solar cells, which are in high demand and available for residential use, were fabricated in Bell Laboratory. Ohl patented these solar cells in 1946 [15], and further, Chapin, Fuller, and Pearson demonstrated these cells in 1954 [16]. It consists of a single c-Si wafer where absorption of light occurs at single c-Si and charge separation occurs at the p-n junction, having a PCE of ~5%. Highly efficient silicon solar cell have an efficiency of 26% [17], and consists of thin silicon wafers also known as mono or single c-Si solar cells. Sharp Corporation of Japan initiated the production of commercial photovoltaic modules in 1963. Mono c-Si photovoltaic cells have the advantages of longevity, high efficiency, inexpensive installation cost, and non-hazardous. Mono c-Si wafers also known as single crystalline wafers comprise approximately 30% of the market and are manufactured by the Czochralski process in the silicon electronics industry. Hence, it is expensive material owing to the high purity of monocrystalline silicon. Another issue with the mono c-Si solar cell is it loses its efficiency as temperature increases above 25°C. In addition to this, mono c-Si solar cells are also fragile in nature, which makes it possible to protect solar cells/ solar panels during natural disasters like storms, etc. In comparison to this, polycrystalline silicon is less pure and less expensive. The first polycrystalline silicon photovoltaic cell was introduced in 1981. The highly efficient polycrystalline photovoltaic cell has an efficiency of 21% [17]. Polycrystalline silicon solar cells can be synthesized by a simple molding process by cooling the melted silicon to the desired crystal structure using a seed crystal. Hence, it avoids wastage of silicon, whereas the mono c-Si photovoltaic cell has the disadvantage of

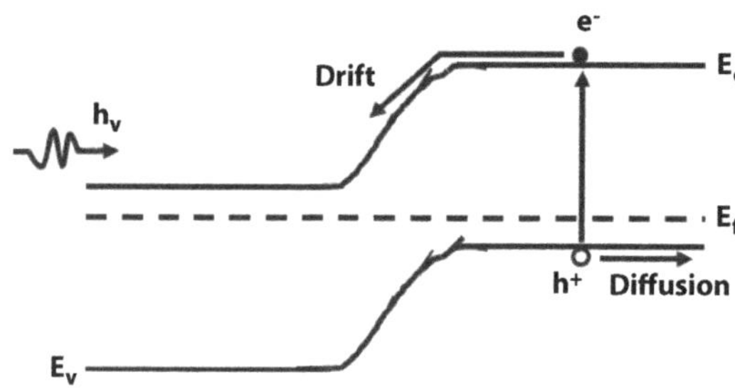

Figure 11.5 Working principle of a wafer-Si solar cell.

silicon wastage during the silicon cutting process. The polycrystalline silicon solar cell also has the advantage of working at high temperatures. It does not lose its efficiency at high temperatures as compared to the mono c-Si photovoltaic cell.

A silicon solar cell consists of two layers: the first layer is p-type and the second layer is n-type. The p-type silicon layer consists of p-doped silicon, where boron is used for doping results in creation of extra holes in the silicon lattice, whereas the n-type silicon layer consists of n-doped silicon, where phosphorus is used for doping results in the creation of extra electrons in the silicon lattice. The p-n junction in the silicon solar cell acts as charge separation. Figure 11.5 demonstrates the basic principle of solar cell operation. Absorption of photons on the surface of the material leads to charge separation that implicates charge carrier diffusion towards the depletion region. The material quality is critical in matters of carrier mobility and minority carrier lifetime for fabricating highly efficient photovoltaic cells.

11.2.1.2 Gallium Arsenide Solar Cells

Silicon is not the one and only material for crystalline solar cells. Gallium arsenide (GaAs) is another alternative semiconducting material having a direct bandgap of 1.43 eV with characteristics of suitable material for solar cell applications [18]. GaAs is a III–V direct bandgap semiconductor having a zinc blend crystal structure, high absorption coefficient, and wider bandgap that makes it a suitable material for solar cell application. GaAs has a crystal structure very much the same as that of mono c-Si, but with alternating Ga and As atoms. It has high electron mobility as compared to silicon and, hence, can be used more efficiently in photovoltaic cells as compared to that of silicon. GaAs photovoltaic cells can be produced by Ga and As layer deposition on single-crystal GaAs with defined orientation. GaAs photovoltaic cells' efficiency is higher than that of mono c-Si as well as polycrystalline Si solar cells and no loss in efficiency at high temperatures makes it possible to use these solar cells in high-temperature areas. The major disadvantage of GaAs photovoltaic cells is the expensive material and manufacturing cost of these photovoltaic cells [19].

Finally, the high manufacturing cost of first-generation solar cells leads to the craving for less expensive smart materials for photovoltaic cell applications. Some additional disadvantages of first-generation solar cells that lead to the desire to explore more materials for solar cell fabrication are as follows: (i) growing and sawing of ingots is a very laborious task, (ii) hitting a left off hole during previous photoexcitation process generates electron in a new molecule, and (iii) heat wastage at the violet and blue end of spectrum where the energy of photon is higher.

11.2.2 Second-Generation Solar Cells

Second-generation solar cells are generally recognized as thin-film solar cells (TFSCs). TFSCs are solar cells consisting of only a few micrometer-thick layers of semiconducting materials. This type of solar cell technology attracts manufacturers because of cost reduction due to the usage of less material and the low cost of manufacturing processes. Mass production of TFSCs can be easily achieved. Solar cell panel fitting on light or flexible materials is possible in this case, whereas heavy support is required for mounting first-generation solar cell panels on the rooftop. Four basic types of TFSCs are amorphous silicon (a-Si), cadmium telluride (CdTe), copper indium gallium diselenide (CIGS), and copper zinc tin sulfide (CZTS). In 2009, TFSCs accounted for approximately 16.8% of the sold panels. a-Si photovoltaic cell has an efficiency of 11.8%, CdTe photovoltaic cells have shown 21.4% efficiency, and the highest efficiency of 21.6% has been shown by CIGS photovoltaic cells. The main fundamental difference between the first-generation and the second-generation solar cell is solar cell design. These TFSCs consist of a few layers of appropriate thickness named the window layer, the absorber layer, and the buffer layer. A thin absorber layer captures a significant amount of solar radiation and efficiently converts that solar radiation into electricity at a lower processing cost [20]. TFSCs can have two types of cell configuration according to the deposition sequence of these layers: superstrate-type and substrate-type configuration (Figure 11.6).

TFSCs can be made either in substrate configuration (Figure 11.6a), where light illumination occurs in from the film side, or in superstrate configuration (Figure 11.6b), where light is illuminated on a transparent substrate like ITO- and FTO-coated glass, and this transparent substrate acts as both support for the growth of layers and a transparent window.

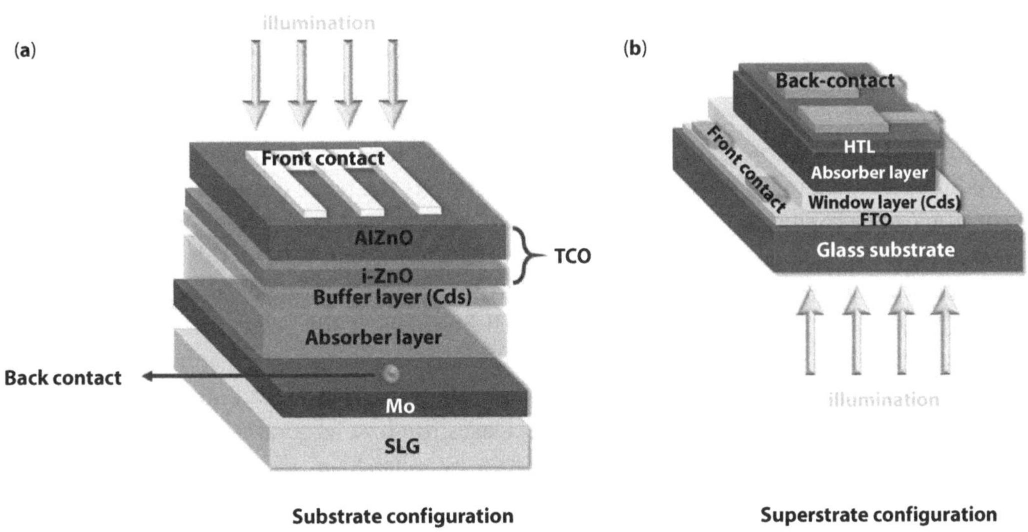

Figure 11.6 Superstrate-type and substrate-type configuration of TFSCs.

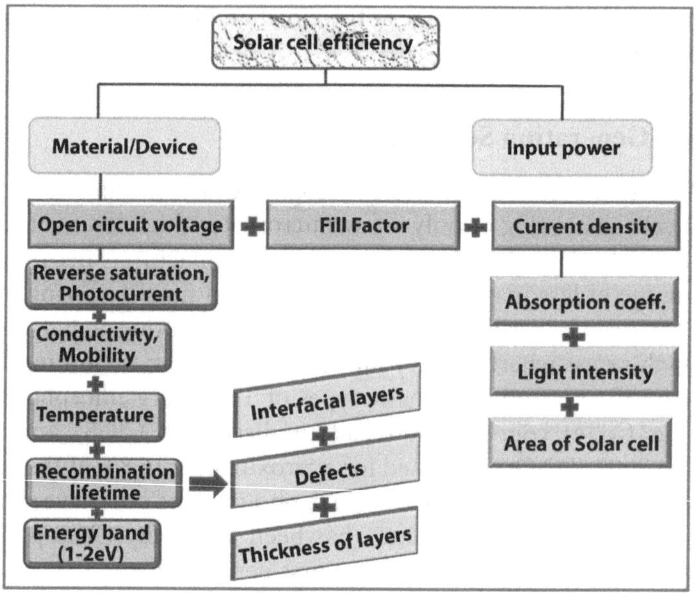

Figure 11.7 The schematic diagram indicating the influence of various parameters on thin-film solar cell efficiency. Reproduced with permission from Science Elsevier 2019 [23].

In substrate configuration, the substrate acts as support for the deposited layers but is not transparent in nature. Hence, metal sheets such as stainless steel or metalized polyimide foils are the main substrates used for this type of configuration in TFSCs. Typically, CIGS solar cells are constructed in substrate configuration, whereas CdTe solar cells are constructed in superstrate configuration [21, 22]. The photovoltaic device parameters such as V_{OC}, J_{SC}, and FF are the measures for TFSC efficiency as indicated in Figure 11.7.

11.2.2.1 Amorphous Silicon (a-Si) Solar Cells

Amorphous silicon (a-Si) solar cell devices are accustomed in small electronic devices like calculators. a-Si solar cells are manufactured by thin-layer deposition of silicon on substrates like metal, glass, or plastic. Hence, fabrication of flexible solar cells is possible in this case, which is not possible in first-generation silicon solar cells where silicon crystal growth is used for solar cell manufacturing. The p-n junction is the basic design for both first- and second-generation silicon photovoltaic cells. The difference between an a-Si solar cell and a c-Si solar cell is that an a-Si solar cell has a direct bandgap (1.7–1.9 eV) and is less efficient per unit area (~12%) [17], whereas a first-generation solar cell has indirect bandgap and highly efficient in nature. An a-Si solar cell has a short life span in comparison with c-Si solar cells. In 1976, Carlson and Wronski reported hydrogenated amorphous silicon (a-Si:H) solar cells [24]. Thereafter, research on a-Si:H solar cells and large-area modules caught more attention. In 1986, Arco Solar releases the first large-area commercial a-Si:H solar module. By now, these modules have been established in the optoelectronic device market. Japanese companies like Sharp, Kaneka, and MHI started the mass production of a-Si solar cell modules on glass. The main reasons for using a-Si for TFSCs are as follows:

a-Si is a non-hazardous and abundantly available material on earth. Fabrication reproducibility and control of the doping effect are the additional benefits of using a-Si in TFSCs. Even though the efficiency of a-Si-based TFSCs is not as high as that of other TFSCs, the self-annealing effect and low-temperature coefficient of a-Si (-0.2%/°C) ensures this material for use at high operating temperature. As illustrated in Figure 11.8, the hydrogenated amorphous silicon (a-Si:H) solar cell can exist in PIN (superstrate) and NIP (substrate) configurations.

High V_{OC} can be achieved by these TFSCs; however, insufficient absorption of solar spectrum leads to limit photocurrent and, hence, efficiency of these photovoltaic cells. The new way to enhance PCE and stability is introduction of amorphous (a-Si:H) and microcrystalline silicon (μc-Si:H) tandem photovoltaic cells. Here, μc-Si:H has an absorption spectrum resembling c-Si. First, μc-Si:H solar cells have been demonstrated in 1994 [26]. μc-Si:H consists of coalescence of small crystalline grains and a-Si:H. During manufacturing of tandem solar cells, the thickness of the μc-Si:H layer is a multiple of a-Si:H, and the deposition rate of both layers becomes a key issue. The best PCE of these solar cells has been analyzed at the transition from microcrystalline to amorphous growth and counts on the substrate surface morphology. Cautious microstructure control during growth is the critical step to getting a highly efficient solar cell. The dilution profiling method is the emerging technique to regulate these microstructures, where either dilution of the silane by hydrogen is continuously changed during the growth of the μc-Si:H active layer for getting a constant crystalline content along the growth axis [16, 37–39] or dilution of hydrogen is performed for stability of the cell against light-induced defects [27]. The light trapping optimization of PIN solar cells strongly depends on the roughness of the TCO layer (ZnO or SnO_2). Hence, TCO improvement and cost-effective fabrication of solar cell requires a lot of research exposure toward the solution of this issue.

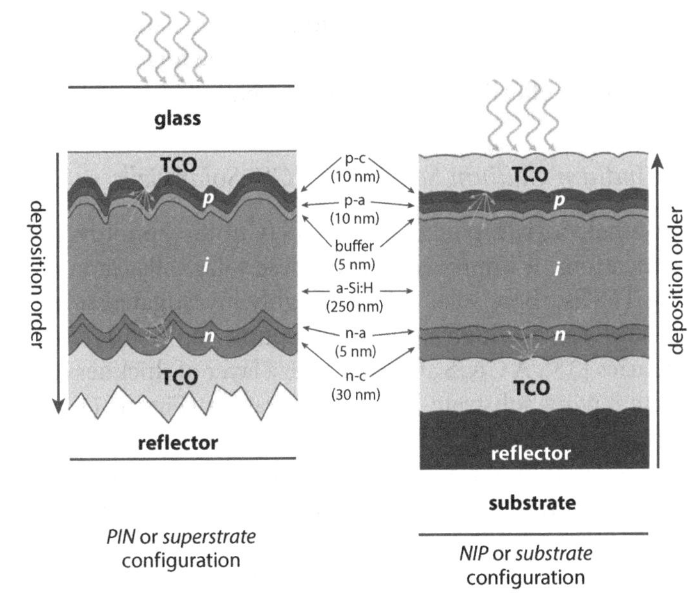

Figure 11.8 Amorphous silicon (a-Si:H) solar cell in PIN (superstrate) and NIP (substrate) configuration. Reproduced with permission from Science Elsevier 2017 [25].

11.2.2.2 *Cadmium Telluride (CdTe) Solar Cells*

The number one producer and seller of photovoltaic cell modules in the world use fabricating cadmium telluride (CdTe) absorber-based photovoltaic cells. These photovoltaic cells are able to reach a PCE of ~20% in the laboratory with the highest achieved efficiency of 22.1% till now. The commercial module PCE of CdTe photovoltaic cells has been disclosed to be 18%. CdTe has an optical bandgap of 1.45 eV with an absorption coefficient of 5×10^{15} cm^{-1}, which makes it suitable for the photon distribution in the solar radiation spectrum for solar cell applications. A CdTe-based photovoltaic cell is a promising material for affordable and flexible photovoltaic cell fabrication as materials used in the fabrication process are non-expensive and easily available. The superstrate-type configuration has been used for CdTe solar cell fabrication where a 2-μm-thick layer of p-type CdTe act as an absorber layer (Figure 11.6b). A CdTe photovoltaic cell consists of a p-n heterojunction, where CdTe performs as a p-type absorber layer and cadmium sulfide (CdS) performs as an n-type buffer layer. CdS is the most effective n-type heterojunction partner layer or window layer used by many researchers. The miscibility of CdTe/CdS and the formation of the $CdS_{1-x}Te_x$ interfacial layer during cell fabrication are the key reasons for the enhancement in the PCE of these solar cells [28]. $CdCl_2$ is used as post-deposition heat treatment of the CdTe, which causes further enhancement in the efficiency. This enhanced efficiency is attributed to recrystallization and grain growth of the CdTe layer and defect passivation/charge carrier lifetime improvement of CdTe [29]. A thin layer of <100 nm is preferred for CdS deposition as this also leads to recombination, which leads to reduced photocurrent because of the poor properties of CdS as its bandgap falls in the wavelength region of below 510 nm. Cu/Au, ZnTe, and doped graphite are used as back contact materials during solar cell fabrication [30, 31]. The back-contact treatment by using surface modification solutions like Br_2/methanol or nitric phosphoric acid mixture is the critical step [31, 32]. This step accelerates the Te content in the CdTe surface to make it Te-rich, which is further responsible for surface conductivity enhancement and tunneling junction formation [32]. During the back contact formation, mostly copper deposition accelerates the formation of a Cu_2Te on CdTe, which is also critical for the effective back contact formation.

11.2.2.3 *Copper Indium Gallium Selenide (CIGS) Solar Cells*

In the 1950s, ternary chalcopyrite-type absorber CIS ($CuInSe_2$) photovoltaic cells have been reported. The modifications or improvements of these solar cells lead to CIGS [Cu(In, Ga) Se_2] TFSCs. Among TFSCs, these solar cells are highly investigated solar cells. CIGS has a tunable bandgap (1.0–1.12 eV), high optical absorption (up to 10^5 cm^{-1}), and somewhat benign grain boundaries [33]. A CIGS absorber with a layer of thickness of 2–2.5 μm deposited on Mo substrate having substrate-type configuration (Figure 11.4a) by vacuum and non-vacuum techniques [34] and CIGS absorber-based photovoltaic cells are a favorable candidate for future-generation power-efficient photovoltaic cells having an efficiency similar to that of crystalline silicon-based photovoltaic cells [34]. Among second-generation photovoltaic cells, CIGS-based photovoltaic cells are non-toxic as compared to CdTe-based solar cells. Easy structural modification of crystalline phase and electronic structure are responsible to enhance the photovoltaic cell efficiency by regulating functionalities of the CIGS absorber layer [35–37]. Nevertheless, CIGS photovoltaic cells are not as extensively commercialized as that of crystalline silicon photovoltaic cells owing to the lower efficiency of CIGS solar modules as compared to efficiency at the lab scale [38]. CIGS solar cell works on the principle of generation of electron–hole pairs after transmitting light from the n-type CdS buffer layer to the CIGS absorber layer. The electrons move away from the p-type CIGS absorber to the n-type CdS buffer layer due to the built-in potential across the interface and collected by the n-type electrode. Likewise, holes move away from the n-type CdS layer to the p-type CIGS layer and collected by a p-type electrode. The back surface field (BSF) has also been generated by a Ga gradient, which leads to the electron reflection towards the p-n junction and is collected by an n-type electrode. This BSF is responsible for the minority carrier recombination declination at the CIGS/Mo interface.

11.2.2.4 Copper Zinc Tin Sulfide (CZTS) Solar Cells

The scarcity of use of toxic metals like indium and gallium in CIGS solar cell motivates researchers to explore more environment-friendly materials as absorber layer in the solar cell. As a result, inexpensive and non-toxic CZTS-based solar cells come into the picture in 1988 [39]. CZTS has two equilibrium states named kesterite (Zn-rich) and stannite (Zn-poor) depending on crystalline structures where the kesterite structure is more thermodynamically stable [40, 41]. The champion efficiency of CZTS solar cells has been recorded to be 12.6% [42]. The direct bandgap, high optical absorption (10^4 cm^{-1}), and optical bandgap of 1.4–1.5 eV make CZTS a promising material for use as an absorber layer in the photovoltaic cell [43]. Owing to these properties, CZTS enables 32.2% power conversion efficiency with a photon energy of 1.1–1.6 eV as depicted by the Shockley–Queisser (SQ) limit [42, 44]. IBM and Tokyo Ohka Kogyo (TOK) jointly achieved the world's highest recorded efficiency and has been verified by Newport Corporation [45–47].

Finally, the efficiencies of TFSCs are inferior to silicon (wafer-based) photovoltaic cells. The enhanced noxiousness owing to hazardous materials like Cd and In, and the lack of broad radiation spectral absorption range and, hence, high carrier motilities lead to exploring more advanced smart materials for solar cell applications. Recently, new emerging earth abundant smart materials such as SnS, SnSe, Sb_2S_3, and Sb_2Se_3 have been explored for TFSC applications [48–51]. Various techniques such as grading in thin-film absorber layer has been explored to achieve the high efficiency of TFSCs [52].

11.2.3 Third-Generation Solar Cells

The third-generation solar cells are solar cells made from a diversity of innovative materials like conducting polymers, quantum dots, organic dyes, perovskites, etc. The extravagance of these first-generation photovoltaic cells and the toxicity of materials in second-generation photovoltaic cells make researchers explore novel smart materials for photovoltaic applications. Most of the research work related to these materials is being performed in laboratories. In this chapter, dye-sensitized solar cells (DSSC), perovskite-based solar cells, and organic solar cells (OPV) are mainly discussed in detail.

11.2.3.1 Dye-Sensitized Solar Cells (DSSCs)

In third-generation solar cells, dye-sensitized solar cells (DSSC) attracted attention because of their advantages such as cheap manufacturing cost, environment-friendly, stability, and operational capability in low light. In 1991, Brian O'Regan and Michael Gratzel fabricated a thin-film device with TiO_2 nanoparticles coated with charge transfer dye on a film 10 μm thick and high surface area and achieved 7.1% light conversion efficiency in the simulated light source. This was the first DSSC [53]. The dye-sensitized solar cell is a thin film-based photoelectrochemical reaction system (Figure 11.9). DSSCs rely on four components for the operation, which are semiconductor electrode (photoanode), dye, electrolyte, and counter electrode. Herein, when light is incident on the photoanode, the dye reaches excitation and introduces the electrons in the conduction band of the photoanode. This electron reaches the external load by diffusing through the working electrode and oxidation takes place at the dye. At the same time, the electrolyte that acts as a mediator in the reaction donates the electron, which takes the place of previously injected electrons, and by the recombination, a reduction reaction takes place in the dye. The oxidized electrolyte ions get their electrons back through the counter electrode who has a continuous supply of electrons from the external load. This whole process takes place in milliseconds [54]. This entire system is inspired by the naturally occurring phenomenon of photosynthesis in plants, also known as Z-scheme [55].

In conventional solar cells, silicon plays two roles, supplying photoelectrons and providing an electric field to generate current by charge separation. However, in the case of DSSCs, semiconductor electrodes only transport charge while the dye is the source for photoelectrons. The theoretical calculations reveal that DSSCs can achieve around 20%

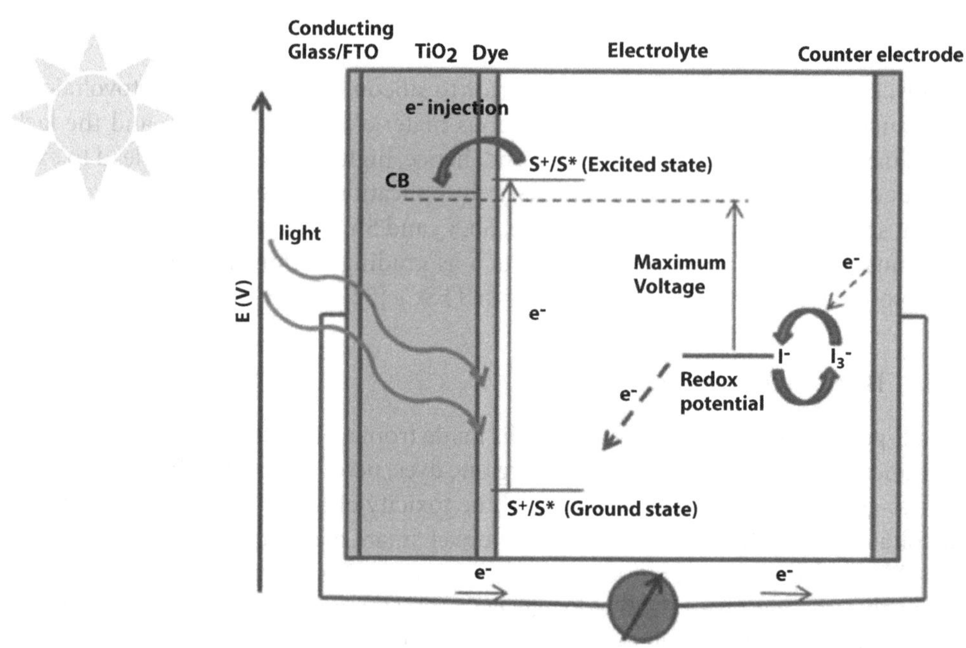

Figure 11.9 Schematic diagram representing the working mechanism of a DSSC solar cell. Reproduced with permission from Springer Nature 2018 [56].

of efficiency, while the highest reposted efficiency achieved by DSSC is around 11% [56]. Types of DSSC are based on components and dyes used in making it, mainly dye, a mediator electrolyte solution, a counter electrode, and a semiconducting working electrode. The most commonly used semiconducting material and electrolyte for DSSC are TiO_2 and iodide-triiodide(I^-/I^-_3), respectively. Semiconductors with a bandgap in the range of 1–2 eV are supposed to be the best photoanodes, but they suffer from corrosion in contact with electrolytes. Therefore, researchers choose materials like TiO_2 and ZnO for this work due to high bandgap values. Still, TiO_2 is better than ZnO due to its properties like highly stable form and non-toxic nature. TiO_2 has a bandgap of 3.2 eV, suitable to exhibit UV-sensitive photocatalytic activity in the 380–480 nm wavelength range. Along with all this, TiO_2 has high carrier mobility, low dielectric constant, and intense hydroxylation nature, which is beneficial for dye absorption [57]. Moreover, TiO_2 quickly transfers injected electrons to ZnO due to the density of states and has a high refractive index that helps to absorb more light [58, 59]. One more fact that affects the dye absorption efficiency of TiO_2 is the morphology. The surface area of porous structure or surface roughness of material increases and leads to absorb more and more dye. This factor improves photoelectron generation [55]; therefore, nanostructures like nano-porous film, nano-leaves, nano-flowers, nano-cones, nano-rods, or a mixture of all these are preferred to be at the active site of photoanode.

Dyes function as the working horse of the DSSC system. After photoexcitation, dye releases photoelectrons to the external load through the semiconducting electrode. The efficiency of the device can affect a lot by the type of dye being sensitized on the surface of the photoanode. Ruthenium-based dyes are in use because of the initiative of the DSSC, which provided really good efficiency because of the compatibility of this dye with TiO_2. Today, the market is full of different kinds of dyes like ruthenium complex dyes, metal-free organic dyes, natural dyes, and porphyrin dyes. To be highly efficient, dyes should have some of the following inherent properties:

- High adsorption coefficient
- Sensitive for the broad visible spectrum
- Strong presence of anchoring groups (such as hydroxyl and methoxy protons)
- Rapid regeneration capability
- Stability at low and high temperature

Along with this, it is better to have dye that is non-hazardous, low-cost, and easy to synthesize, but all of them have some advantages and disadvantages.

11.2.3.1.1 Dyes

11.2.3.1.1.1 Ruthenium Dye

While discussing the pros and cons of dyes, Ruthenium dyes are known for their high efficiency. Polynuclear bipyridyl Ru dye, carboxylate polypyridyl Ru dye, and phosphonate Ru dye are some types of ruthenium-based dyes. When talking about their advantages, Ru-based dyes exhibit fast electron transfer, have a good anchoring with TiO_2, remain in stable oxidation states, and have a high solubility in all solvents. On the contrary, the decline in efficiency occurs after a long operational time due to desorption of dye molecules, which is caused by the continuous penetration of H_2O molecules from electrolytes. Moreover, ruthenium dyes are costly, contain rare and heavy metal elements, and are toxic [60].

11.2.3.1.1.2 Metal-Free Organic Dye

Nowadays, many other dyes have been investigated to replace Ru-complex dye. Metal-free dyes have the capability to replace Ru-based dyes as they have low toxicity and high structural flexibility, are environmental-friendly, have facile synthesis, and are highly molar coefficient. Metal-free sensitized indolines and coumarins are good examples of metal-free dyes available in the market. This type of dye has three important structural parts: donor, linker, and acceptor (D-π-A, respectively), where the linker group (π) forms a bridge between electron donor (D) and electron acceptor groups [61]. Horiuchi *et al.* reported DSSCs with an indoline sensitizer, which showed an efficiency of 6.1% compared to N3 Ru-based dye, which has 6.3% of efficiency in the same experimental condition [62]. However, this dye suffers from long-term stability issues; chemicals used in the dye are rare and costly, and the purity of chemicals affects efficiency a lot. In efforts to improve the performance of the dye, a variety of chromophores can be used, which improves the electron generation and injection capability of the dye.

11.2.3.1.1.3 Natural Dye

Recently, natural sensitizers show up as potential candidates with attractive features. Normally, this kind of dye is prepared from plants, flowers, leaves, and fruits, and some dyes are extracted from the roots of the plants, which makes them non-toxic and low-cost, and easy to synthesize. Here, plant pigments act as photo absorbers in the visible light region. Anthocyanin, chlorophyll, and betalain are commonly known dyes that can be extracted from plants. Anthocyanin is a pigment that is a color provider for plants, flowers, leaves, etc. and chlorophyll is present in every plant where it plays a role in photosynthesis by absorbing light and converting it into energy. The carbonyl and hydroxyl groups from anthocyanin act as an anchorage to create a strong bond with TiO_2 molecules. This is an important point to be noted for improvement in efficiency. Anthocyanin withdrawn from different plants can show different performances. Dye betalain has a high oxidation ratio that helps it to interact well with electrolytes. Even natural dyes have some benefits over synthetic dyes, but still, they have low power conversion efficiency that pulls them on backfoot [54, 55].

11.2.3.1.2 Electrolytes

Photosynthesis is the inspiration of the DSSC. In the photosynthesis process, water is supplied to the electrons for the regeneration of chlorophyll enzyme in the same way electrolytes work in DSSC to have a continuous supply of electrons from the counter electrode to dye for regeneration. The selection of suitable cationic electrolytes is important to improve the output charge. The following properties of electrolytes can make them the best mediators during the reaction:

- Highly conductive with quick redox reactivity
- Chemically inert
- Highly soluble with high concentration so that a fast charge supply is achieved
- Diffusion coefficient should be high to ensure good electron transport

Overall, the dye regeneration should be faster than the recombination of the electrons; also, the oxidized dye should give preference to taking electrons from electrolytes instead of taking them from injected ones. Electrolytes like acetonitrile with dissolved iodine ion and iodine possess such properties, but they have stability issues with concentration and temperature. At low concentrations, the dye does not show good operational capabilities, while at high concentrations, corrosion occurs at the photoanode. At low temperatures, the solution of dye freezes, and the system dies at high temperatures due to the expansion of electrolyte solution leakage, which reduces the V_{OC}. Another problem electrolytes have to face on a regular basis is losing dye molecules from the electrode surface due to penetration of water molecules as discussed previously. Because of this, scientists started working on solid-state electrolytes in search of a high-efficiency and leakage-free system.

11.2.3.1.3 Counter Electrode

The counter electrode is an imperative part of the DSSC system. It acts as a catalyst for redox reactions in DSSC. Various types of counter electrodes like platinum, graphite, activated carbon, carbon black, etc. are available. Some major properties of a good counter electrode are as follows:

- High conductivity
- Good mobility of electrons
- Electrocatalytically active
- Resistive to electrolyte corrosion

The present metal-based counter electrodes are good but they have some drawbacks such as high resistance at high temperatures, expensive, electrolyte corrosion, etc. A new generation of counter electrodes is the answer to this problem. Vertically grown carbon nanotubes on glass substrates possess remarkable stability with notable photoactivity compared to old materials. Carbon nanotube's high surface area provides them extra space for good photocatalytic activity [55].

11.2.3.2 Perovskite-Based Solar Cells

Perovskites are originally discovered in the form of ternary oxides possessing the general formula of ABO_3 ($CaTiO_3$) in 1839 by the Russian scientist Gustav Rose. Based on this formula, metal halide-based perovskites are generated, which exist in the form of ABX_3, where A denotes monovalent organic/inorganic cation ($CH_3NH_3^+$, $CH_5N_2^+$, Cs^+, Rb^+), B denotes divalent metal cation (Pb^+, Sn^+, Ge^+), and X denotes halide anion (Cl^-, Br^-, I^-). The perovskite structure consists of a network of corner-sharing BX6 octahedra. In the last two decades, metal halide perovskite-based solar cells gained tremendous attention from scientists due to their tunable optical bandgap by controlling the chemical composition and exhibiting high absorption coefficient, ambipolar charge transport, low charge recombination rate, high carrier mobility, weak exciton binding energy, and long carrier diffusion lengths [63]. Besides having their major application in photovoltaic devices, the metal halide perovskites also show extensive application in different areas such as photodiodes, LEDs, thermoelectric, lasers, X-ray detectors, and memory devices [64]. Basically, metal halide perovskites can be synthesized by co-precipitation, hot injection, physical deposition, and chemical deposition methods. Therefore, for solar cell application, the most widely utilized deposition (chemical and physical) methods include spin coating, inkjet printing, blade coating, dip coating, thermal evaporation, atomic layer, and chemical vapor deposition [65]. Among them, the spin coating technique shows superior selection over other techniques due to low processing temperature, no requirement of a high vacuum system, low-cost solution processing, and controlled film thickness. These fascinating attributes make perovskite material a potential candidate in solar cell applications. Interestingly, the tunable properties of metal halide perovskite depend on solvent engineering, synthesis, and deposition methods [66, 67]. A variety of parameters such as usage of different types of solvent and antisolvents, deposition time, antisolvent volume, temperature, dipping, additives, spin-coating parameters, surrounding atmosphere, and thickness of deposited layers influence the PCE of metal halide perovskite solar cells [68]. The antisolvent treatment during the spin coating method facilitates solvent removal and initiates perovskite film crystallization. During the film formation step, the perovskite material behavior can be analyzed by using *in situ* and *ex situ* techniques that reveal complex and composition-dependent crystal dynamics of the materials due to different crystallization routes taking place. Perovskite solar cells can be fabricated in both superstrate and substrate configurations. The general structure of a perovskite photovoltaic cell includes an electron transport layer (ETL) protected with an absorber layer of halide perovskite (sometimes called an active layer or material) and a hole transport layer (HTL). The perovskite solar cell structure can be fabricated in a number of ways such as liquid electrolyte sensitized, meso-superstructure, mesoscopic, and planar type as shown in Figure 11.10a–d [69]. Basically, two types of device architectures are available, direct (n-i-p) and inverted (p-i-n), where direct is represented as glass/transparent conductive oxide (TCO)/ETL/perovskite/HTL/metal electrode and inverted is represented as glass/transparent conductive oxide (TCO)/HTL/perovskite/ETL/metal electrode (Figure 11.10e and f).

Pb-based perovskites exhibit a direct bandgap of 1.48–1.62 eV. Recent reports have shown that Pb-based perovskites ($FAPbI_3$) solar cells achieved a maximum PCE up to 25.6% [7, 70]. The Pb-based perovskite layer can be deposited in two ways: one-step deposition and two-step deposition. For example, in the one-step method, $CH_3NH_3PbI_3$ (methylammonium lead iodide) perovskite thin film can be directly deposited using a mixed solution of lead iodide (PbI_2) and methylammonium iodide (MAI) prepared in an organic solvent. In the two-step method, two separate solutions of PbI_2 and MAI are prepared using solvents. Firstly, PbI_2 is deposited followed by MAI deposition to obtain a crystallized thin film of $MAPbI_3$. Despite having higher efficiency, the Pb-based perovskites lack in reversible IV characteristics, long-term stability, and high toxicity hindering the commercial production of these materials. Recently, to overcome the issues with Pb-based perovskites, researchers started exploring Sn- and Ge-based perovskites as these are less toxic and can form various types of cationic perovskite compounds. Notably, Sn and Ge perovskites exhibit similar electronic and optical properties to Pb due to the same group occupancy. The Sn-based perovskites show fascinating properties such as a narrow bandgap, lower exciton binding energy, small ionic radius, and higher charge carrier mobility. Generally, the Sn-perovskite materials possess a bandgap in the range of 1.1–1.4 eV, which is the ideal bandgap for obtaining high PCE. Interestingly, these perovskites decompose into a non-toxic entity such as SnO_2. However, the lack of understanding of basic chemistry of Sn^{2+} compounds

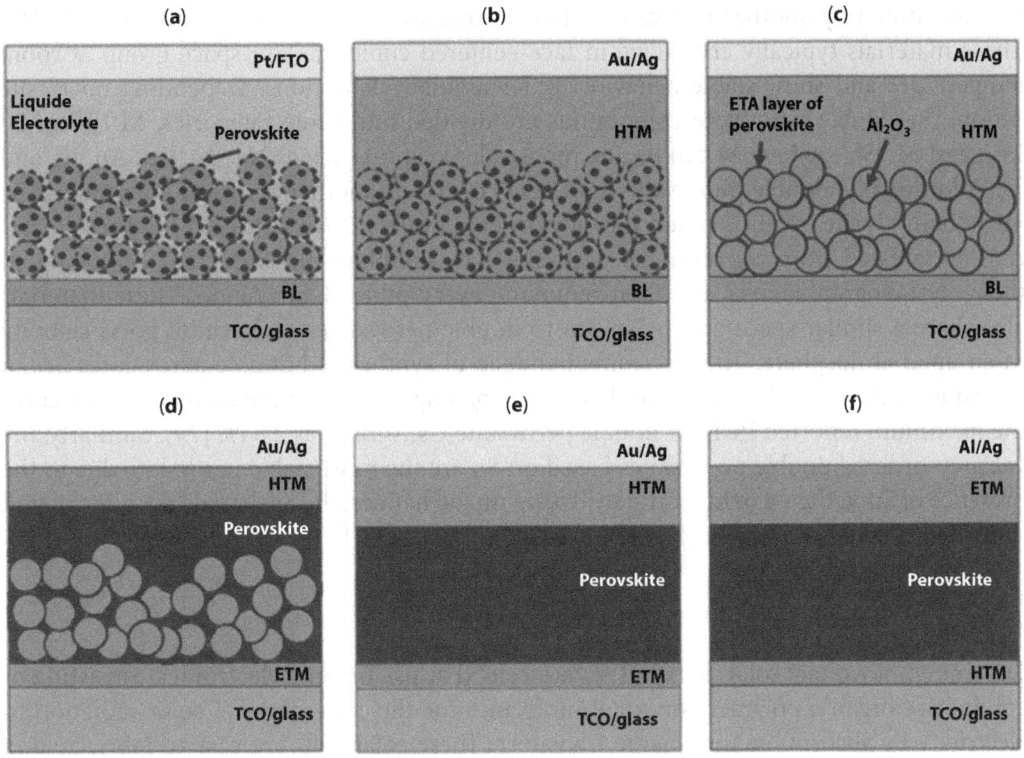

Figure 11.10 Schematic of a perovskite solar cell structure: (a) liquid electrolyte-sensitized (DSSC) structure, (b) mesoscopic structure, (c) meso-superstructure, (d) regular structure, (e) direct n-i-p architecture, and (f) inverted p-i-n architecture. Reproduced with permission from Science Elsevier 2018 [69].

hinders the fabrication of high-quality Sn films. In addition, the stability issue of Sn-based perovskites due to oxidation of Sn^{2+} to Sn^{4+} inhibits its market production. In the current scenario, the maximum power conversion efficiency obtained for $FASnI_3$ and $CsSnI_3$ is 14.6% [71] and 12.96% [72], respectively. Continuous efforts are being made by researchers to improve the stability of these perovskites as these can change the future of the solar industry. Currently, Ge-based perovskites are also being explored by many researchers as germanium belongs to the same Pb group with a smaller ionic radius and narrow bandgap. These materials show a direct bandgap of 1.4–2.2 eV according to the organic or inorganic (A) cation element. These materials are the least explored materials experimentally in the field of perovskites solar cells owing to their inferior efficiency and stability as compared to Sn-based perovskites. To date, Ge-based perovskites $MAGeI_3$ and $CsGeI_3$ have shown a maximum PCE of 0.68% and 0.11%, respectively [73, 74]. However, to overcome the stability issue of these perovskite materials, alloys of Sn and Ge perovskites have been explored, which shows enhanced power conversion efficiency and stability in comparison to pristine Ge-based perovskite. For example, solar cells fabricated using $CsSn_{0.5}Ge_{0.5}I_3$ perovskite materials showed a PCE of 7.11% [75].

Double perovskites with the general formula $A_2BB'X_6$ are another class of potential perovskite materials that are being explored to replace Pb-perovskites. Double perovskites retain a similar structure to conventional perovskites; however, every two B^{2+} site cations are substituted by another two cations having oxidation states from +1 and +3 [76–78]. These materials typically crystallize in face-centered cubic Fm_3m space group at room temperature and show stable behavior under ambient conditions. Depending on B' site cation, the double perovskite compounds are divided into three categories, M^{+}/Bi^{3+}, M^{+}/In^{3+}, and M^{+}/Sb^{3+}, where M can be any monovalent cation such as Na, Ag, Cu, Au, In, and Tl. Additionally, double perovskite materials consist of another series of compounds by introducing B site vacancy called vacancy ordered double perovskites A_2BX_6. These materials are created from conventional perovskites by doubling the ABX_3 unit cell along all their crystallographic axes and then removing every other B-site cation. These materials also share a similar space group Fm-3m to double perovskites and exhibit good stability in an open atmosphere. The electronic bandgap of synthesized double perovskites lies in the range of 2.0–3.4 eV which is not beneficial for single-junction photovoltaics. Currently, the maximum reported PCE for double perovskite $Cs_2AgBiBr_6$ is 2.84% [79]. Similarly, the vacancy-ordered double perovskites based on Sn are the most stable perovskite due to the presence of Sn in the +4 oxidation state. Using mixed halides, the Sn-based vacancy ordered double perovskites $Cs_2SnI_3Br_3$ has achieved a PCE of 3.63% [80].

11.2.3.3 Organic Solar Cells (OPV)

Organic photovoltaic solar cells (OPV) are cells that use organic electronics consisting of conductive organic polymers or small molecules for the conversion of solar radiation to electricity by absorption and charge transport. These solar cells are easy to fabricate and cost-efficient owing to the use of simple techniques for device fabrication such as printing, spray deposition, and spin coating. Large-area, lightweight, and flexible solar cell devices can be fabricated using OPV. Two types of OPV solar cells are small molecule and polymer solar cells. The PCE for polymer solar cells lies nearby 10%. Although the fabrication of OPV is cost-effective and simple, OPV technology is still under development. The theoretical limit of PCE of organic solar cells is low as compared to inorganic solar cells. Although there is no comparison of both organic solar cells and inorganic solar cells, organic solar cell leads in terms of cost, simple fabrication, and novelty. OPV consists of a minimum of four layers on the surface of transparent substrate [81] (Figure 11.11).

Generally, glass or polyethylene terephthalate (PET) are used for superstrate structures. Indium tin oxide (ITO)-coated glass substrate acts as a transparent conductive oxide (TCO) layer. Here, the TCO layer acts as both a transparent window layer and an electrode to collect photogenerated holes. The charge trapping at the charge trap centers leads to device degradation and, hence, the diffusion of anode elements into the active layer demands the protective layer [82]. Hence, poly (3,4-ethylenedioxythiophene) poly (styrene sulfonate) (PEDOT: PSS) is used as a separator between the active layer and anode, and acts as an electron-blocking layer. Additionally, it also assists photogenerated holes' transportation to the anode. Polyaniline doped with camphor sulfonic acid is used as an HTL and it is crucial in the case of bulk devices as compared to bilayer devices [83]. Both donor and acceptor touch the anode and the cathode leads to electrons and holes flowing to the respective electrode. Hence, they degrade the device's performance. Donor and acceptor organic materials comprise the active layer. Both layer-by-layer and mixed or blended organic materials can act as an active layer. In the case of nanocomposite active materials, CNTs are used as acceptors. Light passes through TCO, then through HTL, and gets absorbed by the active layer.

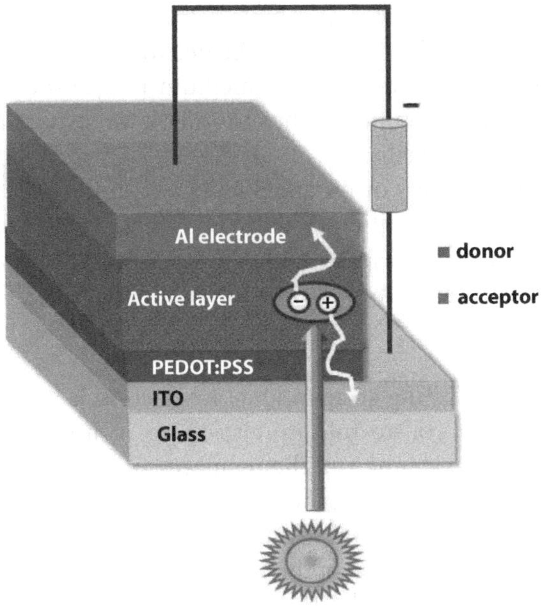

Figure 11.11 A schematic diagram for basic OPV.

Finally, the cathode (aluminum, copper) collects the charge for better electron transport. Sometimes, LiF is introduced as a protective buffer layer between the cathode and active layer. It acts as an electron transport or hole-blocking layer and prevents diffusion of cathode elements to the active layer [82]. The high deposition temperature of these buffer layers is the main issue in OPV. The stability of organic active layers as well as cathode element diffusion into the organic active layer needs much research attention [84]. The working principle of OPV consists of three steps: generation of charge carriers due to light absorption, charge separation, and transportation to the electrodes. The solar radiations get absorbed on the active layer and generation of free charge carriers or excitons occurs. In other words, the charge carrier (electron) hops from the lowest unoccupied molecular orbital (LUMO) into leaving a hole behind in the highest occupied molecular orbital (HOMO), after absorption of the photon leads to exciton generation. The exciton is bound by Coulomb attractions and the binding energy of the exciton is higher than the thermal energy [85]. Hence, exciton dissociation requires a large internal electric field. The formation of the donor–acceptor interface leads to the collection and separation of charge carriers at the interface. The metal electrode collects the charge and converts it into electricity.

11.3 Future Trends and Possibilities for Tackling the Challenges in the Improvement of Smart Materials

11.3.1 Silicon Solar Cells

Multiple stacks of single-junction cells comprise the multijunction solar cells and major research challenges associated with these solar cells are as follows:

(1) Stability enhancement of amorphous semiconducting photovoltaic cells either by improving the ordering of the network or by fabricating the protocrystalline network via developed medium-range order.
(2) Boosting the deposition rate of the absorber layer, especially in the case of μc-Si:H photovoltaic cells.
(3) Developing novel layers or modified layers via doping that are compatible with the advanced deposition systems.
(4) Increasing both V_{OC} and J_{SC} by highly developed light trapping configuration by means of different ways like internal reflectors, dielectric mirrors, and textured surfaces.

Another major challenge during solar module fabrication is an increase in the deposition area either in batchwise, roll-to-roll production, or in-line processes. In addition to this, solutions for low-cost packaging and the requirement of fast monolithic interconnect methods are the main challenges in the solar cell industry.

To overcome these challenges, several research progress works have been implemented. For instance, PECVD using H_2 dilution of the silane feedstock gas has been verified in both p–i–n and n–i–p configured solar cells. This enhanced the remarkable stable cell performance owing to the medium-range order of the protocrystalline silicon. For deposition rate enhancement, various inexpensive and fast techniques including PECVD techniques have been studied. Another successful technique is the use of elevated frequencies in the very high frequency (VHF) domain [86–88], silane depletion conditions under high pressure [89, 90], or a combination of both with showerhead gas distribution [91, 92]. Further, techniques like microwave PECVD [93] or the expanding thermal plasma CVD technique [94] and other progressive electrode designs are being examined, Furthermore, hotwire or catalytic CVD methods are the topic of new research area in this field [95–98].

11.3.2 Thin-Film Solar Cells

Further challenges in CdTe/CdS TFSCs include the following: (i) short minority carrier lifetime owing to the electron–hole pair recombination at the defect centers of absorber layers and at the CdTe/CdS junction, (ii) inadequate TCO transparency and CdS window layers, (iii) scarcity of smooth ohmic contact between back-contact and CdTe absorber layer, and (iv) possible stable way of p-type CdTe absorber layer doping. Mentioned challenges greatly influence the efficiency and cost of solar cell modules of CdTe/CdS TFSCs solar cells and can be resolved in many ways including modification of fabrication technology, choice of back contact, and structural design of the photovoltaic solar cell. Generally, the device fabrication method includes radio frequency sputtering (RF sputtering) [99], chemical bath deposition (CBD) method [100], and close-spaced sublimation (CSS) method [101] for CdS deposition, whereas CdTe thin-film formation can be performed by electrodeposition (ED) [102], CSS [103–105], and screen printing (SP) [106]. CdS deposition by RF sputtering leads to poor-quality films owing to fast layer deposition. Although CdS deposited by the CBD method results in dense and smooth film deposition, the solution waste after deposition requires waste recycling and management, which leads to an increase in the cost. Hence, modification of these fabrication techniques is the current topic of research, which includes RF sputtering in the presence of argon (Ar) and CHF_3 atmosphere, CdS bilayer deposition, low-temperature deposition of CdS and CdTe layer, dry $CdCl_2$ treatment of CdTe absorber layer, and non-toxic chlorine ambient treatment method [107]. Further, ohmic or tunneling back contact is another requirement for fabricating highly efficient solar cells. Nitric-phosphoric (NP) and nitric-acetic (NA) acid treatments are emerging etching methods to enhance solar cell efficiency by forming ohmic contact via the formation of a diffusion layer of Cu_xTe alloy during annealing after copper deposition [107]. Furthermore, modification of CdS/CdTe solar cell configuration also leads to improvement of solar cell efficiency. Nanostructures and tandem structures are the two routes for obtaining favorable formation enthalpy and good chemical stability of the CdTe/CdS interface, which leads to enhancement in solar cell efficiency. A CdS nanopillar has been used as a window layer as it enhances transparency as well as light absorption in the visible range due to the quantum confinement effect [108]. The base tandem solar cell has been discovered in the 1980s and intensive research has been started after 2004 [109–112]. The increment in solar cell efficiency is owing to the absorption of a broad part of the solar radiation spectrum [107].

The crucial concern regarding CIGS solar cells is that CIGS solar cell module efficiency is lower than that of a CIGS single solar cell. Post-deposition treatment with alkali elements, more gallium incorporation in the absorber layer, and fabrication of tandem solar cells having multijunction with perovskites techniques have been used for enhancing the module efficiency of CIGS solar cells [38]. Stability is another issue related to CIGS solar cell. CIGS solar panels are heavier than c-Si solar cells since heavy glass has been used during CIGS solar cell fabrication. Exposure to substrate in an open-air environment while transferring the device from a vacuum system to a non-vacuum system leads to contamination. Moisture trapping in the back contact Mo causes Mo degradation. Stability issues and contamination can be improved by performing device fabrication steps in a vacuum chamber. The long processing time for fabrication of a CIGS solar cell is the main hurdle for industrial-level production of the solar module. Processing time can be reduced by using high-speed deposition and a thin layer of CIGS absorber material.

In the case of CZTS TFSCs, the main challenge towards enhancing the power conversion efficiency is an exploration of low-cost methodology. CZTS solar cells have a tendency to achieve an efficiency of 22.6% by controlling the absorber material composition. Loss of charge carriers by recombination during collection and transportation of charge at the electrode is the main reason for efficiency degradation. Absorber/buffer and absorber/metallic contact interfacial layer directly affects these parameters. The recombination time for the CZTS/CdS interface is very low due to Cd atom diffusion into the absorber layer; hence, the CdS buffer layer is not perfect for this thin-film solar cell. Lattice mismatch problem occurs due to spike type band offset having 0.4 eV energy at the CdS/CZTS interface [113]. Hence, the exploration of a suitable buffer layer for CZTS/CdS is the main requirement for high PCE of CZTS solar cells. The best solution for this problem is the use of carbon nanotube (CNT) and graphene as the buffer layer. In addition to this, the $Mo(S_2/Se_2)$ layer can be used as a buffer layer. However, the maximum series resistance of this material is another hurdle in the enhancement of solar cell efficiency. The best solution to get rid of this is to avoid the annealing process after device fabrication and removal of Mo metal by using high functional materials such as Pt (5.64), NiO (5.2), Ni (5.22), and Pd (5.22) [114]. All the charge carriers can be collected without energy loss for high-work function electrodes due to the presence of these highly functional materials, which leads to efficiency enhancement. Hence, charge carrier extraction probability is higher owing to the formation of ohmic contact.

Second-generation solar cells or TFSCs are cost-effectively manufactured and highly efficient in nature. Most of the materials (indium, cadmium) used to fabricate these cells are either rare, more expensive, or harmful. Several more years are required to resolve this problem.

11.3.3 New Emerging Solar Cells

Overall, perovskite solar cells are the future of the PV industry, which can solve the future energy crisis of power sectors. The long-term stability issue due to heat, moisture, oxygen, and toxicity of lead obstructs the future commercialization of perovskite materials. The toxicity issue in Pb-based perovskites can be solved by using an alternative metal cation in place of lead. The perovskite layer instability due to the moisture and oxygen present in the atmosphere can be overcome by using encapsulation techniques. The perovskite solar cell can be encapsulated using polymeric layers such as ethylene vinyl acetate (EVA), Surlyn, polyurethane (PU), polyolefin, polyethylene glycol (PEG), Teflon, edge seal, etc. via spin coating, UV curing, and atomic layer deposition (ALD) methods [115]. The low-cost and low-temperature encapsulation process minimizes the penetration of moisture and oxygen inside the solar cell and maintains the performance of the perovskite solar cell. The heat and temperature instability can be avoided by adding various additives during the deposition of the perovskite layer. For instance, the addition of ionic liquid in the perovskite materials prevents the ion migration under the illumination of light and heat and maintains the device's performance and long-term stability [116].

However, OPV solar cells show significant PCE in a cost-effective manner, and the commercialization of these solar cells has not been achieved yet. For achieving high efficiency, the V_{OC} of the OPV device is easily intensified by choosing a large ΔE between the D_{LUMO} and A_{LUMO}. This makes it challenging to dissociate exciton efficiently and, hence, affects the resulting electricity in the outside circuit. The mobility and diffusion length of OPV solar cells are lower as compared to the inorganic solar cells [117–119]. Hence, the thickness of the active layer in OPV is limited to tens of nanometers only, where light absorption is intrinsically high [120]. Several approaches to designing OPV materials and devices have been introduced to overcome these inadequacies such as improvement of the crystal structure for enhancing the electrical conductivity of active materials [121], the creation of a large donor–acceptor interface for promotion of more excitons' dissociation, and more photogenerated carrier collection.

11.4 Summary

The decline of non-renewable energy sources leads to the requirement for the usage of renewable energy sources in our daily life. Solar energy is abundant in all energy and easy to utilize for energy storage. Nowadays, solar cells are the much-investigated energy storage devices, which convert solar energy into electricity. The bandgap and optical absorption properties of the material highly impact the solar cell performance. In the present chapter, we discussed the smart materials used for solar cell manufacturing. Further, we have described various types of solar cells in detail including working principles, benefits, and drawbacks. The insight into solar cell mechanism and fundamentals was studied thoroughly.

State-of-the-art solar cells have efficiencies presently reaching ~26% at the minimodule level. Various challenges in the production and development of solar cell devices are pointed out. Future perspectives to develop the quality of materials and, hence, the performance of photovoltaic cells are discussed. Solar cell technology is considered a major technology to explore at a fast pace for use of renewable energy sources in the future.

References

1. Lewis, N. S.; Crabtree, G.; Nozik, A. J.; Wasielewski, M. R.; Alivisatos, P.; Kung, H.; Tsao, J.; Chandler, E.; Walukiewicz, W.; Spitler, M.; Ellingson, R.; Overend, R.; Mazer, J.; Gress, M.; Horwitz, J.; Ashton, C.; Herndon, B.; Shapard, L.; Nault, R. M.; *Basic Research Needs for Solar Energy Utilization. Report of the Basic Energy Sciences Workshop on Solar Energy Utilization,* April 18-21, 2005.
2. Duffie, J. A.; Beckman, W. A.; *Solar Engineering of Thermal Processes*; John Wiley & Sons, Inc.: Hoboken, NJ, USA, 2013.
3. Chu, Y.; Meisen, P.; Review and Comparison of Different Solar Energy Technologies. Global Energy Network Institute (GENI), San Diego, CA, 2011. https://www.academia.edu/35942446/Review_and_Comparison_of_Different_Solar_Energy_Technologies.
4. Joseph, S.; Raymond, F.; Ashok, S.; Solar Cell. www.britannica.com; 2015.
5. Barnett, A.; Rand, J.; Hall, R.; Bisaillon, J.; DelleDonne, E.; Feyock, B.; Ford, D.; Ingram, A.; Mauk, M.; Yaskoff, J.; Sims, P.; High Current, Thin Silicon-on-Ceramic Solar Cell. *Sol. Energy Mater. Sol. Cells,* 2001, 66, 45–50.
6. Paul, H.; Zweibel, K.; *Basic Photovoltaic Principles and Methods*; No. SERI/SP-290-1448, Ed.; Solar Energy Research Inst., Golden, CO (USA), 1982.
7. NREL. https://www.nrel.gov/pv/cell-efficiency.html.
8. Geisz, J. F.; France, R. M.; Schulte, K. L.; Steiner, M. A.; Norman, A. G.; Guthrey, H. L.; Young, M. R.; Song, T.; Moriarty, T.; Six-Junction III–V Solar Cells with 47.1% Conversion Efficiency under 143 Suns Concentration. *Nat. Energy,* 2020, 5, 326–335.
9. Slade, A.; Garboushian, V.; 27.6% Efficient Silicon Concentrator Solar Cells for Mass Production. *Tech. Dig. 15th Int. Photovolt. Sci. Eng. Conf.,* 1–2.
10. Nakamura, M.; Yamaguchi, K.; Kimoto, Y.; Yasaki, Y.; Kato, T.; Sugimoto, H.; Cd-Free Cu(In,Ga)(Se,S) 2 Thin-Film Solar Cell With Record Efficiency of 23.35%. *IEEE J. Photovolt.,* 2019, 9, 1863–1867.
11. First Solar. http://www.firstsolar.com/en-AU/-/media/First-Solar/Technical-Documents/Series- 6-Datasheets/Series-6-Datasheet.ashx.
12. Min, H.; Lee, D. Y.; Kim, J.; Kim, G.; Lee, K. S.; Kim, J.; Paik, M. J.; Kim, Y. K.; Kim, K. S.; Kim, M. G.; Shin, T. J.; Il Seok, S.; Perovskite Solar Cells with Atomically Coherent Interlayers on SnO2 Electrodes. *Nature,* 2021, 598, 444–450.
13. Fritts, C. E.; On a New Form of Selenium Cell, and Some Electrical Discoveries Made by Its Use. *Am. J. Sci.,* 1883, *s3-26,* 465–472.
14. Ranabhat, K.; Patrikeev, L.; Antal'evna-Revina, A.; Andrianov, K.; Lapshinsky, V.; Sofronova, E.; An Introduction to Solar Cell Technology. *Istraz. I Proj. Za Privredu,* 2016, 14, 481–491.
15. Ohl, R. S.; Light-Sensitive Electric Device, 1946.
16. Chapin, D. M.; Fuller, C. S.; Pearson, G. L.; A New Silicon P-n Junction Photocell for Converting Solar Radiation into Electrical Power [3]. *J. Appl. Phys.,* 1954, 25, 676–677.
17. Green, M. A.; Emery, K.; Hishikawa, Y.; Warta, W.; Dunlop, E. D.; Solar Cell Efficiency Tables (Version 48). *Prog. Photovoltaics Res. Appl.,* 2016, 24, 905–913.

18. Sze, S. M.; Irvin, J. C.; Resistivity, Mobility and Impurity Levels in GaAs, Ge, and Si at 300°K. *Solid State Electron,* 1968, 11, 599–602.
19. S Peake.; *Renewable Energy - Power for a Sustainable Future,* 4th ed.; Oxford University Press, 2018.
20. AbuShama, J. A. M.; Johnston, S.; Moriarty, T.; Teeter, G.; Ramanathan, K.; Noufi, R.; Properties of ZnO/CdS/CuInSe2 Solar Cells with Improved Performance. *Prog. Photovoltaics Res. Appl.*, 2004, 12, 39–45.
21. Yan, F.; Metacarpa, D. J.; Sundaramoorthy, R.; Fobare, D.; Haldar, P.; Evaluation of CIGS Cell Interconnection Methods. In *2013 IEEE 39th Photovoltaic Specialists Conference (PVSC)*; IEEE, 2013; pp. 2064–2067.
22. Bermudez, V.; Perez-Rodriguez, A.; Understanding the Cell-to-Module Efficiency Gap in Cu(In,Ga)(S,Se)2 Photovoltaics Scale-Up. *Nat. Energy,* 2018, 3, 466–475.
23. Pal, K.; Singh, P.; Bhaduri, A.; Thapa, K. B.; Current Challenges and Future Prospects for a Highly Efficient (<20%) Kesterite CZTS Solar Cell: A Review. *Sol. Energy Mater. Sol. Cells,* 2019, 196, 138–156.
24. Carlson, D. E.; Wronski, C. R.; Amorphous Silicon Solar Cell. *Appl. Phys. Lett.*, 1976, 28, 671–673.
25. Stuckelberger, M.; Biron, R.; Wyrsch, N.; Haug, F.-J.; Ballif, C.; Review: Progress in Solar Cells from Hydrogenated Amorphous Silicon. *Renew. Sustain. Energy Rev.*, 2017, 76, 1497–1523.
26. Meier, J.; Flückiger, R.; Keppner, H.; Shah, A.; Complete Microcrystalline p-i-n Solar Cell—Crystalline or Amorphous Cell Behavior? *Appl. Phys. Lett.*, 1994, 65, 860–862.
27. Matsui, T.; Matsuda, A.; Kondo, M.; High-Rate Plasma Process for Microcrystalline Silicon: Over 9% Efficiency Single Junction Solar Cells. *MRS Proc.*, 2004, *808*, A8.1-A8.1.
28. Ferekides, C. S.; Marinskiy, D.; Viswanathan, V.; Tetali, B.; Palekis, V.; Selvaraj, P.; Morel, D. L.; High Efficiency CSS CdTe Solar Cells. *Thin Solid Films,* 2000, *361–362*, 520–526.
29. Levi, D. H.; Moutinho, H. R.; Hasoon, F. A.; Keyes, B. M.; Ahrenkiel, R. K.; Al-Jassim, M.; Kazmerski, L. L.; Birkmire, R. W.; Micro through Nanostructure Investigations of Polycrystalline CdTe: Correlations with Processing and Electronic Structures. In *Proceedings of 1994 IEEE 1st World Conference on Photovoltaic Energy Conversion - WCPEC (A Joint Conference of PVSC, PVSEC and PSEC)*; IEEE; Vol. 1, pp. 127–131.
30. Gessert, T. A.; Sheldon, P.; Li, X.; Dunlavy, D.; Niles, D.; Sasala, R.; Albright, S.; Zadler, B.; Studies of ZnTe Back Contacts to CdS/CdTe Solar Cells. In *Conference Record of the Twenty Sixth IEEE Photovoltaic Specialists Conference - 1997*; IEEE; pp. 419–422.
31. McCandless, B. E.; Qu, Y.; Birkmire, R. W.; A Treatment to Allow Contacting CdTe with Different Conductors [Solar Cells]. In *Proceedings of 1994 IEEE 1st World Conference on Photovoltaic Energy Conversion - WCPEC (A Joint Conference of PVSC, PVSEC and PSEC)*; IEEE; Vol. 1, pp. 107–110.
32. Levi, D. H.; Woods, L. M.; Albin, D. S.; Gessert, T. A.; Niles, D. W.; Swartziender, A.; Rose, D. H.; Ahrenkiel, R. K.; Sheldon, P.; Back Contact Effects on Junction Photoluminescence in CdTe/CdS Solar Cells. In *Conference Record of the Twenty Sixth IEEE Photovoltaic Specialists Conference - 1997*; IEEE; pp. 351–354.
33. Schock, H.; Noufi, R.; CIGS-based Solar Cells for the next Millennium. *Prog. Photovoltaics Res. Appl.*, 2000, 8, 151–160.
34. Mufti, N.; Amrillah, T.; Taufiq, A.; Sunaryono; Aripriharta; Diantoro, M.; Zulhadjri; Nur, H.; Review of CIGS-Based Solar Cells Manufacturing by Structural Engineering. *Sol. Energy,* 2020, 207, 1146–1157.
35. Asaduzzaman, M.; Hasan, M.; Bahar, A. N.; An Investigation into the Effects of Band Gap and Doping Concentration on Cu(In,Ga)Se2 Solar Cell Efficiency. *Springerplus,* 2016, 5, 578.

36. Chirilă, A.; Reinhard, P.; Pianezzi, F.; Bloesch, P.; Uhl, A. R.; Fella, C.; Kranz, L.; Keller, D.; Gretener, C.; Hagendorfer, H.; Jaeger, D.; Erni, R.; Nishiwaki, S.; Buecheler, S.; Tiwari, A. N.; Potassium-Induced Surface Modification of Cu(In,Ga)Se2 Thin Films for High-Efficiency Solar Cells. *Nat. Mater.*, 2013, 12, 1107–1111.
37. Sun, Y.; Lin, S.; Li, W.; Cheng, S.; Zhang, Y.; Liu, Y.; Liu, W.; Review on Alkali Element Doping in Cu(In,Ga)Se 2 Thin Films and Solar Cells. *Engineering,* 2017, 3, 452–459.
38. Ramanujam, J.; Singh, U. P.; Copper Indium Gallium Selenide Based Solar Cells – a Review. *Energy Environ. Sci.*, 2017, 10, 1306–1319.
39. Nakazawa, K. I.; Electrical and Optical Properties of Stannite-Type Quaternary Semiconductor Thin Films, *Jpn. J. Appl. Phys.*, 1988, *27.* 2094.
40. Schäfer, W.; Nitsche, R.; Tetrahedral Quaternary Chalcogenides of the Type Cu2·II·IV·S4(Se4). *Mater. Res. Bull.*, 1974, 9, 645–654.
41. Schorr, S.; Structural Aspects of Adamantine like Multinary Chalcogenides. *Thin Solid Films* 2007, 515, 5985–5991.
42. Wang, W.; Winkler, M. T.; Gunawan, O.; Gokmen, T.; Todorov, T. K.; Zhu, Y.; Mitzi, D. B.; Device Characteristics of CZTSSe Thin-Film Solar Cells with 12.6% Efficiency. *Adv. Energy Mater.*, 2014, 4, 1301465.
43. Mitzi, D. B.; Gunawan, O.; Todorov, T. K.; Wang, K.; Guha, S.; The Path towards a High-Performance Solution-Processed Kesterite Solar Cell. *Sol. Energy Mater. Sol. Cells,* 2011, 95, 1421–1436.
44. Shockley, W.; Queisser, H. J.; Detailed Balance Limit of Efficiency of P-n Junction Solar Cells. *J. Appl. Phys.*, 1961, 32, 510–519.
45. Yao, L.; Ao, J.; Jeng, M.-J.; Bi, J.; Gao, S.; Sun, G.; He, Q.; Zhou, Z.; Sun, Y.; Chang, L.-B.; A CZTSe Solar Cell with 8.2% Power Conversion Efficiency Fabricated Using Electrodeposited Cu/Sn/Zn Precursor and a Three-Step Selenization Process at Low Se Pressure. *Sol. Energy Mater. Sol. Cells,* 2017, 159, 318–324.
46. Zoppi, G.; Forbes, I.; Miles, R. W.; Dale, P. J.; Scragg, J. J.; Peter, L. M.; Cu 2 ZnSnSe 4 Thin Film Solar Cells Produced by Selenisation of Magnetron Sputtered Precursors. *Prog. Photovoltaics Res. Appl.*, 2009, 17, 315–319.
47. Bag, S.; Gunawan, O.; Gokmen, T.; Zhu, Y.; Mitzi, D. B.; Hydrazine-Processed Ge-Substituted CZTSe Solar Cells. *Chem. Mater.*, 2012, 24, 4588–4593.
48. Cho, J. Y.; Kim, S.; Nandi, R.; Jang, J.; Yun, H.-S.; Enkhbayar, E.; Kim, J. H.; Lee, D.-K.; Chung, C.-H.; Kim, J.; Heo, J.; Achieving over 4% Efficiency for SnS/CdS Thin-Film Solar Cells by Improving the Heterojunction Interface Quality. *J. Mater. Chem. A,* 2020, 8, 20658–20665.
49. Nandi, R.; Pawar, P. S.; Neerugatti, K. E.; Cho, J. Y.; Kim, S.; Cho, S. H.; Lee, Y. S.; Heo, J.; Vapor-Transport-Deposited Orthorhombic-SnSe Thin Films: A Potential Cost-Effective Absorber Material for Solar-Cell Applications. *Sol. RRL,* 2022, 6, 2100676.
50. Pawar, P. S.; Nandi, R.; Rao Eswar Neerugatti, K.; Sharma, I.; Kumar Yadav, R.; Tae Kim, Y.; Yu Cho, J.; Heo, J.; Atomic-Layer-Deposited TiO2 and SnO2 Coupled with CdS as Double Buffer Layers for HTL-Free Sb2S3 Thin-Film Solar Cells. *Sol. Energy,* 2022, 246, 141–151.
51. Karade, V. C.; Jang, J. S.; Kumbhar, D.; Rao, M.; Pawar, P. S.; Kim, S.; Gour, K. S.; Park, J.; Heo, J.; Dongale, T. D.; Kim, J. H.; Combating Open Circuit Voltage Loss in Sb2Se3 Solar Cell with an Application of SnS as a Back Surface Field Layer. *Sol. Energy,* 2022, 233, 435–445.
52. Sharma, I.; Pawar, P. S.; Kumar Yadav, R.; Nandi, R.; Heo, J.; Review on Bandgap Engineering in Metal-Chalcogenide Absorber Layer via Grading: A Trend in Thin-Film Solar Cells. *Sol. Energy,* 2022, 246, 152–180.
53. O'Regan, B.; Grätzel, M.; A Low-Cost, High-Efficiency Solar Cell Based on Dye-Sensitized Colloidal TiO2 Films. *Nature,* 1991, 353, 737–740.

54. Jamalullail, N.; Mohamad, I. S.; Norizan, M. N.; Baharum, N. A.; Mahmed, N.; Short Review: Natural Pigments Photosensitizer for Dye-Sensitized Solar Cell (DSSC). In *2017 IEEE 15th Student Conference on Research and Development (SCOReD)*; IEEE, 2017; pp 344–349.
55. Prabavathy, N.; Shalini, S.; Balasundaraprabhu, R.; Velauthapillai, D.; Prasanna, S.; Muthukumarasamy, N.; Enhancement in the Photostability of Natural Dyes for Dye-Sensitized Solar Cell (DSSC) Applications: A Review. *Int. J. Energy Res.*, 2017, 41, 1372–1396.
56. Sharma, K.; Sharma, V.; Sharma, S. S.; Dye-Sensitized Solar Cells: Fundamentals and Current Status. *Nanoscale Res. Lett.* 2018, 13, 381.
57. Gupta, S. M.; Tripathi, M.; A Review of TiO2 Nanoparticles. *Chinese Sci. Bull.*, 2011, 56, 1639–1657.
58. Listorti, A.; O'Regan, B.; Durrant, J. R.; Electron Transfer Dynamics in Dye-Sensitized Solar Cells. *Chem. Mater.*, 2011, 23, 3381–3399.
59. Ravidhas, C.; Anitha, B.; Arivukarasan, D.; Venkatesh, R.; Christy, A. J.; Jothivenkatachalam, K.; Nithya, A.; Moses Ezhil Raj, A.; Ravichandran, K.; Sanjeeviraja, C.; Tunable Morphology with Selective Faceted Growth of Visible Light Active TiO2 Thin Films by Facile Hydrothermal Method: Structural, Optical and Photocatalytic Properties. *J. Mater. Sci. Mater. Electron.*, 2016, 27, 5020–5032.
60. Hamadanian, M.; Safaei-Ghomi, J.; Hosseinpour, M.; Masoomi, R.; Jabbari, V.; Uses of New Natural Dye Photosensitizers in Fabrication of High Potential Dye-Sensitized Solar Cells (DSSCs). *Mater. Sci. Semicond. Process.*, 2014, 27, 733–739.
60. Devadiga, D.; Selvakumar, M.; Shetty, P.; Santosh, M. S.; Recent Progress in Dye Sensitized Solar Cell Materials and Photo-Supercapacitors: A Review. *J. Power Sources,* 2021, 493, 229698.
62. Horiuchi, T.; Miura, H.; Uchida, S.; Highly-Efficient Metal-Free Organic Dyes for Dye-Sensitized Solar Cells. *Chem. Commun.*, 2003, No. 24, 3036.
63. Hao, F.; Stoumpos, C. C.; Guo, P.; Zhou, N.; Marks, T. J.; Chang, R. P. H.; Kanatzidis, M. G.; Solvent-Mediated Crystallization of CH 3 NH 3 SnI 3 Films for Heterojunction Depleted Perovskite Solar Cells. *J. Am. Chem. Soc.*, 2015, 137, 11445–11452.
64. Kim, H.; Han, J. S.; Choi, J.; Kim, S. Y.; Jang, H. W.; Halide Perovskites for Applications beyond Photovoltaics. *Small Methods,* 2018, 2, 1700310.
65. Basumatary, P.; Agarwal, P.; A Short Review on Progress in Perovskite Solar Cells. *Mater. Res. Bull.*, 2022, 149, 111700.
66. Mathies, F.; Eggers, H.; Richards, B. S.; Hernandez-Sosa, G.; Lemmer, U.; Paetzold, U. W.; Inkjet-Printed Triple Cation Perovskite Solar Cells. *ACS Appl. Energy Mater.*, 2018, 1, 1834–1839.
67. Howard, I. A.; Abzieher, T.; Hossain, I. M.; Eggers, H.; Schackmar, F.; Ternes, S.; Richards, B. S.; Lemmer, U.; Paetzold, U. W.; Coated and Printed Perovskites for Photovoltaic Applications. *Adv. Mater.*, 2019, 31, 1806702.
68. Taylor, A. D.; Sun, Q.; Goetz, K. P.; An, Q.; Schramm, T.; Hofstetter, Y.; Litterst, M.; Paulus, F.; Vaynzof, Y.; A General Approach to High-Efficiency Perovskite Solar Cells by Any Antisolvent. *Nat. Commun.*, 2021, 12, 1878.
69. Watthage, S. C.; Song, Z.; Phillips, A. B.; Heben, M. J.; Evolution of Perovskite Solar Cells. In *Perovskite Photovoltaics*; Elsevier, 2018; pp. 43–88.
70. Jeong, J.; Kim, M.; Seo, J.; Lu, H.; Ahlawat, P.; Mishra, A.; Yang, Y.; Hope, M. A.; Eickemeyer, F. T.; Kim, M.; Yoon, Y. J.; Choi, I. W.; Darwich, B. P.; Choi, S. J.; Jo, Y.; Lee, J. H.; Walker, B.; Zakeeruddin, S. M.; Emsley, L.; Rothlisberger, U.; Hagfeldt, A.; Kim, D. S.; Grätzel, M.; Kim, J. Y.; Pseudo-Halide Anion Engineering for α-FAPbI3 Perovskite Solar Cells. *Nature,* 2021, 592, 381–385.
71. Jiang, X.; Li, H.; Zhou, Q.; Wei, Q.; Wei, M.; Jiang, L.; Wang, Z.; Peng, Z.; Wang, F.; Zang, Z.; Xu, K.; Hou, Y.; Teale, S.; Zhou, W.; Si, R.; Gao, X.; Sargent, E. H.; Ning, Z.; One-Step Synthesis of SnI

2 (DMSO) x Adducts for High-Performance Tin Perovskite Solar Cells. *J. Am. Chem. Soc.*, 2021, 143, 10970–10976.
72. Chen, L.-J.; Lee, C.-R.; Chuang, Y.-J.; Wu, Z.-H.; Chen, C.; Synthesis and Optical Properties of Lead-Free Cesium Tin Halide Perovskite Quantum Rods with High-Performance Solar Cell Application. *J. Phys. Chem. Lett.*, 2016, 7, 5028–5035.
73. Kopacic, I.; Friesenbichler, B.; Hoefler, S. F.; Kunert, B.; Plank, H.; Rath, T.; Trimmel, G.; Enhanced Performance of Germanium Halide Perovskite Solar Cells through Compositional Engineering. *ACS Appl. Energy Mater.*, 2018, 1, 343–347.
74. Krishnamoorthy, T.; Ding, H.; Yan, C.; Leong, W. L.; Baikie, T.; Zhang, Z.; Sherburne, M.; Li, S.; Asta, M.; Mathews, N.; Mhaisalkar, S. G.; Lead-Free Germanium Iodide Perovskite Materials for Photovoltaic Applications. *J. Mater. Chem. A*, 2015, *3*, 23829–23832.
75. Chen, M.; Ju, M.-G.; Garces, H. F.; Carl, A. D.; Ono, L. K.; Hawash, Z.; Zhang, Y.; Shen, T.; Qi, Y.; Grimm, R. L.; Pacifici, D.; Zeng, X. C.; Zhou, Y.; Padture, N. P.; Highly Stable and Efficient All-Inorganic Lead-Free Perovskite Solar Cells with Native-Oxide Passivation. *Nat. Commun.*, 2019, 10, 16.
76. Slavney, A. H.; Hu, T.; Lindenberg, A. M.; Karunadasa, H. I.; A Bismuth-Halide Double Perovskite with Long Carrier Recombination Lifetime for Photovoltaic Applications. *J. Am. Chem. Soc.*, 2016, *138*, 2138–2141.
77. McClure, E. T.; Ball, M. R.; Windl, W.; Woodward, P. M.; Cs 2 AgBiX 6 (X = Br, Cl): New Visible Light Absorbing, Lead-Free Halide Perovskite Semiconductors. *Chem. Mater.*, 2016, 28, 1348–1354.
78. Wei, F.; Deng, Z.; Sun, S.; Xie, F.; Kieslich, G.; Evans, D. M.; Carpenter, M. A.; Bristowe, P. D.; Cheetham, A. K.; The Synthesis, Structure and Electronic Properties of a Lead-Free Hybrid Inorganic–Organic Double Perovskite (MA) 2 KBiCl 6 (MA = Methylammonium). *Mater. Horizons,* 2016, 3, 328–332.
79. Jokar, E.; Chien, C.-H.; Tsai, C.-M.; Fathi, A.; Diau, E. W.-G.; Robust Tin-Based Perovskite Solar Cells with Hybrid Organic Cations to Attain Efficiency Approaching 10%. *Adv. Mater.*, 2019, 31, 1804835.
80. Kaltzoglou, A.; Antoniadou, M.; Perganti, D.; Siranidi, E.; Raptis, V.; Trohidou, K.; Psycharis, V.; Kontos, A. G.; Falaras, P.; Mixed-Halide Cs2SnI3Br3 Perovskite as Low Resistance Hole-Transporting Material in Dye-Sensitized Solar Cells. *Electrochim. Acta,* 2015, 184, 466–474.
81. Verploegen, E.; Mondal, R.; Bettinger, C. J.; Sok, S.; Toney, M. F.; Bao, Z.; Effects of Thermal Annealing Upon the Morphology of Polymer-Fullerene Blends. *Adv. Funct. Mater.,* 2010, 20, 3519–3529.
82. Jing, X.; Zhenbo, D.; Chunjun, L.; Denghui, X.; Ying, X.; Dong, G.; Effect of LiF Buffer Layer on the Performance of Organic Electroluminescent Devices. *Phys. E Low-Dimensional Syst. Nanostructures,* 2005, 28, 323–327.
83. Abdulrazzaq, O.; Bourdo, S. E.; Saini, V.; Bairi, V. G.; Dervishi, E.; Viswanathan, T.; Nima, Z. A.; Biris, A. S.; Optimization of the Protonation Level of Polyaniline-Based Hole-Transport Layers in Bulk-Heterojunction Organic Solar Cells. *Energy Technol.*, 2013, 1, 463–470.
84. Jørgensen, M.; Norrman, K.; Krebs, F. C.; Stability/Degradation of Polymer Solar Cells. *Sol. Energy Mater. Sol. Cells,* 2008, 92, 686–714.
85. Barth, S.; Bässler, H.; Intrinsic Photoconduction in PPV-Type Conjugated Polymers. *Phys. Rev. Lett.*, 1997, 79, 4445–4448.
86. Curtins, H.; Wyrsch, N.; Shah, A. V.; High-Rate Deposition of Amorphous Hydrogenated Silicon: Effect of Plasma Excitation Frequency. *Electron. Lett.*, 1987, 23, 228–230.
87. Keppner, H.; Kroll, U.; Torres, P.; Meier, J.; Fischer, D.; Goetz, M.; Tscharner, R.; Shah, A.; Scope of VHF Plasma Deposition for Thin-Film Silicon Solar Cells. In *Conference Record of the Twenty Fifth IEEE Photovoltaic Specialists Conference - 1996*; IEEE, 1996; pp. 669–672.

88. Vetterl, O.; Finger, F.; Carius, R.; Hapke, P.; Houben, L.; Kluth, O.; Lambertz, A.; Mück, A.; Rech, B.; Wagner, H.; Intrinsic Microcrystalline Silicon: A New Material for Photovoltaics. *Sol. Energy Mater. Sol. Cells,* 2000, 62, 97–108.
89. Kondo, M.; Matsuda, A.; An Approach to Device Grade Amorphous and Microcrystalline Silicon Thin Films Fabricated at Higher Deposition Rates. *Curr. Opin. Solid State Mater. Sci.,* 2002, 6, 445–453.
90. Rech, B.; Kluth, O.; Repmann, T.; Roschek, T.; Springer, J.; Müller, J.; Finger, F.; Stiebig, H.; Wagner, H.; New Materials and Deposition Techniques for Highly Efficient Silicon Thin Film Solar Cells. *Sol. Energy Mater. Sol. Cells,* 2002, 74, 439–447.
91. Kondo, M.; Microcrystalline Materials and Cells Deposited by RF Glow Discharge. *Sol. Energy Mater. Sol. Cells,* 2003, 78, 543–566.
92. Mai, Y.; Klein, S.; Carius, R.; Wolff, J.; Lambertz, A.; Finger, F.; Geng, X.; Microcrystalline Silicon Solar Cells Deposited at High Rates. *J. Appl. Phys.*, 2005, 97, 114913.
93. Saito, K.; Sano, M.; Ogawa, K.; Kajita, I.; High Efficiency A-Si:H Alloy Cell Deposited at High Deposition Rate. *J. Non. Cryst. Solids,* 1993, *164–166,* 689–692.
94. Smets, A. H. M.; Kessels, W. M. M.; van de Sanden, M. C. M.; Temperature Dependence of the Surface Roughness Evolution during Hydrogenated Amorphous Silicon Film Growth. *Appl. Phys. Lett.,* 2003, 82, 865–867.
95. Schropp, R. E. I.; Veen, M. K. van; Werf, C. H. M. van der; Williamson, D. L.; Mahan, A. H.; Protocrystalline Silicon at High Rate from Undiluted Silane. *MRS Proc.,* 2004, *808,* A8.4-A8.4.
96. Schropp, R. E. I.; Present Status of Micro- and Polycrystalline Silicon Solar Cells Made by Hot-Wire Chemical Vapor Deposition. *Thin Solid Films,* 2004, *451–452,* 455–465.
97. Matsumura, H.; Catalytic Chemical Vapor Deposition (CTC–CVD) Method Producing High Quality Hydrogenated Amorphous Silicon. *Jpn. J. Appl. Phys.,* 1986, *25,* L949–L949.
98. Klein, S.; Finger, F.; Carius, R.; Stutzmann, M.; Deposition of Microcrystalline Silicon Prepared by Hot-Wire Chemical-Vapor Deposition: The Influence of the Deposition Parameters on the Material Properties and Solar Cell Performance. *J. Appl. Phys.,* 2005, 98, 24905.
99. Romeo, N.; Bosio, A.; Canevari, V.; Terheggen, M.; Vaillant Roca, L.; Comparison of Different Conducting Oxides as Substrates for CdS/CdTe Thin Film Solar Cells. *Thin Solid Films,* 2003, *431–432,* 364–368.
100. Khrypunov, G.; Romeo, A.; Kurdesau, F.; Batzner, D.; Zogg, H.; Tiwari, A.; Recent Developments in Evaporated CdTe Solar Cells. *Sol. Energy Mater. Sol. Cells,* 2006, 90, 664–677.
101. Oman, D. M.; Dugan, K. M.; Killian, J. L.; Ceekala, V.; Ferekides, C. S.; Morel, D. L.; Reduction of Recombination Current in CdTe/CdS Solar Cells. *Appl. Phys. Lett.* 1995, 67, 1896–1898.
102. Morris, G.C.; Das, S.K.; Some Fabrication Procedures For Electro Deposited CdTe Solar Cells. *Int. J. Sol. Energy,* 1992, *12,* 95–108.
103. Chu, T. L.; Chu, S. S.; High Efficiency Thin Film CdS/CdTe Solar Cells†. *Int. J. Sol. Energy,* 1992, 12, 121–132.
104. Romeo, N.; Bosio, A.; Tedeschi, R.; Romeo, A.; Canevari, V.; A Highly Efficient and Stable CdTe/CdS Thin Film Solar Cell. *Sol. Energy Mater. Sol. Cells,* 1999, 58, 209–218.
105. Britt, J.; Ferekides, C.; Thin-film CdS/CdTe Solar Cell with 15.8% Efficiency. *Appl. Phys. Lett.,* 1993, 62, 2851–2852.
106. Ikegami, S.; CdS/CdTe Solar Cells by the Screen-Printing-Sintering Technique: Fabrication, Photovoltaic Properties and Applications. *Sol. Cells,* 1988, 23, 89–105.
107. Fang, Z.; Wang, X. C.; Wu, H. C.; Zhao, C. Z.; Achievements and Challenges of CdS/CdTe Solar Cells. *Int. J. Photoenergy,* 2011, 2011, 1–8.
108. Liu, P.; Singh, V. P.; Jarro, C. A.; Rajaputra, S.; Cadmium Sulfide Nanowires for the Window Semiconductor Layer in Thin Film CdS–CdTe Solar Cells. *Nanotechnology,* 2011, 22, 145304.

109. Nakada, T.; Hirabayashi, Y.; Tokado, T.; Ohmori, D.; Mise, T.; Novel Device Structure for Cu(In,Ga)Se2 Thin Film Solar Cells Using Transparent Conducting Oxide Back and Front Contacts. *Sol. Energy,* 2004, 77, 739–747.
110. Tiwari, A. N.; Khrypunov, G.; Kurdzesau, F.; Bätzner, D. L.; Romeo, A.; Zogg, H.; CdTe Solar Cell in a Novel Configuration. *Prog. Photovoltaics Res. Appl.*, 2004, 12, 33–38.
111. Romeo, A.; Khrypunov, G.; Galassini, S.; Zogg, H.; Tiwari, A. N.; Bifacial Configurations for CdTe Solar Cells. *Sol. Energy Mater. Sol. Cells,* 2007, 91, 1388–1391.
112. Marsillac, S.; Parikh, V. Y.; Compaan, A. D.; Ultra-Thin Bifacial CdTe Solar Cell. *Sol. Energy Mater. Sol. Cells,* 2007, 91, 1398–1402.
113. Minemoto, T.; Matsui, T.; Takakura, H.; Hamakawa, Y.; Negami, T.; Hashimoto, Y.; Uenoyama, T.; Kitagawa, M.; Theoretical Analysis of the Effect of Conduction Band Offset of Window/CIS Layers on Performance of CIS Solar Cells Using Device Simulation. *Sol. Energy Mater. Sol. Cells,* 2001, 67, 83–88.
114. Michaelson, H. B.; The Work Function of the Elements and Its Periodicity. *J. Appl. Phys.*, 1977, 48, 4729–4733.
115. Li, J.; Xia, R.; Qi, W.; Zhou, X.; Cheng, J.; Chen, Y.; Hou, G.; Ding, Y.; Li, Y.; Zhao, Y.; Zhang, X.; Encapsulation of Perovskite Solar Cells for Enhanced Stability: Structures, Materials and Characterization. *J. Power Sources,* 2021, 485, 229313.
116. Bai, S.; Da, P.; Li, C.; Wang, Z.; Yuan, Z.; Fu, F.; Kawecki, M.; Liu, X.; Sakai, N.; Wang, J. T.-W.; Huettner, S.; Buecheler, S.; Fahlman, M.; Gao, F.; Snaith, H. J.; Planar Perovskite Solar Cells with Long-Term Stability Using Ionic Liquid Additives. *Nature,* 2019, 571, 245–250.
117. Halls, J. J. M.; Pichler, K.; Friend, R. H.; Moratti, S. C.; Holmes, A. B.; Exciton Diffusion and Dissociation in a Poly(p -phenylenevinylene)/C 60 Heterojunction Photovoltaic Cell. *Appl. Phys. Lett.*, 1996, 68, 3120–3122.
118. Halls, J. J. M.; Friend, R. H.; The Photovoltaic Effect in a Poly(p-Phenylenevinylene)/Perylene Heterojunction. *Synth. Met.*, 1997, 85, 1307–1308.
119. Kerp, H. R.; Donker, H.; Koehorst, R. B. M.; Schaafsma, T. J.; van Faassen, E. E.; Exciton Transport in Organic Dye Layers for Photovoltaic Applications. *Chem. Phys. Lett.*, 1998, 298, 302–308.
120. Hoppe, H.; Sariciftci, N. S.; Organic Solar Cells: An Overview. *J. Mater. Res.*, 2004, 19, 1924–1945.
121. Aziz, E. F.; Vollmer, A.; Eisebitt, S.; Eberhardt, W.; Pingel, P.; Neher, D.; Koch, N.; Localized Charge Transfer in a Molecularly Doped Conducting Polymer. *Adv. Mater.*, 2007, 19, 3257–3260.

12

Mixed-Dimensional 2D–3D Perovskite Solar Cells: Origin, Development, and Applications

Vani Pawar*, Bhumika Sharma and Sushobhan Avasthi†

Centre for Nano Science and Engineering, Indian Institute of Science, Bangalore, Karnataka, India

Abstract

Photovoltaics has emerged as the most promising renewable energy source to displace fossil fuels. Conventional silicon-based solar cells are widely used, but their limitation in efficiency (Shockley–Queisser limit) and functionality creates opportunities for new materials. Among the alternatives, 3D perovskites (ABX_3) the most promising. Constant efforts and improvements enabled a power conversion efficiency (PCE) of more than 25%, comparable to silicon. However, there are many critical barriers to commercialization, for example, stability. Typical lead-halide perovskites (3D) are unstable. They degrade with light, heat, and moisture, so the state-of-the-art T80 is still only around 5000 hours. Reduced-dimensional 2D–3D halide perovskite semiconductors are inherently more stable. However, the bandgap of 2D–3D halide perovskite is higher than 3D perovskite, so efficiencies were initially lower. Various innovations have since been made to enhance the photovoltaic performance, e.g., additives, replacement of A-site cations, mixed-halide compositions, novel transport layers, and passivating interlayers. The combination of these advances has enabled 2D–3D perovskite solar cells that are not just stable but also efficient. The fabrication process of 2D–3D perovskite is still very simple and compatible with substrates like glass, plastic, and steel. The 2D–3D perovskite PV will lead to new types of solar panels, like flexible solar cells, indoor solar cells, and self-powered smart tags. Due to unique features like tunable bandgap, 2D–3D perovskite cells can be engineered to be selectively semitransparent. Semitransparent solar cells harvest a part of the solar spectrum for energy but transmit the rest. Such cells can be exploited for dual-use opportunities like agrivoltaics and building-integrated PV, where solar cells do not just produce energy but also have a second functionality. In this chapter, we introduce 3D perovskites, their advantages, and limitations. This motivates the need for 2D and 2D–3D perovskites. We shall discuss the trend of photophysical properties as we move from pure 3D perovskites to mixed 2D–3D and then pure 2D perovskites. This trend has been studied and explored with insights into how the device performance changes while using them as an absorber layer. Further, we will discuss the various techniques used to enhance the performance of 2D–3D perovskite solar cells. At the end of this chapter, we talk about the dual-purpose semitransparent perovskite solar cells and their usefulness in architecture and agriculture.

***Keywords*:** Low-dimensional perovskites, 2D-3D perovskites solar cells, RP-DJ perovskites, additive engineering, compositional engineering, semi-transparent perovskite, solar cells

**Corresponding author*: vanipawar@iisc.ac.in

†*Corresponding author*: savasthi@iisc.ac.in

Upendra Kumar and Piyush Kumar Sonkar (eds.) Smart Materials for Science and Engineering, (221–268) © 2024 Scrivener Publishing LLC

12.1 Introduction

It is said that the price of everything is "dirt plus energy". Energy is a core sector that impacts all facets of our lives. The world's energy demand is continuously increasing, rising from 9,702 TWh in 1990 to 25,343 TWh in 2021 (refer to Figure 12.1a) [1]. We cannot depend solely on fossil fuels to fuel this growth, firstly because fossil fuels are non-renewable and, to a large extent, finite, and secondly, because fossil fuels are the leading source of greenhouse emissions and the primary cause of climate change [2]. Considering this, we must scale and substitute fossil fuels with renewable energy [3].

Solar photovoltaics has the most potential compared to other renewable energy sources, such as wind, hydro, and geothermal. The sun is an abundant source delivering about 1.8×10^{14} kW of power. Even a fraction of this is enough to provide all the world's energy requirements [5]. The challenge is to harvest this energy, which falls over large areas and concentrate it to usable levels. Solar photovoltaics has emerged as the most cost-effective method for this conversion. The attractive techno-economics has enabled an exponential increase in solar photovoltaics, from 179 TWh to 1000 TWh in the last decade (Figure 12.1b) [4].

At a fundamental level, solar cells are relatively simple electronic devices: a p-n or p-i-n junction diode made with a light-absorbing semiconductor. The choice of the absorber is a key factor that dictates the solar cell's efficiency and stability. The theoretical maximum efficiency possible for a single-absorber solar cell is given by the Shockley–Queisser (SQ) limit, calculated using the concepts of thermodynamics and detailed balance. The SQ limit varies with the bandgap of the absorber (Figure 12.2). For instance, if an absorber material has a bandgap of 1.34 eV, it can achieve a maximum efficiency of 33.7% at an incident light intensity of 1000 W/m^2 (see Figure 12.2) [6].

Due to its technological maturity, silicon has been the most common absorber. The first practical solar cell was demonstrated on 25 April 1954 at Bell Labs [7]. After decades of research and optimization, the technology improved, and today crystalline silicon solar cells dominate the market. Commercial silicon solar modules are the cheapest source of energy. They support efficiency ranging from 18 to 22.8% (based on technology).

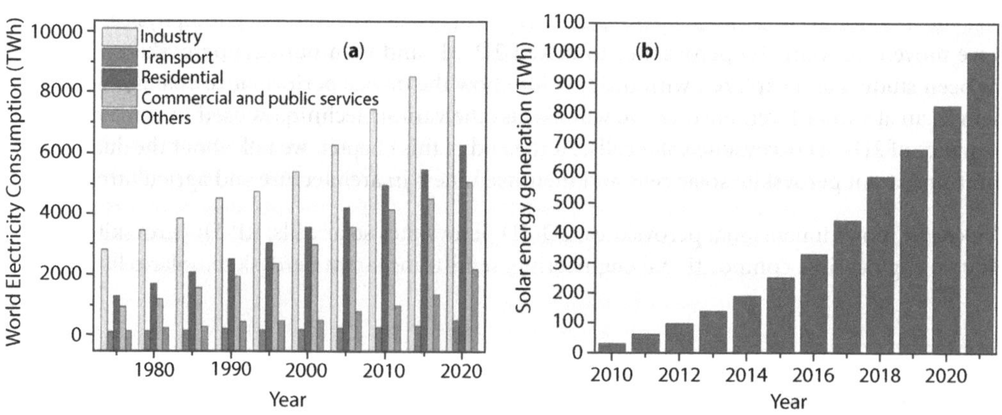

Figure 12.1 (a) World electricity consumption from different sectors [1] and (b) solar energy generation data from 2010 to 2021 [4].

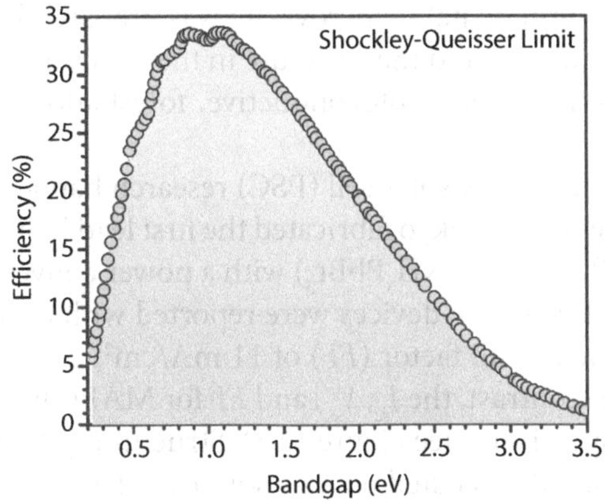

Figure 12.2 Theoretical Shockley–Queisser Limit for the efficiency of a solar cell as a function of the bandgap. Sun's intensity is not a constant. Thus, such calculations are done for standardized sun spectrum. This curve is for AM 1.5 [6].

Conventional silicon solar cells require a high-quality monocrystalline silicon wafer. The manufacture of silicon wafers requires significant technological and energy inputs. Researchers have proposed alternatives where wafers are replaced by thin films deposited on low-cost glass, plastic, or steel substrates. Several thin films have been suggested for solar cell fabrication, notably amorphous silicon (a-Si), cadmium telluride (CdTe), copper indium gallium selenide (CIGS), copper zinc tin sulfide (CZTS), conjugated organic semiconductors, and polymers. Some of these were scaled up, e.g., CIGS, and some were commercialized, e.g., a-Si. However, most proved to be commercial dead-ends due to issues like efficiency and manufacturability. Currently, *First Solar* is the only large-scale thin-film PV manufacturer selling CdTe modules, with a market share of just 3% that continues to shrink [8].

In the last decade, lead halide perovskites thin-film PV has emerged as the most promising challenger to silicon's dominance. The raw materials for synthesizing the absorber are dirt cheap. Lead-halide perovskites can be deposited on virtually any substrate, with equipment closer to a newsprint printer than a semiconductor deposition tool. The absence of wafers enables migration from a batch process to a roll-to-roll process, potentially with higher throughput and lower capital expenditure. The family of lead-halide perovskites is diverse, with many different bandgaps. This allows engineers to select suitable bandgap, as per the intended applications. However, significant challenges remain, like stability, lead toxicity, upscaling, and market definition.

12.2 Perovskite Solar Cells (PSCs)

- **Perovskites: Origin and Evolution**

Perovskites were discovered in 1839 by Gustav Rose in the form of a naturally occurring mineral, $CaTiO_3$, and named in honor of Russian mineralogist Lev von Perovski. Perovskite now refers to the family of oxides with the chemical formula of ABO_3. Many perovskites are known, such as $PbTiO_3$, $BiFeO_3$, $BaTiO_3$, $SrTiO_3$, etc. The halide perovskite discussed here differs from oxide perovskites, with a chemical formula of ABX_3 (X refers to any halide, Cl,

Br, I, etc.) [9]. The first synthetic halide perovskite was prepared by H. L. Wells using lead and caesium. C.K. Møller discovered the structure in the 1950s. It is noteworthy that Møller found the halide perovskite to be photoconductive, foreshadowing their importance for solar cells [10].

The breakthrough in perovskite solar cell (PSC) research happened in 2009. Kojima and Miyasaka from the University of Tokyo fabricated the first lead halide PSC by using $MAPbI_3$ ($CH_3NH_3PbI_3$) and $MAPbBr_3$ ($CH_3NH_3PbBr_3$) with a power conversion efficiency (PCE) of 3.81% and 3.13%, respectively. The devices were reported with a short-circuit current (J_{sc}), open-circuit voltage (V_{oc}), and fill factor (FF) of 11 mA/cm^2, 0.61 V, and 0.57, respectively, for $MAPbI_3$ solar cells. In contrast, the J_{sc}, V_{oc}, and FF for $MAPbBr_3$-based cell was 5.57 mA/cm^2, 0.96 V, and 0.59, respectively (see Figure 12.3). Structurally, these were electrochemical cells with perovskite as the "dye" coated on a mesoporous TiO_2 scaffold [11].

Soon, researchers realized that perovskite is not just a dye but an ambipolar semiconductor with excellent transport and photophysical properties. PSCs have since evolved to a solid-state architecture [12] with a perovskite absorber layer sandwiched between electron and hole transport layers (Figure 12.4). Depending on the order of the layers and direction of light incidence, two architectures are possible: (a) n-i-p (conventional) and (b) p-i-n (inverted). In the former, the electron transport layer (ETL) is light facing, while in the latter, the hole transport layer (HTL) is light facing [13].

N. G. Park *et al.* reported the first solid-state mesoscopic heterojunction PSC having a PCE of 9.7%. Methylammonium iodide (MAI) and lead iodide (PbI_2) were made to react on a submicron-thick mesoscopic film of TiO_2, leading to the synthesis of

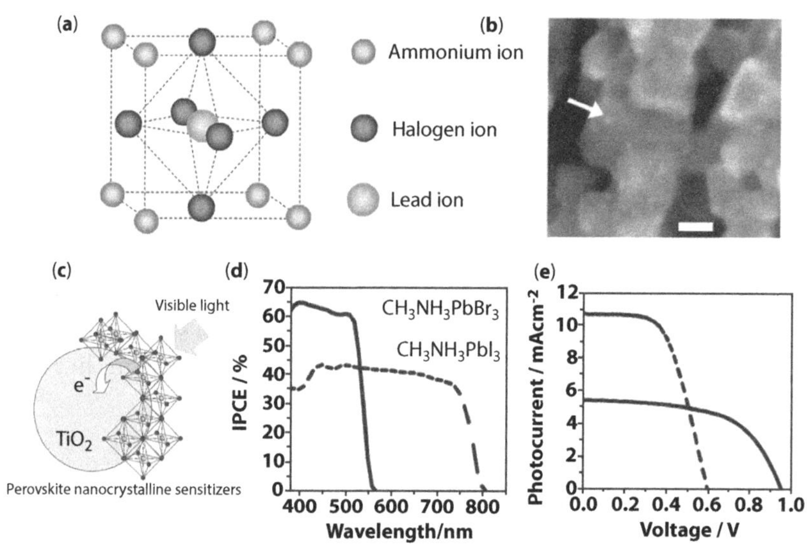

Figure 12.3 (a) Crystal structure of perovskite compound. (b) SEM micrograph of $CH_3NH_3PbBr_3$ nanoparticles (represented by an arrow) deposited on TiO_2 surface (scale bar = 10 nm). (c) Schematic of perovskite compound as a sensitizer of TiO_2 for visible-light conversion. (d) Incident photon-to-current conversion efficiency (IPCE) spectra for photoelectrochemical cells using $CH_3NH_3PbI_3/TiO_2$ (dashed line) and $CH_3NH_3PbBr_3/TiO_2$ (solid line). (e) Photocurrent–voltage spectra of solar cells under 1 sun illumination utilizing $CH_3NH_3PbI_3/TiO_2$ (dashed line) and $CH_3NH_3PbBr_3/TiO_2$ (solid line) perovskites. Reproduced with permission from [11]. Copyright 2009, American Chemical Society.

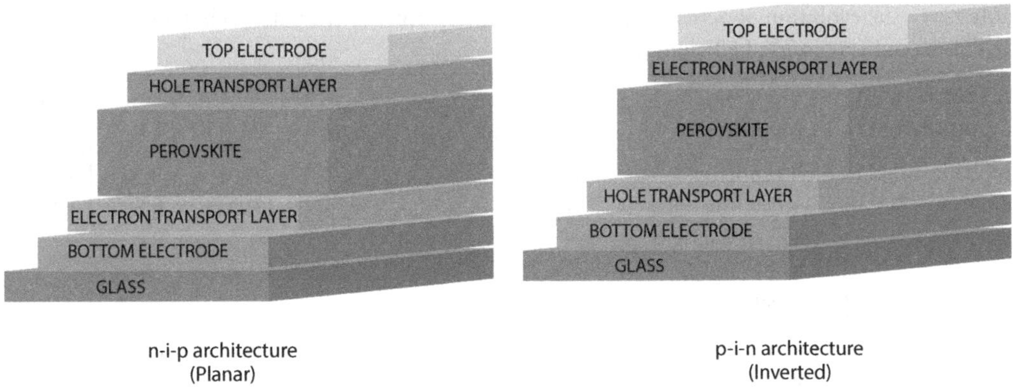

Figure 12.4 Schematic representation of planar and inverted device architecture of solar cells.

perovskite nanoparticles. The device parameters for this cell were reported to be 17.6 mA/cm^2, 0.88 V, and 0.62 for J_{sc}, V_{oc}, and *FF*, respectively (see Figure 12.5a–d). Later, Lee *et al.* reported a solar cell in 2012, which was relatively cheaper and had an efficiency of 10.9%. They used a mesoporous alumina film as a chemically inactive layer over TiO_2. The alumina film provided a better transport path for electrons. The champion devices exhibited J_{sc}, V_{oc}, and *FF* of 17.8 mA/cm^2, 0.98 V, and 0.63, respectively [14].

The perovskite synthesis till then was done in the form of nanostructures for solar cells. Breaking this compulsion in 2013, Snaith *et al.* reported solar cells in which thermal deposition of perovskite was done using a dual source. This simple planar heterojunction solar cell exhibited an efficiency of 15.4% and a V_{oc} of 1.07 V. The J_{sc} and *FF* for this champion device were 21.5 mA/cm^2 and 0.67, respectively. Hence, the scientific community shifted to fabricating thin-film perovskite-based solar cells [15].

The current efficiency record for lead halide PSCs is 25.6%, as reported by Jeong *et al.* The champion device exhibited J_{sc}, V_{oc}, and *FF* of 26.35 mA/cm^2, 1.18 V, and 0.81, respectively. Formamidinium formate (FAHCOO) was used as an additive for the perovskite ($FAPbI_3$) precursor. It has a formate anion that is small enough to fit in the anion (iodine) vacancy,

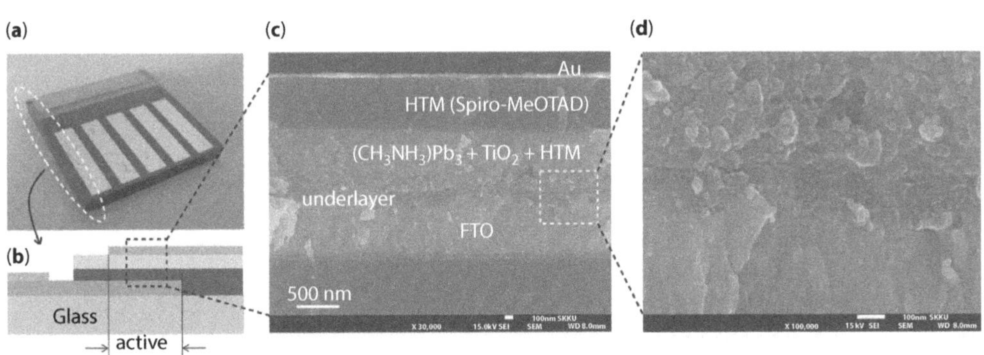

Figure 12.5 (a) Original solid-state device. (b) Cross-sectional view of the device structure. (c) Cross-sectional SEM micrograph of the device showing various interlayers. (d) SEM micrograph of FTO-underlayer-active layer interfacial junction. Reproduced with permission from [14]. Copyright 2012, Springer Nature.

Table 12.1 The evolution of perovskite solar cells (PSCs) from the very first cell made to the current maximum reported efficiency.

Year of publication	Key feature	Perovskite	J_{sc} (mA/cm^2)	V_{oc} (V)	FF	PCE (%)
2009	First reported perovskite solar cell used as a visible-light sensitizer [17]	$MAPbI_3$ $MAPbBr_3$	11 5.57	0.61 0.96	0.57 0.59	3.81 3.13
2012	Chemical deposition of $MAPbI_3$ nanoparticles on a submicron-thick mesoscopic film of TiO_2 [14]	$MAPbI_3$	17.6	0.88	0.62	9.7
2012	A chemically inactive layer of mesoporous alumina was used over TiO_2 for better structuring and carrier transport [18]	$MAPbI_3$	17.8	0.98	0.63	10.9
2013	First simple planar heterojunction solar cell by thermal evaporation of perovskite using a dual source [19]	$MAPbI_3$	21.5	1.07	0.67	15.4
2013	Sequential deposition of PbI_2 and MAI to form $MAPbI_3$ perovskite phase on nanoporous TiO_2 [20]	$MAPbI_3$	20.0	0.99	0.73	15.0
2017	Moderate residual of PbI_2 compound in perovskite can provide stability and high efficiency without any hysteresis [21]	$(FAPbI_3)_{1-x}(MAPbBr_3)_x$	23.5	1.13	0.78	20.9

(Continued)

Table 12.1 The evolution of perovskite solar cells (PSCs) from the very first cell made to the current maximum reported efficiency. (*Continued*)

Year of publication	Key feature	Perovskite	J_{sc} (mA/cm^2)	V_{oc} (V)	FF	PCE (%)
2016	Triple cation (A-site) and multi halide (X-site) perovskite: Thermally stable, highly pure, less effective to processing conditions [22]	$Cs_x(MA_{0.17}FA_{0.83})_{(100-x)}Pb(I_{0.83}Br_{0.17})_3$	23.5	1.14	0.78	21.17
2019	Stabilizer (MACl and $MAPbBr_3$) assisted growth of high-quality and phase-stable FA-based perovskite [23]	$(FAPbI_3)_{95}(MAPbBr_3)_5$	24.47	1.12	0.82	22.5
2022	Crystallinity was improved by anion engineering. Pseudo-halide anion format was used to suppress anion-vacancy defects, which were present at grain boundaries and perovskite film surface [16]	α-$FAPbI_3$	26.35 26.25	1.18 1.17	0.810 .81	25.6 25.2

hence passivating these defects. The defects present at the grain boundaries and surface are effectively passivated, leading to improved V_{oc} and *FF*. Also, the film's crystallinity and grain size are enhanced because they attain better performance and device stability (450 hours of long-term operational stability) [16]. Table 12.1 shows the list of evolution of PSCs from the very first cell made to the current maximum reported efficiency.

Improvement in efficiency, from 3% to 25.7% [16], happened over just 10 years. The field of PSCs has grown significantly over this time (refer to Figure 12.6). The research focus has also evolved. Initially, the focus was on improving the quality of the deposited perovskite absorber since light absorption and transport were the major challenges [24]. Various methods were tried to obtain a film that was dense, pinhole-free, and defect-free [25]. The importance of larger grain size and higher crystallinity led to the development of treatments such as two-step deposition [26], anti-solvent dripping [27], additives [28], and solvent-assisted vapor annealing multi-cation perovskites [29–32]. As the losses in bulk are reduced, losses at the interface become important, prompting work on interfacial

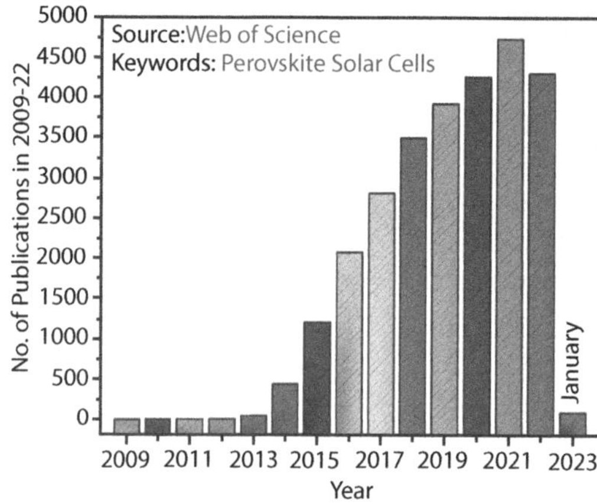

Figure 12.6 Yearly publications in perovskite solar cells (PSCs) from January 2009 to January 2023 (Keyword: Perovskite solar cells, Source: Web of Science).

passivation [33], novel transport layers [34], passivating layers [35, 36], etc. With efficiency sorted, the focus has shifted to the harder problem of stability and manufacturability. These priorities have expanded the library of lead halide perovskites. Many permutations of "A", "B", and "X" in ABX_3 are under study. A newer family of low-dimensional perovskites with related properties but different structures are now being studied [37–39].

- **Perovskites: Structure and Chemistry**

Perovskites are represented by the general formula ABX_3, where "A" and "B" are cations, and X is a halide (Cl^-, Br^-, and I^-). "A" usually is a large monovalent cation such as MA^+, FA^+, and Cs^+. "B" is a bivalent metal cation, generally Pb^{2+}, but others, like Sn and Ge, are also possible. Structurally, the halide atoms occupy the corners of an octahedron (BX_6). The cations sit either at the octahedra centre or in the voids formed by the octahedra in the lattice [38, 39] (see Figure 12.7).

The structural stability of perovskites is often estimated by Goldschmidt's Tolerance Factor (t), which measures the distortion in the perovskite structure:

$$t = \frac{r_A + r_X}{\sqrt{2}(r_B + r_X)}$$

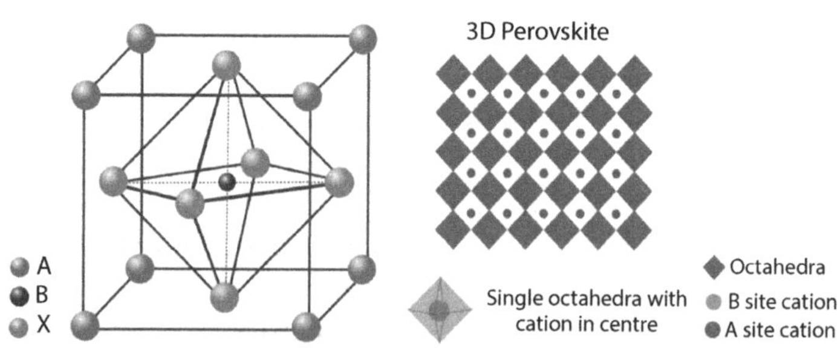

Figure 12.7 Crystal structure of perovskite solar cell (left), pictorial depiction of the 3D perovskite structure.

r_A, r_B, and r_X are the ionic radii of A, B, and X ions. For stable 3D perovskites such as $MAPbI_3$ and $FAPbI_3$, the t value ranges between ~ $0.8 < t < 1$. Any deviation from this range leads to distortion in the face-shared and edge-shared octahedra of the perovskite [40].

While PSCs yield excellent efficiencies, they are not very stable. Perovskite films degrade rapidly in the presence of moisture, oxygen, light, and heat by decomposition into metal halide and organic halide constituents. The possible mechanism of decomposition in moisture may be as follows [41–44]:

$$CH_3NH_3PbI_3 \leftrightarrow PbI_2\ (s) + CH_3NH_3I\ (aq.)\ [\text{In presence of } H_2O]$$

$$CH_3NH_3I\ (aq.) \leftrightarrow CH_3NH_2\ (aq) + HI\ (aq.)$$

$$HI\ (aq.) + O_2 \leftrightarrow 2I_2\ (s) + 2H_2O\ (l)$$

$$2HI\ (aq.) \leftrightarrow H_2\ (g) + I_2\ (s)\ [\text{in the presence of } h\nu]$$

One way to prevent degradation is to reduce exposure to water and oxygen by encapsulating the devices in impermeable membranes or incorporating getters. Another approach is to improve the intrinsic stability of the perovskite film by passivating surfaces and grain boundaries [42]. Additives in trace amounts are also known to improve stability, provided they do not modify the perovskite structure to the extent that the desirable optoelectronic properties degrade. Despite these advances, the stability of PSCs continues to be a serious concern [45–48].

Low-dimensional perovskites are potentially a viable solution for intrinsically stable perovskite. When the organic cation was too big to be accommodated within the octahedral cage of 3D perovskite, it led to distortions in the geometry. The initial 3D tetragonal structure conversion to a 2D layered structure energetically stabilized the structure [37, 49, 50].

12.3 Low-Dimensional (2D or 2D–3D Mixed) Perovskites

Low-dimensional perovskites (2D or 2D–3D) are represented by the general formula $A'_mA_{n-1}B_nX_{3n+1}$, in which A' represents a bulky organic cation. These perovskite structures can be considered inorganic octahedral sheets separated by bulky organic cations (Figure 12.8). These bulky organic cations effectively shield inorganic cages from moisture and oxygen exposure, thus acting as a blocking layer and slowing down degradation. They also prevent cations from escaping the octahedral cage (by inhibiting ionic movement), hence preventing structural distortion to a greater extent [39, 51, 52].

In the chemical formula, "n" represents the number of inorganic octahedral layers between two subsequent layers of organic cations. For instance, $n = 1$ low-dimensional perovskite has one sheet of inorganic octahedra between subsequent layers of organic cations. Higher-order ($n > 1$) low-dimensional perovskites can be systematically created by adding additional layers of octahedra. Just like the 3D perovskite structure can be cut along its three crystallographic planes, <100>, <110>, and <111>, the 2D perovskite family can also be synthesized with three different orientations: <100>, <110>, and <111>.

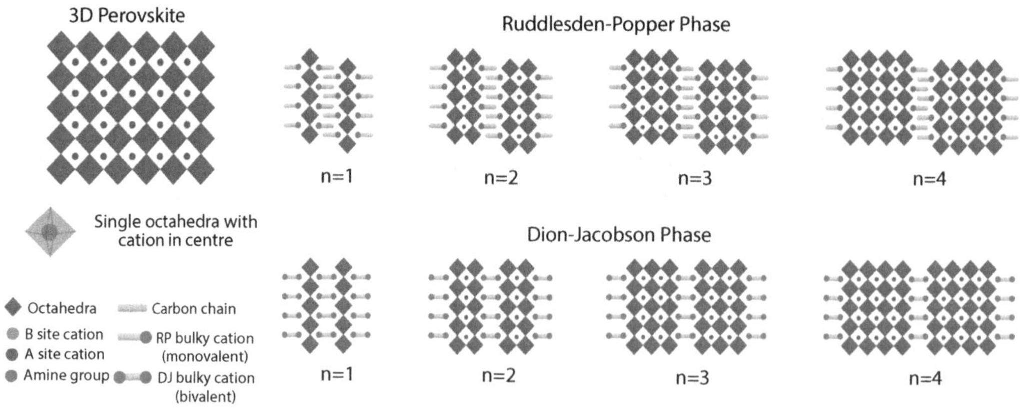

Figure 12.8 Pictorial depictions of 3D and low-dimensional perovskites.

The <100>-oriented perovskites are flat inorganic sheets. The <100> perovskites are the most studied 2D halide perovskite and are most frequently used in solar cells as absorbers [39].

Further, sub-classification of <100>-oriented perovskites are the Ruddlesden–Popper (RP), Dion–Jacobson (DJ), and the phases with alternating cations in the interlayer space (ACI). Depending on the value of "*m*", the low-dimensional perovskites can be categorized as Ruddlesden-Popper (RP) perovskites for $m = 2$ and Dion–Jacobson (DJ) perovskites for $m = 1$. The general formula hence becomes $A'_2A_{n-1}B_nI_{3n+1}$ for RP and $A'A_{n-1}B_nI_{3n+1}$ for DJ perovskites. Another class of low-dimensional perovskites is the alternating cation interlayer (ACI). In this, two cations (A' and A) alternate between two layers of inorganic octahedra. The general formula for this is $A'A_nB_nX_{3n+1}$ [40, 53].

It is well understood that the organic chains of low-dimensional perovskites enhance stability but at the cost of efficiency. Figure 12.9 shows the trend in bandgap from $n = 1$ 2D perovskite to 3D perovskite. The bandgap increases from 1.52 eV in 3D perovskites to 2.3 eV in pure 2D perovskite. As shown in Figure 12.2, due to the larger bandgap, the efficiency potential of 2D perovskite is much lower than 3D perovskite. The organic groups

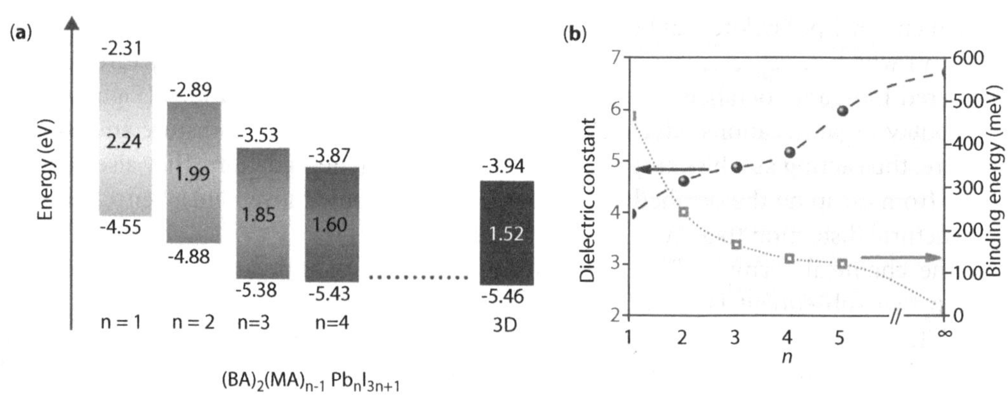

Figure 12.9 (a) Schematic band diagram for $(BA)_2(MA)_{n-1}Pb_nI_{3n+1}$ perovskite family [54]. (b) Variation of exciton binding energy (E_b) and dielectric constant of the $(BA)_2(MA)_{n-1}Pb_nI_{3n+1}$ perovskite family as a function of the layer number *n*. The $n = \infty$ values correspond to $MAPbI_3$ perovskites. Data obtained from Ref. [55].

also increase the exciton binding energy (Figure 12.9b). As a result, charge separation and carrier collection probability are worsened. The organic cations typically used in low-dimensional perovskites are insulating, which reduces mobility and charge collection. The problem is even worse for RP perovskites, in which the two subsequent bulky cations are connected only by weak Van der Waals forces. Preventing these losses is the major challenge for low-dimensional perovskites [39].

12.4 Ruddlesden–Popper (RP) Perovskites

Ruddlesden–Popper (RP) perovskites have the general formula $A'_2A_{n-1}B_nX_{3n+1}$, where A' is aryl or alkyl ammonium cations. Typical cations are butylammonium (BA^+) and phenylethylammonium (PEA^+). RP perovskites have a monovalent organic spacer, which can attach to the inorganic octahedral at only one end. Weak Van der Waals forces bond two adjacent spacers. The inorganic layers are often offset by one octahedral unit, presenting in-plane displacement, as shown in Figure 12.8 [56].

As discussed in the previous section, the wider bandgap of pure 2D perovskites (~2.3 eV) makes them less desirable for photovoltaics. One of the first approaches to overcome this issue was by Karunadasa *et al.* in 2014. They mixed 2D perovskite with conventional 3D perovskites to form a single absorber layer to get the best of both worlds. The low-dimensional perovskite was RP phase with a composition of $(PEA)_2MA_2Pb_3I_{10}$ [$(PEA)_2(MA)_{n-1}(Pb)_n(I)_{3n+1}$ for $n = 3$]. The ($n = 3$) 2D perovskite manifests a wider bandgap value of 2.10 eV compared to 1.63 eV for conventional 3D $MAPbI_3$. The schematic of the $(PEA)_2MA_2Pb_3I_{10}$ ($n = 3$) crystal structure is shown in Figure 12.10, where the 3D perovskite layer (inorganic sheet) is separated by bulky PEA^+ monovalent cation to form a layered structure.

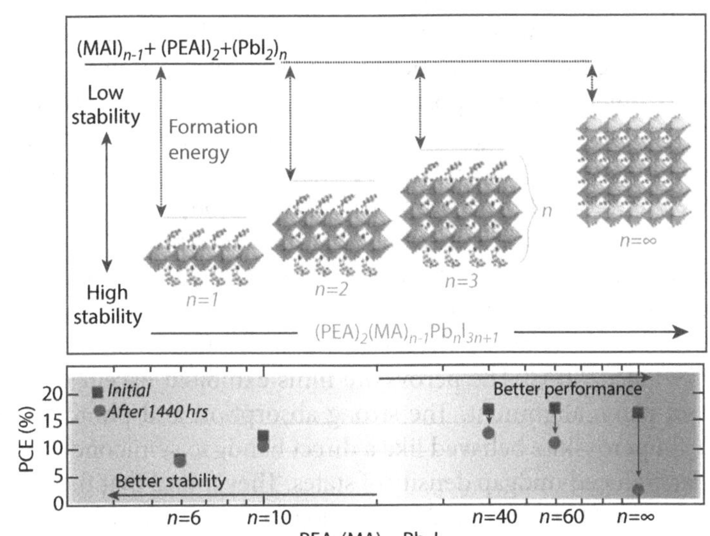

Figure 12.10 Formation energy, stability, and unit cell structure variation of $(PEA)_2(MA)_{n-1}Pb_nI_{3n+1}$ perovskites with different "*n*" values, displaying the evolution of dimensionality from 2D ($n = 1$) to 3D ($n = \infty$). Reproduced with permission from [58]. Copyright 2016, American Chemical Society.

PSCs were fabricated with compact TiO_2 and spiro-OMeTAD as the electron (ETM) and hole transport materials (HTM), forming an FTO/c-TiO_2/perovskite/Spiro-OMeTAD/Au structure. The champion 2D perovskite device attained a modest PCE of 4.73% but a high open-circuit voltage (V_{oc}) of 1.18 V under AM 1.5G. They further reported an improvement in stability. While the $n = 3$ film survived for 46 days in the air (Relative Humidity (RH) = 52%), the 3D $MAPbI_3$ degraded completely, turning from brown to yellow. Unfortunately, the stability studies were only based on material characterizations like X-ray diffraction (XRD) and UV-visible (UV-Vis) spectroscopy, but no evidence was reported regarding the stability of the device with time [57].

In 2016, Sargent *et al.* reported the stability and device performance of higher "*n*" members of $(PEA)_2(MA)_{n-1}Pb_nI_{3n+1}$ mixed dimensional perovskites, where $n = 6$, 10, 40, 60, and ∞. The theoretical studies using the ab initio density functional theory calculation (DFT) suggested that bulkier organic moieties in PEA-based perovskites reinforced Van der Waals interactions between the inorganic layers. This led to the increment of formation energy and enhancement in the stability of mixed dimensional perovskites (Figure 12.10). Also, the energy required to detach PEAI (phenylethylammonium iodide) from $(PEA)_2(MA)_{n-1}Pb_nI_{3n+1}$ perovskite matrix was 0.36 eV higher as compared to MAI (methylammonium iodide), which could be important for the decomposition rate of the material. Motivated by theory, the authors fabricated dimensionally tuned PSC devices using single-step spin coating. The characterizations confirmed the expected enhanced stability of $n = 10$, 40, and 60 (quasi-2D) perovskites compared to 3D perovskites. On the other hand, they showed (Figure 12.10) that lower "*n*" perovskites closer to 2D were the most stable ones. Still, their photovoltaic performance was inferior to higher "*n*" (i.e., approaching 3D). The lower photovoltaic performances mainly originated from poor charge carrier transport and a wider bandgap. The best-performing device of $(PEA)_2(MA)_{n-1}Pb_nI_{3n+1}$ perovskite matrix was obtained from $n = 60$, demonstrating the V_{oc} and PCE of 1.08 V and 15.3%, respectively. Unfortunately, these devices showed a decreased PCE (11.3%) when exposed to low humidity conditions (RH ~ 55%). This suggests that higher members of the RP perovskite phase (dimensionality approaching pure 3D perovskite) had relatively poor stability compared to lower members (dimensionality approaching pure 2D perovskite). This report showed a trade-off in "*n*" value between a high PCE and more stable devices. It could be achieved by optimizing mixed dimensional 2D–3D perovskites or by further studying different organic moieties in perovskite-based devices to improve stability [58].

Continuing the momentum, Mohite *et al.* demonstrated a novel "hot-casting" technique (HC) to boost the efficiency and stability of 2D–3D perovskite structures ($(BA)_2(MA)_3Pb_4I_{13}$; $n = 4$). Here, the perovskite precursor solution was spin-coated on the preheated substrate (fluorine-doped tin oxide (FTO)/poly(3,4-ethylenedioxythiophene) polystyrene sulfonate (PEDOT: PSS)) at ~150°C. The perovskite films exhibited excellent crystallinity with preferential out-of-plane alignment. The strong absorption and photoluminescence (PL) indicated that $n = 4$ perovskite behaved like a direct bandgap semiconductor with very few traps and disorder-induced midgap density of states. They found that the perovskite growth direction lay along the (111) and (202) crystallographic planes, which benefited carrier collection efficiency. The solar cell demonstrated a PCE of 12.52% with $V_{oc} = 1.01$ V, $J_{sc} = 16.76$ mA/cm^2, and $FF = 74.13\%$. Unencapsulated devices were exposed to continuous light illumination (AM 1.5G) for 2250 h (~94 days). The 2D devices retained 80% of their initial efficiency after 200 h and then slowly decreased to 70% after another 2050 h. In contrast,

3D perovskite devices degraded 40% of their initial PCE within 24 h and followed the slow degradation of around 10% of its initial values over the next 2250 h. 2D perovskites were also reported to have higher tolerance towards 65% relative humidity than 3D perovskites. The same group published another report on higher members of the RP phase of $(BA)_2(MA)_{n-1}Pb_nI_{3n+1}$ family, i.e., $(BA)_2(MA)_4Pb_5I_{16}$ (n = 5). The PSC fabricated using a similar HC approach exhibited a PCE of 8.7% with a V_{oc} of 1.0 V. The main objective of this report was to investigate the physical properties and stability of higher members of the RP perovskite phase [51].

The vertical orientation of perovskite film depends on the organic spacer or the thickness of the inorganic layer. Kanatzidis and co-workers introduced a new custom-made molecule named phenyl-2-propenammonium (PPA) to investigate the role of spacers on the preferred crystal growth direction. This chemical derivative served as an organic conjugated counter-cation spacer in the RP perovskite structure. The high-quality single crystals of $(PPA)_2(FA_{0.5}MA_{0.5})_{n-1}Pb_nI_{3n+1}$ (n = 2 and 3) perovskite were fabricated and characterized by single crystal X-ray diffraction. The results suggested that the corresponding material belonged to the RP perovskites family. It was ascertained as the structure consisted of $[PbI_6]^{4-}$ octahedra of inorganic layers separated by two layers of PPA^+ cations, and the adjacent inorganic layers were shifted by ½ unit cell length. The optical studies indicated the reduction in the bandgap with increasing "n" due to less quantum and dielectric confinement. After this, the p-i-n PSC devices were fabricated by a low-temperature synthesis of 2D PPA-RP perovskites. The stoichiometry was $(PPA)_2(Cs_{0.05}(FA_{0.88}MA_{0.12})_{0.95})_{n-1}Pb_n(I_{0.88}Br_{0.12})_{3n+1}$ for n = 2, 3, and 4 with the device stack as shown in Figure 12.11. The champion device for n = 2, 3, and 4 displayed the PCE of 3.35%, 9.60%, and 14.76%, respectively, with negligible hysteresis. The encapsulated PPA-RP device was tested up to 600 h at 85 °C with an RH of 85%. The devices maintained ~94% of their initial value of PCE as compared to ~85% for 3D perovskite. This work concluded that as the "n" value increased, the perovskite films became less crystalline and less oriented. As "n" varied from 2 to 4, the (020) plane diminished gradually and indicated the suppressed horizontal orientation. In higher "n" films (n = 4), the intensity of the (111) plane showed a larger percentage of the 3D-like phase, which implied a higher degree of vertical orientation [59].

Some of the previous studies on 3D perovskite-based optoelectronic devices illustrated that modulating residual strain could effectively alter the carrier mobility, band structure, ion migration activation energy, and defect formation energy. Hence, it could improve the stability and photovoltaic properties of devices. Taking all this effect into consideration, in a very recent report (in 2022) by Zhou *et al.*, the impact of strain relaxation on 2D RP PSC was studied. They combined the rigid PEA^+ and flexible amylammonium (AA) cations to investigate the residual strain by producing $(PEA_{1-x}AA_x)_2MA_4Pb_5I_{16}$ (x = 0, 0.1, 0.2, 0.3) matrix. They found that perovskite had a positive value of residual strain that indicated the tensile strain in an out-of-plane direction. They evaluated Young's modulus of perovskite films, and its value decreased from 9.307 GPa (control film; x = 0) to 4.816 GPa (x = 0.2), confirming the efficient strain relaxation and improved mechanical stability. They reported that the organic spacer cations are arranged in an edge-to-face manner in the pure PEA-based film, which hampers the conjugation between the layered spacer cations. This tilted the inorganic framework of the perovskite structure and brought out tensile strain in the out-of-plane direction [60].

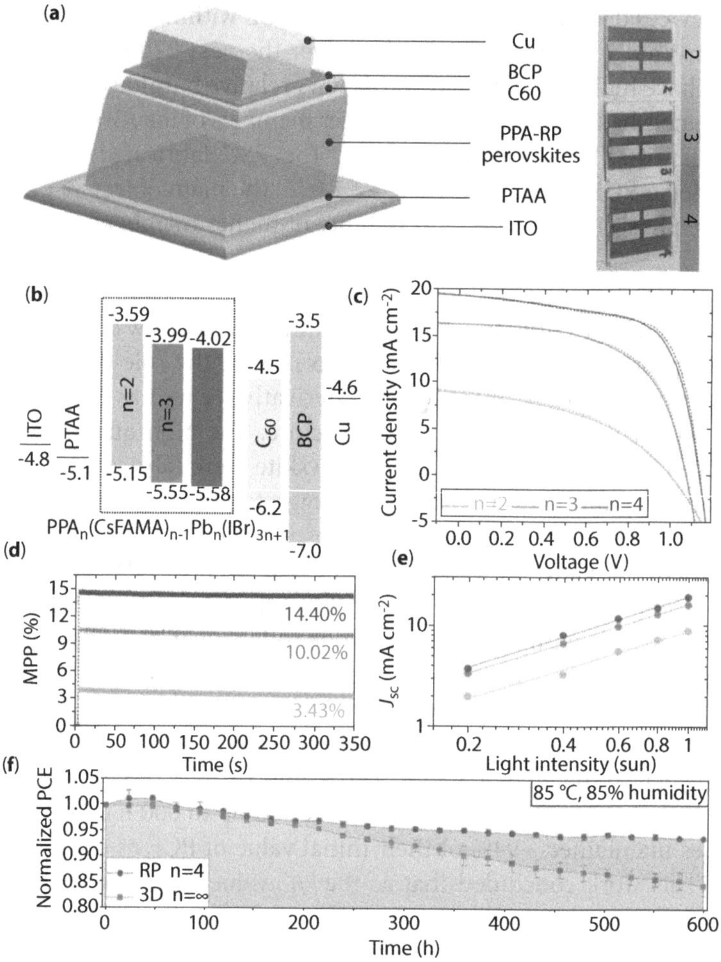

Figure 12.11 (a) Device architecture and digital photographs of p-i-n solar cell based on $(PPA)_2(Cs_{0.05}(FA_{0.88}MA_{0.12})_{0.95})_{n-1}Pb_n(I_{0.88}Br_{0.12})_{3n+1}$ for $n = 2$, 3, and 4 perovskite films. (b) Band diagram of fabricated devices. (c) *J-V* curve of champion devices of $n = 2$, 3, and 4 solar cells. (d) Maximum power point (MPP) tracking following the voltage at PCE. (e) Light-intensity-dependent J_{sc} of the best $n \leq 4$ solar cells. (f) Stability of encapsulated devices based on $n = 4$ (PPA-RP) and 3D ((Cs/FA/MA)Pb(I/Br)) perovskites tested at 85°C (humidity = 85%). Reproduced with permission from [59]. Copyright 2020, American Chemical Society.

On the other hand, in the alkyl AA incorporated perovskite system, authors suspected the reduction in steric hindrance as the spacer cations were arranged in parallel stacking. This could improvise the conjugation and release the tensile strain along the out-of-plane direction. The champion device of $(PEA_{0.8}AA_{0.2})_2MA_4Pb_5I_{16}$ (referred to as mix device) perovskite exhibited the J_{sc}, V_{oc}, and *FF* of 19.10 mA/cm^2, 1.228 V, and 76.88%, respectively, with the highest PCE of 18.03%. These devices showed an improved photovoltaic performance along with a decrement in hysteresis value compared to the control device ($(PEA)_2MA_4Pb_5I_{16}$). The decrement in the ideality factor (n) from 1.66 for the control device to 1.24 for the mix device represented the reduced trap density and suppressed recombination. From the dark *I–V* measurements based on the space-charge-limited-current (SCLC) method, the ratio of electron and hole mobility (μ_e/μ_h) for the mix device decreased from 3.86 to 1.78, representing a more balanced charge carrier transport. Remarkably, the strain-released devices (mix one) maintained 86% of their initial PCE for 1080 h when exposed to ambient with an RH of 85%. Overall, this work revealed that strain relaxation was crucial in improvising the device performance of quasi-2D perovskites [60].

Numerous bulky cations such as ammonium n-butylamine (BA) [49], 4-fluorophenethylammonium (F-PEA) [61], 3-phenyl-2-propenammonium (PPA) [59], 4-chloro-phenylformamidinium (CPFA+) [62], benzimidazolium (Bn) [51], 2-thiophene-methylammonium (ThMA+) [63], 2-furfurylammonium (FuMA+) [64], 1-amino-3-butene hydrochloride (BEA) [49], 3-butyn-1-amine hydrochloride (BYA) [49], benzamidine (PhFA) and para-fluorobenzamidine (p-FPhFA) [65], guanidinium (GA+) [66], amylammonium (AA) [60], etc. have been introduced by various researchers to incorporate into 2D and 2D–3D mixed RP perovskite structures and later into photovoltaic devices. Table 12.2 below summarizes the usage of different spacers for RP perovskites, the "*n*" formed, the device architecture, stability, and efficiency (perovskite abbreviated as PSK).

Table 12.2 RP perovskite solar cells with spacer used, device architecture, PCE and stability.

Spacer cation	Perovskite formula	Device structure	PCE (%)	Stability
PEA (*n*=3)	$(PEA)_2(MA)_2Pb_3I_{10}$	FTO/c-TiO_2/PSK/Spiro-OMeTAD/Au	4.73	Structural stability of 46 days, exposed to air, RH of 52% [57]
PEA (*n*=5)	$(PEA)_2(MA)_4Pb_5I_{16}$	FTO/TiO_2/PSK/Spiro-OMeTAD/Ag	9.74	1440 h moisture and temperature stable [67]
BA (*n*=4)	$(BA)_2(MA)_3Pb_4I_{13}$	FTO/PEDOT:PSS/PSK/PCBM/Al	12.52	Retained 80% of its original PCE after 200 h and slowly degraded to about 70% after another 2050 h [56]
BA (*n*=4)	$(BA)_2(MA)_3Pb_4I_{13}$	ITO/SnO_2/PSK/Spiro-OMeTAD/MoO_3/Au	16.25	Retained 96.2% of the initial PCE after continuous illumination at AM 1.5G for 1100 h [68]

(*Continued*)

Table 12.2 RP perovskite solar cells with spacer used, device architecture, PCE and stability. (*Continued*)

Spacer cation	Perovskite formula	Device structure	PCE (%)	Stability
PEA/GA (*n*=5)	$(PEA,GA)_2MA_4Pb_5I_{16}$	ITO/PTAA/PSK/PCBM/PEI/Ag	15.27	80% of its initial PCE after 800 h (RH of 55%–60%) [69]
F-PEA (*n*=5)	$(F\text{-}PEA)_2MA_4Pb_5I_{16}$	FTO/c-TiO_2/PSK/Spiro-OMeTAD/Au	13.64	65% of their initial PCE after 576 h (ambient environment at 70°C) [61]
PEA (*n*=5)	$(PEA)_2(MA)_4Pb_5I_{16}$	ITO/PTAA/PSK/PCBM/PEI/Ag	18.04	PCE maintained 97.7% at 80°C over 180 h, retaining 96.1% of initial efficiency after 8-month storage in the glove box [70]
Bn (*n*=10)	$(Bn)_2(MA)_9Pb_{10}I_{31}$	ITO/SnO_2/PSK/Spiro-OMeTAD/MoO_3/Ag	15.12	Maintained 92.1% of its original performance in the air (RH 40–45%) for 720 h, 87.9% of initial performance after 528 h of heating at 85°C [51]
PEA (*n*=40)	$(PEA)_2MA_{39}Pb_{40}I_{121}$	FTO/c-TiO_2-SnO_2/PSK/Spiro-OMeTAD/Au	15.07	Retention of PCE ~88% over 168 h (RH < 40%) [71]
PEA (*n*=5)	$PEA_2MA_4Pb_5I_{16}$	ITO/PEDOT:PSS/PSK/PCBM/BCP/Ag	15.14	Retained 98.6% of its original PCE after 1440 h [72]

(*Continued*)

Table 12.2 RP perovskite solar cells with spacer used, device architecture, PCE and stability. (*Continued*)

Spacer cation	Perovskite formula	Device structure	PCE (%)	Stability
PEA (*n*=60)	$PEA_2MA_{59}Pb_{60}I_{181}$	FTO/TiO_2/PSK/Spiro-OMeTAD/Au	15.30	Declined to 11.3% PCE after 60 days under low-humidity atmosphere [58]
CPFA (*n*=9)	$CPFA_2MA_8Pb_9(I_{0.857}Cl_{0.143})_{28}$	FTO/c-TiO_2/Meso-TiO_2/PSK/Spiro-OMeTAD/Au	14.78	Retained ~80% of its original PCE after 2000 h storage under ambient conditions [62]
BA (*n*=4)	$BA_2MA_3Pb_4I_{13}$	ITO/PEDOT:PSS-CuCl/PSK/PCBM/C_{60}/LiF/Cu	13.36	85% of initial PCE after 480 h (humidity of 85%) [73]
PEA/AA (*n*=5)	$(PEA_{1-x}AA_x)_2MA_4Pb_5I_{16}$	ITO/PTAA/PSK/C_{60}/BCP/Ag	18.03	Maintained 86% of the initial PCE in the air with RH 85% and 86% after continuous heating at 85°C for 1080 h [60]
FuMA (*n*=5)	$(FuMA)_2(MA)_4Pb_5I_{16}$	ITO/SnO_2/PSK/Spiro-OMeTAD/MoO_3/Au	15.66	Retained 94.0% of its initial value after 1640 min of ageing (RH 5–10%) [64]
F-PEA/GA (*n*=4)	$(F\text{-}PEA_{0.8}GA_{0.2})_2MA_3Pb_4I_{13}$	ITO/PTAA/PSK/C_{60}/BCP/Cu	17.50	Maintained 87% of its starting efficiency after 500 aging hours in ambient air [66]

(*Continued*)

Table 12.2 RP perovskite solar cells with spacer used, device architecture, PCE and stability. (*Continued*)

Spacer cation	Perovskite formula	Device structure	PCE (%)	Stability
ThMA (*n*=3)	$(ThMA)_2(MA)_2Pb_3I_{10}$	ITO/ PEDOT:PSS/ PSK/PCBM/ BCP/Ag	15.42	Retained 90% of its initial efficiency after 1000 h (RH 30 ± 10%) [63]
BA, BYA, BEA (*n*=4)	$(BA)_2MA_3Pb_4I_{13}$, $(BYA)_2MA_3Pb_4I_{13}$, $(BEA)_2MA_3Pb_4I_{13}$	ITO/ PEDOT:PSS/ PSK/PCBM/ LiF/Al	13.8, 15.1, 16.1	Without encapsulation in the glovebox for 3500 h remained 68.1%, 77.2%, and 83.1% of original PCE for BA, BYA, and BEA devices, respectively. [49]
PhFA, p-FPhFA (*n*=5)	$(PhFA)_2MA_{n-1}Pb_nI_{3n+1-x}Cl_x$, $(p\text{-}FPhFA)_2MA_{n-1}Pb_nI_{3n+1-x}Cl_x$	ITO/ PEDOT:PSS/ PSK/PCBM/ BCP/Ag	12.92, 17.37	Unencapsulated p-FPhFA-Pb and PhFA-Pb devices maintained 99% and 68% of their original PCE after 3000 h stored in an N_2-filled glovebox, respectively. [65]
PPA (*n* = 4)	$(PPA)_2(Cs_{0.05}(FA_{0.88}MA_{0.12})_{0.95})_3Pb_4(I_{0.88}Br_{0.12})_{13}$	ITO/PTAA/ PSK/C_{60}/ BCP/Cu	14.76	93.8 ± 0.25% PCE after 600 h at 85°C (RH=85%) with encapsulation [59]

12.5 Dion–Jacobson (DJ) Perovskites

As discussed above, Dion–Jacobson (DJ) perovskites have the general formula $A'A_{n-1}B_nX_{3n+1}$. Unlike R-P perovskite, DJ perovskites contain cations that are bivalent organic spacers that attach to the inorganic octahedral cages at both ends. As a result, the issues caused by the Van der Waals forces in the RP phases are eliminated. The scientific community has widely explored DJ perovskites. Some significant examples are described briefly below.

Ma *et al.* used smaller-chain cations to reduce the influence of quantum confinement and improve charge transport in lower-dimensional perovskites. Instead of butane-1,4-diammonium iodide ($BDAI_2$), propane-1,3-diammonium iodide ($PDAI_2$) was used as a spacer. The transport properties were compared for $n = 4$ perovskites for both spacers. XRD was used to measure the interlayer distance, which was nearly 7 Å for (BDA)$MA_3Pb_4I_{13}$, but was reduced to around 2 Å for (PDA)$MA_3Pb_4I_{13}$. The SCLC plots proved the hypothesis, showing that PDA-based perovskites have higher carrier mobility than BDA-based perovskites. The PDA-based device has the higher PCE of 13%, with a J_{sc}, V_{oc}, and *FF* of 18 mA/cm^2, 0.97 V, and 0.74, respectively. The stability of the devices was also compared because smaller cations typically impair stability. After 1000 h, the PCE of BDA-based devices declined to 40% of their initial value, while PDA-based devices retained 90% of their initial efficiency. Hence, the small-chain cations improved both efficiency and stability [74].

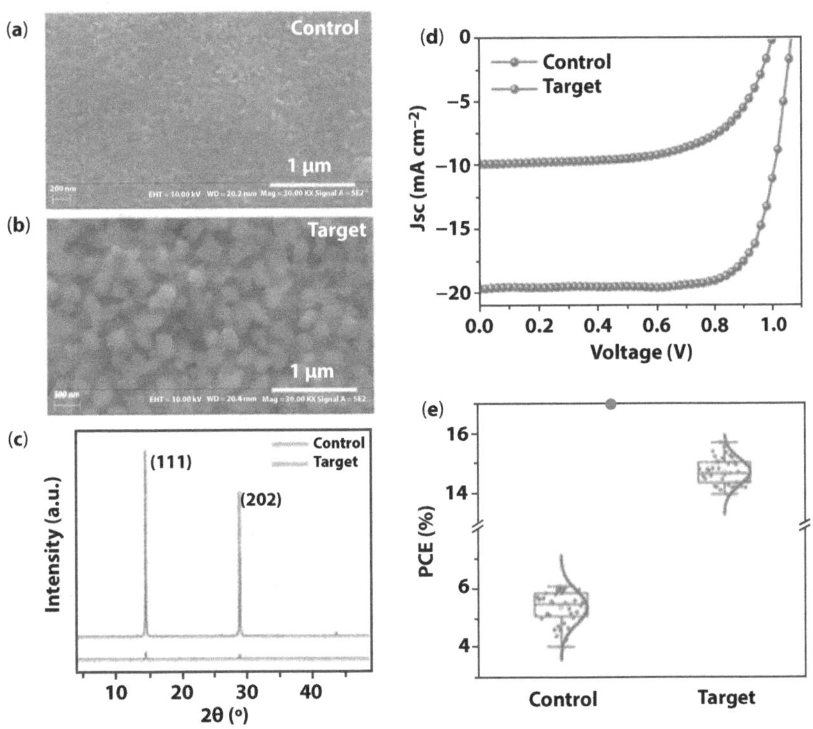

Figure 12.12 SEM of (a) control and (b) target film of (ThDMA)$(MA)_4Pb_5I_{16}$ DJ perovskite. (c) XRD, (d) light *J–V* curve, and (e) statistics plot for efficiency of control and target films. Reproduced with permission from [75]. Copyright 2020, American Chemical Society.

Lu *et al.* used an aromatic spacer, 2,5-thiophenedimethylammonium (ThDMA), to fabricate $n = 5$ DJ perovskite-based solar cells with an efficiency of 15.75% (Figure 12.12d). They claimed that film crystallinity, grain orientation, trap density, and stability all improved with one part of DMSO with 9 parts of DMF in the precursor solution (Figure 12.12). DMSO coordinates strongly with PbI_2, which reduces crystallization rate leading to larger grains. The DMSO devices showed J_{sc}, V_{oc}, *FF*, and PCE values of 9.83 mA/cm², 1.00 V, 0.62, and 6.11%, respectively. Devices without DMSO showed J_{sc}, V_{oc}, *FF*, and PCE of just 19.55 mA/cm², 1.07 V, 0.75, and 15.75%, respectively. Also, the unencapsulated cells with DMSO retained 95% of their initial efficiency in nitrogen ambient after 1655 h [75].

He *et al.* claimed that an inhomogeneous energy landscape is a major reason for V_{oc} losses in quasi-2D perovskites. The carriers photogenerated in the higher bandgap semiconductor first relax into the band edges of the lower gap semiconductor via thermal relaxation and dissociate (Figure 12.13a). The thermalization losses are higher, and the V_{oc} of the device is limited by the smallest bandgap in the 2D/3D hybrid absorber. The inhomogeneous energy bands also affected minority carrier transport and carrier generation of band tail states.

The energy landscape could be homogenized by carefully engineering material structure and crystallization rate. It was proposed that enhanced octahedral tilting and larger interlayer distance result in the pushing away of bands, increasing the bandgap. For lower V_{oc} losses, we need a flatter band structure, lower octahedral tilting, and lesser interlayer distance. MAMP (meta-(aminomethyl) piperidinium) was one such immobile cation with a short interlayer halide–halide distance. The colloidal MA^+ and PbI_x^- complexes are the nucleating sites that lead to the high crystallization rate of conventional perovskites. An acidic additive, APSA (2-aminophenol-4-sulfonic acid), was used to suppress the formation

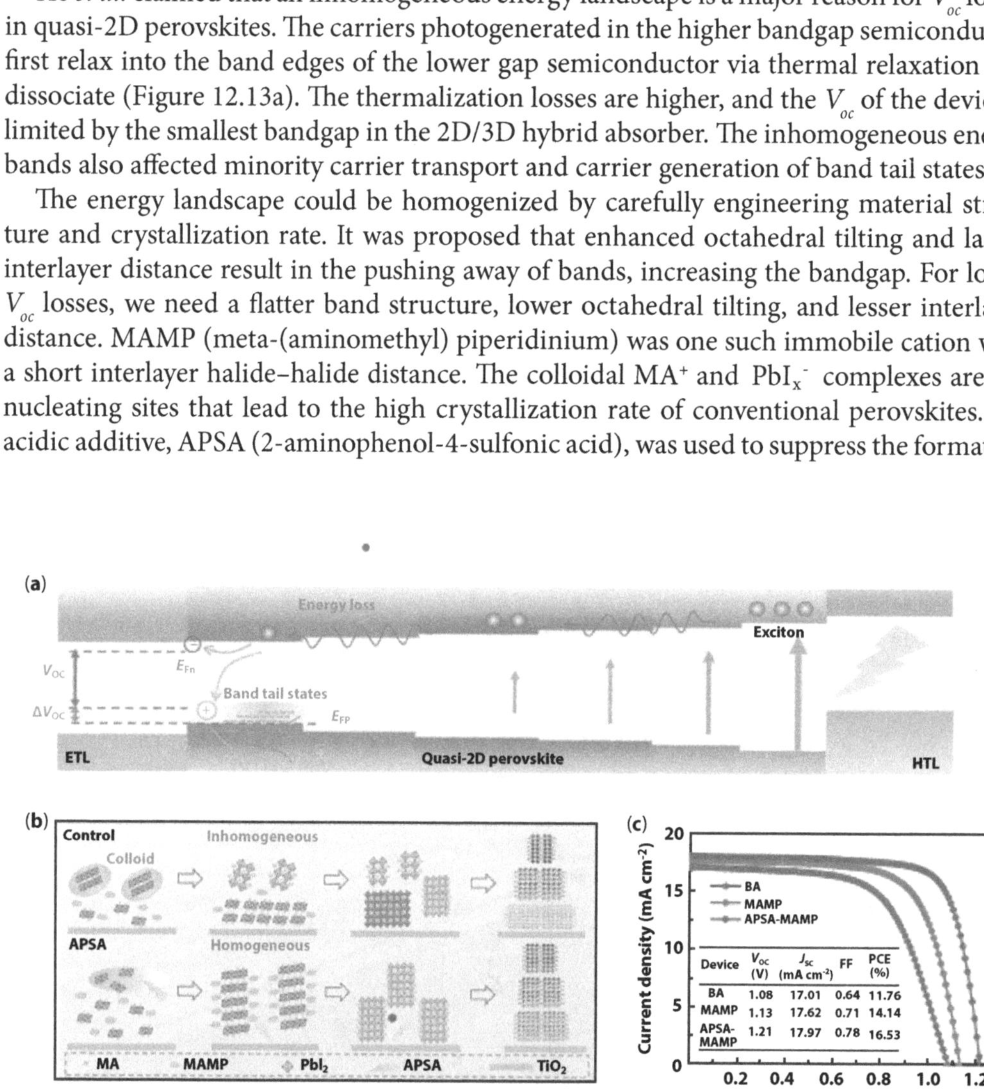

Figure 12.13 (a) Charge carrier transport in Quasi-2D perovskites along with loss mechanisms and V_{oc} of device. (b) Pictorial depiction of crystallization of films for control and APSA-treated perovskite precursor. (c) Cyclic light *J*–*V* curve for devices when spacers used were BA, MAMP, and MAMP when the precursor was treated with APSA. Reproduced with permission from [76]. Copyright 2020, Springer Nature.

of the complexes, reducing the crystallization rate (Figure 12.13b). The resulting films yielded a solar cell efficiency of 16.53%, with a high V_{oc} of 1.21 V (Figure 12.13c). The devices also exhibited enhanced stability, retaining 92% of the efficiency after 1000 h of storage [76].

Lv *et al.* suggested spacers with more non-covalent interactions with the perovskite octahedra. They compared two spacers for $n = 4$ DJ perovskite: 4-phenylenedimethanammonium (PhDMA) (non-fluorinated) and 2,3,5,6-tetrafluoro-1,4-phenylenedimethanammonium 4F-PhDMA (fluorinated). The fluorinated spacer-based perovskite had a higher dissociation energy because of multiple non-covalent interactions between the organic spacer and adjacent octahedra, like CH-F and F-I. Dynamic Light Scattering (DLS) showed that the particle size in the perovskite precursor solution increased from 400 to 800 nm for the non-fluorinated spacer to 1000–2000 nm for the fluorinated spacer. The large size of the colloid was speculated to be responsible for the larger grain size and high quality of the film in Figure 12.14a. The non-fluorinated spacer-based devices showed a J_{sc}, V_{oc}, *FF*, and PCE of 14.63 mA/cm^2, 1.03 V, 0.66, and 10.11%, respectively. Performance for the fluorinated spacer-based devices was better with a J_{sc}, V_{oc}, *FF*, and PCE of 19.58 mA/cm^2, 1.10 V, 0.77, and 16.62%, respectively (Figure 12.14b). The fluorinated spacer also enhanced stability. The fluorinated spacer-based devices retained more than 94% of their initial efficiency after 350 h when kept at 80°C in nitrogen ambient in the dark, while the non-fluorinated ones retained only about 66% of their initial PCE under the same conditions [77].

Jin *et al.* proposed an effective method to reduce surface defects with a secondary antisolvent, specifically MAI. DJ PSCs ($n = 5$) from the single-antisolvent method showed a J_{sc}, V_{oc}, and *FF* of 20.49 mA/cm^2, 1.10 V, and 0.74, respectively, i.e., an efficiency of 16.89%. Devices with secondary antisolvent showed a J_{sc}, V_{oc}, and *FF* of 21.37 mA/cm^2, 1.16 V, and 0.78, respectively, i.e., an efficiency of 19.55%. The carrier lifetime also improved. The secondary antisolvent treatment forms a lower-n phase ($n = 2$) at the interface, which passivates the surface and improves performance. When stored in the ambient at a relative humidity of

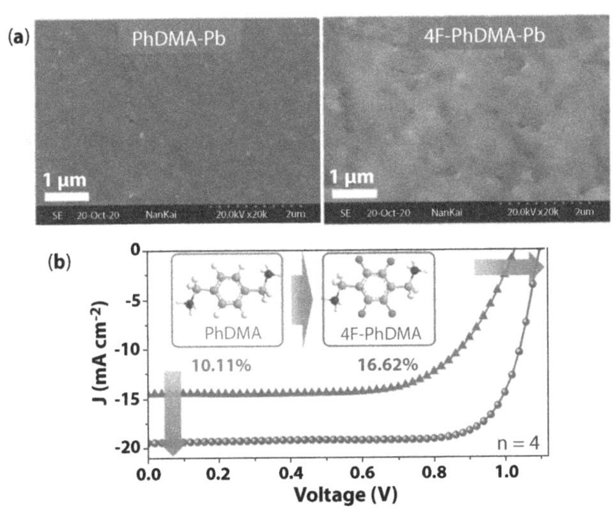

Figure 12.14 (a) SEM images for the non-fluorinated (left) and fluorinated (right) spacer-based perovskite. (b) Light *J–V* characteristics for non-fluorinated and fluorinated spacer-based perovskite solar cells. Reproduced with permission from [77]. Copyright 2021, American Chemical Society.

35–50%, devices from single antisolvent treatment retained 74.63% of the initial efficiency after 576 h. Devices from secondary antisolvent retained 83.44% of the efficiency. Hence, an enhancement of both stability and efficiency was observed with secondary antisolvent treatment. Table 12.3 below summarizes the usage of different spacers for DJ perovskites, the "*n*" formed, the device architecture, and the efficiency [78].

Table 12.3 DJ perovskite solar cells with spacer used, device architecture, PCE and stability (perovskite abbreviated as PSK).

Spacer cation	Perovskite formula	Device structure	PCE (%)	Stability
AMP (*n*=4)	(3AMP) $(MA_{0.75}FA_{0.25})_3Pb_4I_{13}$	FTO/PEDOT:PSS/ PSK/C_{60}/BCP/Ag	12.04	Retained nearly 22% efficiency under continuous AM 1.5 G illumination in ambient air with humidity of 50–70% [79].
BPDA (*n*=4)	$(PDA)MA_3Pb_4I_{13}$	ITO/PEDOT:PSS/ PSK/C_{60} (20 nm)/ BCP (8 nm)/Ag (70 nm)	13.0	Devices kept in dark at 70°C, and RH 85% retained 99% efficiency after 100 h [74].
PDA (*n*=4)	$(PDA)(FA)_3Pb_4I_{13}$	ITO/SnO_2/PSK/ Spiro-OMeTAD/ Ag	13.8	At room temperature and RH of 30–40%, cells retained 85% of their initial efficiency after 540 h [80].
BDA (*n*=5)	(BDA) $(MA)_4Pb_5I_{16}$	ITO/PCBM/PSK/ PCBM/BCP/Ag	14.53	Under RH 50±5%, the cells retained 85% of their initial efficiency for 900 h [81].
PXD (*n*=3)	(PXD) $(MA)_2Pb_3I_{10}$	ITO/PEDOT:PSS/ PSK/PCBM/BCP/ Ag	15.6	The perovskite started showing a PbI_2 peak after 65 days at RH 60-80% in dark [82].
ThDMA (*n*=5)	(ThDMA) $(MA)_4Pb_5I_{16}$	ITO/PEDOT:PSS/ PSK/PCBM/BCP/ Ag	15.75	The unencapsulated cells retained 95% of their initial efficiency in nitrogen ambient after 1655 h [75].

(Continued)

Table 12.3 DJ perovskite solar cells with spacer used, device architecture, PCE and stability (perovskite abbreviated as PSK). (*Continued*)

Spacer cation	Perovskite formula	Device structure	PCE (%)	Stability
PDMA (n=4)	(PDMA) $(MA)_3Pb_4I_{13}$	FTO/TiO_2/PSK/ Spiro-OMeTAD/ Au	15.81	The cells retained 89% of their initial efficiency in the air under RH 30% after 744 h [83].
AMP (n=4)	(3AMP) $(MA_{0.75}FA_{0.25})_3$ Pb_4I_{13}	ITO/PTAA/PFN-Br/ PSK/PCBM/BCP/ Ag	16.26	The unencapsulated devices retained 90% of their initial stability after 800 h of air exposure at an RH of 55±5% [84].
MAMP (n=4)	(MAMP) $MA_3Pb_4I_{13}$	FTO/c-TiO_2/PSK/ Spiro-OMeTAD/ Au	16.53	The unencapsulated cells retained 92% of their initial efficiency after 1000 h of storage at RH 40–50% [76].
4F-PhDMA (n=4)	(4F-PhDMA) $(MA)_3Pb_4I_{13}$	ITO/PEDOT:PSS/ PSK/PCBM/BCP/ Ag	16.62	The devices maintained more than 93% of their initial efficiency after 1839 h when stored in dark in nitrogen ambient [77].
BDA (n=5)	(BDA) $(MA)_4Pb_5I_{16}$	ITO/PEDOT:PSS/ PSK/PCBM/LiF/ Au	17.91	The unencapsulated devices retained 84% of their initial efficiency in ambient air [85].
BDA (n=5)	(BDA) $(Cs_{0.1}FA_{0.9})_4$ Pb_5I_{16}	FTO/cp-TiO_2/ mpTiO_2/PSK/ Spiro-OMeTAD/ Au	18.2	The unencapsulated device retained more than 80% of the initial efficiency after 800 h [86].
4AMP (n=3)	(4AMP) $MA_2Pb_3I_{10}$	ITO/PEDOT:PSS/ PSK/PCBM/BCP/ Al	18.3	N/A [87]

(*Continued*)

Table 12.3 DJ perovskite solar cells with spacer used, device architecture, PCE and stability (perovskite abbreviated as PSK). (*Continued*)

Spacer cation	Perovskite formula	Device structure	PCE (%)	Stability
3AMP (n=4)	(3AMP)$(MA_{0.75}FA_{0.25})_3Pb_4I_{13}$	ITO/(NiO_x/PTAA)/PSK/PCBM/BCP/Ag	18.67	The devices retained 90% of their initial efficiency at RH 45±5% in air for 60 days [88].
TTDMA (n=4)	(TTDMA)$(MA)_3Pb_4I_{13}$	ITO/PEDOT:PSS/PSK/PCBM/BCP/Ag	18.82	The cells retained 99% of their initial efficiency in nitrogen ambient after 4400 h [89].
BDA (n=5)	(BDA)$(MA_{0.07}FA_{0.93})_4Pb_5I_{16}$	ITO/SnO_2/PSK/Spiro-OMeTAD/Au	19.55	The unencapsulated devices retained 74.89% of their initial efficiency after 240 h at 60°C at RH 25–35% [78].

12.6 Alternating Cation Interlayers

Soe *et al.*, in 2017, first reported 2D perovskites with alternating cations guanidinium (GA^+, $C(NH_2)_3$) and methylammonium (CH_3NH_3, MA^+) in the interlayer space (ACI). The distinct ordering of spacer cations leads to the evolution of a unique structure type that combines the structural feature of DJ perovskites with the chemical formula of RP perovskites. Currently, $(GA)(MA)_nPb_nI_{3n+1}$ compositions are the most studied ACI perovskites, represented in Figure 12.15a. The 2D nature is visible in n = 3 ($GAMA_3Pb_3I_{10}$) with its plate-like crystal morphology (Figure 12.15b). The higher crystal symmetry and incorporation of MA cation in the spacer favors the reduction in bandgap value (1.73 eV), and smaller dielectric confinement of these quantum well-like systems in ACI perovskite phase leads to lower binding energy, which is useful for photovoltaic applications. Therefore, the planar solar cells (device stack in Figure 12.15c) comprising the n = 3 perovskite as an absorber layer demonstrated a PCE of 7.26% with a V_{oc} of 0.974 V and a striking FF of ~80% (Figure 12.15d) [90].

Zhao and co-workers focused on the further advancement in GA-based 2D ACI PSCs. They applied the advantages of hot-casting (HC) and anti-solvent engineering techniques to control the kinetic transformation from disordered sol–gel colloidal precursor to the crystalline phase. They demonstrated the influence of the dynamical process on the resulting perovskite thin film ($GAMA_3Pb_3I_{10}$) and its photovoltaic properties. The outcomes from *in situ* GIWAXS revealed the emergence of an intermediate phase, including

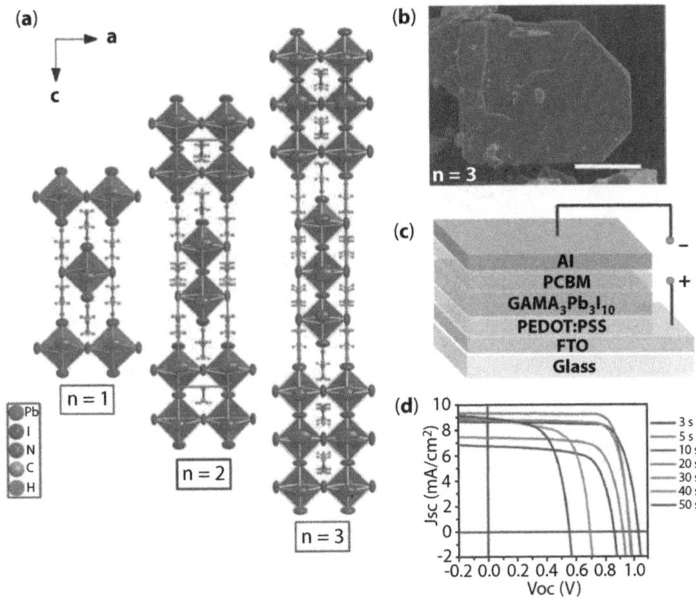

Figure 12.15 (a) Unit cell of $(GA)(MA)_nPb_nI_{3n+1}$ (n = 1–3) perovskites along the crystallographic b-axis. (b) SEM image of n = 3 ($GAMA_3Pb_3I_{10}$) perovskite crystal displaying the plate-like morphology. (c) The device structure of the fabricated solar cell. (d) J–V characteristics of $GAMA_3Pb_3I_{10}$ perovskite as a function of antisolvent (chlorobenzene) dropping delay time. Reprinted with permission from [90]. Copyright 2017, American Chemical Society.

an intermediate crystalline solvate and the 2D GA_2PbI_4 perovskite (represented in Figure 12.16a, b), followed by the final transformation into the desired ACI 2D perovskites. They proposed that the reason behind the formation of the intermediate phase can be attributed to the intricate interactions between the solvent, lead halide, and organic components. The impact of the deposition technique on the nature of the thin film was examined by scanning electron microscopy (SEM), shown in Figure 12.16g–i. The crystal structure of as-formed perovskite is depicted in Figure 12.16d. All the ACI films exhibited elongated micrometre-sized grains, strikingly different from a polycrystalline film of 3D $MAPbI_3$, which has densely packed granular grains. Consequently, the PCE of 14.69% was higher for antisolvent-assisted devices (device stack Figure 12.16c) than standard (PCE = 12.91%) and hot casted (PCE = 13.87%) devices, represented in Figure 12.16e. The unencapsulated antisolvent processed device exhibited retention of 88% of its original efficiency after 240 days (Figure 12.16f) of ambient exposure to dark (RH ~ 30–40% and T = 25°C). Later, they demonstrated the enhancement in PCE from 15.79% for the pristine case to 18.48% for the MACl incorporated case. A better film quality with more prominent grains and a smoother surface was obtained by introducing the MACl as an additive into $GAMA_3Pb_3I_{10}$ perovskite [91].

In 2022, N. G. Park and a co-worker modified the surface of GA-based quasi-2D perovskites with amphoteric imidazolium iodide (ImI) and achieved over 22% PCE.

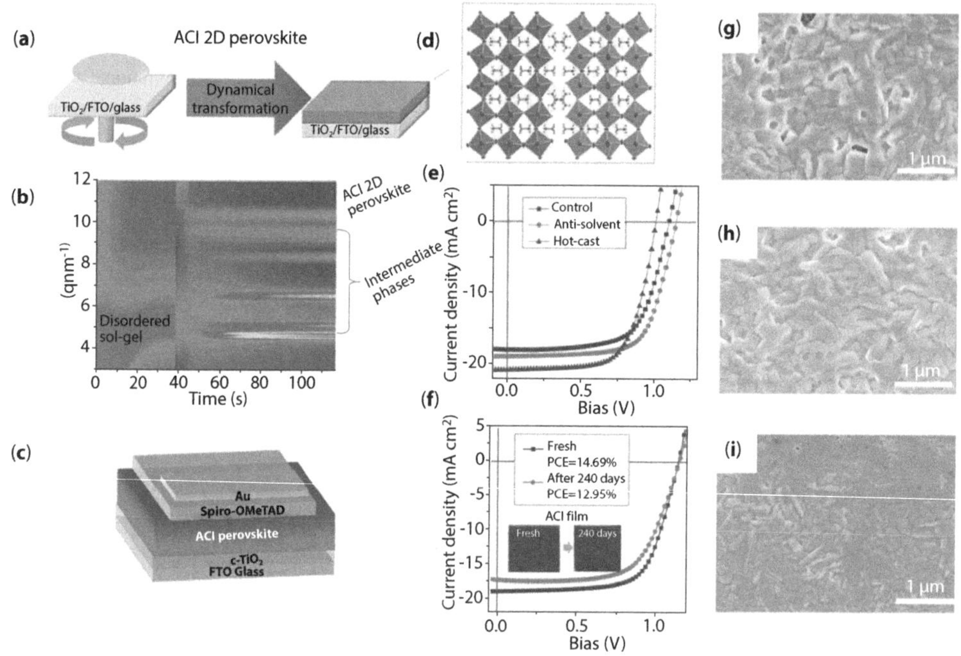

Figure 12.16 (a) Pictorial depiction of kinetic transformation of $(GA)(MA)_nPb_nI_{3n+1}$ ACI perovskites from disordered sol-gel to the pure crystalline phase. (b) *In situ* GIWAXS analysis representing the dynamical transformation from intermediate phase to control perovskite phase processed by antisolvent assisted technique. (c) Fabricated device and (d) crystal structure of $n = 3$ $(GA)(MA)_3Pb_3I_{10}$ ACI 2D perovskites. (e) *J–V* curves of the three best devices processed by control, anti-solvent, and hot-casting methods. (f) *J–V* curve and PCE of the unencapsulated antisolvent processed device before and after 240 days of ambient exposure with a relative humidity of ~30–40%. (g–i) SEM images of control, anti-solvent processed and hot-casting method used perovskite films. Reprinted with permission from [91]. Copyright 2019, American Chemical Society.

They used ImI as a post-treatment material for $GA(MA)_5Pb_5I_{16}$ perovskite because ImI was expected to interact with the perovskite layer via ionic bonding and the spiro-MeOTAD layer via π–π interaction. They concluded that a low concentration of ImI (less than 2 mg/ml) can improve crystallinity and effective surface defect passivation, which led to an increase in the carrier lifetime and improved V_{oc} and *FF*. On the other hand, if ImI was used in high concentration, it further deteriorated the photovoltaic performance by forming an insulating layer of $(Im)_2PbI_4$ (Figure 12.17a). The residual ImI on the perovskite surface also hampered the interfacial charge transfer process. ImI-treated devices were also found to be more stable for 50 days (1200 h) under ambient conditions (RH ~ 25%, T = 25°C) as they maintained a higher value of PCE than their pristine counterpart (refer Figure 12.17) [92].

Some of the evolving works in ACI-based PSCs have been listed in Table 12.4, along with their efficiency and stability. With the combination of properties of RP and DJ perovskites, ACI perovskites have proven to be a competent material for absorbers for solar cells.

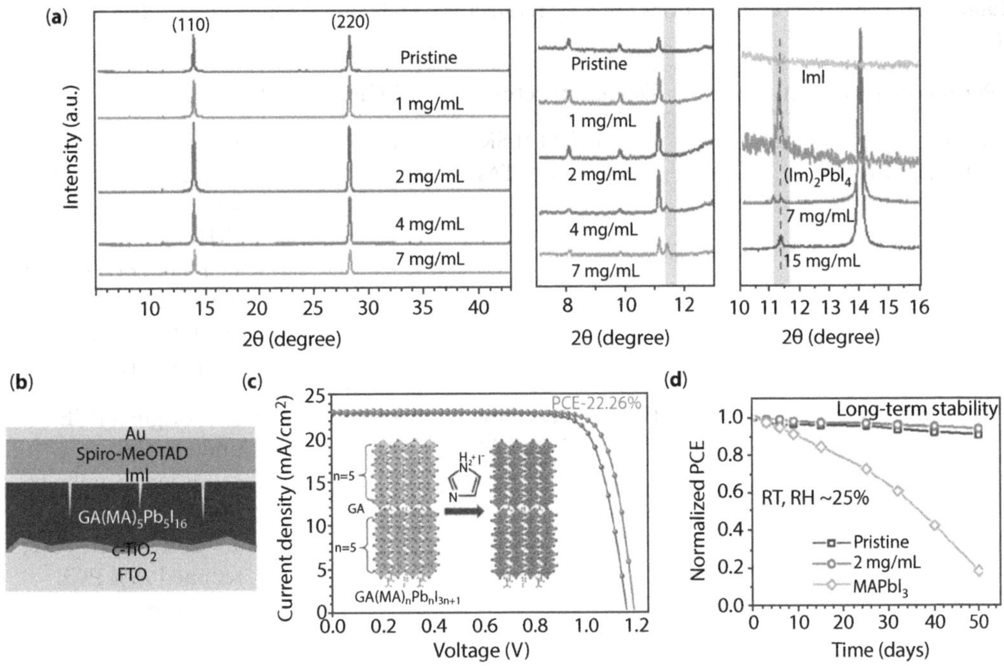

Figure 12.17 (a) Left: XRD patterns of $GA(MA)_5Pb_5I_{16}$ perovskite films treated with different concentrations of ImI. Middle: XRD patterns of low angles from 7° to 13°. Right: XRD of ImI, $(ImI)_2PbI_4$, and perovskite treated with a high concentration of ImI (7 and 15 mg/ml). (b) The device structure of $GA(MA)_5Pb_5I_{16}$ 2D ACI perovskite-based solar cells. (c) *J–V* characteristics of champion devices; the inset shows the crystal structure of $n = 5$ ($GA(MA)_5Pb_5I_{16}$) ACI perovskites. (d) Long-term stability of PSCs without (standard) and with 2 mg/mL ImI treatment, along with $MAPbI_3$-based 3D PSC. Reproduced with permission from [92]. Copyright 2022, American Chemical Society.

Table 12.4 Perovskite compositions, device architecture, PCE, and stability of ACI devices.

Perovskite composition	**Device structure**	**PCE (%)**	**Stability**
$(GA)(MA)_3Pb_3I_{10}$ ($n = 3$)	FTO/PEDOT:PSS/ PSK/PCBM/Al	7.26	N/A [90]
$(GA)(MA)_4Pb_4I_{13}$ ($n = 4$)	ITO/PEDOT:PSS/ PSK/PTAA/Ag	14.3	92% of its initial PCE after 600 h, under ambient conditions (devices without encapsulation) [28]
$(GA)(MA)_3Pb_3I_{10}$ ($n = 3$)	ITO/PEDOT:PSS/ PSK/PCBM/LiF:Al	16.32	80% of its initial PCE after 2400 h, stored in a glove box (without encapsulation) [93]

(*Continued*)

Table 12.4 Perovskite compositions, device architecture, PCE, and stability of ACI devices. (*Continued*)

Perovskite composition	Device structure	PCE (%)	Stability
$(BEA)_{0.5}Cs_{0.15}(FA_{0.83}MA_{0.17})_{2.85}Pb_3(I_{0.83}Br_{0.17})_{10}$ (n = 3)	FTO/c-TiO_2/PSK/Spiro-OMeTAD/Au	17.39	90% of its initial PCE after 2400 h, stored in ambient conditions (RH ~35%); 93% of PCE under continuous illumination after 500 h (at room temperature) [94]
$(GA)(MA)_3Pb_3I_{10}$ (n = 3)	FTO/c-TiO_2/PSK/Spiro-OMeTAD/Au	18.48	95% of original PCE under a dark condition in ambient air (RH ~ 30–40%, T = 25°C) for 31 days; retained 79% PCE after continuous heating at 80 °C for 65 h in an inert atmosphere [95]
$(GA)(MA)_3Pb_3I_{10}$ (n = 3)	FTO/TiO_2/PSK/Spiro-OMeTAD/Au	14.69	88% of initial PCE after 240 days, exposed to ambient air (RH ~ 30–40%, T = 25 °C) [91]
$(GA)(MA)_4Pb_4I_{13}$ (n = 4)	ITO/PEDOT:PSS/PSK/PCBM/BCP/Ag	15.27	75% of its original PCE after 480 h without encapsulation (RH ~ 50%, T = 42 °C) [96]
$GA(MA)_nPb_nI_{3n+1}$ (n = 5)	FTO/c-TiO_2/PSK/ImI/Spiro-OMeTAD/Au	22.26	94% of its initial PCE after 1200 h, under ambient conditions, light soaking (RH = 25%, T = 25 °C) [92]

Low-dimensional perovskites are known for their enhanced stability, but higher efficiency is always challenging to achieve because of quantum confinement by the large organic spacers. Several efforts were made to carefully tailor the properties of these low-dimensional perovskites such that they had both better efficiency and stability. Two of the most common approaches to this were using additive and compositional engineering, which exhibited the enhancement in efficiency without compromising their stability. These two approaches are described in the section below, with some examples.

12.7 Additive Engineering

In 2D–3D mixed-dimensional layered perovskites, the movement of charge carriers is restricted along the directions of insulating organic layers. Consequently, the efficiency of these low-dimensional PSCs is lower than expected. Generally, two different strategies are adopted to reduce the insulating effects of the organic layers and improve the photovoltaic performance of the devices: (a) decrease the content of the insulating organic layer, and (b) vertically orient the metal halide layers to favor the transport of photocurrent (Figure 12.18c). Additives in precursor solutions are known to control the crystallization of 3D perovskites. A similar approach is envisioned to modify the crystallization of 2D–3D-based perovskites. Table 12.5 summarized the use of different additives for perovskite solar cells.

Chen *et al.* first added NH_4SCN (ammonium thiocyanate) into the $(BA)_2(MA)_2Pb_3I_{10}$ perovskite compound, and the films showed the larger grains that are vertically oriented perpendicular to the substrate and reduced grain boundaries, leading to enhanced carrier lifetime and transport mobility. The device performance improved from 1.31% (without NH_4SCN) to 6.89% (with NH_4SCN), indicating the improved charge transport and extraction due to the oriented growth of perovskite films. The unsealed devices manifested high stability with almost consistent PCE during the storage period in the N_2 glove box. Later, they incorporated NH_4SCN additive into $(PEA)_2(MA)_4Pb_5I_{16}$ (n = 5) films and fabricated

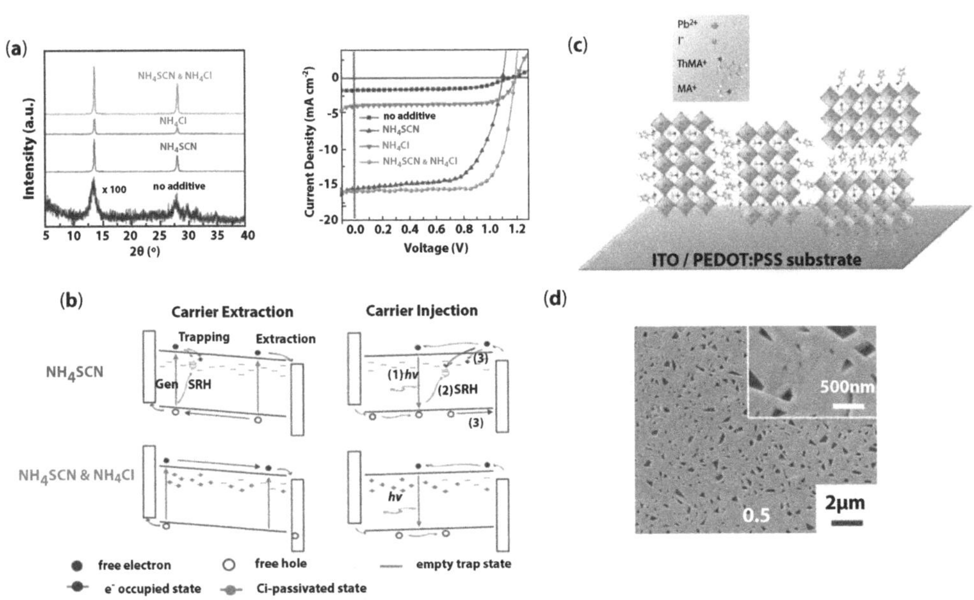

Figure 12.18 (a) XRD pattern and J–V spectra of $(PEA)_2(MA)_4Pb_5I_{16}$-based (n = 5) perovskite films and devices processed with various additives. (b) Schematic carrier injection and carrier extraction mechanism in the devices with NH_4SCN and NH_4SCN+NH_4Cl as additives. Reproduced with permission from [98]. Copyright 2018, American Chemical Society. (c) Schematic diagram of proposed packing structures of $(ThMA)_2(MA)_2Pb_3I_{10}$-based 2D perovskite with MACl/MAI weight ratio of 0.3, 0.7, and 1.0. (d) SEM image of $(ThMA)_2(MA)_2Pb_3I_{10}$-based 2D perovskite with a MACl/MAI weight ratio of 0.5. Reproduced with permission from [99]. Copyright 2018, American Chemical Society.

Table 12.5 Use of different additives for perovskite solar cells. Here, the abbreviation for Device **A:** ITO/PEDOT:PSS/PSK/PCBM/BCP/Ag, Device **B:** ITO/SnO_2/PSK/Spiro/MoO_3/Ag.

Additive	**Perovskite**	**Device structure**	**Stability**	**PCE (%)**
NH_4SCN	$(BA)_2(MA)_2Pb_3I_{10}$ ($n = 3$)	Device A	100% PCE after 25 days (stored in N_2 filled glove box)	6.82 [100]
NH_4SCN	$(PEA)_2(MA)_4Pb_5I_{16}$ ($n = 5$)	Device A	78.5% of its initial PCE after 160 h storage in air (RH 55 ± 5%)	11.01 [97]
NH_4SCN +NH_4Cl	$(PEA)_2(MA)_4Pb_5I_{16}$ ($n = 5$)	Device A	90% PCE after 45 days, stored in ambient air, dark, RH ~ 30%	14.1 [98]
PEAI	$BA_2MA_4Pb_5I_{16}$ ($n = 5$)	Device A	90% of PCE after 1000 h exposure to air (RH 25 ± 5%)	14.09 [101]
MACl	$(ThMA)_2(MA)_2Pb_3I_{10}$ ($n = 3$)	Device A	90% of PCE after 1000 h, stored in the dark (RH 30 ± 10%)	15.42 [99]
GASCN + MACl	$(GA)(MA)_4Pb_4I_{13}$ ($n = 4$)	Device A	Retained 75% of PCE for 480 h (RH 50 ± 5%)	15.27 [96]
GAI/MACl	$BA_2MA_4Pb_5I_{16}$ ($n = 5$)	Device A	90% of PCE after 800 h, exposure to air (RH 55 ± 5%)	16.26 [84]
KI	$(Bn)_2(MA)_9Pb_{10}I_{31}$ ($n = 10$)	Device B	92.1% of initial PCE after 720 h, kept in the air (RH 40–45%)	15.12 [51]

the devices by one-step spin coating technique. With the optimized value of NH_4SCN for $n = 5$, the PCE of the device boosted from 0.56% (original) to 11.01% (optimized) [97].

In another report by Jen *et al.*, NH_4Cl was combined with NH_4SCN as an additive for $(PEA)_2(MA)_4Pb_5I_{16}$-based perovskite devices. They found that the addition of NH_4SCN and NH_4Cl improved the crystallinity of the quasi-2D perovskite film (Figure 12.18a) and stimulated the preferred orientation along (0k0) planes perpendicular to the electrode. Moreover, they observed that the addition of NH_4Cl increases the V_{oc} value from 1.11 V (with only NH_4SCN additive) to 1.21 V, which confirmed the achievement of low nonradiative V_{oc} loss (~0.16 V) for the $(PEA)_2(MA)_4Pb_5I_{16}$ perovskite systems. This reduction in nonradiative recombination arose from the passivation of electron traps by Cl^- near the PEDOT:PSS/perovskite hole collecting interface, which compensated for the halide vacancies and contributed to improvement in charge carrier collection. The schematic of this mechanism is represented in Figure 12.18b. The planar structured PSCs with the device architecture ITO/PEDOT:PSS/$(PEA)_2(MA)_4Pb_5I_{16}$/$PC_{61}BM$/BCP/Ag exhibited the PCE of 14.1% (average value 12.9 ± 0.8%) with a high V_{oc} of 1.21 ± 0.01 V. Additionally, the encapsulated device retained 90% of its original PCE after 45 days of exposure into ambient with a humidity of ~30%. In contrast, the 3D perovskite devices retain only ~50% of their initial PCE after 10 days of exposure under the same conditions, indicating that the addition of additives into 2D–3D perovskites material not only enhanced the photovoltaic performance but also strengthened the long-term stability of devices [98].

Lai *et al.* proposed the use of MACl (methylammonium chloride)-assisted process to produce the highly oriented thin films of $(ThMA)_2(MA)_2Pb_3I_{10}$ (2-thiophenemethylammonium ($ThMA^+$)). The resulting 2D perovskites dramatically improved the PCE from 1.74% to 15.42%. The perovskite precursor solution was synthesized by varying the weight ratio of MACl and MAI from 0 to 1. Perovskite films were casted on ITO/PEDOT:PSS substrate by combining antisolvent treatment. A dense nano-rod-like morphology was formed along with improved crystallinity, increased crystal size, and out-of-plane crystal orientation (Figure 12.18c and d). The higher device performance was mainly due to the increased charge carrier mobility and charge carrier lifetime [99].

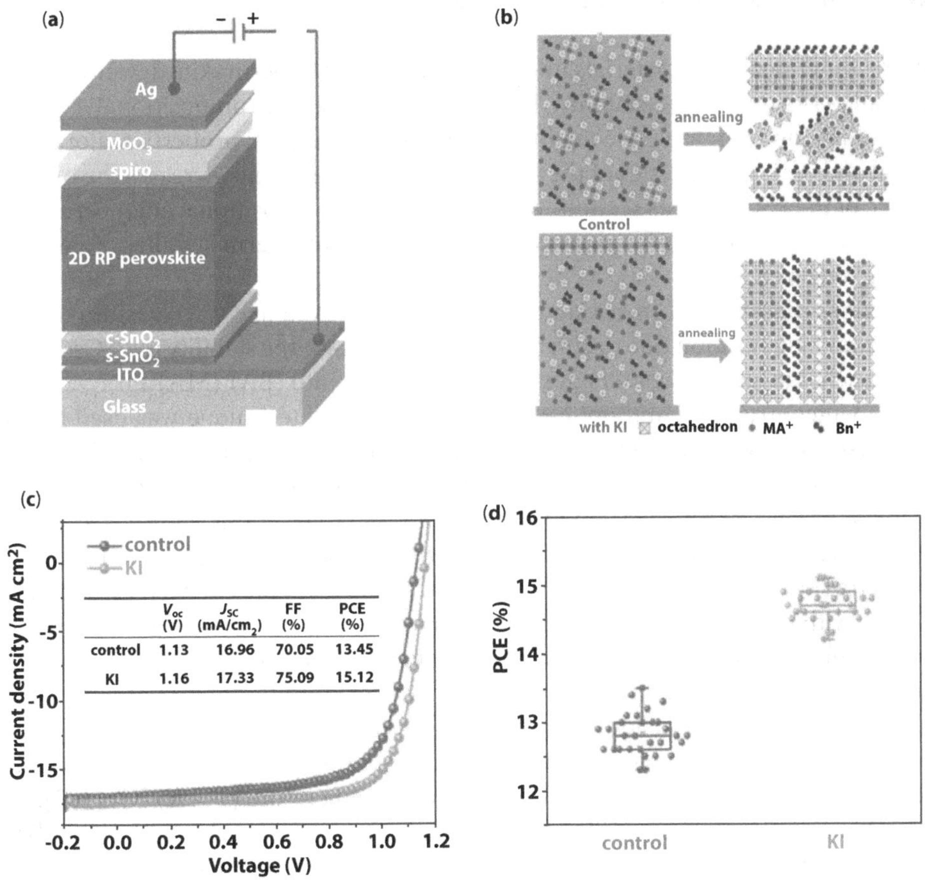

Figure 12.19 (a) The fabricated device structure of $(Bn)_2(MA)_9Pb_{10}I_{31}$ ($n = 10$) 2D RP PSCs. (b) Schematic diagram of perovskite crystal growth without or with KI additive condition. (c) *J–V* characteristics of control (without) and with KI additive perovskite champion devices. (d) Statistics of PCEs for the corresponding devices. Reproduced with permission from [51]. Copyright 2021, American Chemical Society.

Recently, Li *et al.* reported a multifunctional strategy by adding KI as an additive into benzimidazolium iodide (BnI)-based $(Bn)_2(MA)_9Pb_{10}I_{31}$ ($n = 10$) 2D RP perovskites. The high-quality 2D RP perovskite films were fabricated by the hot-casting free and antisolvent-free spin-coated method. Enhanced charge carrier transport within the 2D RP perovskite layer was achieved due to the delocalization of π-electrons in the entire conjugated backbone of Bn^+. Consequently, higher values of PCE, 13.42% and 13.45%, were obtained for $(Bn)_2(MA)_{n-1}Pb_nI_{3n-1}$ with $n = 5$ and $n = 10$, respectively. In addition, for (n = 10) 2D RP perovskite layer, additive engineering was performed using KI, which improved the crystal orientation by inducing nucleates to grow along (111) and (202) crystal planes. As a result, an increased PCE of 15.12% was achieved as well as higher stability when kept in air for 720 h at 40–45% of RH (shown in Figure 12.19) [51].

Similar successful approaches were also followed with additives such as PEAI [99], GAI [84], and GASCN [96] in pursuit of higher PCE and long-term stability of these low-dimensional 2D–3D PSCs.

12.8 Compositional Engineering

Including functional nanomaterials into various layers of the perovskite device can enhance the overall performance and stability. This is achieved through compositional engineering. These modifications improve the intrinsic physical and chemical properties, along with the electrical and optical properties of the materials, which ultimately increases the overall performance of the PSCs. In this section, various aspects of compositional engineering are discussed, and their effects on the material properties and long-term stability of PSC devices have been analyzed.

In previous reports on 3D perovskites, it has been observed that Cs^+ (caesium) cation can improve the crystallization due to entropy gain and enhance the stability of PSCs. Following this approach, Liu and co-workers doped the Cs^+ cation into $(BA)_2(MA)_3Pb_4I_{13}$ 2D RP perovskites to alter structural and optical properties. The HC technique was used to prepare the precursor solution of $(BA)_2Cs_{3x}(MA)_{3-3x}Pb_4I_{13}$ perovskite by changing x values from 0% to 5%, 10%, and 15%. The results indicated that Cs^+ doping influenced the crystallization behavior of perovskite films, leading to the growth of larger grains with phase purity and crystal domains oriented perfectly with respect to the substrate. The optimized 5% Cs^+-doped perovskite film was used to fabricate PSCs. The champion device exhibited a PCE of 13.68% with $V_{oc} = 1.08$ V, $J_{sc} = 19.95$ mA/cm^2, and $FF = 63.47\%$. The improvement in optoelectronic properties in terms of charge carrier mobility, trap density, and charge transfer kinetics are mainly attributed to the doping of Cs^+ cation on the site of MA^+ cation, as displayed in Figure 12.20. In addition, these devices maintained 89% of their original PCE after 1400 h of exposure under ambient conditions (RH ~ 30%) and showed enhanced tolerance to high humidity (between 65% to 85%) and temperature (T = 85 °C) [102].

Later, Shao *et al.* incorporated the large guanidinium cation (GA^+) with the bulky PEA cation to form a mix $(PEA, GA)_2MA_4Pb_5I_{16}$ 2D RP perovskite. The mixed cations (PEA+GA) led to a denser and smoother morphology without the cracks and pinholes seen in standard PEA films. GA cations were responsible for the increased nucleation density and controlled crystallization process, leading to improved film. The bandgaps for standard PEA and mixed PEA+GA-based devices were about 1.63 eV and 1.58 eV, respectively, so the

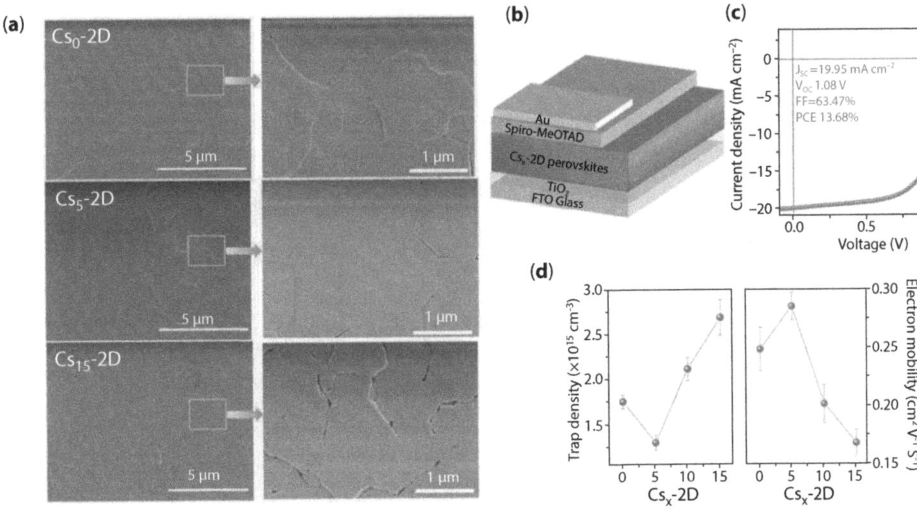

Figure 12.20 (a) SEM micrographs of low- and high-magnification Cs^+-doped 2D $(BA)_2Cs_{3x}(MA)_{3-3x}Pb_4I_{13}$ perovskite films. (b) Device architecture of fabricated solar cells. (c) *J–V* spectra of champion device. (d) The variation of trap density and electron mobility in the Cs_x-2D perovskite films. Reproduced with permission from [102]. Copyright 2017, The Royal Society of Chemistry.

PEA+GA devices were more absorptive. Compared to $(PEA)_2MA_4Pb_5I_{16}$, the devices with $(PEA_{0.8}GA_{0.2})_2MA_4Pb_5I_{16}$ perovskite showed a higher PCE: 11.58% versus 15.27%. Both J_{sc} and *FF* improved. The charge carrier mobility measured by the SCLC method showed that the electron and hole mobility of the mixed (PEA+GA) films was higher than standard PEA films. Conclusively, they found that the incorporation of GA^+ cation strengthened the stability of 2D devices, which maintained over 80% of their original efficiency after storing for 800 h under the ambient condition (represented in Figure 12.21a–c) [69].

Recently, Song and co-workers used 1,4-butanediamine (BEA) as an insulating organic cation to synthesize a new type of 2D low-dimensional perovskite with the stoichiometry of $(BEA)_{0.5}MA_nPb_nI_{3n+1}$ ($n = 3$), which comprised monoammonium (MA^+) and diammonium (BEA^{2+}) alternating cations in the interlayer space (B-ACI). Contrary to conventional ACI perovskite, B-ACI perovskites had more stability due to hydrogen bonds between BEA and inorganic slabs. This structure combined the advantage of both DJ and ACI perovskites. The $(BEA)_{0.5}MA_3Pb_3I_{10}$ perovskite film manifested full coverage and single crystal-like morphology along with the large grain. The thickness of the films was around 350 nm, which matched well with the charge carrier diffusion length. Highly textured films with excellent crystallinity confirmed the crystal orientation parallel to the substrate. The B-ACI perovskite provided an efficient charge carrier localization process and enabled a balanced carrier transport pathway to enhance mobility. The devices based on $(BEA)_{0.5}MA_3Pb_3I_{10}$ perovskite as an active layer reached a performance of 14.86%, which further improved to 17.39% by incorporating the mixed cations such as FA, Cs, and Br into the composition ($(BEA)_{0.5}Cs_{0.15}(FA_{0.83}MA_{0.17})_{2.85}Pb_3(I_{0.83}Br_{0.17})_{10}$). Conclusively, encapsulated triple cation device retained 90% of the original PCE when kept in ambient atmospheric conditions in the dark for up to 2400 h (RH ~ 35%). No sign of degradation under constant illumination for over 500 h was observed at room temperature [94].

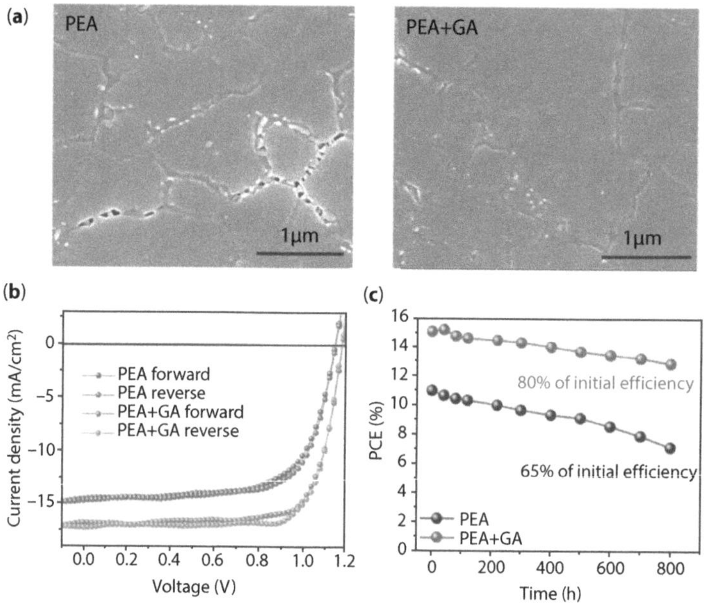

Figure 12.21 (a) SEM images of films, (b) *J–V* characteristics, and (c) stability under humidity of PEA and PEA+GA ($(PEA, GA)_2MA_4Pb_5I_{16}$)-based 2D RP PSCs. Reproduced with permission from [69]. Copyright 2020, AIP Publishing.

A similar approach was made by Zhou *et al.*, in which they incorporated FA^+ cation into $(BA)_2(MA,FA)_3Pb_4I_{13}$ 2D perovskites and reported that FA^+ cation modulated the crystallization kinetics of perovskite, which decreased the nonradiative recombination centers to obtain high-quality films. The PSC devices based on this perovskite consisted of a maximum PCE of 12.81% and increased moisture stability. Table 12.6 shows the list of various cations used as absorber material in 2D PSCs by compositional engineering method. These findings of compositional engineering benefit the understanding of doping behavior in low-dimensional perovskites in the context of device performance and stability [103].

The above sections dealt with how we can enhance the performance of perovskite-based solar cells. A salient feature of low-dimensional perovskites is having a large bandgap. Hence, they can be a potential candidate for applications that do not require absorption of the entire visible spectrum. One such area is semitransparent solar cells. The upcoming section discusses various possible applications relevant to semitransparent perovskites.

12.9 Functional Perovskite Photovoltaics

Photovoltaics is already the cheapest source of renewable energy in many markets. However, affordability is just a necessary, not sufficient condition for market penetration. Consumers often demand additional attributes, like functionality, aesthetics, and integration. If the goal is ubiquitous photovoltaics, then we have to cater to the need for niche applications like building-integrated photovoltaics and agrivoltaics. Many of these functionalities can be enabled by semitransparent solar cells. Due to wide compositional variety, tunable bandgap, and performance, 2D–3D perovskites are uniquely positioned to dominate this application space.

Table 12.6 Various representative cations used as absorber material in 2D PSCs by compositional engineering method.

Perovskite	Device structure	PCE (%)	Stability
$(PEA_{0.8}GA_{0.2})_2(MA)_4Pb_5I_{16}$	ITO/PTAA/PSK/ PCBM/PEI/Ag	15.27	80% of PCE after 800 h, dark, ambient air, RH ~ 55–60% [69]
$F\text{-}PEA_2MA_2Pb_3I_{10}$	FTO/NiO_x/PSK/ $PC_{61}BM$/BCP/Ag	5.8	N/A [104]
$(4\text{-}AEP)_2MA_4Pb_5I_{13}$	FTO/C_{60}/PSK/ Spiro-OMeTAD/ Au	11.68	95% of PCE after 1000 h, air, RH ~ 30% [105]
$(t\text{-}BA)_2(FA_{0.85}Cs_{0.15})_4(Pb_{0.6}Sn_{0.4})_5I_{16}$	ITO/PEDOT:PSS/ PSK/PCBM/BCP/ Ag	10.6	45% of PCE after 10 h, under air, 1 sun illumination, 65°C [106]
$(F\text{-}PEA)_2(MA)_4Pb_5I_{16}$ ($n = 5$)	ITO/PEDOT:PSS/ PSK/$PC_{61}BM$/ BCP/Ag	14.5	90% of PCEs after 40 days, dark, RH ~ 40–50% [107]
$(BEA)_{0.5}Cs_{0.15}(FA_{0.83}MA_{0.17})_{2.85}Pb_3(I_{0.83}Br_{0.17})_{10}$	FTO/c-TiO_2/PSK/ Spiro-OMeTAD/ Au	17.39	90% of its initial PCE after 2400 h, stored in ambient conditions (RH ~35%); 93% of PCE under continuous illumination after 500 h (at room temperature) [94]
$BA_2(Cs_{0.02}MA_{0.64}FA_{0.34})_4Pb_5I_{16}$	FTO/c-TiO_2/PSK/ spiro MeOTAD/ Au	14.23	74% of PCE after 600 h, 80°C, RH ~ 15–20% [108]
$(BA)_2Cs_{0.15}(MA)_{2.85}Pb_4I_{13}$	FTO/TiO_2/PSK/ Spiro-MeOTAD/ Au	13.68	89% of PCE after 1400 h, ambient condition, 30% RH [102]
$(BA)_2(MA_{0.8}FA_{0.2})_3Pb_4I_{13}$	ITO/PEDOT:PSS/ PSK /PCBM/BCP/ Ag	12.81	88% of PCE after 1300 h, ambient air, RH ~ 40–60% [103]
$(3AMPY)(MA)_3Pb_4I_{13}$	FTO/PEDOT:PSS/ PSK/C_{60}/BCP/Ag	9.20	N/A [109]
$(F\text{-}PEA)_2MA_4Pb_5I_{16}$	FTO/c-TiO_2/PSK/ spiro MeOTAD/ Au	13.64	65% of PCE after 576 h, ambient environment, 70°C [61]

Semitransparent perovskite solar cells (ST-PSCs), as the name suggests, are cells that transmit (not absorb) a certain percentage of incident light. Semi-transparency can be achieved by reducing the thickness of the absorber or by tuning the bandgap of the absorber. The absorbed light is used to produce electricity. The transmitted light can be used for other purposes, such as lighting, aesthetics, tandem cells, agriculture, etc. We shall specifically discuss two exciting applications: Building Integrated Photovoltaics (BIPV) and Agrivoltaics.

Conventional silicon solar panels are often bulky, unwieldy, and ugly. On buildings, they are mostly installed after construction, on frames and roofs, with little regard to functionality. BIPV systems propose a different model in which parts of a building's structure are replaced by solar cells for energy generation. The energy generated by cells can help buildings achieve "net-zero" status. More importantly, the system is seamlessly integrated into the structure. The module form factor can vary, such as flexible steel foils for curved surfaces; shingles for roofs and facades; PV-integrated corrugated steel sheets; and glazing products for windows, facades, and glass ceilings. Unfortunately, conventional silicon technology is woefully impractical for many of these applications. Technologies like CIGS are too uneconomical. Perovskites, especially 2D/3D perovskites [110], have a good chance of capturing the BIPV market with economic and aesthetic products.

Solar cells compete for land with applications like agriculture. This is often a flashpoint in countries with subsistence farming. Agrivoltaics assumes importance because it integrates agriculture and photovoltaics so that both share the same land. The combination not only increases farmer income manifold but also saves water, improves yield, increases productivity, and enables farming of high-value crops. The harvested energy can be consumed by the local economy or fed back to the grid. The current state of the art is a sparse array of silicon solar cells mounted on a tracker. A better solution is a selectively transparent module that transmits photons needed for photosynthesis and absorbs the rest [111]. Lower-dimensional perovskites are inherently semitransparent due to the bandgap, making them an excellent candidate for agrivoltaics. Printable perovskite modules have already been demonstrated for the pilot demonstration of agrivoltaics.

These applications, as stated above, require either partial or selective absorption for semitransparent solar cells. This requirement can be well served by lower-dimensional perovskites due to the following reasons.

1. It is generally noted that the light transmission of lower-dimensional perovskites in the absorption range is better than their 3D counterparts (due to a lesser absorption coefficient). Hence, they can prove to be comparatively better suited for semitransparent applications.
2. Lower-dimensional perovskites have bandgaps higher than conventional 3D perovskites, because of which their absorption range is comparatively lesser. This makes a larger portion of the visible spectrum available for transmission, making semitransparent applications feasible.

In order to make semitransparent solar cells, it is essential that, along with the perovskite, the transport layers and contacts should not absorb the transmitted spectrum. The transport layers are chosen such that they do not absorb in the visible spectrum. Also, they are usually thinner when compared to absorber layers, so the absorption, if any, is insignificant.

The contacts, however, need to be made deliberately thin so that they shall not absorb the incident light. The schematic of the device stack for an ST-PSC is shown in Figure 12.22.

The semi-transparency of cells is analyzed by a parameter known as Average Visible Transmittance (AVT). The calculation for AVT is done by using the below-mentioned formula where λ is the wavelength, S(λ) is solar energy flux for AM 1.5 at a specific wavelength, T(λ) is transmittance at that wavelength, and P(λ) is the photopic response of human eye at the same wavelength [112].

$$AVT = \frac{\int T(\lambda)P(\lambda)S(\lambda)d\lambda}{\int P(\lambda)S(\lambda)d\lambda}$$

The photopic response is defined as the response of the human eye to the visible spectrum under well-lit conditions. This response is mainly dependent on the response of rods and cones present in the eye. The human eye has a higher response in the middle of the spectrum while it is comparatively less towards the ends (red and blue wavelengths). The expected AVT for cells to be suitable for semitransparent application lies in the range of a minimum of 20–30% [113].

Recent research in ST-PSCs has been focused on using 3D perovskites as absorbers, while modifications are mainly done for the transport layers. Chang *et al.* proposed using low-temperature ALD (<100 °C) for depositing a thin layer of ZnO as a cathode buffer layer in order to fabricate ST-PSCs (Figure 12.23a). It was observed that the work function of the top electrode was fine-tuned due to the presence of the ZnO layer, as a result of which the electron transport ability was improvised. The film coverage was excellent, along with better charge selectivity due to the presence of the buffer layer. The transparent top electrode for the cell was composed of Ag nanowires that yielded an efficiency of 10.8% with an AVT of 25.5% (Figure 12.23a). To enhance the ambient stability of devices, a 50-nm encapsulating layer of ALD-deposited Al_2O_3 was used. The layer exhibited an excellent barrier performance against both oxygen and water vapor by decreasing the oxygen transmission rate to 1.9×10^{-3} cm^3 m^{-2} day^{-1} and the water vapor transmission rate to 9×10^{-4} g m^{-2} day^{-1}.

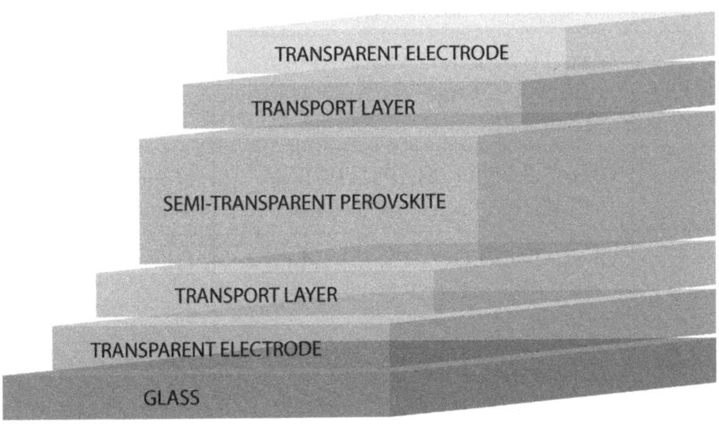

Figure 12.22 Pictorial depiction of device architecture for a semitransparent perovskite solar cell (ST-PSC).

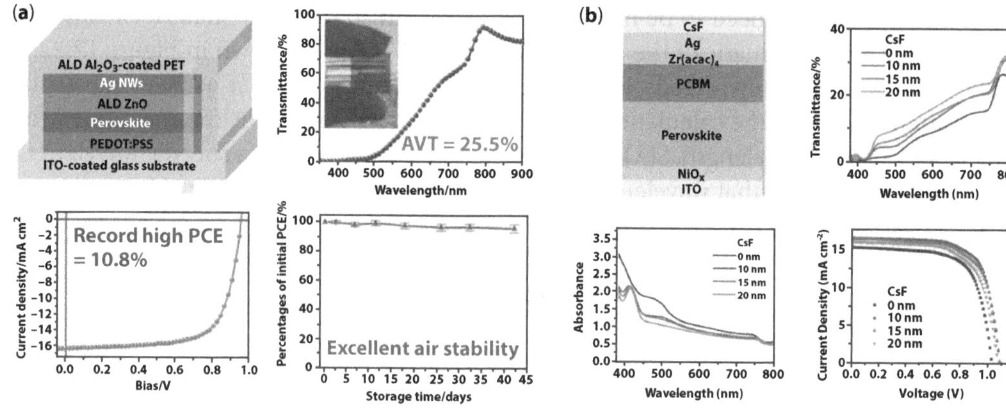

Figure 12.23 (a) Device architecture, transmittance curve, light *J–V* curve, and stability curve for devices when ALD-deposited ZnO and Al_2O_3 were used as a buffer layer and encapsulating layer, respectively. Reproduced with permission from [114]. Copyright 2015, American Chemical Society. (b) Device architecture, transmittance curve, absorbance curve, and light *J-V* curve of devices when BiPy-I was used as a green antisolvent for perovskite fabrication. Reproduced with permission from [115]. Copyright 2022, Elsevier.

Table 12.7 ST-PSCs device performance along with transparent electrodes and AVT.

Top electrode	Device structure	PCE (%)	AVT (%)
Ag nanowires	Glass/ITO/PEDOT:PSS/mixed-halide PSK/PCBM/ZnO/AgNWs	10.81, 9.55	23, 29 [116]
Au (10 nm)/MoO_3 (5 nm)	Glass/ITO/C_{60}/$MAPbI_{3-x}Br_x$/spiro-OMeTAD/MoO_3 (5 nm)/Au (10 nm)/MoO_3 (5 nm)	11.03	21 [117]
Ag (20 nm)/CsF	Glass/ITO/NiO_x/$MAPbI_3$/PCBM/Zr(acac)4/Ag (20 nm)/CsF	11.74	23 [115]
Cu (1 nm)/Au (7 nm)	Glass/ITO/SnO_2/$CsPbI_2Br$/PTAA/BCP/Cu (1 nm)/Au (7 nm)	14.01, 10.36	31.7, 40.9 [118]
AgNWs/ALD Al_2O_3-coated PET	Glass/ITO/PEDOT:PSS/$MAPbI_3$/ALD-ZnO/ AgNWs/ALD Al_2O_3-coated PET	10.8	25.5 [114]
Ag (8 nm) or Cu (8 nm)	Glass/ITO/NiO_x/$MAPbI_3$/PCBM/PEIE/Ag (8 nm) or Cu (8 nm)	11.95	20 [119]
Ag (8 nm)	Glass/ITO/PEDOT:PSS/$MAPbI_3$/$PC_{61}BM$/ MUTAB (10 nm)/Ag (8 nm)	11.8	20.8 [120]
Ag (20 nm)	Glass/ITO/CuSCN/$MAPbI_3$/$PC_{61}BM$/Bis-C_{60}/Ag (20 nm)	10.22	25.7 [121]

Hence, by using ALD-deposited ZnO as a cathode buffer layer and Al_2O_3 as an encapsulating layer, both the performance and stability of ST-PSC were enhanced [114].

Zhang *et al.* used a greener anti-solvent rather than using chlorobenzene to fabricate good coverage semitransparent films for ST-PSCs. The green anti-solvent used was 1-propyl-[4,4′-bipyridin]-1-ium iodide (BiPy-I) and was believed to confine the growth of perovskite crystals. As a result, the thickness was not decreased much (206 ± 3 nm), but the transparency was well in the range of cells being qualified for ST-PSCs (Figure 12.23b, absorbance). For the anti-solvent, a 0.001 M solution of BiPy-I was prepared in NMP (N-methyl pyrrolidone). A 2 vol% concentration of this was added to the perovskite precursor and was stirred for 30 min at 70 °C inside the glove box. The spin coating of the film was a two-step process in which ethyl acetate and perovskite precursor were sequentially dripped onto the HTL-coated substrate and then annealed for perovskite crystallization. The devices exhibited an efficiency of 11.74% with an AVT of 23%. The devices also retained 91% of their initial efficiency when stored in ambient for 240 h. Some of such works have been summarized in Table 12.7, along with their cell performance and AVT [115].

Lower-dimensional perovskites have not been explored much when it comes to ST-PSCs. However, their selective absorption, as well as larger bandgap (lesser absorption in the visible region), makes them a potential candidate for ST-PSC absorber layers. Hence, there is a good chance that they might be explored in this field in the coming years.

12.10 Conclusion and Future Outlook

Perovskites have been an active field of research for more than a decade. The PCE of 3D perovskites is excellent, and they seem amenable for low-cost and high-throughput fabrication. The main challenge has been long-term stability, especially in the presence of oxygen and moisture. Lower-dimensional perovskites show excellent stability. However, the challenge is to improve charge transport. Several efforts have been made by using additives, compositional engineering, interlayers, etc. which made it feasible to attain high efficiencies without compromising stability. However, the fundamental aspects of their growth, along with developing an understanding of their properties, still need to be explored. Engineering the spacers is an important direction in low-dimensional perovskites. Owing to the chemical richness of the perovskite family, it is crucial to select a spacer that reduces the quantum confinement and maintains stability. Spacers that are linear, have a π–π conjugation, or have non-covalent bonding with the inorganic layers may be a better choice.

Thermodynamics and kinetics of growth need to be well understood for depositing high-quality films. The thermodynamically stable phases tend to crystallize in a purer form. It has been observed that purer phases of lower-dimensional perovskites with a single "*n*" can result in better efficiencies. However, it is questionable if the presence of several "*n*", which provides a cascade for carrier transport, is better. A better understanding of thermodynamics could aid in answering these questions. The crystallization kinetics have also been a mystery. Kinetics controls the crystallization sequence of different "*n*", influencing the device performance. To better understand kinetics, single-crystal studies of lower-dimensional perovskites could be significant. This is because single crystals form a strict pure phase, which is not achievable with thin films.

The comparatively higher bandgap of lower-dimensional perovskites creates a need for novel transport layers. The conventional transport layers are well suited for 3D perovskites but often lead to unfavorable band alignments for lower-dimensional counterparts. The exploration of better-aligned transport layers could prove to enhance the performance of cells. With these understandings becoming clearer and explorations being done, the community could lead to a path where lower-dimensional PSCs could be increasingly made more efficient. Eventually, they could also serve for commercialization along with other range applications available such as building-integrated photovoltaics, agrivoltaics, solar-powered automotive, wearable electronics, etc.

References

1. "IEA (2021), Electricity Information: Overview, IEA, Paris, https://www.iea.org/reports/electricity-information-overview, License: CC BY 4.0."
2. P. Nejat, F. Jomehzadeh, M. M. Taheri, M. Gohari and M. Z. Muhd, "A global review of energy consumption, CO2 emissions and policy in the residential sector (with an overview of the top ten CO2 emitting countries)," *Renewable Sustainable Energy Rev.*, vol. 43, pp. 843–862, 2015, doi: 10.1016/j.rser.2014.11.066.
3. P. A. Owusu and S. Asumadu-Sarkodie, "A review of renewable energy sources, sustainability issues and climate change mitigation," *Cogent Eng.*, vol. 3, no. 1, pp. 1–14, 2016, doi: 10.1080/23311916.2016.1167990.
4. IEA, "IEA (2022), Solar PV, IEA, Paris, https://www.iea.org/reports/solar-pv, License: CC BY 4.0."
5. T. C. Sum and N. Mathews, "Advancements in perovskite solar cells: Photo-physics behind the photovoltaics," *Energy Environ. Sci.*, vol. 7, no. 8, pp. 2518–2534, 2014, doi: 10.1039/c4ee00673a.
6. W. Shockley and H. J. Queisser, "Detailed balance limit of efficiency of p-n junction solar cells," *J. Appl. Phys.*, vol. 32, no. 3, pp. 510–519, 1961, doi: 10.1063/1.1736034.
7. D. Chapin and C. Fuller, "The Silicon Solar Cell Turns 50."
8. First Solar, "Revisiting First Solar's Current Competitiveness A Decade After The Polysilicon Bubble," Jun. 2019.
9. A. K. Jena, A. Kulkarni and T. Miyasaka, "Halide Perovskite Photovoltaics: Background, Status, and Future Prospects," *Chem. Rev.*, vol. 119, no. 5, pp. 3036–3103, 2019, doi: 10.1021/acs.chemrev.8b00539.
10. A. Krishna, S. Gottis, M. K. Nazeeruddin and F. Sauvage, "Mixed Dimensional 2D/3D Hybrid Perovskite Absorbers: The Future of Perovskite Solar Cells?," *Adv. Funct. Mater.*, vol. 29, no. 8. Wiley-VCH Verlag, Feb. 21, 2019. doi: 10.1002/adfm.201806482.
11. Tsutomu M. Akihiro Kojima, Kenjiro Teshima, Yasuo Shirai, "Organometal Halide Perovskites as Visible- Light Sensitizers for Photovoltaic Cells," *J. Am. Chem. Soc.*, vol. 131, no. October, pp. 6050–6051, 2009, doi: 10.1021/ja809598r.
12. J. Even *et al.*, "Solid-state physics perspective on hybrid perovskite semiconductors," *J. Phys. Chem. C*, vol. 119, no. 19, pp. 10161–10177, 2015, doi: 10.1021/acs.jpcc.5b00695.
13. E. Elahi, G. Dastgeer, A. S. Siddiqui, S. A. Patil, M. W. Iqbal and P. R. Sharma, "A review on two-dimensional (2D) perovskite material-based solar cells to enhance the power conversion efficiency," *Dalton Transactions*, vol. 51, no. 3. *R. Soc. Chem.*, pp. 797–816, Jan. 21, 2022. doi: 10.1039/d1dt02991f.
14. H. S. Kim *et al.*, "Lead iodide perovskite sensitized all-solid-state submicron thin film mesoscopic solar cell with efficiency exceeding 9%," *Sci. Rep.*, vol. 2, 2012, doi: 10.1038/srep00591.

15. H. J. Snaith, "Perovskites: The Emergence of a New Era for Low-Cost, High-Efficiency solar cells," *J. Phys. Chem. Lett.*, 4, 21, 3623-3630, 2013.
16. J. Jeong *et al.*, "Pseudo-halide anion engineering for α-FAPbI3 perovskite solar cells," *Nature*, vol. 592, no. 7854, pp. 381–385, Apr. 2021, doi: 10.1038/s41586-021-03406-5.
17. A. Kojima, K. Teshima, Y. Shirai and T. Miyasaka, "Organometal halide perovskites as visible-light sensitizers for photovoltaic cells," *J. Am. Chem. Soc.*, vol. 131, no. 17, pp. 6050–6051, May 2009, doi: 10.1021/ja809598r.
18. Michael M. Lee, Joël Teuscher, Tsutomu Miyasaka, Takurou N. Murakami, and Henry J. Snaith, "Efficient Hybrid Solar Cells Based on Meso-Superstructured Organometal Halide Perovskites," *Science* (1979), vol. 338, pp. 643–647, 2012.
19. M. Liu, M. B. Johnston and H. J. Snaith, "Efficient planar heterojunction perovskite solar cells by vapour deposition," *Nature*, vol. 501, no. 7467, pp. 395–398, 2013, doi: 10.1038/nature12509.
20. J. Burschka *et al.*, "Sequential deposition as a route to high-performance perovskite-sensitized solar cells," *Nature*, vol. 499, no. 7458, pp. 316–319, 2013, doi: 10.1038/nature12340.
21. Q. Jiang *et al.*, "Planar-Structure Perovskite Solar Cells with Efficiency beyond 21%," *Adv. Mater.*, vol. 29, no. 46, Dec. 2017, doi: 10.1002/adma.201703852.
22. M. Saliba *et al.*, "Cesium-containing triple cation perovskite solar cells: Improved stability, reproducibility and high efficiency," *Energy Environ. Sci.*, vol. 9, no. 6, pp. 1989–1997, Jun. 2016, doi: 10.1039/c5ee03874j.
23. G. Yang, H. Zhang, G. Li and G. Fang, "Stabilizer-assisted growth of formamdinium-based perovskites for highly efficient and stable planar solar cells with over 22% efficiency," *Nano Energy*, vol. 63, Sep. 2019, doi: 10.1016/j.nanoen.2019.06.031.
24. C. Motta, F. el Mellouhi, S. Kais, N. Tabet, F. Alharbi, and S. Sanvito, "Revealing the role of organic cations in hybrid halide perovskites CH3NH3PbI3," *Nat. Commun.*, vol. 6, no. May 2014, pp. 1–7, 2014, doi: 10.1038/ncomms8026.
25. P. Liu *et al.*, "Interfacial electronic structure at the CH3NH3PbI3/MoOx interface," *Appl. Phys. Lett.*, vol. 106, no. 19, 2015, doi: 10.1063/1.4921339.
26. H. Chen, "Two-Step Sequential Deposition of Organometal Halide Perovskite for Photovoltaic Application," *Adv. Funct. Mater.*, vol. 27, 1605654, 2017, doi: 10.1002/adfm.201605654.
27. A. D. Taylor *et al.*, "A general approach to high-efficiency perovskite solar cells by any antisolvent," *Nat. Commun.*, vol. 12, no. 1, Dec. 2021, doi: 10.1038/s41467-021-22049-8.
28. P. Yan, C. Xiang, Z. Wu, C. Wang, H. Li and C. Sheng, "Methylammonium chloride or guanidinium chloride as an additive to improve performance of 2D alternating cation perovskite solar cells: A direct comparison," *J. Alloys Compd.*, vol. 933, p. 167671, Feb. 2023, doi: 10.1016/j.jallcom.2022.167671.
29. J. B. Patel *et al.*, "Influence of Interface Morphology on Hysteresis in Vapor-Deposited Perovskite Solar Cells," *Adv. Electron. Mater.*, vol. 3, 1600470, pp. 1–6, 2017, doi: 10.1002/aelm.201600470.
30. R. Sheng *et al.*, "Methylammonium Lead Bromide Perovskite-Based Solar Cells by Vapor-Assisted Deposition," *J. Phys. Chem. C.*, vol. 119, no. 7, pp. 3545-3549, 2015, doi: 10.1021/jp512936z.
31. L. K. Ono, M. R. Leyden, S. Wang and Y. Qi, "Organometal halide perovskite thin films and solar cells by vapor deposition," *J. Mater. Chem. A Mater*, vol. 4, no. 18, pp. 6693–6713, 2016, doi: 10.1039/c5ta08963h.
32. Z. Ding, X. Hu, P. L. Yue, G. Q. Lu and P. F. Greenfield, "Synthesis of anatase TiO2 supported on porous solids by chemical vapor deposition," *Catal. Today*, vol. 68, no. 1–3, pp. 173–182, 2001, doi: 10.1016/S0920-5861(01)00298-X.
33. H. Tan *et al.*, "Efficient and stable solution-processed planar perovskite solar cells via contact passivation," *Science*, (1979), vol. 355, no. 6326, pp. 722–726, 2017, doi: 10.1126/science.aai9081.

34. K. Mahmood, S. Sarwar and M. T. Mehran, "Current status of electron transport layers in perovskite solar cells: materials and properties," *RSC Adv.*, vol. 7, no. 28, pp. 17044–17062, 2017, doi: 10.1039/c7ra00002b.
35. J. Lian, B. Lu, F. Niu, P. Zeng and X. Zhan, "Electron-Transport Materials in Perovskite Solar Cells," *Small Methods*, vol. 2, no. 10, p. 1800082, 2018, doi: 10.1002/smtd.201800082.
36. H. Liu, Z. Huang, S. Wei, L. Zheng, L. Xiao and Q. Gong, "Nano-structured electron transporting materials for perovskite solar cells," *Nanoscale*, vol. 8, no. 12, pp. 6209–6221, 2016, doi: 10.1039/c5nr05207f.
37. L. Chao, Z. Wang, Y. Xia, Y. Chen and W. Huang, "Recent progress on low dimensional perovskite solar cells," *J. Energy Chem.*, vol. 27, no. 4. Elsevier B.V., pp. 1091–1100, Jul. 01, 2018. doi: 10.1016/j.jechem.2017.10.013.
38. F. Zhang, H. Lu, J. Tong, J. J. Berry, M. C. Beard and K. Zhu, "Advances in two-dimensional organic-inorganic hybrid perovskites," *Energy Environ. Sci.*, vol. 13, no. 4. *R. Soc. Chem.*, pp. 1154–1186, Apr. 01, 2020. doi: 10.1039/c9ee03757h.
39. C. Ortiz-Cervantes, P. Carmona-Monroy and D. Solis-Ibarra, "Two-Dimensional Halide Perovskites in Solar Cells: 2D or not 2D?," *ChemSusChem*, vol. 12, no. 8, pp. 1560–1575, 2019. doi: 10.1002/cssc.201802992.
40. P. Liu, S. Yu and S. Xiao, "Research progress on two-dimensional (2D) halide organic-inorganic hybrid perovskites," *Sustain. Energy Fuels.*, vol. 5, no. 16, pp. 3950–3978, 2021. doi: 10.1039/d1se00589h.
41. D. Wang, M. Wright, N. K. Elumalai and A. Uddin, "Stability of perovskite solar cells," *Sol. Energy Mater. Sol. Cells*, vol. 147, pp. 255–275, 2016, doi: 10.1016/j.solmat.2015.12.025.
42. X. Zhao and N.-G. Park, "Stability Issues on Perovskite Solar Cells," *Photonics*, vol. 2, no. 4, pp. 1139–1151, 2015, doi: 10.3390/photonics2041139.
43. J. Cao, J. Yin, S. Yuan, Y. Zhao, J. Li and N. Zheng, "Thiols as interfacial modifiers to enhance the performance and stability of perovskite solar cells," *Nanoscale*, vol. 7, pp. 9443–9447, 2015, doi: 10.1039/c5nr01820j.
44. M. I. Asghar, J. Zhang, H. Wang and P. D. Lund, "Device stability of perovskite solar cells – A review," *Renewable Sustainable Energy Rev.*, vol. 77, no. July 2016, pp. 131–146, 2017, doi: 10.1016/j.rser.2017.04.003.
45. G. Niu, X. Guo and L. Wang, "Review of recent progress in chemical stability of perovskite solar cells," *J. Mater. Chem. A Mater.*, vol. 3, no. 17, pp. 8970–8980, 2015, doi: 10.1039/c4ta04994b.
46. S. Emami, "Recent progress in long-term stability of perovskite solar cells," *Adv. Sci.*, vol. 5, 1700387 vol. 2, pp. 52–62, 2018, doi: 10.1002/advs.201700387
47. F. Li and M. Liu, "Recent efficient strategies for improving the moisture stability of perovskite solar cells," *J. Mater. Chem. A Mater.*, vol. 5, no. 30, pp. 15447–15459, 2017, doi: 10.1039/c7ta01325f.
48. B. Li, Y. Li, C. Zheng, D. Gao and W. Huang, "Advancements in stability of perovskite solar cells: degradation mechanisms and improvement approaches," *RSC Adv.*, vol. 6, pp. 38079–38091, 2016, doi: 10.1039/c5ra27424a.
49. L. Chao, T. Niu, Y. Xia, X. Ran, Y. Chen and W. Huang, "Efficient and Stable Low-Dimensional Ruddlesden-Popper Perovskite Solar Cells Enabled by Reducing Tunnel Barrier," *J. Phys. Chem. Lett.*, vol. 10, no. 6, pp. 1173–1179, Mar. 2019, doi: 10.1021/acs. jpclett.9b00276.
50. M. M. Davy, T. M. Jadel, C. Qin, B. Luyun and G. Mina, "Recent progress in low dimensional (quasi-2D) and mixed dimensional (2D/3D) tin-based perovskite solar cells," *Sustain. Energy Fuels*, vol. 5, no. 1. *R. Soc. Chem.*, pp. 34–51, Jan. 07, 2021. doi: 10.1039/d0se01520b.
51. W. Yang, Y. Zhan, F. Yang and Y. Li, "Hot-Casting and Anti-solvent Free Fabrication of Efficient and Stable Two-Dimensional Ruddlesden-Popper Perovskite Solar Cells," *ACS Appl. Mater. Interfaces*, vol. 13, no. 51, pp. 61039–61046, Dec. 2021, doi: 10.1021/acsami.1c17169.

52. Sum T. C. and Mathews N., "Advancements in perovskite solar cells: Photophysics behind the photovoltaics," *Energy Environ. Sci.*, vol. 7, pp. 2518–2534, 2014.
53. H. Tsai *et al.*, "High-efficiency two-dimensional ruddlesden-popper perovskite solar cells," *Nature*, vol. 536, no. 7616, pp. 312–317, Jul. 2016, doi: 10.1038/nature18306.
54. C. M. Myae Soe *et al.*, "Structural and thermodynamic limits of layer thickness in 2D halide perovskites," *Proc. Natl. Acad. Sci. U S A*, vol. 116, no. 1, pp. 58–66, Jan. 2019, doi: 10.1073/pnas.1811006115.
55. J. C. Blancon *et al.*, "Scaling law for excitons in 2D perovskite quantum wells," *Nat. Commun.*, vol. 9, no. 1, Dec. 2018, doi: 10.1038/s41467-018-04659-x.
56. H. Tsai *et al.*, "High-efficiency two-dimensional ruddlesden-popper perovskite solar cells," *Nature*, vol. 536, no. 7616, pp. 312–317, Jul. 2016, doi: 10.1038/nature18306.
57. I. C. Smith, E. T. Hoke, D. Solis-Ibarra, M. D. McGehee and H. I. Karunadasa, "A Layered Hybrid Perovskite Solar-Cell Absorber with Enhanced Moisture Stability," *Angew. Chem. Int. Ed.*, vol. 53, no. 42, pp. 11232–11235, Oct. 2014, doi: 10.1002/anie.201406466.
58. L. N. Quan *et al.*, "Ligand-Stabilized Reduced-Dimensionality Perovskites," *J. Am. Chem. Soc.*, vol. 138, no. 8, pp. 2649–2655, Mar. 2016, doi: 10.1021/jacs.5b11740.
59. J. Xi *et al.*, "Alternative Organic Spacers for More Efficient Perovskite Solar Cells Containing Ruddlesden-Popper Phases," *J. Am. Chem. Soc.*, vol. 142, no. 46, pp. 19705–19714, Nov. 2020, doi: 10.1021/jacs.0c09647.
60. Q. Cheng *et al.*, "Impact of Strain Relaxation on 2D Ruddlesden–Popper Perovskite Solar Cells," *Angew. Chem. Int. Ed.*, vol. 61, no. 36, Sep. 2022, doi: 10.1002/ anie.202208264.
61. F. Zhang *et al.*, "Enhanced Charge Transport in 2D Perovskites via Fluorination of Organic Cation," *J. Am. Chem. Soc.*, vol. 141, no. 14, pp. 5972–5979, Apr. 2019, doi: 10.1021/jacs.9b00972.
62. X. Li *et al.*, "Phenylformamidinium-enabled quasi-2D Ruddlesden-Popper perovskite solar cells with improved stability," *J. Energy Chem.*, vol. 66, pp. 680–688, Mar. 2022, doi: 10.1016/j.jechem.2021.09.026.
63. H. Lai *et al.*, "Two-dimensional ruddlesden-popper perovskite with nanorod-like morphology for solar cells with efficiency exceeding 15%," *J. Am. Chem. Soc.*, vol. 140, no. 37, pp. 11639–11646, Sep. 2018, doi: 10.1021/jacs.8b04604.
64. Y. Zheng, S. C. Chen, Y. Ma and Q. Zheng, "Furfurylammonium as a Spacer for Efficient 2D Ruddlesden–Popper Perovskite Solar Cells," *Sol. RRL*, vol. 6, no. 8, Aug. 2022, doi: 10.1002/solr.202200221.
65. Q. Li *et al.*, "Fluorinated Aromatic Formamidinium Spacers Boost Efficiency of Layered Ruddlesden-Popper Perovskite Solar Cells," *ACS Energy Lett.*, vol. 6, no. 6, pp. 2072–2080, Jun. 2021, doi: 10.1021/acsenergylett.1c00620.
66. J. Xu *et al.*, "Organic spacer engineering of Ruddlesden-Popper perovskite materials toward efficient and stable solar cells," *Chem. Eng. J.*, vol. 453, Feb. 2023, doi: 10.1016/j. cej.2022.139790.
67. F. Huang *et al.*, "Controlled crystallinity and morphologies of 2D Ruddlesden-Popper perovskite films grown without anti-solvent for solar cells," *Chem. Eng. J.*, vol. 394, Aug. 2020, doi: 10.1016/j.cej.2020.124959.
68. C. Liang *et al.*, "Two-dimensional Ruddlesden–Popper layered perovskite solar cells based on phase-pure thin films," *Nat. Energy*, vol. 6, no. 1, pp. 38–45, Jan. 2021, doi: 10.1038/s41560-020-00721-5.
69. J. Shi, X. Jin, Y. Wu and M. Shao, "Mixed bulky cations for efficient and stable Ruddlesden-Popper perovskite solar cells," *APL Mater.*, vol. 8, no. 10, Oct. 2020, doi: 10.1063/5.0024135.
70. J. Zhang *et al.*, "Uniform Permutation of Quasi-2D Perovskites by Vacuum Poling for Efficient, High-Fill-Factor Solar Cells," *Joule*, vol. 3, no. 12, pp. 3061–3071, Dec. 2019, doi: 10.1016/j.joule.2019.09.020.

71. V. Romano *et al.*, "Two-Step Thermal Annealing: An Effective Route for 15 % Efficient Quasi-2D Perovskite Solar Cells," *ChemPlusChem*, vol. 86, no. 8, pp. 1044–1048, Aug. 2021, doi: 10.1002/cplu.202000777.
72. F. Zhu *et al.*, "Pressure-Enhanced Vertical Orientation and Compositional Control of Ruddlesden-Popper Perovskites for Efficient and Stable Solar Cells and Self-Powered Photodetectors," *ACS Appl. Mater. Interfaces*, vol. 14, no. 1, pp. 1526–1536, Jan. 2022, doi: 10.1021/acsami.1c18522.
73. X. Zhu *et al.*, "PEDOT:PSS/CuCl Composite Hole Transporting Layer for Enhancing the Performance of 2D Ruddlesden-Popper Perovskite Solar Cells," *J. Phys. Chem. Lett.*, vol. 13, no. 26, pp. 6101–6109, Jul. 2022, doi: 10.1021/acs.jpclett.2c01399.
74. C. Ma, D. Shen, T. W. Ng, M. F. Lo and C. S. Lee, "2D Perovskites with Short Interlayer Distance for High-Performance Solar Cell Application," *Adv. Mater.*, vol. 30, no. 22, May 2018, doi: 10.1002/adma.201800710.
75. D. Lu, G. Lv, Z. Xu, Y. Dong, X. Ji and Y. Liu, "Thiophene-Based Two-Dimensional Dion-Jacobson Perovskite Solar Cells with over 15% Efficiency," *J. Am. Chem. Soc.*, vol. 142, no. 25, pp. 11114–11122, Jun. 2020, doi: 10.1021/jacs.0c03363.
76. T. He *et al.*, "Reduced-dimensional perovskite photovoltaics with homogeneous energy landscape," *Nat. Commun.*, vol. 11, no. 1, Dec. 2020, doi: 10.1038/s41467-020-15451-1.
77. G. Lv *et al.*, "Multiple-Noncovalent-Interaction-Stabilized Layered Dion-Jacobson Perovskite for Efficient Solar Cells," *Nano Lett.*, vol. 21, no. 13, pp. 5788–5797, Jul. 2021, doi: 10.1021/acs.nanolett.1c01505.
78. L. Jin *et al.*, "Secondary Anti-Solvent Treatment for Efficient 2D Dion–Jacobson Perovskite Solar Cells," *Small*, 19, 2205088, 2023, doi: 10.1002/smll.202205088.
79. W. Ke *et al.*, "Compositional and Solvent Engineering in Dion–Jacobson 2D Perovskites Boosts Solar Cell Efficiency and Stability," *Adv. Energy Mater.*, vol. 9, no. 10, Mar. 2019, doi: 10.1002/aenm.201803384.
80. L. Cheng *et al.*, "Highly Thermostable and Efficient Formamidinium-Based Low-Dimensional Perovskite Solar Cells," *Angew. Chem. Int. Ed.*, vol. 60, no. 2, pp. 856–864, Jan. 2021, doi: 10.1002/anie.202006970.
81. F. Li *et al.*, "Vertical Orientated Dion–Jacobson Quasi-2D Perovskite Film with Improved Photovoltaic Performance and Stability," *Small Methods*, vol. 4, no. 5, May 2020, doi: 10.1002/smtd.201900831.
82. J. Wang *et al.*, "Suppressing the Excessive Solvated Phase for Dion–Jacobson Perovskites with Improved Crystallinity and Vertical Orientation," *Sol. RRL*, vol. 4, 2000371, 2020, doi: 10.1002/solr.202000371
83. X. Zhang, T. Yang, X. Ren, L. Zhang, K. Zhao and S. Liu, "Film Formation Control for High Performance Dion–Jacobson 2D Perovskite Solar Cells," *Adv. Energy Mater.*, vol. 11, no. 19, May 2021, doi: 10.1002/aenm.202002733.
84. X. Lian *et al.*, "Solvation effect in precursor solution enables over 16% efficiency in thick 2D perovskite solar cells," *J. Mater. Chem. A Mater.*, vol. 7, no. 33, pp. 19423–19429, 2019, doi: 10.1039/c9ta06009j.
85. T. Niu *et al.*, "Reduced-Dimensional Perovskite Enabled by Organic Diamine for Efficient Photovoltaics," *J. Phys. Chem. Lett.*, vol. 10, no. 10, pp. 2349–2356, May 2019, doi: 10.1021/acs.jpclett.9b00750.
86. X. Zhao, T. Liu, A. B. Kaplan, C. Yao and Y. L. Loo, "Accessing Highly Oriented Two-Dimensional Perovskite Films via Solvent-Vapor Annealing for Efficient and Stable Solar Cells," *Nano Lett.*, vol. 20, no. 12, pp. 8880–8889, Dec. 2020, doi: 10.1021/acs.nanolett.0c03914.

87. W. Li *et al.*, "Light-activated interlayer contraction in two-dimensional perovskites for high-efficiency solar cells," *Nat. Nanotechnol.*, vol. 17, no. 1, pp. 45–52, Jan. 2022, doi: 10.1038/s41565-021-01010-2.
88. H. Wu *et al.*, "Merged interface construction toward ultra-low: Voc loss in inverted two-dimensional Dion-Jacobson perovskite solar cells with efficiency over 18%," *J. Mater. Chem. A Mater.*, vol. 9, no. 21, pp. 12566–12573, Jun. 2021, doi: 10.1039/d1ta02015c.
89. Z. Xu, D. Lu, X. Dong, M. Chen, Q. Fu and Y. Liu, "Highly Efficient and Stable Dion–Jacobson Perovskite Solar Cells Enabled by Extended π-Conjugation of Organic Spacer," *Adv. Mater.*, vol. 33, no. 51, 2021, doi: 10.1002/adma.202105083.
90. C. M. M. Soe *et al.*, "New Type of 2D Perovskites with Alternating Cations in the Interlayer Space, (C(NH2)3)(CH3NH3)nPbnI3n+1: Structure, Properties, and Photovoltaic Performance," *J. Am. Chem. Soc.*, vol. 139, no. 45, pp. 16297–16309, 2017, doi: 10.1021/jacs.7b09096.
91. Y. Zhang *et al.*, "Dynamical Transformation of Two-Dimensional Perovskites with Alternating Cations in the Interlayer Space for High-Performance Photovoltaics," *J. Am. Chem. Soc.*, vol. 141, no. 6, pp. 2684–2694, 2019, doi: 10.1021/jacs.8b13104.
92. Y. Zhang and N. G. Park, "Quasi-Two-Dimensional Perovskite Solar Cells with Efficiency Exceeding 22%," *ACS Energy Lett.*, vol. 7, no. 2, pp. 757–765, 2022, doi: 10.1021/acsenergylett.1c02645.
93. H. Gu *et al.*, "Nanoscale hybrid multidimensional perovskites with alternating cations for high performance photovoltaic," *Nano Energy*, vol. 65, 2019, doi: 10.1016/j.nanoen.2019.104050.
94. P. Li *et al.*, "Low-Dimensional Perovskites with Diammonium and Monoammonium Alternant Cations for High-Performance Photovoltaics," *Adv. Mater.*, vol. 31, no. 35, 2019, doi: 10.1002/adma.201901966.
95. T. Luo *et al.*, "Compositional Control in 2D Perovskites with Alternating Cations in the Interlayer Space for Photovoltaics with Efficiency over 18%," *Adv. Mater.*, vol. 31, no. 44, 2019, doi: 10.1002/adma.201903848.
96. Y. Zhang *et al.*, "Highly Efficient Guanidinium-Based Quasi 2D Perovskite Solar Cells via a Two-Step Post-Treatment Process," *Small Methods*, vol. 3, no. 11, 2019, doi: 10.1002/smtd.201900375.
97. X. Zhang *et al.*, "Orientation Regulation of Phenylethylammonium Cation Based 2D Perovskite Solar Cell with Efficiency Higher Than 11%," *Adv. Energy Mater.*, vol. 8, no. 14, May 2018, doi: 10.1002/aenm.201702498.
98. W. Fu *et al.*, "Two-Dimensional Perovskite Solar Cells with 14.1% Power Conversion Efficiency and 0.68% External Radiative Efficiency," *ACS Energy Lett.*, vol. 3, no. 9, pp. 2086–2093, Sep. 2018, doi: 10.1021/acsenergylett.8b01181.
99. H. Lai *et al.*, "Two-dimensional Ruddlesden-Popper Perovskite with Nanorod-like Morphology for Solar Cells with Efficiency Exceeding 15%," *J. Am. Chem. Soc.*, vol. 140, 37, pp. 11639–11646, 2018, doi: http://pubs.acs.org
100. X. Zhang *et al.*, "Vertically Oriented 2D Layered Perovskite Solar Cells with Enhanced Efficiency and Good Stability," *Small*, vol. 13, no. 33, Sep. 2017, doi: 10.1002/smll.201700611.
101. X. Lian *et al.*, "The Second Spacer Cation Assisted Growth of a 2D Perovskite Film with Oriented Large Grain for Highly Efficient and Stable Solar Cells," *Angew. Chem.*, vol. 131, no. 28, pp. 9509–9513, Jul. 2019, doi: 10.1002/ange.201902959.
102. X. Zhang *et al.*, "Stable high efficiency two-dimensional perovskite solar cells via cesium doping," *Energy Environ. Sci.*, vol. 10, no. 10, pp. 2095–2102, Oct. 2017, doi: 10.1039/c7ee01145h.
103. N. Zhou *et al.*, "Exploration of Crystallization Kinetics in Quasi Two-Dimensional Perovskite and High Performance Solar Cells," *J. Am. Chem. Soc.*, vol. 140, no. 1, pp. 459–465, Jan. 2018, doi: 10.1021/jacs.7b11157.

104. H. Pan, X. Zhao, X. Gong, Y. Shen and M. Wang, "Atomic-Scale Tailoring of Organic Cation of Layered Ruddlesden-Popper Perovskite Compounds," *J. Phys. Chem. Lett.*, vol. 10, no. 8, pp. 1813–1819, 2019, doi: 10.1021/acs.jpclett.9b00479.
105. Y. Li, H. Cheng, K. Zhao and Z. S. Wang, "4-(Aminoethyl)pyridine as a Bifunctional Spacer Cation for Efficient and Stable 2D Ruddlesden-Popper Perovskite Solar Cells," *ACS Appl. Mater. Interfaces*, vol. 11, no. 41, pp. 37804–37811, 2019, doi: 10.1021/acsami.9b13951.
106. D. Ramirez, M. Alejandro, J. F. Montoya and F. Jaramillo, "Understanding the Role of the Mesoporous Layer in the Thermal Crystallization of a Meso-Superstructured Perovskite Solar Cell," *J. Phys. Chem. C*, vol. 120, 16, pp. 8559–8567, doi: 10.1021/acs.jpcc.6b02808.
107. W. Fu, H. Liu, X. Shi, L. Zuo, X. Li and A. K. Y. Jen, "Tailoring the Functionality of Organic Spacer Cations for Efficient and Stable Quasi-2D Perovskite Solar Cells," *Adv. Funct. Mater.*, vol. 29, no. 25, 2019, doi: 10.1002/adfm.201900221.
108. L. Gao *et al.*, "Enhanced Charge Transport by Incorporating Formamidinium and Cesium Cations into Two-Dimensional Perovskite Solar Cells," *Angew. Chem. Int. Ed.*, vol. 58, no. 34, pp. 11737–11741, 2019, doi: 10.1002/anie.201905690.
109. X. Li *et al.*, "Two-Dimensional Dion-Jacobson Hybrid Lead Iodide Perovskites with Aromatic Diammonium Cations," *J. Am. Chem. Soc.*, vol. 141, no. 32, pp. 12880–12890, 2019, doi: 10.1021/jacs.9b06398.
110. J. Ponchai *et al.*, "Modified colored semi-transparent perovskite solar cells with enhanced stability," *J. Alloys Compd.*, vol. 875, 2021, doi: 10.1016/j.jallcom.2021.159781.
111. A. Weselek, A. Ehmann, S. Zikeli, I. Lewandowski, S. Schindele and P. Högy, "Agrophotovoltaic systems: applications, challenges, and opportunities. A review," *Agron. Sustain. Dev.*, vol. 39, no. 4. Springer-Verlag France, 2019. doi: 10.1007/s13593-019-0581-3.
112. T. Sano, S. Inaba and V. Vohra, "Ternary Active Layers for Neutral Color Semitransparent Organic Solar Cells with PCEs over 4%," *ACS Appl. Energy Mater.*, vol. 2, no. 4, pp. 2534–2540, 2019, doi: 10.1021/acsaem.8b02144.
113. Y. Zhu, L. Shu and Z. Fan, "Recent Progress on Semi-transparent Perovskite Solar Cell for Building-integrated Photovoltaics," *Chem. Res. Chin. Univ.*, vol. 36, no. 3. Springer Berlin Heidelberg, pp. 366–376, 2020. doi: 10.1007/s40242-020-0105-3.
114. C. Y. Chang, K. T. Lee, W. K. Huang, H. Y. Siao and Y. C. Chang, "High-Performance, Air-Stable, Low-Temperature Processed Semitransparent Perovskite Solar Cells Enabled by Atomic Layer Deposition," *Chem. Mater.*, vol. 27, no. 14, pp. 5122–5130, 2015, doi: 10.1021/ acs.chemmater.5b01933.
115. Y. W. Zhang, P. P. Cheng, W. Y. Tan and Y. Min, "Balance the thickness, transparency and stability of semi-transparent perovskite solar cells by solvent engineering and using a bifunctional additive," *Appl. Surf. Sci.*, vol. 537, 2021, doi: 10.1016/j.apsusc.2020.147908.
116. C. O. Ramírez Quiroz *et al.*, "Pushing efficiency limits for semi-transparent perovskite solar cells," *J. Mater. Chem. A Mater.*, vol. 3, no. 47, pp. 24071–24081, 2015, doi: 10.1039/c5ta08450d.
117. L. Yuan *et al.*, "Semi-transparent perovskite solar cells: Unveiling the trade-off between transparency and efficiency," *J. Mater. Chem. A Mater.*, vol. 6, no. 40, pp. 19696–19702, 2018, doi: 10.1039/c8ta07318j.
118. M. Wang, F. Cao, M. Wang, K. Deng and L. Li, "Intermediate-Adduct-Assisted Growth of Stable CsPbI2Br Inorganic Perovskite Films for High-Efficiency Semitransparent Solar Cells," *Adv. Mater.*, vol. 33, no. 10, 2021, doi: 10.1002/adma.202006745.
119. H. J. Lee, S. P. Cho, S. @ in Na and S. S. Kim, "Thin metal top electrode and interface engineering for efficient and air-stable semitransparent perovskite solar cells," *J. Alloys Compd.*, vol. 797, pp. 65–73, 2019, doi: 10.1016/j.jallcom.2019.05.051.

120. C. Y. Chang, Y. C. Chang, W. K. Huang, K. T. Lee, A. C. Cho and C. C. Hsu, "Enhanced Performance and Stability of Semitransparent Perovskite Solar Cells Using Solution-Processed Thiol-Functionalized Cationic Surfactant as Cathode Buffer Layer," *Chem. Mater.*, vol. 27, no. 20, pp. 7119–7127, 2015, doi: 10.1021/acs.chemmater.5b03137.
121. J. W. Jung, C. C. Chueh and A. K. Y. Jen, "High-Performance Semi-transparent Perovskite Solar Cells with 10% Power Conversion Efficiency and 25% Average Visible Transmittance Based on Transparent CuSCN as the Hole-Transporting Material," *Adv. Energy Mater.*, vol. 5, no. 17, 2015, doi: 10.1002/aenm.201500486.

13

Advanced Materials in Energy Conversion Devices: Fuel Cells and Biofuel Cells

Amit Kumar Verma[1], Prerna Tripathi[1], Akhoury Sudhir Kumar Sinha[2] and Shikha Singh[1]*

[1]*Department of Sciences and Humanities, Rajiv Gandhi Institute of Petroleum Technology, Jais, Amethi, India*

[2]*Department of Chemical Engineering and Biochemical Engineering, Rajiv Gandhi Institute of Petroleum Technology, Jais, Amethi, India*

Abstract

A fuel cell is, in its most basic sense, a device that transforms chemical energy into electrical energy, whereas a biofuel cell is a subcategory of a fuel cell that uses a biocatalyst. The ideal candidate to replace current energy technologies is thought to be fuel cells. Fuel cell systems have a wide range of potential uses, including stationary power generation, transportation, and renewable energy storage when used in reverse as electrolyzer. Additionally, carbon capture and utilization (molten carbonate fuel cells) and power to gas processes involving the co-electrolysis of CO_2 and H_2O are good candidates for high-temperature fuel cells (solid oxide fuel cells and electrolyzer). Low-temperature biofuel cells are a great example of bio-compatible energy systems. Only highly developed and effective catalytic materials in both the anodic and cathodic positions enable the use of various fuel types (gas or liquid, hydrogen or hydrocarbons), an extension of lifetime, and a reduction in operating temperature. A new era of metal-free and/or nanostructured electrodes has significantly advanced fuel cell technology. In this regard, the current chapter offers comprehensive information on the cutting-edge materials utilized in fuel cell technology. Additionally, a classification and overview of fuel cells are given in this chapter, along with details on their major uses and the equations governing their primary reactions. In the end, the goals for further development of advanced materials for both mobile and stationary fuel cell applications, as well as the problems with cost and reliability, have been thoroughly discussed.

***Keywords*:** Fuel cells, devices, energy conversion, nanomaterials, efficiency, activity, biomimetic

13.1 Introduction

For a wide range of applications, fuel cell technology has been regarded as the preferred energy source. Since hydrogen has a higher energy density per unit weight than any other fuel [1–6], a power plant fueled by hydrogen might be more efficient on a weight basis compared to a conventional power plant that burns various fuels [7]. Fuel cell systems can

**Corresponding author*: shikhas@rgipt.ac.in

Upendra Kumar and Piyush Kumar Sonkar (eds.) Smart Materials for Science and Engineering, (269–286) © 2024 Scrivener Publishing LLC

produce energy at an efficiency of up to 60%, and a traditional combustion-based power plant normally generates electricity at 33 to 35% [8]. The petrol engine in a typical car today is less than 20% efficient at converting the chemical energy in petrol into power that drives the vehicle [9]. Moreover, fuel cells are suitable for a wide range of applications, own fewer moving parts, and run silently [10, 11]. Recently, nearly all the big manufacturers invested a sizable amount of money in the advancement of fuel cells. The development of fuel cells is receiving more attention than ever. Nearly all the major automakers have internal fuel cell development initiatives, or are collaborating closely with other businesses to build power plants for use in transportation [12]. The price of this technology has clearly been the biggest barrier [13]. A cathode, an anode, and an ion conducting electrolyte are the three essential components of a fuel cell. The electrodes/electrocatalysts are essential to the operation of a fuel cell. Transport of ions, electrons, and gaseous (or liquid) species must all be made possible within the electrodes, and at the triple-point (or triple-line) boundaries. The electrocatalysts possess a tendency to oxidize the fuel at the cathode and reduce the oxygen at the anode, respectively. Absolutely to accomplish this, the electrodes must be electrochemically active, and have large surface areas together with decent electrical and ionic conductivity [14]. The majority of widely used fuel cells, including DMFC, PAFC, and PEM, heavily rely on platinum to speed up the reaction rate. The cost of the catalysts used by a fuel cell accounts for around 26% of the whole cost [15]. The expansion of advancements in fuel cell technology is anticipated to be constrained in the coming years by the high cost of catalysts [16]. Recently, biofuel cells have been advanced, which partly differ from traditional fuel cells as they use naturally occurring proteins or microorganisms as biocatalysts instead of expensive metallic catalyst materials for respective anodic and cathodic electrochemical reactions [17–21]. The hunt for novel enhanced electrocatalyst materials with better activity, improved stability, cheap cost, and excellent electrical conductivity is now an important field of study that would benefit the development of fuel cell industry. This book chapter explores the fascinating potential of fuel cells with a focus on the state of development of the various advanced materials, a comparison of their operational traits, and the current techno-economic roadblocks to their advancement and the outlook for the future.

13.1.1 Electrochemical and Thermodynamic Principles of Fuel Cells

Different thermodynamic and electrochemical factors, such as pressure, temperature, current density, reactant consumption, gas concentration, etc., have an impact on fuel cell operation. These factors also directly affect the value of cell potential and voltage losses [22]. A typical fuel cell reaction that occurs in fuel cells is given below (Atkins 1986):

$$H_2 + \frac{1}{2}O_2 = H_2O \tag{13.1}$$

The maximum amount of work produced by the aforementioned reaction, as determined by a thermodynamic analysis, is correlated with the reaction's change in free energy.

The above reaction is thermodynamically and spontaneously preferred. Equation 13.2 shows the fuel cell reaction's standard free energy change.

$$\Delta G = -nFE \tag{13.2}$$

where F is Faraday's constant, n is the number of moles of electrons involved, ΔG is the free energy change, and E is the reversible potential. The value of ΔG corresponding to Equation 13.1 is −229 kJ/mol. The total heat produced by a fuel cell reaction is represented by the enthalpy change (ΔH, under constant pressure). The full potential of fuel cell in accordance with ΔH is defined as the thermo-neutral potential (E_t) (Equation 13.3). For the reaction indicated by Equation 13.1, E_t has a value of 1.48 V. The generalized perspective of the electrode reactions that occur in fuel cells with different feedstocks are briefly summarized in Table 13.1.

$$\Delta H = -nFE_t \tag{13.3}$$

The electrochemical processes in fuel cells are listed in Table 13.2 along with the Nernst equations associated [23]. These equations can be used to compute the ideal voltage at various temperatures and pressures if the ideal potential under standard conditions is known [24].

A fuel cell produces electrical energy when a current is applied, but due to irreversible losses caused by numerous factors, the real cell potential is lower than its equilibrium potential (Figure 13.1).

Many parameters are attributed to the irreversible losses in a real fuel cell. Gas concentration polarization, activation polarization, and Ohmic polarization are the main causes of the losses, which are commonly referred to as polarization or overpotential (Chase *et al.* 1985). The activation polarization can be described as (Equation 13.4)

Table 13.1 Electrochemical reactions in different fuel cells.

S. no.	Fuel cell type	Anode reaction	Cathode reaction
1	Acidic Fuel Cells	$H_2 \rightarrow 2H^+ + 2e^-$	$1/2\ O_2 + 2H^+ + 2e^- \rightarrow H_2O$
2	Alkaline Fuel Cells	$H_2 + 2(OH)^- \rightarrow 2H_2O + 2e^-$	$1/2\ O_2 + 2H_2O + 2e^- \rightarrow 2(OH)^-$
3	Oxide Fuel Cells	$H_2 + O2^- \rightarrow H_2O + 2e^-$ $CO + O2^- \rightarrow CO_2 + 2e^-$ $CH_4 + 4\ O2^- \rightarrow 2H_2O + CO_2 + 8e^-$	$1/2\ O_2 + 2e^- \rightarrow O2^-$
4	Molten Carbonate Fuel Cells	$H_2 + CO_32^- \rightarrow H_2O + CO_2 + 2e^-$	$1/2\ O_2 + CO_2 + 2e^- \rightarrow CO_32^-$

Table 13.2 Nernst equation for different fuel cell reactions [The symbol E represents the equilibrium potential, Eo the standard potential, P the gas pressure, R the universal gas constant, F Faraday's constant and T the absolute temperature].

S. no.	Fuel cell reaction	Anode reaction
1	$H_2 + 1/2\ O_2 \rightarrow 2H_2O$	$E = E° + (RT/nF)\ln[P_{H2}/P_{H2O}] + (RT/nF)\ln[P_{O2}^{1/2}]$
2	$H_2 + 1/2\ O_2 + CO_{2\,(cathode)} \rightarrow H_2O + CO_{2\,(anode)}$	$E = E° + (RT/nF)\ln[P_{H2}/P_{H2O}\,(P_{CO2\ anode})] + (RT/nF)\ln[P_{O2}^{1/2}\,(P_{CO2\ cathode})]$
3	$CO + 1/2\ O_2 \rightarrow CO_2$	$E = E° + (RT/nF)\ln[P_{CO}/P_{CO2}] + (RT/nF)\ln[P_O 2^{1/2}]$
4	$CH_4 + 2O_2 \rightarrow 2H_2O + CO_2$	$E = E° + (RT/8F)\ln[P_{CH4}/P_{H2O}^{2}P_{CO2}] + (RT/8F)\ln[P_{O2}^{2}]$

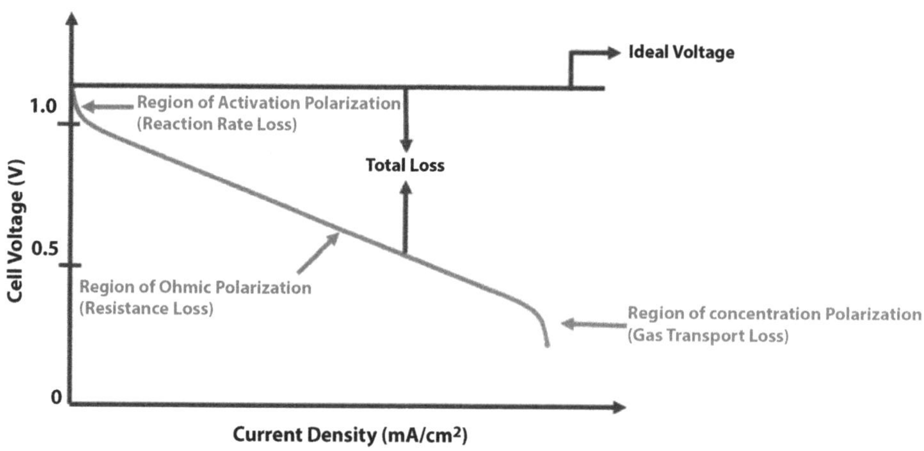

Figure 13.1 Actual fuel cell performance in relation to the potential current response.

$$\eta_{act} = \frac{RT}{\alpha nF} ln(\frac{i}{i_0}) \tag{13.4}$$

where R is the universal gas constant, η_{act} is the activation polarization, α is the charge transfer coefficient, T is the temperature, n is the number of electrons involved, i is the current density, F is the Faraday constant, and i_0 is the exchange current density. The sluggish electrochemical reactions at the electrode surface are the basis of activation polarization [25]. The rate of oxidation or reduction of the fuel or oxidant is closely correlated with activation polarization. The reactive species in fuel cell reactions must get through the activation barrier. The ohmic polarization is depicted by Equation 13.5, where R_c is the cell resistance and η_{ohm} is the ohmic polarization. It moves through resistance to the movement of ions in the electrolyte, the passage of electrons through the electrodes, and the movement of current in the external electrical circuit [24]. The electrolyte accounts for the majority of ohmic loss, which is minimized by reducing electrode separation, increasing the electrolyte's ionic conductivity, and altering its physical characteristics.

$$\eta_{ohm} = iR_c \tag{13.5}$$

Importantly, at practical current densities, sluggish reactant transport to the electrode and slow product removal from the reaction site are important causes of concentration polarization. The concentration polarization is represented by Equation 13.6, where i_l is the limiting current density and η_{con} is the concentration polarization. Many factors contribute to the formation of the concentration polarization.

$$\eta_{con} = \frac{RT}{nF} ln(1 - \frac{i}{i_l}) \tag{13.6}$$

13.1.2 Fuel Cell Efficiency

The effectiveness of any energy conversion apparatus is determined by the ratio of useful energy production to energy input (Equations 13.7 and 13.8). In fuel cell, the energy input is the hydrogen's enthalpy, or higher heating value (HHV), and the practical energy output is the electrical energy generated. The fuel cell's maximum theoretical efficiency, assuming that 100% of the Gibbs free energy can be converted into electrical energy (the reaction is reversible), is 83% [26].

$$\eta = \frac{actual\ electic\ work}{maximum\ available\ work} \tag{13.7}$$

$$\eta = \frac{\Delta G}{\Delta H} = \frac{237.34}{286.02} \times 100\% = 83\% \tag{13.8}$$

The fuel cell efficiency at lower heating value for hydrogen's (LHV) is calculated to be 94.5% [27],

$$\eta = \frac{\Delta G}{\Delta H} = \frac{228.74}{241.98} \times 100\% = 94.5\% \tag{13.9}$$

Due to the fact that the fuel cell's reversible efficiency declines as operating temperature rises, the LHV is more efficient than the HHV [28, 29].

13.2 Fuel Cell Types and Advancement in Electrode Materials

William Grove, a British physicist, employed platinum electrodes and hydrogen and oxygen as catalysts in 1839. Nowadays, numerous fuel cell varieties have been identified. Herein,

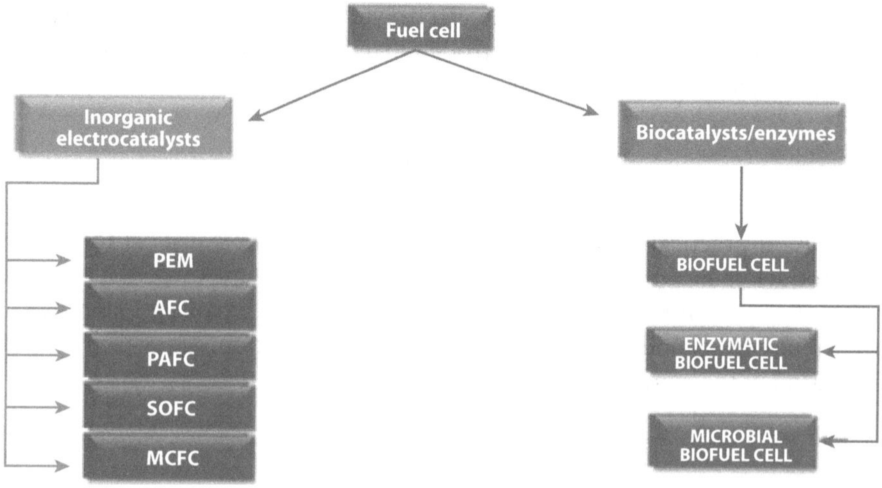

Figure 13.2 Classification of fuel cells (PEM, Proton-exchange membrane fuel cells; AFC, Alkaline fuel cells; PAFC, Phosphoric acid fuel cells; SOFC, solid oxide fuel cell; MCFC, Molten carbonate fuel cells).

to have a more simplified view, we have classified the fuel cell into two categories based on the type of material used: inorganic and biocatalyst/enzyme type (Figure 13.2). The recent advancement in the design and electrocatalyst materials for various fuel cell types will be briefly discussed below.

13.2.1 Advanced Materials for PEM, AFC, PAFC, SOFC, and MCFC Fuel Cells

13.2.1.1 Platinum Group Metals (PGM)

Regardless of whether hydrogen is being reduced or oxidized, platinum (Pt) is the electrocatalytic material that operates at its peak [30]. For the hydrogen oxidation reaction (HOR) or for PEMFCs, very durable and adaptable WO_3 and evenly adorned Pt nanoparticles (20 wt%) that supported WO_3 nanoplatelets were produced [31]. The chronoamperometry (CA) investigations that were carried out for 24 hours confirmed the robustness and stability of this HOR electrocatalyst. A maximum power density (p.d.) of 430 mW cm^{-2} for Pt/WO_3-550 at 50°C is noted. By using cyclic voltammetry and square wave pulse electrochemical technique, Pt-Ni bimetallic nanoparticles were electro-deposited on a novel carbon fiber paper filled with reduced graphene oxide (rGO) [32]. The electrocatalytic mass activity of the Pt-Ni-CV(A)-rGO/CFP electrode was 5.3 ampere per mg of Pt, which was approximately 5.6 times that of the standard Pt/C catalyst consisted of 20 wt% Pt nanoparticles. For use as an integrated electrode in proton exchange membrane fuel cells (PEM), the Pt-Ni-rGO/CFPs composite electrode looked promising as it was being manufactured. For use in DMFC, ultra-stable self-supported Pt-Cu nanowires (NWs) with lots of Cu vacancies were developed, where the OH adsorption energy was found to be stronger by the vacancy defect through dealloying and the CO adsorption energy is weakened by the doping of Cu elements [33]. The well-designed Pt-Cu NWs provide exceptional activity for the MOR, with a specific activity 7.5 times higher than that for the bare Cu. For MOR, graphene oxide (GO), reduced graphene oxide (r-GO), and graphitic nitride (GCN) hybrid composites were used as a stable and electron-rich substrate for Pt-Sn bimetallic nanoparticles [34]. Similar to this, the porous Pt-Ni nanobelt arrays were made by sputtering Pt and Ni using a magnetron on a sacrificial template [35]. In H_2/air and at 40% relative humidity, the porous Pt-Ni nanobelt array-based catalyst layer showed improved fuel cell performance. The MEA with porous Pt-Ni nanobelt arrays has a performance of 366.6 and 597.6 mW cm^{-2}. Nitrogen-doped active carbon (N-doped Black Pearl, NBP) was used to deposit an atomically dispersed Pt-based heterogeneous electrocatalyst. A Pt content of 5% (Pt1/NBP) using the hydrothermal ethanol reduction method was achieved [36]. The highly efficient platinum single-atom catalysts for oxygen reduction reaction (ORR) were investigated. The produced Pt@NBP catalyst exhibits outstanding ORR stability and performance together with quick ORR kinetics at a high potential region. At a cell voltage of 0.9 V, it exhibits 8.7 times more mass activity than the standard Pt/C as a cathode catalyst of a PEM fuel cell. Pt electrocatalysts coated with Co_{100-x}-Ni_x (x = 0, 30, 70, and 100) were produced [37]. A two-stage procedure consisting of an adaptable ball milling and the galvanic displacement approach was used to generate the electrocatalysts. Two types of systems were produced as a result: BM-Pt/C ($Co_{30}Ni_{70}$-Pt/C and $Co_{70}Ni_{30}$-Pt/C) and M-Pt/C (Co_{100}-Pt/C and Ni_{100}-Pt/C). The final compositions of each electrocatalyst were as follows: 10 weight percent Pt, 40 weight percent $Co_{100-x}Ni_x$, and 50 weight percent Vulcan carbon. During 3000 cycles of an accelerated deterioration test, the $Co_{30}Ni_{70}$-Pt/C electrocatalyst showed the best stability compared to the others, while the Ni_{100}-Pt/C electrocatalyst displayed the highest mass activity. Ni_{100}-Pt/C and $Co_{30}Ni_{70}$-Pt/C electrocatalysts were manufactured and were used as cathode in two membrane electrode assemblies (MEAs) for PEM fuel cells.

13.2.1.2 Platinum Group Metal Free Materials

Platinum group metal (PGM) free materials have received a lot of interest as electrocatalysts, but there are still issues with their application due to their low catalytic activity and instability. Materials made of transition metals are frequently employed because of their inexpensive cost and comparatively high activity. The stability and activity of the catalytic process may be improved by the electrical, compositional, and geometric impacts of the metal–support interaction. Co-N-P-C PGM free electrocatalysts were immobilized on MXene (Mo_2TiC_2) [38], which possess strong corrosion resistance and outstanding electrical conductivity. Co-N-P-C@Mo_2TiC_2 demonstrates exceptional stability in single-cell and ORR testing, with only loss of 18 mV after 10,000 cycles. Co-N-P-C@Mo_2TiC_2 has a half-wave potential of 0.78 V and a peak power density (p.d.) of 880 mV cm^{-2}. Similarly, Fe-N-C PGM free electrocatalysts for PEM fuel cell application were produced by pyrolyzing N or N/S co-doped artificial polymers impregnated with ferric chloride [39]. The performance of Fe-N-C-based cathode catalyst layers (CCLs) in H_2-air PEM fuel cells was examined in relation to the initial $FeCl_3$ content, sulfur co-doping, and addition of phenanthroline. In an H_2-air PEMFC, the top-performing CCL delivered a peak power d. of less than 0.15 $W.cm^{-2}$. The development of highly effective, reliable, and reasonably priced electrocatalysts with a super aerophobic feature for the hydrazine oxidation reaction (HzOR) is one of the key difficulties in the expansion of direct hydrazine fuel cells (DHzFCs). Cobalt phosphide nanowire array (CoP-NWA@CP) was developed on a carbon paper substrate for the HzOR in alkaline media [40]. The catalyst demonstrates a substantial HzOR performance with an onset potential of -61 mV (vs RHE) and an ultralow overpotential of -32 mV@10 mA cm^{-2}. The array also revealed reliable stability with little c.d. loss over 10,000 s at an overpotential of 0.42 V (vs RHE). The CoP-NWA@CP catalyst's astounding HzOR performance is ascribed to both its nanowire array architecture and high intrinsic activity in addition to its high intrinsic activity. Reversal-tolerant anodes (RTAs) for PEMFCs were created by uniformly depositing niobium-doped IrOX nanoparticles over TiO_2 ($IrO_X/Nb_{0.1}Ti_{0.9}O_2$) using the sol-gel method [41]. The uniformly dispersed IrO_X nanoparticles boost the oxygen evolution reaction and, additionally, the use of a non-carbon support to prevent the carbon oxidation reaction. The iron/nitrogen co-doped carbon that was developed by converting the ZIF-8 regular dodecahedron into a hierarchical porous structure (Fe-ZIF/CN-UC) demonstrated impressive ORR activity in an acidic solution [42]. The constructed Fe-ZIF/CN-UC electrocatalyst was used as the cathode catalyst of H_2-O_2 PEMFCs and exhibits impressive performance with a c.d. of 2 A cm^{-2} at 0.2 V vs. RHE and a maximum power d. of 0.484 Wcm^{-2}. NGM free Mo carbide composite with N-doped C (MoCx-NC) [43] for DHzFCs application was produced by the rapid pyrolysis method. Among the electrocatalysts as-prepared, the best MoCx-NC electrocatalyst (900°, 5%) has the highest electrocatalytic performance (3.56 mA cm^{-2}) and cycling stability (86.13%) for the HzORs. Long-term operational stability for 12 hours, an open circuit voltage of 0.96 V (80°), and a maximum power d. of 158.26 mW cm^{-2} were the features of the MoCx-NC electrocatalyst. Cobalt (II) phthalocyanine (CoPc) was investigated as a potential platinum (Pt) substitute as cathode in PEM fuel cells [44]. A CVD graphene-based CoPc catalyst was developed using the impregnation process. It was found that electrochemical activity and stability of CoPc/CVD graphene combinations have improved with higher heat-treatment temperatures. A maximum power density (p.d.) of 186 mW cm^{-2} was attained by CoPc-Graphene at 80°C, 25 psi backpressure, and 100% RH.

13.2.1.3 Carbonaceous Materials

Heteroatom-doped (N, P, S, C and O) porous carbon materials (HPCNs) derived from plant polyphenols (tannic acid) through a simple template free one-step approach was found to have admirable biocompatibility and great conductivity, making them a good candidate for MFCs [45]. The real power density (RPD) recorded for HPCN-decorated carbon cloth-based MFCs was 1.72 W m^{-2} and the c.d. was 4.52 A m^{-2}, which were 1.82 times and 1.44 times higher than the bare carbon cloth (CC) as anode, respectively. The extracellular electron transfer (EET) and cell survival of the electroactive biofilms were markedly enhanced by the suggested anode. Recently, to generate carbon structures doped with nitrogen and phosphorus as well as single cobalt active sites, tofu containing bimetallic zinc and cobalt ions was used [46]. The electrocatalytic activity of the prepared tofu-derived catalyst displays outstanding ORR, with a half-wave potential of 0.86 V, an onset potential of 0.981 V, with enhanced durability. A 3D heterojunction catalyst composed of hollow carbon nanospheres and graphitic C_3N_5 was developed for ORR [47]. It was reported that the existence of conductive heterojunction allowed for quick electron/charge transfers between g-C_3N_5 and HCNs. The HCNs@g-C_3N_5 composite showed significantly increased ORR activity compared to standalone HCNs and g-C_3N_5, with a greater positive half-wave potential and quicker kinetics in the alkaline medium. Negatively doped poly(benzimidazobenzophenanthroline) (BBL) was investigated as a stable and effective oxygen cathode material [48].

The BBL provides electrocatalysis that entirely converts dioxygen into water through a (2 + 2)-electron transfer pathway with hydrogen peroxide (H_2O_2) as a middle step. The deoxygenated graphene oxide that formed on the carbon brush's surface (RGO-CB) was investigated for use as a bio-anode in MFCs [49]. The electrophoretic deposit of graphene oxide on the surface of the carbon brush boosted the power d. by a factor of more than 10 (from 33 mW m^{-2} to 381 mWm-^{2}). Similar to that, a microbial fuel cell (MFC) anode made of polydopamine-deposited reduced oxide graphene (rGO@PDA) on carbon cloth (CC) [50]. The anodes provided superhydrophilicity, high conductivity, improved biocompatibility, and long-term stability since they contained both inorganic and organic components. A new covalent organic framework (NDI-COF) based on naphthalene diimide was developed for metal-free fuel cells [51]. In aqueous settings, the naphthalene diimide units serve as an efficient electrocatalyst for O reduction. Highly graphitic nitrogen-doped carbon nano-onions (N-CNOs) were formed using a one-step, direct *in situ* flame synthesis procedure, and their potential for application as oxygen reduction process (ORR) catalysts in a microbial fuel cell (MFC) was evaluated [52]. N-CNO achieved a power density of 49.6 mW m^{-2}, which was twice as efficient as an undoped CNO-MFC.

13.2.1.4 *Perovskite Materials*

The SOFC is the fuel cell type where materials having a perovskite structure are used and researched the most. $Sr_2Ti_{0.8}Co_{0.2}FeO_{6-\delta}$ perovskite ceramic was used as perovskite anode with *in situ* exsolved nanocatalysts [53]. It has the characteristic of helping to reduce carbon deposition and boost anode catalytic activity (SOFCs). Using hydrogen, methane, methanol, and ethanol, respectively, the peak power densities of the cells were measured to be 0.82, 0.59, 0.43, and 0.33 W cm^{-2} at 800°C. Similarly, $SrCo_{0.8}Zr_{0.2}O_{3-\delta}$ was evaluated as the anode for SOFCs [54]. The cubic $SrZr_{0.8}Co_{0.2}O_{3-\delta}$ (SZC, Pm_3m, 30% weight%) and hexagonal $SrCo_{0.9}Zr_{0.1}O_{3-\delta}$ (SCZ, R_3cH, 70% weight%) perovskite phases were discovered co-assembled at the nanoscale area. At 750–550°C, the SZC–SCZ nanocomposite cathode cell has a superior p.d. of 1.48–0.41 W cm^{-2} compared to a single cubic SZC cathode cell (0.24–0.06 W cm^{-2}) and a conventional hexagonal $SrCoO_{3-\delta}$ cathode cell (1.25-0.32 W cm^{-2}). The impact of the $La_{0.6}Sr_{0.4}CoO_{3-\delta}$-$Gd_{0.1}Ce_{0.9}O_{1.95}$ (LSC-GDC) perovskites composite cathode on the performance of SOFCs was examined [55]. $LiNi_{0.8}Co_{0.15}Al_{0.05}O_{2-\delta}$ (NCAL) perovskites seemed to provide an excellent electrode for solid oxide fuel cells, having both hydrogen oxidation and oxygen reduction catalytic activity (SOFCs) [56]. For the anode application in a solid oxide fuel cell (SOFC), porous $Gd_{0.1}Ce_{0.9}O_{2-\delta}/SrMg_{0.1}Mo_{0.9}O_{3-\delta}$ perovskites were explored [57]. The electrochemical performance obtained was discovered to be significantly better than the electrochemical performance of the other Ni-free ceramic anodes reported before, highlighting the potential of the $0.7Gd_{0.1}Ce_{0.9}O_{2-\delta}/SrMg_{0.1}Mo_{0.9}O_{3-\delta}$ for the SOFC anode. A-site-deficient $(Ba_{0.9}La_{0.1})_{0.95}Co_{0.7}Fe_{0.2}Nb_{0.1}O_{3-\delta}$ perovskite oxide with *in situ* exsolved Co-Fe nanoparticles, designated (BL95CFN), was created and researched for SOFCs [58]. Cobalt and iron were more easily released from the perovskite matrix of BL95CFN when compared to stoichiometric BLCFN due to a minor A-site shortage (5%). Excellent electrochemical performance was displayed by this hybrid anode catalyst. At 750°C, the polarization resistance of the BL95CFN anode is 0.11 cm^2, which is 42% less than that of the BLCFN anode. Overall, the hybrid electrocatalyst has a promising future for use in SOFCs due to its high catalytic activity and coking tolerance. Perovskite

$PrBaFe_2O_{5+\delta}$ appeared as a promising electrode material for symmetrical solid oxide fuel cells (Sym-SOFCs) [59]. Fe was partially replaced with W to generate a composition of $(PrBa)_{0.95}(Fe_{0.95}W_{0.05})_2O_{5+\delta}$ with the goal of improving Sym-SOFC performance. Doping W maintains the stability of $PrBaFe_2O_{5+\delta}$ after exposure to high temperatures in both hydrogen and air atmospheres, and reduces the thermal expansion coefficient from 17.11×10^{-6} to 14.59×10^{-6} K^{-1}. Doping W significantly improves SOFCs electrochemical performance. In order to create exsolution systems for solid oxide fuel cell anodes and catalytic applications, Ni-doped perovskites, $(Pr_{0.5}Ba_{0.5})_{1-x/2}Mn_{1-x/2}Ni_{x/2}O_{3^-\delta}$ with $x = 0$, 0.05, 0.1, and 0.2 (S-PBMNx), were synthesized [60]. $La_{2-x}Ce_xCuO_4$ (LCCO) layer-perovskite oxide was used as a cathode, anode, and electrolyte materials for SOFCs [61]. Due to the LCCO's versatility, it may be used to streamline the production process of SOFCs. Although H_2 in the anode is seen to lower LCCO, the cell performance remains largely steady when operated repeatedly. For use in an electrolyte, LCCO is combined in a variety of weight ratios with the ionic conductor $Ce_{0.8}Sm_{0.2}O_{2-\delta}$ (SDC). The electroactive electrode $Sr_{0.93}(Ti_{0.3}Fe_{0.56}Ni_{0.07}Co_{0.07})O_{3-\delta}$ perovskite (STFNC) (STFNC) was investigated for symmetrical SOFCs (S-SOFCs) application [62]. When the NiCoFe phase was exsolved *in situ* at 700°C, the anode polarization resistance of the STFNC electrode was 1.12 cm^2 and the cathode polarization resistance was 0.054 and 0.042 cm^2, before and after exsolution, respectively. This electrode has an intriguing potential as an S-SOFC electrode. For intermediate-temperature symmetrical solid oxide fuel cells (IT-SSOFC), cathode and anode materials made of $La_{0.7}Sr_{0.3}Ti_{0.1}Fe_{0.6}Ni_{0.3}O_{3^-\delta}$ (LSTFN) were used [63]. The produced perovskite material exhibits a low polarization resistance (Rp) of 0.047 cm^2 with air and 0.201 cm^2 with dry H_2 at a temperature of 800°C. The constructed symmetrical cell showed excellent stability with no noticeable deterioration over the 300 hours of testing in air and 100 hours of testing with dry H_2.

13.2.2 Advanced Materials for Biofuel Cells

This group of fuel cells comprises those that can transform chemical energy into electrical energy through electrochemical reactions involving biological pathways when a biocatalytic component is linked to effective electrodes [64]. Enzymatic biofuel cells (EFCs) are bio-electronic devices that convert energy into electricity by oxidizing an organic substrate and/or reducing oxygen or peroxide using oxidoreductase enzymes as electrocatalysts [65]. Microbial fuel cells (MFCs) are technologies that oxidize organic and inorganic materials using microorganisms as catalysts to produce current [66]. In advance, polypyrrole nanotubes (PPy-NT) appeared as an effective anode catalyst in single-chamber microbial fuel cells (SMFC) [67]. With the hydrothermal process, chitosan (CHI) and nickel oxide (NiO/CHI and PPy-NT) was subsequently coated on PPy-NT to further improve electrochemical efficiency. In comparison to PPy-NT, cyclic voltammetry characterization reveals high oxidation peaks for CHI and PPy-NT (-0.0282 mA) and NiO/CHI and PPy-NT (-0.0286 mA) (0.00860 mA). A power density of 755 mW m^{-2} was produced by covering of stainless-steel mesh (SSM) anode with NiO/CHI and PPy-NT, and it was 4.4 times more powerful than a stainless-steel anode that was not coated (170 mW m^{-2}). Similarly, the fabrication of MFC prototypes using stainless-steel meshes with rGO and sNiO was made [68]. Structured nickel oxide (sNiO), which serves as the air cathode, and reduced graphene oxide (rGO), which has a high electrical conductivity, are utilized in the generation of energy and the treatment of domestic sewage. To significantly enhance the anode's efficiency in microbial fuel cells (MFCs), β-FeOOH was coated on porous carbon monolith (WGC) [69]. The MFCs with WGC and β-FeOOH /WGC anodes had maximum power densities of 913.9 and 1355.1 mW/m^2, respectively. For use in MFCs, carbon nanotubes (CNTs) were merged with electrospun porous–α-Fe_2O_3 nanofibers [70]. Due to its excellent electricity, extremely high porosity, and three-dimensional interpenetrated network of produced CNTs-Fe_2O_3, the decorated anode was able to enrich active bacteria and improve the effective EET rate. CNTs/-Fe_2O_3 nanofibers as the anode surpass -Fe_2O_3 anode (940 mW/m^2; 81.66%) and bulk carbon cloth anode (432 mW/m^2; 65.83%), achieving an outstanding power density of 1959 mW/m^2 and a high COD elimination efficiency of 85.04%. To enhance the anode characteristics of MXene, the superhydrophilic cationic polymer poly(diallyldimethylammonium chloride) (PDDA) was used [71]. MXene@PDDA/carbon cloth (CC) was then created as an MFC anode. The enrichment of electrochemically active bacteria was made easier by the anode, and the EET process was made more effective. The maximum output voltage and power for the system are 585 mV and 811 mW m^{-2}, respectively. After 12 hours, the COD and methyl orange decolorization efficiency were up to 79% and 84%, respectively. A hydrophobic Fe-N_4 macromolecular catalyst was directly mounted on the surface of activated carbon (AC) (Fe-N_4/AC cathode) [72] for MFCs application. A maximum p.d. of 2.5 W m^{-2} was attained with a 10 wt% Fe-N_4/AC cathode, which was 25% more powerful than standard AC cathodes (2.0 W m^{-2}). For the improvement of MFCs, it was also investigated whether altering catalyst hydrophobicity would concurrently increase power density and coulombic efficiency.

13.3 Current Application Status

The use of fuel cell technology is extremely diverse. Fuel cells, as opposed to batteries, can be quickly recharged, saving time and money on battery changes. It has been proven that fuel cells can be used for stationary, automotive, backup, or portable power applications [73]. Particularly in the telecommunications industry, fuel cells offer a potentially practical solution for backup power [74]. The United Nations Development Program (UNDP) and Global Environment Facility (GEF) are supporting and coordinating an international program to install fuel cell buses in significant cities throughout the world, including Mexico City, New Delhi, Sao Paolo, Shanghai, Cairo, and Beijing [75]. Some of these buses have already been developed and demonstrated [76]. Currently, the US and South Korean markets, which account for about 95% of installed capacity, are leading the world in the deployment of large-scale fuel cells [77]. The global fuel cell market is anticipated to raise at a compound annual growth rate (CAGR) of 26.0% from 2022 to 2027, from USD 2.9 billion to USD 9.1 billion [78].

13.4 Challenges

Fuel cells are perfect for industrial applications because they represent a renewable energy source with no toxic by-products that pollute the environment. Unfortunately, several obstacles prevent the widespread use of fuel cells. Over the past two decades, Pt concentration in PEM fuel cells has dramatically decreased, although total Pt eradication is still a goal.

In the majority, high-cost contributing components are the catalyst (precious metal, Pt, or Pt-alloys) and the ionomer membrane. Further improvements in material development and production techniques are essential for cost and complexity reduction success. Although carbonaceous materials are frequently used in fuel cell applications, carbon monoxide poisoning, the longevity of carbonaceous materials, and carbon corrosion are some of the key elements that influence a fuel cell's performance. In order to make SOFCs viable, absolutely active cathode materials and highly selective anode materials that are unable to catalyze carbon deposition must be developed. Additional effort is required to improve the performance of low-temperature operating nanostructure materials. The MFC technique's development is currently constrained by inefficiencies related to electrode material characteristics including charge transfer, surface area, catalytic behavior, and cost. Investigating innovative functional materials with high electrical and ionic conductivity, porosity, and biocompatibility is crucial to proliferate the growth of biofilm on its surface. Future research should focus on scaling up the effective electrodes for MFCs as well as creating inventive combinations of anode and membrane or cathode and membrane to create the optimum membrane electrode assembly.

13.5 Conclusion

Fuel cells are a wonderful technology that is easy to use, effective, and low polluting. These factors enable numerous fuel cell applications on their own. There is an effort to produce as much electricity as is feasible while being sustainable. Now more than ever, it is critical to develop fresh approaches to meet our rising electricity demand ethically and safely. By developing novel, extremely efficient electrocatalysts, fuel cells' efficiency, output, and adaptability would all be improved. If all the related or affecting factors are thoroughly considered and put into practice, fuel cells can actively compete with other storage technologies and may even outperform them.

References

1. Møller, K. T.; Jensen, T. R.; Akiba, E.; Li, H.-w.; Hydrogen - a Sustainable Energy Carrier. *Prog. Nat. Sci.: Mater. Int.*, 2017, 27, 34-40.
2. Rude, L. H., *et al.*; Tailoring Properties of Borohydrides for Hydrogen Storage: A Review. *Phys. Status Solidi (a)*, 2011, 208, 1754-1773.
3. Lai, Q., *et al.*; Hydrogen Storage Materials for Mobile and Stationary Applications: Current State of the Art. *ChemSusChem*, 2015, 8, 2789-2825.
4. Rosen, M. A.; Koohi-Fayegh, S.; The Prospects for Hydrogen as an Energy Carrier: An Overview of Hydrogen Energy and Hydrogen Energy Systems. *Energy Ecol. Environ.*, 2016, 1, 10-29.
5. Ramachandran, R.; Chen, T.-W.; Veerakumar, P.; Anushya, G.; Chen, S.-M.; Kannan, R.; Mariyappan, V.; Chitra, S.; Ponmurugaraj, N.; Boominathan, M.; Recent Development and Challenges in Fuel Cells and Water Electrolyzer Reactions: An Overview. *RSC Adv.*, 2022, 12, 28227-28244.
6. Fan, L.; Tu, Z.; Chan, S. H.; Recent Development of Hydrogen and Fuel Cell Technologies: A Review. *Energy Rep.*, 2021, 7, 8421-8446.

7. Edwards, P. P.; Kuznetsov, V. L.; David, W. I. F.; Brandon, N. P.; Hydrogen and Fuel Cells: Towards a Sustainable Energy Future. *Energy Policy,* 2008, 36, 4356-4362.
8. Yazici, M. S.; In Nanoscale Materials for Hydrogen and Fuel Cell Systems, *Functionalized Nanoscale Materials, Devices and Systems*, Dordrecht, 2008//; Vaseashta, A.; Mihailescu, I. N.; Eds. Springer Netherlands: Dordrecht, 2008; *pp* 283-290.
9. Leach, F.; Kalghatgi, G.; Stone, R.; Miles, P.; The Scope for Improving the Efficiency and Environmental Impact of Internal Combustion Engines. *Transp. Eng.*, 2020, 1, 100005.
10. Zhou, S.; Fan, L.; Zhang, G.; Gao, J.; Lu, Y.; Zhao, P.; Wen, C.; Shi, L.; Hu, Z.; A Review on Proton Exchange Membrane Multi-Stack Fuel Cell Systems: Architecture, Performance, and Power Management. *Appl. Energy,* 2022, 310, 118555.
11. Gencoglu, M. T.; Ural, Z.; Design of a Pem Fuel Cell System for Residential Application. *Int. J. Hydrog. Energy,* 2009, 34, 5242-5248.
12. Fan, L.; Tu, Z.; Chan, S. H.; Recent Development of Hydrogen and Fuel Cell Technologies: A Review. *Energy Rep.*, 2021, 7, 8421-8446.
13. Perry, M. L.; Fuller, T. F.; A Historical Perspective of Fuel Cell Technology in the 20th Century. *J. Electrochem. Soc.*, 2002, 149, S59.
14. Haile, S. M.; Materials for Fuel Cells. *Mater. Today,* 2003, 6, 24-29.
15. Du, L.; Prabhakaran, V.; Xie, X.; Park, S.; Wang, Y.; Shao, Y.; Low-Pgm and Pgm-Free Catalysts for Proton Exchange Membrane Fuel Cells: Stability Challenges and Material Solutions. *Adv. Mater.*, 2021, 33, 1908232.
16. Fuel Cell Market Study (Markets & Markets; May 2022, Code- EP 2762
17. Ghassemi, Z.; Slaughter, G.; Biological Fuel Cells and Membranes 2017, DOI: 10.3390/membranes7010003.
18. Yin, S.; Varun, P.; Chunlei, W.; Recent Development of Miniatured Enzymatic Biofuel Cells 657. In *Biofuel's Engineering Process Technology*, Marco Aurélio dos Santos, B.; Ed. IntechOpen: Rijeka, 2011; *p* Ch. 28
19. Jia, H.; Wang, P.; Biocatalytic Generation of Power from Biofuels: Biofuel Cells. *Biofuels,* 2010, 1, 479-492.
20. Kaur, R.; Marwaha, A.; Chhabra, V. A.; Kim, K.-H.; Tripathi, S. K.; Recent Developments on Functional Nanomaterial-Based Electrodes for Microbial Fuel Cells. *Renew. Sustain. Energy Rev.*, 2020, 119, 109551.
21. Peera, S. G.; Maiyalagan, T.; Liu, C.; Ashmath, S.; Lee, T. G.; Jiang, Z.; Mao, S.; A Review on Carbon and Non-Precious Metal Based Cathode Catalysts in Microbial Fuel Cells. *Int. J. Hydrog. Energy,* 2021, *46*, 3056-3089
22. Lindiwe, K.; Fuel Cell Thermodynamics. In *Thermodynamics and Energy Engineering*, Petrică, V., Ed. IntechOpen: Rijeka, 2019; *p* Ch. 1.
23. Pilatowsky, I.; Romero, R. J.; Isaza, C. A.; Gamboa, S. A.; Sebastian, P. J.; Rivera, W.; Thermodynamics of Fuel Cells. In *Cogeneration Fuel Cell-Sorption Air Conditioning Systems*, Eds. Springer London: London, 2011; *pp* 25-36.
24. Chapter 2: Thermodynamics of Fuel Cells. *Fuel Cell Fundamentals*, edited by S.-W. Cha, W. Colella and F.B. Prinz, R. O'Hayre.
25. Birss, V.; El Sawy, E.; Ketabi, S.; Keyvanfar, P.; Li, X.; Young, J.; Electrochemical Energy Production Using Fuel Cell Technologies. In *Handbook of Industrial Chemistry and Biotechnology*, Kent, J. A.; Bommaraju, T. V.; Barnicki, S. D.; Eds. Springer International Publishing: Cham, 2017; pp. 1729-1779.
26. Pavelka, M.; Maršík, F.; Detailed Thermodynamic Analysis of Polymer Electrolyte Membrane Fuel Cell Efficiency. *Int. J. Hydrog. Energy,* 2013, 38, 7102-7113.
27. Lutz, A. E.; Larson, R. S.; Keller, J. O.; Thermodynamic Comparison of Fuel Cells to the Carnot Cycle. *Int. J. Hydrog. Energy,* 2002, 27, 1103-1111.

28. Reimer, U.; Lehnert, W.; Holade, Y.; Kokoh, B.; Chapter 2 - Irreversible Losses in Fuel Cells. In *Fuel Cells and Hydrogen*, Hacker, V.; Mitsushima, S., Eds. Elsevier: 2018; *pp* 15-40.
29. *Cybern. Phys.*, VOL. 1, NO. 1, 2012, 67–72
30. Sheng, W.; Gasteiger, H. A.; Shao-Horn, Y.; Hydrogen Oxidation and Evolution Reaction Kinetics on Platinum: Acid Vs Alkaline Electrolytes. *J. Electrochem. Soc.*, 2010, *157*, B1529.
31. A R, D.; Ganguly, D.; Sundara, R.; High Temperature Annealed (002) Oriented Wo3 Nanoplatelets with Uniform Pt Decoration as Durable Carbon Free Anode Electrocatalyst for Pemfc Application. *Int. J. Hydrogen Energy*, 2022, *47*, 24978-24990.
32. Li, Y.; Zhu, E.; Yu, J.; Yu, J.; Yang, J.; Liu, W.; Cui, X.; Yang, X.; Zhang, Y.; Xu, M.; Pt–Ni Nanoparticles Electrodeposited on Rgo/Cfp as High-Performance Integrated Electrode for Methanol Oxidation. *Int. J. Hydrog. Energy*, 2022, 47, 23957-23970.
33. Wang, K.; Huang, D.; Guan, Y.; Liu, F.; He, J.; Ding, Y.; Fine-Tuning the Electronic Structure of Dealloyed Ptcu Nanowires for Efficient Methanol Oxidation Reaction. *ACS Catal.*, 2021, 11, 14428-14438.
34. Poudyal, D. C.; Dugani, R.; Dash, B. S.; Dhavale, M.; Satpati, A. K.; Haram, S. K.; Γ-Ray-Assisted Synthesis of a Pt–Sn Bimetallic Composite Loaded on Graphene–Graphitic Carbon Nitride Hybrid: A Cocktail Electrocatalyst for the Methanol Oxidation Reaction. *ACS Omega*, 2021, 6, 13579-13587.
35. Yao, D.; Yu, H.; Song, W.; Gao, X.; Sun, X.; Jiang, G.; Hao, J.; Shao, Z.; Porous Pt-Ni Nanobelt Arrays with Superior Performance in H2/Air Atmosphere for Proton Exchange Membrane Fuel Cells. *ACS Appl. Energy Mater.*, 2021, 4, 10703-10712.
36. Liu, J.; Bak, J.; Roh, J.; Lee, K.-S.; Cho, A.; Han, J. W.; Cho, E.; Reconstructing the Coordination Environment of Platinum Single-Atom Active Sites for Boosting Oxygen Reduction Reaction. *ACS Catal.*, 2021, 11, 466-475.
37. Flores-Rojas, E.; Reyes-Rodríguez, J. L.; Cruz-Martínez, H.; Rojas-Chávez, H.; Samaniego-Benítez, J. E.; Solorza-Feria, O.; Applications of Cathodic Co100-Xnix (X = 0, 30, 70, and 100) Electrocatalysts Chemically Coated with Pt for Pem Fuel Cells. *Int. J. Hydrog. Energy*, 2020, 45, 13726-13737.
38. Ren, W.; Chang, Z.; Zhang, J.; Wang, L.; Xu, C.; Nonprecious Co-N-P-C@Mo2tic2 Catalyst for the High-Performance Oxygen Reduction Reaction in Pemfcs. *ACS Appl. Energy Mater.*, 2022, 5, 15828-15833.
39. Kiciński, W.; Dyjak, S.; Gratzke, M.; Tokarz, W.; Błachowski, A.; Platinum Group Metal-Free Fe–N–C Catalysts for Pem Fuel Cells Derived from Nitrogen and Sulfur Doped Synthetic Polymers. *Fuel*, 2022, 328, 125323.
40. Sun, H.; Gao, L.; Kumar, A.; Cao, Z.; Chang, Z.; Liu, W.; Sun, X.; Superaerophobic Cop Nanowire Arrays as a Highly Effective Anode Electrocatalyst for Direct Hydrazine Fuel Cells. *ACS Appl. Energy Mater.*, 2022, 5, 9455-9462.
41. Liao, J.; Wang, Y.; Chen, M.; Wang, M.; Fan, J.; Li, H.; Wang, H.; Zeng, L.; Zhao, T.; Irox Supported onto Niobium-Doped Titanium Dioxide as an Anode Reversal Tolerant Electrocatalyst for Proton Exchange Membrane Fuel Cells. *ACS Appl. Energy Mater.*, 2022, 5, 3259-3268.
42. Zhang, L.; Li, L.; Gao, Z.; Guo, L.; Li, M.; Su, J.; Porous Hierarchical Iron/Nitrogen Co-Doped Carbon Etched by G-C3n4 Pyrolysis as Efficient Non-Noble Metal Catalysts for Pem Fuel Cells. *ChemElectroChem*, 2022, 9, e202101681.
43. Deng, J.; Li, X.; Imhanria, S.; Chen, K.; Deng, X.; Wang, W.; Molybdenum Carbide-Nitrogen Doped Carbon Composites as Effective Non-Precious Electrocatalyst for Direct Hydrazine Fuel Cell. *Electrochim. Acta*, 2021, 384, 138417.
44. Dursun, S.; Akay, R. G.; Yazici, M. S.; Cvd Graphene Supported Cobalt (Ii) Phthalocyanine as Cathode Electrocatalyst for Pem Fuel Cells. *Int. J. Hydrog.Energy*, 2020, 45, 34837-34844.

45. Zhu, K., *et al.*; Heteroatom-Doped Porous Carbon Nanoparticle-Decorated Carbon Cloth (Hpcn/Cc) as Efficient Anode Electrode for Microbial Fuel Cells (Mfcs). *J. Cleaner Prod.*, 2022, 336, 130374.
46. Im, K.; Choi, K. H.; Park, B. J.; Yoo, S. J.; Kim, J.; Tofu-Derived Heteroatom-Doped Carbon for Oxygen Reduction Reaction in an Anion Exchange Membrane–Fuel Cell. *Energy Convers. Manage.*, 2022, 265, 115754.
47. Li, Q.; Song, S.; Mo, Z.; Zhang, L.; Qian, Y.; Ge, C.; Hollow Carbon Nanospheres@Graphitic C3n5 Heterostructures for Enhanced Oxygen Electroreduction. *Appl. Surf. Sci.*, 2022, 579, 152006.
48. Vagin, M.; Gueskine, V.; Mitraka, E.; Wang, S.; Singh, A.; Zozoulenko, I.; Berggren, M.; Fabiano, S.; Crispin, X.; Negatively-Doped Conducting Polymers for Oxygen Reduction Reaction, *Adv. Energy Mater.*, 2021, *11*, 2002664.
49. Sayed, E. T.; Alawadhi, H.; Olabi, A. G.; Jamal, A.; Almahdi, M. S.; Khalid, J.; Abdelkareem, M. A.; Electrophoretic Deposition of Graphene Oxide on Carbon Brush as Bioanode for Microbial Fuel Cell Operated with Real Wastewater. *Int. J. Hydrog. Energy*, 2021, 46, 5975-5983.
50. Liu, H.; Zhang, Z.; Xu, Y.; Tan, X.; Yue, Z.; Ma, K.; Wang, Y.; Reduced Graphene Oxide@ Polydopamine Decorated Carbon Cloth as an Anode for a High-Performance Microbial Fuel Cell in Congo Red/Saline Wastewater Removal. *Bioelectrochemistry*, 2021, 137, 107675.
51. Royuela, S.; Martínez-Periñán, E.; Arrieta, M. P.; Martínez, J. I.; Ramos, M. M.; Zamora, F.; Lorenzo, E.; Segura, J. L.; Oxygen Reduction Using a Metal-Free Naphthalene Diimide-Based Covalent Organic Framework Electrocatalyst. *Chem. Commun.*, 2020, 56, 1267-1270.
52. Han, T. H.; Mohapatra, D.; Mahato, N.; Parida, S.; Shim, J. H.; Nguyen, A. T. N.; Nguyen, V. Q.; Cho, M. H.; Shim, J.-J.; Effect of Nitrogen Doping on the Catalytic Activity of Carbon Nano-Onions for the Oxygen Reduction Reaction in Microbial Fuel Cells. *J. Ind. Eng. Chem.*, 2020, 81, 269-277.
53. Wang, T.; Wang, R.; Xie, X.; Chang, S.; Wei, T.; Dong, D.; Wang, Z.; Robust Direct Hydrocarbon Solid Oxide Fuel Cells with Exsolved Anode Nanocatalysts. *ACS Appl. Mater. Interfaces*, 2022, 14, 56735-56742.
54. Qi, H.; Tu, B.; Cheng, M.; Zhang, T.; Investigation of in Situ Co-Assembled Sr(Co,Zr)O3–Δ-Based Perovskite Nanocomposite Cathode for Intermediate-Temperature Solid Oxide Fuel Cells. *ACS Appl. Energy Mater.*, 2022, 5, 14881-14890.
55. Solovyev, A. A.; Shipilova, A. V.; Rabotkin, S. V.; Smolyanskiy, E. A.; Shmakov, A. N.; Magnetron Sputtered Lsc-Gdc Composite Cathode Interlayer for Intermediate-Temperature Solid Oxide Fuel Cells. *Int. J. Hydrogen Energy*, 2022, 47, 37967-37977.
56. Wang, G.; Wu, J.; Li, S.; Cai, H.; Yu, Z.; Wang, X.; Xia, C.; Wang, B.; Wang, H.; Dong, W.; Effect of the Online Reaction Byproducts of Lini0.8co0.15al0.05o2-Δ Electrodes on the Performance of Solid Oxide Fuel Cells. *Int. J. Hydrog. Energy*, 2022, 47, 33850-33860.
57. Das, A.; Kumar, S.; Jana, B.; Suresh, M. B.; Prashanthi, C.; Omar, S.; Electrochemical Performance of Srmg0.1mo0.9o3-Based Composites for Solid Oxide Fuel Cell Anodes. *ACS Appl. Energy Mater.*, 2022, 5, 1607-1617.
58. Hu, H.; Li, M.; Min, H.; Zhou, X.; Li, J.; Wang, X.; Lu, Y.; Ding, X.; Enhancing the Catalytic Activity and Coking Tolerance of the Perovskite Anode for Solid Oxide Fuel Cells through in Situ Exsolution of Co-Fe Nanoparticles. *ACS Catal.*, 2022, 12, 828-836.
59. Zhang, B.; Wan, Y.; Hua, Z.; Tang, K.; Xia, C.; Tungsten-Doped Prbafe2o5+Δ Double Perovskite as a High-Performance Electrode Material for Symmetrical Solid Oxide Fuel Cells. *ACS Appl. Energy Mater.*, 2021, 4, 8401-8409.

60. Managutti, P. B., *et al.*; Exsolution of Ni Nanoparticles from a-Site-Deficient Layered Double Perovskites for Dry Reforming of Methane and as an Anode Material for a Solid Oxide Fuel Cell. *ACS Appl. Mater. Interfaces,* 2021, 13, 35719-35728.
61. Liu, Q.; Xiao, Z.; Xie, H.; Gao, J.; Yuan, M.; Dong, W.; Multifunctional Layer-Perovskite Oxide La2-Xcexcuo4 for Solid Oxide Fuel Cell Applications. *Int. J. Hydrog. Energy,* 2021, 46, 9818-9825.
62. Santaya, M.; Troiani, H. E.; Caneiro, A.; Mogni, L. V.; Ternary Ni–Co–Fe Exsolved Nanoparticles/ Perovskite System for Energy Applications: Nanostructure Characterization and Electrochemical Activity. *ACS Appl. Energy Mater.,* 2020, 3, 9528-9533.
63. Hanif, M. B.; Gao, J.-T.; Shaheen, K.; Wang, Y.-P.; Yasir, M.; Zhang, S.-L.; Li, C.-J.; Li, C.-X.; Performance Evaluation of Highly Active and Novel La0.7sr0.3ti0.1fe0.6ni0.3o3-Δ Material Both as Cathode and Anode for Intermediate-Temperature Symmetrical Solid Oxide Fuel Cell. *J. Power Sources,* 2020, 472, 228498.
64. Yadav, M.; Sehrawat, N.; Singh, M.; Kumar, V.; Sharma, A. K.; Kumar, S.; Chapter 11 - Thermophilic Microbes-Based Fuel Cells: An Eco-Friendly Approach for Sustainable Energy Production. In *Bioremediation for Environmental Sustainability,* Kumar, V.; Saxena, G.; Shah, M. P., Eds. Elsevier: 2021; *pp.* 235-246.
65. Xiao, X.; Xia, H.-q.; Wu, R.; Bai, L.; Yan, L.; Magner, E.; Cosnier, S.; Lojou, E.; Zhu, Z.; Liu, A.; Tackling the Challenges of Enzymatic (Bio)Fuel Cells. *Chem. Rev.,* 2019, 119, 9509-9558.
66. Zhou, M.; Chi, M.; Luo, J.; He, H.; Jin, T.; An Overview of Electrode Materials in Microbial Fuel Cells. *J. Power Sources,* 2011, 196, 4427-4435.
67. Geetanjali; Dhillon, S. K.; Kundu, P. P.; Development of Polypyrrole Nanotube Coated with Chitosan and Nickel Oxide as a Biocompatible Anode to Enhance the Power Generation in Microbial Fuel Cell. *J. Power Sources,* 2022, *539,* 231595.
68. González Vázquez, O. F.; Frausto Reyes, C.; Ortiz Morales, M.; Kamaraj, S.-K.; Moreno Virgen, M. d. R.; Hernández Montoya, V.; Facile Scalable Manufacture of Improved Electrodes Using Structured Surface Coatings of Nickel Oxide as Cathode and Reduced Graphene Oxide as Anode for Evaluation in a Prototype Development on Microbial Fuel Cells. *Int. J. Hydrogen Energy,* 2022, 47, 30248-30261.
69. Tang, X.; Liu, L.; Cui, Y.; B-Feooh Modified Carbon Monolith Anode Derived from Wax Gourd for Microbial Fuel Cells. *Int. J. Hydrog. Energy,* 2022, 47, 4838-4845.
70. Liu, Y.; Zhang, X.; Li, H.; Peng, L.; Qin, Y.; Lin, X.; Zheng, L.; Li, C.; Porous A-Fe2o3 Nanofiber Combined with Carbon Nanotube as Anode to Enhance the Bioelectricity Generation for Microbial Fuel Cell. *Electrochim. Acta,* 2021, 391, 138984.
71. Xing, C.; Jiang, D.; Tong, L.; Ma, K.; Xu, Y.; Xie, K.; Wang, Y.; Mxene@Poly(Dially ldimethylammonium Chloride) Decorated Carbon Cloth for Highly Electrochemically Active Biofilms in Microbial Fuel Cells. *ChemElectroChem,* 2021, 8, 2583-2589.
72. Yang, W.; Wang, X.; Son, M.; Logan, B. E.; Simultaneously Enhancing Power Density and Coulombic Efficiency with a Hydrophobic Fe–N4/Activated Carbon Air Cathode for Microbial Fuel Cells. *J. Power Sources,* 2020, 465, 228264.
73. Saikia, K.; Kakati, B. K.; Boro, B.; Verma, A.; Current Advances and Applications of Fuel Cell Technologies. In *Recent Advancements in Biofuels and Bioenergy Utilization,* Sarangi, P. K.; Nanda, S.; Mohanty, P.; Eds. Springer Singapore: Singapore, 2018; *pp.* 303-337.
74. Wilberforce, T.; Alaswad, A.; Palumbo, A.; Dassisti, M.; Olabi, A. G.; Advances in Stationary and Portable Fuel Cell Applications. *Int. J. Hydrog. Energy,* 2016, 41, 16509-16522.
75. U.S. Department of Energy Efficiency & Renewable Energy; Oct 8, 2019.
76. Fan, L.; Tu, Z.; Chan, S. H.; Recent Development of Hydrogen and Fuel Cell Technologies: A Review. *Energy Rep.,* 2021, 7, 8421-8446.

77. Weidner, E., Ortiz Cebolla, R. and Davies, J., Global deployment of large capacity stationary fuel cells – Drivers of, and barriers to, stationary fuel cell deployment, EUR 29693EN-2109 (Doi:10.2760/372263, JRC115923).
78. Fuel Cell Market Report (Markets and markets; May 2022, Code-EP 2762

14

Smart Materials in Energy Storage Devices: Fuel Cells and Biofuel Cells

Baliram Gurunath Rathod and Venkata Giridhar Poosarla*

Department of Microbiology and FST (Food Science & Technology), GITAM School of Science, GITAM (Deemed to be University), Visakhapatnam, Andhra Pradesh, India

Abstract

With a regulated external stimulus, electric power/biohydrogen/other biofuels can be synthesized by several mechanisms involving innovative smart materials such as microbial fuel cells (MFCs) or biofuel cells. The MFC process drives the oxidation of organic compounds with microbial-mediated catalysis at the anode (donor). This mechanism generates bioelectric power and biohydrogen due to electron transfer to an external circuit system's electron acceptor (cathode). A simple digital multimeter can be used to measure the electric power, and MFCs' operations initially report low power density but generally reaches a maximum at ~30 days of operation. This book chapter mainly focuses on the exploitation of various MFC cathode and anode materials, proton exchange systems (membranes), electrode catalysts, involvement of consortia of microorganisms, different waste utilization, a variety of electron shuttles, sludge materials etc., to derive electric power, hydrogen, and biofuels.

***Keywords*:** Microbial fuel cells (MFCs), cathode, anode, proton exchange membrane, electric power, biohydrogen, bioethanol

14.1 Introduction

The surge in the requirement for sustainable, clean, and efficient energy sources has motivated research into novel energy storage devices. Biofuel and microbial fuel cells (MFCs) have gained widespread recognition because of their environmental benefits and greater energy density [1]. MFCs and BFCs utilize organic materials to convert chemical energy into electrical energy using microbes. The organisms consume organic matter in the anode section of MFCs, producing electrons, which flow to the cathode region via an external circuit. The electrons are then oxidized by oxidizing agents (like O_2) in the cathode compartment, generating electrical energy [2]. In BFCs, enzymes act as biocatalysts to oxidize organic matter, generating electrons, and electrical energy [3].

The use of smart materials in MFCs and BFCs has the potential to revolutionize energy storage and provides several advantages over traditional materials. According to a report,

**Corresponding author*: gpoosarl@gitam.edu

Upendra Kumar and Piyush Kumar Sonkar (eds.) Smart Materials for Science and Engineering, (287–304) © 2024 Scrivener Publishing LLC

the smart materials marketplace is anticipated to grow from USD 33 billion in 2016 to USD 111 billion by 2027, at a CAGR of 13% from 2016 to 2027 [4]. The unique properties of smart materials make them ideal for energy storage devices, as they can be programmed to respond to specific conditions, enabling the devices to perform optimally under different operating conditions [5].

Smart materials, like conductive polymers, carbon nanotubes, and graphene, are being actively investigated as potential components for MFCs and BFCs to enhance their performance and extend their lifespan [6–8]. However, despite these advantages, applying smart materials in MFCs and BFCs is challenging. One major challenge is the need for chemically and biologically stable materials, as the harsh operating conditions in MFCs and BFCs can degrade the materials over time. Another challenge is the need for materials that can effectively transport electrons, as the efficiency of BFCs and MFCs is highly reliant on electron transfer efficiency [9]. Additionally, the cost of smart materials is often a barrier to their widespread use, as some materials are still expensive to produce in large quantities [10]. However, further study is necessary to face the challenges of using smart materials, such as their stability, efficiency, and cost. This chapter will provide an in-depth overview of smart materials in MFCs and BFCs, including the latest advances in materials development and performance improvement.

14.2 Relation of Smart Materials and MFCs

Smart materials, also known as intelligent or functional materials, can alter their properties in reaction to external stimuli like light, temperature, magnetic field, or chemical changes. These components highly fascinate researchers in the field of MFCs as they can enhance the stability and efficiency of these devices [11]. One of the key areas where smart materials have been used in MFCs is the design and fabrication of electrodes. Electrodes are essential in MFCs, as the interface between microorganisms and electrical circuits. The selection and arrangement of electrodes greatly influence the implementation and efficiency of MFCs [12].

Conductive polymers, such as polyaniline (PANI) and polypyrrole (PPy), have been extensively used as smart materials in MFC electrodes because of their higher conductivity and the capacity to alter their electrical characteristics in response to environmental changes. For example, PPy has been used as a cathode material in MFCs, raising the oxygen reduction ratio and enhancing the devices' efficiency. Similarly, in MFCs, PANI is utilized as an anode material, enhancing bacterial growth and improving substrate oxidation rate [13].

In MFCs, smart materials, like graphene and carbon nanotubes, were used as sensing tools, as they are highly conductive and sensitive to environmental changes. For example, graphene has been used as a sensing material in MFCs for detecting changes in the anode compartment's pH and dissolved oxygen levels, which can offer valuable data on the condition of the microbial community and the conditions for operating the devices [7, 8]. In contrast, for energy storage devices in MFCs, smart materials like supercapacitors have been in use. They can store the electrical energy generated by the devices and provide a stable power source, even when the conditions of the MFCs change. For example, supercapacitors have been used as energy storage devices in MFCs to provide a steady power source during periods of low microbial activity, which can occur when the substrate concentration in the MFCs decreases [14].

Another important application of smart materials in MFCs is developing sensing and monitoring systems. These systems are essential for monitoring the performance of MFCs and detecting changes in the operating conditions that may affect the efficiency and stability of the devices. Additionally, smart materials in MFCs are developing energy storage systems. Energy storage systems are essential for stabilizing the output of MFCs and providing a continuous power source, even when the operating conditions of the devices change [15]. Smart materials thus provide significant advantages over traditional methods by enabling enhanced functionality and increased efficiency across a wide range of applications related to MFCs development. As research continues, further innovation will likely lead us towards novel ways of utilizing these versatile compounds to maximize our energy harvesting capabilities while minimizing the environmental impact caused by current generation technologies.

14.3 MFCs and Their Mechanism

MFCs are bio-electrochemical devices where microbial activity produces electrical energy from chemical energy using organic materials. MFC is an environmentally beneficial technology that generates bioelectricity while also managing waste because it uses organic materials like wastewater [16]. Figure 14.1 depicts a simple MFC schematic diagram [17]. In MFC, a proton exchange membrane (PEM) physically segregates the cathode and anode chambers, which are electrodes. The cathode is abiotic, and the anode contains electrochemically active microbes. As a biocatalyst, microorganisms stimulate the breakdown of organic molecules to create electrons that flow to the cathode side of the electric circuit. Exo- stands for exocellular, and electrogens refer to their capacity to transmit electrons directly to substances or materials that are not immediate electron acceptors [16]. Electrons

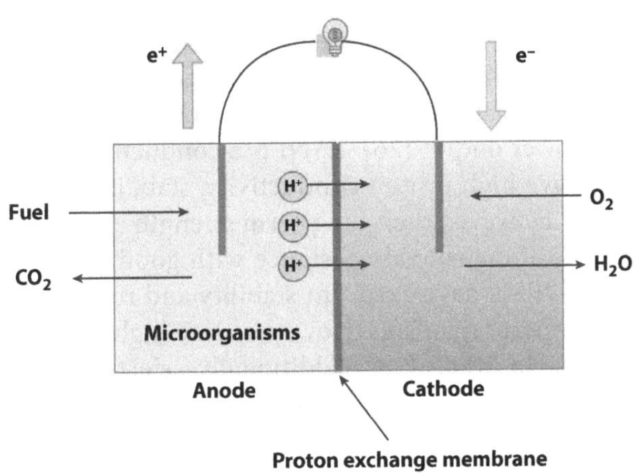

Figure 14.1 The schematic diagram of a simple MFC.

migrate through an external circuit to reach the cathode. At the same time, H^+ ions migrate to the cathode and react with O_2 to generate H_2O in the internal circuit. Therefore, the MFCs are shown to be a possible contender for green electricity [18–20].

During microbial respiration, a substrate transforms into CO_2 through an anodic reaction (anode chamber), which releases electrons and protons. The electrons are directed toward the anode from the cell membrane through direct contact with an electron mediator, which acts as an electrochemically active carrier. The electrons then continue their journey to the cathode via an external circuit [21], and each electron transmitted results in a matching proton, which maintains charge neutrality by mitigating over a membrane (proton-permeable) to the cathodic chamber [22].

Electrons from the anode compartment are transferred to the cathode region through a conducting wire and a water-filled cathodic chamber (oxidation). Additionally, water molecules are formed through a reduction process when protons and oxygen react with electrons. These protons travel to the anode from the cathode through a membrane but not electrons. As a result, the cathodic reaction requires oxygen to be successful. Power is produced by moving electrons to the cathode from the anode, aided through an external circuit. Since microorganisms serve as catalysts in the cathodic reaction, certain metals like palladium or platinum are frequently utilized as catalysts. Figure 14.1 shows how the cathode and anode compartments of the PEM are separated from one another. The PEM functions as a wall that prevents oxygen flow from the cathode to the anode while residuals are permeable to proton flow from the anode to the cathode. Additionally, it permits proton flow while blocking the passage of electrons, substrate, and oxygen from the cathodic chamber [23].

Selecting a PEM is critical for achieving high power output and ensuring the MFC's long-term stability. Several types of PEMs have been used in MFCs, i.e., Nafion, polyethylene oxide (PEO), chitosan, PANI, and graphene oxide (GO). Nafion is a perfluorinated polymer widely used in MFCs as PEM. It has higher proton conductivity and good thermal and chemical stability, and prevents O_2 collisions between the cathode and anode. However, its higher cost and restricted availability have led to the exploration of alternative materials [24]. PEO is a water-soluble polymer employed in MFCs as a PEM because of its higher proton conductivity, ease of modification, and low cost. However, PEO PEMs can be prone to swelling and have limited stability under acidic conditions [25]. Chitosan is a natural polymer derived from the shells of crustaceans. Chitosan PEMs have good mechanical strength, biocompatibility, and low cost. However, their relatively low proton conductivity can limit power output [26]. PANI is a conducting polymer used as a PEM in MFCs. PANI PEMs have high proton conductivity, stability under acidic conditions, and resistance to fouling. However, their long-term strength and durability are still being studied [27]. GO is a two-dimensional substance with good proton conductivity and a higher surface area. GO PEMs have excellent stability and resistance to fouling and are compatible with various microorganisms. However, their high cost and complex preparation have limited their use in MFCs [28]. Additionally, selecting a PEM for an MFC is a critical decision that depends on various factors, including proton conductivity, stability, durability, biocompatibility, and cost. The continued development of new PEM materials will be necessary for improving the performance and scalability of MFCs for various applications.

14.4 Classification of MFCs

MFCs are divided mainly depending on the electron transfer mechanism to the electrode, i.e., mediator MFCs and non-mediator MFCs [16]. The goal of mediator MFCs is to use bacterially mediated electron transfer to migrate electrons to the anode [23, 29]. This kind of MFCs utilizes chemical mediators, also called electron shuttles, to assist in the electron flow to the anode, including anthraquinone-2,6-disulfonate, humic acid, neutral red, and others. The typical mediators were claimed to include electroactive metabolites like methyl viologen, potassium ferricyanide, humic acids, and neutral red [30]. However, these mediators are so expensive and harmful. Therefore, new strategies are needed to increase power production and lower capital cost. Anaerobic digestion is crucial for mediator MFC. Oxygen takes away electrons when present, interfering with the mediator's activity because oxygen is less electronegative. The mediator enters the cell during the electron transfer process, receiving and releasing electrons before giving them to the electron acceptor (anode). The mediator deposits its electrons and returns to its original condition by being oxidized [23].

Electricity can be produced without the use of external mediators by utilizing microorganisms. This type of MFC operates without the addition of any mediators in the system. Wastewater bacteria can transfer electrons to electrodes through nanowires, producing electricity. Compared to MFCs that use mediators, mediator-less MFCs are low-cost and non-toxic [29, 30]. According to [31, 32], most MFCs without mediators are run using the dissimilatory metal-reducing bacterium *Clostridium butyricum*. The efficacy of mediator-less MFCs relies on several factors, including the redox enzymes' presence, the efficiency of transmission of electrons to the anode, the oxidation of the anode's fuel, the circuit's external resistance, the decrease of O_2 at the cathode, and proton transmission to the cathode across the membrane [23]. These factors can impact the generation of electricity, limiting its output.

Plant Microbial Fuel Cell (PMFC) is something, in addition to utilizing microorganisms, where MFCs without any mediator can directly extract energy from specific plants, like algae, lupines, tomatoes, cord grass, rice, and sweet reed grass. There are several types of mediator-less MFCs, including Microbial Electrolysis Cells (MECs), Soil-Based Microbial Fuel Cells (SB-MFCs), and Phototrophic Biofilm Microbial Fuel Cells (PB-MFCs). These designs offer a more sustainable and environmentally friendly approach to generating electricity [16].

14.5 Microorganisms Involved in MFCs

The belief that only a few microorganisms could produce electricity has recently been dispelled. Research has shown that most microorganisms can be utilized in MFCs. The idea of MFCs was first demonstrated in 1910 when *Saccharomyces* sp. and *E. coli* were employed to produce electricity with the help of platinum electrodes. The concept was improved in the early 1980s by using electron mediators to enhance electricity generation significantly. Many microorganisms can transmit electrons from organic and inorganic matter metabolism to an anode. Wastewater, soil, activated sludge, marine sediment, and freshwater sediment are all abundant sources of these microorganisms. Numerous studies have focused on screening and identifying microbes and constructing MFCs using microorganisms capable of generating electricity from organic matter degradation [33].

In MFCs, microorganisms break down organic substances and move electrons to the electrode. The oxidized organic matter releases protons and electrons, leading to electricity similar to that produced in a CFC (chemical fuel cell), with the microorganism serving as catalysts at the anode. Unlike traditional catalysts, which are not altered or do not receive energy from the reaction, the microbes in MFC gain energy via oxidation for their growth, resulting in energy loss. These organisms can acquire all the energy and carbon they need to grow from the oxidation of complex organic material, making the MFC's system self-sufficient [34].

Organisms that can produce electricity in MFCs include members of the Acidobacteria and Firmicutes, four classes of Proteobacteria, and yeast strains such as *Hansenula anomala* and *Saccharomyces cerevisiae*. These microbes interrelate with the anode in various direct and indirect ways, leading to varying levels of the current generation. The most well-researched organism qualified to produce maximum power densities in MFCs was *Geobacter sulfurreducens*, declared as the model for bacterial processes in MFCs. This organism is an electricigen, meaning it conserves energy for growth by fully oxidizing organic substances to CO_2 and moving electrons directly to the anode [33, 35].

Other terms used to describe microorganisms having the potential to migrate electrons to an electrode include electrogenic microorganisms (like anodophiles, exoelectrogens, anode-respiring bacteria, and electrochemically active bacteria). Organisms capable of oxidizing the organic substrate to CO_2 in an MFC system have been reported to have higher coulombic efficiencies. Bacteria capable of oxidizing an organic compound in an MFCs technology include *Rhodoferax ferrireducens, Geothrix fermentans*, and *G.* sp. [35–37]. Table 14.1 shows the microorganisms involved in MFCs [38–48].

Table 14.1 Microorganisms involved in MFCs.

Microorganism	Type of organism	Substrate	Voltage (mV)	Ref.
Bacillus cereus	Facultative anaerobic	Model sewage	463	[47]
Bacillus subtilis	Aerobic	Glucose, sucrose, starch, and acetate	166–298	[46]
Bacillus violaceus	Facultative anaerobic	Asparagine	320	[45]
Clostridium beijerinckii	Obligate anaerobic	Glucose, starch	759	[44]
Clostridium butyricum	Obligate anaerobic	Glucose, starch	473	[44]
Desulfuromonas acetoxidans	Facultative anaerobic	Sodium acetate	-	[37]
Enterobacter spp.	Anaerobic	Wastewater	400	[43]

(*Continued*)

Table 14.1 Microorganisms involved in MFCs. (*Continued*)

Microorganism	Type of organism	Substrate	Voltage (mV)	Ref.
Escherichia coli	facultative anaerobic	Glucose, sucrose, starch, and acetate	187–307	[46]
Geobacter metallireducens	Anaerobic	Aromatic compounds (benzoate, toluene)	200	[37]
Geobacter sulfurreducens	Anaerobic	Methane	273	[48]
Gluconobacter oxydans	Aerobe	Ethanol	360	[38]
Klebsiella pneumoniae	Facultative anaerobic	Lactose	453	[42]
Klebsiella oxytoca	Facultative anaerobic	Palm oil mill effluent (POME)	207	[41]
Methanosarcina acetivorans	Anaerobic	Methane	273	[48]
Proteus vulgaris	Facultative anaerobic	Sucrose	350	[40]
Paracoccus denitrificans	Facultative anaerobic	Methane	-	[48]
Pseudomonas mendocina	Aerobic	Glucose	627	[39]
Saccharomyces cerevisiae	Facultative anaerobic	Glucose, sucrose, and starch	125–183	[46]
Shewanella oneidensis	Facultative anaerobic	Synthetic medium	500	[38]
Stenotrophomonas acidaminiphila	Aerobic	Glucose	445	[39]

14.6 MFC Systems

MFC systems are presented as a tool for utilizing waste material in wastewater and sewage systems. Initially, MFC units were employed as equipment or reactors to treat contaminated water and soil sources. Further, the combination of MFC units can be installed in series or parallel to enhance their performance. Through series or parallel configurations, MFCs have been found to generate electricity through microbial-mediated electricity production, resulting in increased voltage or current. For instance, six independent continuous MFC units can be stacked in a hexacyanoferrate cathode system, with hourly intervals, to generate maximum power [49]. The higher current flow output and voltage created by coupled MFC units provide a powerful external circuit system.

Furthermore, individual MFCs have shown a tripling of the short-term power output values because of a decrease in the transfer of mass limitations and internal resistance. A lack of correlation between MFCs microbial compositions and electrochemical performances was also noticed. This MFC strategy is promising for energy generation from waste organic or biodegradable sources [50].

Power Management System's (PMS) application can convert step-up voltage processes by substituting the unstable stacking arrangement of the MFC technology, which might be helpful for the practical applications of technology of MFCs. This strategy has enabled maximum power extraction from individual units of MFCs. According to the PMS operation study, it also increases the power output performance with minimal power consumption [51]. The PMS reported that the highest power extracted was not delivered to the application load at the point of maximum power extraction. Compared to a typical maximum power point tracking (MPPT) system, an individual MFC produced higher electrical energy for a given applied load. By providing substantial output power over a prolonged period of operation, the optimum power delivery system can establish excellent favorable power conditions for various micro-energy harvesting systems compared to an MPPT system [51, 52]. Mild operating conditions and different biodegradable carbon substrates are used in MFCs. Single-chamber MFC systems have anode and cathode compartments. They are helpful as power generators in small devices and are actively involved in degrading biodegradable waste substrates using bio-electric power generation (like biosensors). Most MFC designs still struggle with low power or current density as a practical barrier to energy output. However, modifications to various MFC components or parts (like cathode, anode, or materials required for membrane design) are possible using cutting-edge materials or advancements that can help control or solve practical challenges in this field [53].

14.7 Design of MFCs

Many factors affect the MFC's production and the design of the reactor systems. The functionality of the electrodes that can be useful for developing MFCs/reactor systems has been discussed. Additionally, various materials or configurations indicated the increased performance of created MFCs with lower material costs during testing. Researchers also investigated the influence of surface changes on anodic electrodes. Facilitating the passage of electrons to electrodes from the microbial cell's surface can enhance bacterial adherence. For both electrodes in the MFC system, recent developments in material designs, selection, or configuration are necessary. The only benefits or downsides mentioned when choosing electrode materials or arrangements are conductivity, surface characteristics, biocompatibility, or cost. The MFC's performance with commercial use can be enhanced by technological change in the anodic electrodes [54].

Scaling-up of MFC systems is possible by knowing the critical roles of several aspects in their parts, reactor geometry, and surface area of the electrode concerning solution condition (substrate concentration or conductivity). The quantity of the substrate can affect the anode by not changing the role of the cathode. Even so, the conductivity of the solution can significantly impact the cathode without impairing the anode electrode's performance. In household wastewater treatment procedures, the cathode electrode's doubled surface area can contribute to a 62% increase in power output quantity or value.

However, the doubled anode electrode's size only boosted the power output by 12%. Cathode-specific area, i.e., the ratio of the cathode's surface area to the reactor's volume, has a linear relationship with volumetric power density. The influence of cathode size on electricity generation depends on the circuit system's substrate conductivity and strength, which was reduced by the chemical oxygen demand value. A cathode-specific surface area was the most crucial component in process optimization in MFC systems to produce maximum power densities [55–57].

When paired with the production of renewable electricity, microbial electrosynthesis (MES) processes can employ electrons to convert carbon dioxide (CO_2) into organic compounds with long chains (generated mainly from wind and solar sources). Interruptions and fluctuations of varying lengths can negatively impact the MES performance, and it can be tested for the first time under various interruption scenarios. The acetate production rate was shown to be affected by this interruption research. After studying all the impacts of disconnections, the synthesis of acetate and acetic acids might be restored by giving the reduction current (between 7 and 16 h to reach a stable state). In times of interruptions, the community of microbial cells on the MES system can offer resilience and aid in recovering the electro-autotrophic activity with the power supply [58, 59].

MFC systems have reportedly encountered numerous difficulties during wastewater treatment procedures. It has been discovered that they primarily lost performance with time because of the cathode electrode fouling. By adopting an *in situ* method to clean air cathodes by scraping off the biofilm growth or deposition with magnets, this problem can be reduced or resolved. In this method, the magnet's power output was cleaned after 1 month of operation at the electrodes' value of 1.32×10^2 $mW.m^{-2}$. Compared to the MES control system, its power output is larger (42%) without magnets.

Additional advanced research is being conducted on no-separator (NS) systems in conjunction with MES systems. It demonstrated an improved (10%) power output compared to the separator system control (SP). Using magnets during operation with a cleaned cathode displayed reduced biofilm formation. These MES design modifications have enhanced the working efficiency with time by slowing the rate at which biofouling accumulates in the air cathode system [60, 61].

Microorganisms (a variety of anaerobic microbial systems) catalyze the oxidation of organic molecules at the anode as part of the operational systems of MFCs, which also generate energy and electron transfer. Additionally, it is obtained via processing secondary sludge's organic materials in an external circuit system. By treating organic acids or carbohydrate-rich effluents, MFC design and operation can function more effectively and produce significant energy [62]. Without specialized equipment, controlling the MFC electrode potential can be accomplished with good performance using an external load (R_{ext}) system. At the ohmic range (10–1000 Ω), the effect of the R_{ext} system was studied on biofilm growth or activity. Their MFC's optimal R_{ext} value was reported to be 50. The maximum current value with the R_{ext} effect is produced by biofilm that has matured under 10 Ω. Extracellular polymer substances of the biofilm layer can be crucial for producing electricity and biofuels [63].

In MFCs, methane generation and the procedures that inhibit it (methanogenesis) showed high power efficiency. It was operated at the least (150 Ω) R_{ext} value and fed with acetate substrate. A critical component affecting the variety of the bacterial community is external resistance. The relationship between the applied R_{ext} load during operation (after the phase of maturing) and the design function of MFCs in terms of energy or fuel supply displayed the finest results when MFCs are operated with close load to internal resistance load control [64, 65].

14.8 Functions/Operations of MFCs

The function of MFCs has been studied by feeding three different species of thermally treated *Chlorella vulgaris* algae into two-compartment MFC systems. The system's performance was monitored over extended periods (more than one to three months) to determine the sustainability of fuel generation. The hydrolysis yield, algal biomass, and thermal treatment at 25, 55, or 95°C of the algae were considered, resulting in an enhancement in the efficiency of the MFC technology. Three MFC systems were tested using algal suspensions (hydrolyzed) from the different thermal treatments, proving to be a reliable fuel source for the electrochemical reaction. The different thermal treatments had varying impacts on the fuel supply for MFC system function over extended operating times and showed the best performance.

Additionally, the polarization curves show that these MFCs are fed with algal suspension, produced following hydrolysis at the lowest temperature treatment [66, 67]. It was evident that modified microbial cells have been used and that the polymer Ppy's conductive properties were used to alter the microbial cells. As a result of the electrons' increased conductivity, MFCs now operate better. Changes in microbial cell's biochemical or metabolic reaction processes (known to be environmentally friendly) have been linked to the reported synthesis of Ppy [68, 69]. The performance of *Aspergillus niger* or *Rhizoctonia* sp. has changed because of the Ppy polymer biosynthesis in the cell components, which improved the efficiency of the current transfer from the fungal cells to electrodes. Two electrochemical techniques are used to assess this electrical charge transfer efficiency, including scanning electrochemical microscopy, which measures the electrochemical activity of Ppy-modified or unmodified MFCs [68–72]. Mediator compounds 1,10-phenanthroline-5,5-dione or 9,10–phenanthrenequinone and potassium ferricyanide were used to determine the electrical charge transfer capacities of microbial cells. The amperometric signals of the Ppy-modified MFCs were six times higher than those from unaltered cells. The cutting-edge structure of the Ppy-modified cell walls had a higher electrical conductivity, which improved the permeability and charge transfer mechanism. These mediator compounds effectively transferred charges through the cell walls and membranes of the microbial cells [70, 73–76]. It could be inferred that the environmentally safe synthetic Ppy polymer approach can assist inefficient methods of enhancing the functionality of electrochemical operations in devices (MFCs or cell-based biosensors) [68, 69, 71–76]. Figure 14.2 shows wastewater treatment using polymer Ppy synthesizing MFCs [16].

With MFCs as a vital component, the membrane's electroconductive nature demonstrated a high potential in the future generation of wastewater treatment facilities. The engineered electro-conductive membrane can improve its functioning while requiring minor maintenance in flux, permissive selectivity, or optimal separation. Recent advancements in these procedures for creating electro-conductive membranes or their mechanisms of

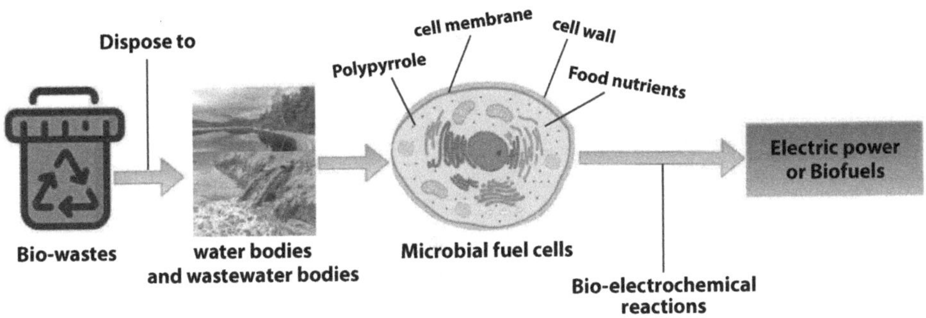

Figure 14.2 Treatment of wastewater using polymer Ppy synthesizing MFCs.

responses are susceptible to the influence of the outside electric voltage. It can assist in better membrane infusion or reduce fouling growth in altered MFCs. Other membrane polymers with electroconductive characteristics can be created by incorporating PANI, Ppy, or carbon nanotubes. Due to electrostatic repulsion, the creation of H_2O_2, or electrochemical oxidation processes, an increase in the electro-conductive nature in membranes can aid in enhancing penetration or fouling mitigation rate [70, 77, 78].

The hydrophilic characteristics of polyvinylidene fluoride (PVDF) are utilized in ultrafiltration membranes, and Ppy coated with multiwall carbon nanotubes (MWCNTs) (raw and oxidized) as an electroconductive polymer nanocomposite was prepared for membrane modification. XRD, FTIR, or SEM techniques have been used to characterize this prepared nanocomposite membrane. This manufactured membrane's hydrophilic qualities were assessed by measuring the water contact angles. Compared to a bare membrane, a mixed matrix membrane demonstrated decreased contact angle and higher hydrophilic properties. The enhanced antifouling feature of the created or formulated membranes increased resistance values for all hybrid membranes. The effectiveness of the membranes can be increased by raising the porosity of the surface and mean pore size [70, 79, 80].

14.9 Components of MFCs

MFCs are composed of different components (Figure 14.3). It has two chambers (cathodic and anodic chambers, consisting of anode or cathode electrodes) separated by a PEM. Utilized as a tool or piece of machinery, the MFC aids in converting chemical energy released through the oxidation of complex organic carbon sources into electrical energy. Additionally, chemical energy is produced when compatible microbial cell systems oxidize organic substrates to generate sustainable energy and fuel. Microbiological metabolic processes collected to maintain a consistent power density supply without producing carbon particles are reported to release electron currents. Numerous aspects are engaged in developing MFC technology by aiding in power generation [70].

Moreover, it has been discovered that these crucial factors can be used as evaluation criteria for the efficiency of device parameters, such as rate of removal of chemical oxygen

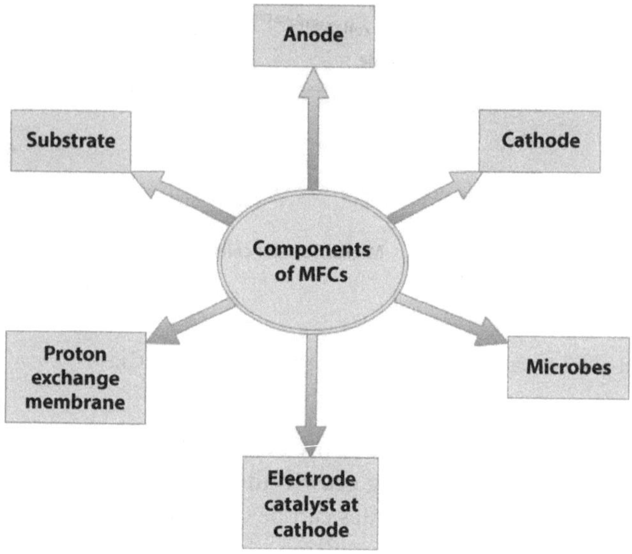

Figure 14.3 Components of MFCs.

demand, coulombic efficiencies, or maximum power density. MFC devices' bioremediation processes generate electricity simultaneously using the appropriate and efficient microbial cell system [53]. This efficient idea may be applied to many industrial, municipal, or agricultural waste management systems. It is said that the MFC system's efficiency in power generation is low/less valuable/used in the present. The MFC's power output can still be significantly increased with the help of technical or advanced modifications to the operation profiles, components, or design, which can increase the likelihood that MFCs will be used in a variety of applications like bioremediation processes, biosensors, or wastewater treatment [53, 79, 80].

14.10 Energy from MFCs

MFCs have been found to produce various types of energy. These cells can produce both electricity and biofuels, such as hydrogen or alcohol, which can contribute to fulfilling the growing demand for sustainable energy sources. MFCs are highly beneficial as they produce little to no harmful byproducts or gases, making them an environmentally friendly energy source. MFCs form electrical energy through the different types of microbial cells, including yeast (*S. cerevisiae*) and various bacteria by converting the stored chemical energy from organic matter, such as wastewater. Furthermore, glucose as a reactant in a hydrogen and air fuel cell, with *Clostridium butyricum* as the fermenting agent at the anode, has produced H^+. Overall, MFCs offer a promising solution to the increasing need for sustainable and clean energy, as they can convert organic matter into electricity and biofuels with minimal environmental impact [70, 81–83].

14.11 Recent Developments and Challenges in Smart Materials for Energy Storage Devices

Recent developments in smart materials for energy storage devices in MFCs have seen significant advancements in terms of efficiency of energy and by being economical. The most commonly used smart materials in MFCs are conductive polymers and nanomaterials, like nanotubes of carbon and graphene. These materials contain unique electrical and thermal conductivity properties that make them best for energy storage devices. Additionally, using graphene and carbon nanotubes can enhance the stability and mechanical strength of the electrodes, which can further increase the effectiveness of the MFCs [7, 8].

One of the key benefits of using smart materials in MFCs is their ability to improve the energy efficiency of fuel cells. Conductive polymers, for example, can be utilized as electrodes in MFCs for promoting the transfer of an electron between the electrode surface and the microbe. This leads to more efficient conversion into electrical energy from chemical energy. Another significant benefit of using smart materials in MFCs is their ability to decrease the cost of fuel cells. For example, using conductive polymers can reduce the MFC's cost by removing the need for expensive catalysts and improving the effectiveness of fuel cells. Similarly, using graphene and carbon nanotubes can decrease the cost of fuel cells by eliminating the need for costly catalysts and increasing the stability of the electrodes [7, 8, 13].

In addition to these benefits, recent developments in smart materials for energy storage devices in MFCs have also seen the use of advanced materials, e.g., metal-organic frameworks (MOFs) and ionic liquids. These materials have unique properties that make them ideal for MFCs, including their ability to store and release energy, high thermal stability, and low cost.

Additionally, when ionic liquids as electrolytes in MFCs are used, further improvement in the energy effectiveness of the fuel cells by reducing the resistance to electron transfer and increasing the stability of the electrodes was observed [84].

Recent developments in smart materials for energy storage devices in MFCs have led to significant energy efficiency and cost-effectiveness advancements. Conductive polymers, nanomaterials, advanced materials like MOFs and ionic liquids, and other smart materials, when used, can revolutionize the energy storage industry and provide a highly cost-efficient and sustainable solution to our energy needs. However, more research and development are needed to realize these smart materials' potential in MFCs fully. In particular, more work is required to optimize the properties of these materials to ensure that they perform optimally in MFCs and reduce the fuel cells' cost, and to understand the long-term stability and durability of these materials in MFCs and their impact on the environment [7, 8, 13, 84].

14.12 Future Perspectives

The MFC technology can potentially be a sustainable energy recycling and waste treatment system and a way to valorize biomass. Although laboratory-scale MFCs have achieved some success in electricity generation, COD removal, and value-added product production, there are still challenges in scaling and commercializing the technology. The low output power of MFCs is one of the significant limitations, which is affected by the electrogenicity of the organism. Screening high-performance electrogenic strains and genetic alteration can help improve performance. Developing electrodes with improved surface area, electrical conductivity, and microbial affinity can also enhance electricity generation. High electrode material costs also hinder commercialization, but using waste and renewable natural feedstocks to produce high-performance electrodes could be a solution. Exploiting 3D technology to improve electrode structure is also promising [85].

The substrate utilization by microbes is crucial for MFC operation. Substrates need to be selected for their utilization efficiency by specific strains. Synthetic wastewater has been studied for its capability to treat wastewater and generate power with MFCs; however, more research is needed on different strains' adaptability to complex practical wastewater environments. Pretreatment methods to improve the availability of lignocellulosic biomass substrates are also required, as well as determining the performance of strains tolerant to cytotoxic substances. The substrate and strain selection should align with the specific function and application of MFCs, whether it is electricity generation or producing value-added products. Metabolic regulation could maximize the function of strains for particular substrates and applications. The stability of internal and external conditions is crucial for efficient MFC operation. The temperature is a significant external factor, and technical improvement is needed for stable performance in a broad temperature range. Experimental design and mathematical modeling can optimize intrinsic factors like pH and cell biomass concentration. The maintenance of MFC devices is essential for long-term energy production efficiency. However, there are still challenges with maintaining anodic electroactive microbes and removing non-electroactive biofilms on the cathode. Convenient maintenance methods are needed to overcome these challenges [85].

14.13 Conclusion

Smart materials have been identified as critical in developing energy storage devices, particularly in FCs and MFCs. The unique properties of conductive polymers, nanomaterials, metal-organic frameworks, and ionic liquids are poised to revolutionize the energy storage industry by providing a sustainable and cost-efficient solution to meet our energy needs. Using smart materials in FCs and MFCs will enhance the energy efficiency, stability, and cost-effectiveness. FCs incorporating smart materials have shown promise in delivering higher energy densities and longer operational lifetimes, reducing maintenance costs and increasing reliability. MFCs using smart materials can also operate at higher efficiency levels by improving electron transfer rates and reducing fuel crossover. Despite these advantages, there is still a need for additional research and development to fully realize the potential of these smart materials in energy storage devices. Optimizing their properties and understanding their environmental impact is crucial to ensure their long-term success and viability. In addition, integrating smart materials with existing energy systems and infrastructure remains a significant challenge that must be overcome to maximize their benefits.

References

1. Moradian, J.M., Z. Fang, and Y.-C. Yong, Recent advances on biomass-fueled microbial fuel cell. *Bioresour. Bioprocess.*, 2021. 8(1): p. 14.
2. Ucar, D., Y. Zhang, and I. Angelidaki, An Overview of Electron Acceptors in Microbial Fuel Cells. *Front. Microbiol.*, 2017. 8.
3. Jia, H. and P. Wang, Biocatalytic generation of power from biofuels: biofuel cells. *Biofuels*, 2010.1(3): p. 479-492.
4. IndustryARC, Smart Materials Market – Forecast (2022-2027). CMR 30928, 2022.
5. Qader, I., *et al.*, A Review of Smart Materials Researches and Applications. *El-Cezeri J. Sci. Eng.*, 2019. 6: p. 755-788.
6. Andriukonis, E., *et al.*, Yeast-assisted synthesis of polypyrrole: Quantification and influence on the mechanical properties of the cell wall. *Colloids Surf., B*, 2018. 164: p. 224-231.
7. Zhou, S., *et al.*, Biosynthetic graphene enhanced extracellular electron transfer for high performance anode in microbial fuel cell. *Chemosphere*, 2019. 232.
8. Mink, J., *et al.*, Vertically Grown Multiwalled Carbon Nanotube Anode and Nickel Silicide Integrated High Performance Microsized (1.25 ??L) Microbial Fuel Cell. *Nano. Lett.*, 2012. 12: p. 791-5.
9. Logan, B.E. and J.M. Regan, Microbial fuel cells--challenges and applications. *Environ. Sci. Technol.*, 2006. 40 17: p. 5172-80.
10. Cai, T., *et al.*, Application of advanced anodes in microbial fuel cells for power generation: Areview. *Chemosphere*, 2020. 248: p. 125985.
11. Kelly, T.R., *Book Review of Intelligent Materials*. 2008, ACS Publications.
12. Kalathil, S., S. Patil, and D. Pant, Microbial Fuel Cells: Electrode Materials. 2017.
13. Ghosh, S., S. Das, and M.E. Mosquera, Conducting polymer-based nanohybrids for fuel cell application. *Polymers*, 2020. 12(12): p. 2993.
14. Poli, F., *et al.*, Boosting microbial fuel cell performance by combining with an external supercapacitor: an electrochemical study. *ChemElectroChem*, 2020. 7(4): p. 893-903.
15. Santoro, C., *et al.*, Microbial Fuel Cells, Concept, and Applications, in *Handbook of Cell Biosensors*, G. Thouand, Editor. 2022, Springer International Publishing: Cham. p. 875-909.
16. Obileke, K., *et al.*, Microbial fuel cells, a renewable energy technology for bio-electricity generation: A mini-review. *Electrochem. Commun.*, 2021. 125: p. 107003.
17. He, Z., *et al.*, An upflow microbial fuel cell with an interior cathode: assessment of the internal resistance by impedance spectroscopy. *Environ. Sci. Technol.*, 2006. 40 17: p. 5212-7.
18. Zhang, Q., J. Hu, and D.-J. Lee, Microbial fuel cells as pollutant treatment units: Research updates. *Bioresour. Technol.*, 2016. 217: p. 121-128.
19. Pant, D., *et al.*, A review of the substrates used in microbial fuel cells (MFCs) for sustainable energy production. *Bioresour. Technol.*, 2010. 101(6): p. 1533-1543.
20. Xu, X., *et al.*, Biodegradation of organic matter and anodic microbial communities analysis in sediment microbial fuel cells with/without Fe(III) oxide addition. *Bioresour. Technol.*, 2017. 225: p. 402-408.
21. Li, Y. The Current Response of a Mediated Biological Fuel Cell with Acinetobacter calcoaceticus: The Role of Mediator Adsorption and Reduction Kinetics. 2013.
22. Kim, I.S., *et al.*, Microbial Fuel Cells: Recent Advances, Bacterial Communities and Application Beyond Electricity Generation. *Environ. Eng. Res.*, 2008. 13(2): p. 51-65.
23. Flimban, S.G.A., *et al.* Overview of Microbial Fuel Cell (MFC) Recent Advancement from Fundamentals to Applications: MFC Designs, Major Elements, and Scalability. 2018.

24. Lunyang, L., W. Chen, and Y. Li, An Overview of the Proton Conductivity of Nafion Membranes Through a Statistical Analysis. *J. Membr. Sci.*, 2016. 504.
25. Tylkowski, B., *et al.*, Polymers application in proton exchange membranes for fuel cells (PEMFCs). *Phys. Sci. Rev.*, 2017. 2.
26. Rosli, N.A.H., K.S. Loh, and W.Y. Wong, Review of Chitosan-Based Polymers as Proton Exchange Membranes and Roles of Chitosan-Supported Ionic Liquids. 2020. 21(2).
27. Wu, W., *et al.*, Polyaniline/Carbon Nanotubes Composite Modified Anode via Graft Polymerization and Self-Assembling for Microbial Fuel Cells. *Polymers*, 2018. 10(7): p. 759.
28. Vinothkannan, M., *et al.*, Sulfonated graphene oxide/Nafion composite membranes for high temperature and low humidity proton exchange membrane fuel cells. *RSC Adv.*, 2018. 8(14): p. 7494-7508.
29. Dharmalingam, S., V. Kugarajah, and M. Sugumar, Membranes for Microbial Fuel Cells. *Microb. Electrochem. Technol.*, 2019.
30. Logan, B.E., *et al.*, Microbial fuel cells: methodology and technology. *Environ. Sci. Technol.*, 2006. 40 17: p. 5181-92.
31. Park, H.S., *et al.*, *A* novel electrochemically active and Fe(III)-reducing bacterium phylogenetically related to Clostridium butyricum isolated from a microbial fuel cell. *Anaerobe*, 2001. 7: p. 297-306.
32. Oh, S.-E. and B.E. Logan, Proton exchange membrane and electrode surface areas as factors that affect power generation in microbial fuel cells. *Appl. Microbiol. Biotechnol.*,2006. 70(2): p. 162-169.
33. Hagoss, Y. and A. Demeke, Review on microbial fuel cell. *Basic Res. J. Microbiol*, 2015. 2: p. 13.
34. Franks, A.E. and K.P. Nevin, Microbial Fuel Cells, A Current Review. *Energies*, 2010. 3: p. 899-919.
35. Chang, I.S., H. Moon, and O. Bretschger, Electrochemically Active Bacteria (EAB) and Mediator-Less Microbial Fuel Cells. *J. Microbiol. Biotechnol.*, 2006. 16: p. 163-177.
36. Nevin, K.P., *et al.*, Power output and columbic efficiencies from biofilms of Geobacter sulfurreducens comparable to mixed community microbial fuel cells. *Environ. Microbiol.*, 2008. 10 10: p. 2505-14.
37. Bond, D.R. and D.R. Lovley, Electricity Production by Geobacter sulfurreducens Attached to Electrodes. *Appl. Environ. Microbiol.*, 2003. 69: p. 1548 - 1555.
38. Konovalova, E.Y., *et al.* The microorganisms used for working in microbial fuel cells. in *AIP Conf. Proc.* 2018. AIP Publishing LLC.
39. Kumari, S., N. Mangwani, and S. Das, Low-voltage producing microbial fuel cell constructs using biofilm-forming marine bacteria. *Curr. Sci.*, 2015: p. 925-932.
40. Park, D.H. and J.G. Zeikus, Electricity generation in microbial fuel cells using neutral red as an electronophore. *Appl. Environ. Microbiol.*, 2000. 66(4): p. 1292-1297.
41. Islam, M.A., *et al.* Performance of Klebsiella oxytoca to generate electricity from POME in microbial fuel cell. In *MATEC Web of Conferences*. 2016. EDP Sciences.
42. Dalvi, A.D., *et al.*, Microbial fuel cell for production of bioelectricity from whey and biological waste treatment. *Int. J. Adv. Biotechnol. Res*, 2011. 2(2): p. 263-268.
43. Tkach, O., L. Liu, and A. Wang, Electricity generation by Enterobacter sp. of single-chamber microbial fuel cells at different temperatures. *J. Clean Energy Technol.*, 2016. 4(1): p. 36-42.
44. Niessen, J., U. Schröder, and F. Scholz, Exploiting complex carbohydrates for microbial electricity generation–a bacterial fuel cell operating on starch. *Electrochem. Commun.*, 2004. 6(9): p. 955-958.

45. Potter, M.C., Electrical effects accompanying the decomposition of organic compounds. *Proc. R. Soc. London, Ser. B, containing papers of a biological character,* 1911. 84(571): p. 260-276.
46. Suresh, K.S. and B. Bulchandani, Comparative study of various substrates and microorganisms in a laboratory designed microbial fuel cell. *Int. J. Res. Chem. Environ. (IJRCE),* 2012. 2(3): p. 168-174.
47. Konovalova, E.Y., *et al.* Using microbial fuel cells for utilization of industrial wastewater. in *Proceedings of the 2014 International Conference on Industrial, Mechanical and Manufacturing Science.* 2015. ICIMMS.
48. McAnulty, M.J., *et al.*, Electricity from methane by reversing methanogenesis. *Nat. Commun.,* 2017. 8(1): p. 15419.
49. Cheng, S., H. Liu, and B.E. Logan, Increased power generation in a continuous flow MFC with advective flow through the porous anode and reduced electrode spacing. *Environ. Sci. Technol.,* 2006. 40 7: p. 2426-32.
50. Aelterman, P., *et al.*, Continuous electricity generation at high voltages and currents using stacked microbial fuel cells. *Environ. Sci. Technol.*, 2006. 40 10: p. 3388-94.
51. Yeo, J., *et al.*, Practical Maximum-Power Extraction in Single Microbial Fuel Cell by Effective Delivery through Power Management System. *Energies,* 2018.
52. Costilla-Reyes, A., *et al.*, A Time-Interleave-Based Power Management System with Maximum Power Extraction and Health Protection Algorithm for Multiple Microbial Fuel Cells for Internet of Things Smart Nodes. *Appl. Sci.*, 2018.
53. Rahimnejad, M., *et al.*, Microbial fuel cell as new technology for bioelectricity generation: A review. *Alexandria Eng. J.*, 2015. 54: p. 745-756.
54. Wei, J., P. Liang, and X. Huang, Recent progress in electrodes for microbial fuel cells. *Bioresour. Technol.*, 2011. 102 20: p. 9335-44.
55. Cheng, S., P.D. Kiely, and B.E. Logan, Pre-acclimation of a wastewater inoculum to cellulose in an aqueous-cathode MEC improves power generation in air-cathode MFCs. *Bioresour. Technol.*, 2011. 102 1: p. 367-71.
56. Cheng, S. and B.E. Logan, Increasing power generation for scaling up single-chamber air cathode microbial fuel cells. *Bioresour. Technol.*, 2011. 102 6: p. 4468-73.
57. Obileke, K., *et al.*, Microbial fuel cells, a renewable energy technology for bio-electricity generation: A mini-review. *Electrochem. Commun.*, 2021. 125: p. 107003.
58. del Pilar Anzola Rojas, M., *et al.*, Effect of the electric supply interruption on a microbial electrosynthesis system converting inorganic carbon into acetate. *Bioresour. Technol.*, 2018. 266: p. 203-210.
59. Mao, Z., *et al.*, Enhancing stability and resilience of electromethanogenesis system by acclimating biocathode with intermittent step-up voltage. *Bioresour. Technol.*, 2021. 337: p. 125376.
60. Mukherjee, A., *et al.*, Effective power management system in stacked microbial fuel cells for onsite applications. *J. Power Sources,* 2022.
61. Rossi, R., *et al.*, *In situ* biofilm removal from air cathodes in microbial fuel cells treating domestic wastewater. *Bioresour. Technol.*, 2018. 265: p. 200-206.
62. Passos, V.F.d., *et al.*, Energy generation in a Microbial Fuel Cell using anaerobic sludge from a wastewater treatment plant. *Sci. Agric.*, 2016. 73: p. 424-428.
63. Xiao, Y., *et al.*, Extracellular polymeric substances are transient media for microbial extracellular electron transfer. *Sci. Adv.*, 2017. 3.
64. Aelterman, P., *et al.*, Loading rate and external resistance control the electricity generation of microbial fuel cells with different three-dimensional anodes. *Bioresour. Technol.*, 2008. 99 18: p. 8895-902.

65. Jung, S.P. and J.M. Regan, Influence of External Resistance on Electrogenesis, Methanogenesis, and Anode Prokaryotic Communities in Microbial Fuel Cells. *Appl. Environ. Microbiol.*, 2010. 77: p. 564 - 571.
66. Elshobary, M.E., *et al.*, Recent insights into microalgae-assisted microbial fuel cells for generating sustainable bioelectricity. *Int. J. Hydrog. Energy*, 2020.
67. Fernández-Marchante, C.M., *et al.*, Thermally-treated algal suspensions as fuel for microbial fuel cells. *J. Electroanal. Chem.*, 2018. 814: p. 77-82.
68. Formoso, P., *et al.*, Electro-Conductive Membranes for Permeation Enhancement and Fouling Mitigation: A Short Review. *Membranes*, 2017. 7.
69. Kisieliūtė, A., *et al.*, Towards microbial biofuel cells: Improvement of charge transfer by self-modification of microoganisms with conducting polymer – Polypyrrole. *Chem. Eng. J.*, 2019.
70. Srivastava, R.K., R. Boddula, and R. Pothu, Microbial fuel cells: Technologically advanced devices and approach for sustainable/renewable energy development. *Energy Convers. Manage.: X*, 2022. 13: p. 100160.
71. Heydorn, R.L., *et al.* Strategies for the Targeted Improvement of Anodic Electron Transfer in Microbial Fuel Cells. 2020.
72. Sharma, S., *et al.*, Recent Trends and Developments in Conducting Polymer Nanocomposites for Multifunctional Applications. *Polymers*, 2021. 13.
73. Morkvėnaitė-Vilkončienė, I., A. Ramanaviciene, and A. Ramanavičius, 9,10-Phenanthrenequinone as a redox mediator for the imaging of yeast cells by scanning electrochemical microscopy. *Sens. Actuators B: Chem.*, 2016. 228: p. 200-206.
74. Rožėnė, J., *et al.*, Yeast-based microbial biofuel cell mediated by 9,10-phenantrenequinone. *Electrochim. Acta*, 2021. 373: p. 137918.
75. Rožėnė, J., *et al.*, Baker's Yeast-Based Microbial Fuel Cell Mediated by 2-Methyl-1,4-Naphthoquinone. *Membranes*, 2021. 11.
76. Zor, E., *et al.*, 1,10-Phenanthroline-5,6-dione and 9,10-phenanthrenequinone as redox mediators for amperometric glucose biosensors. *J. Solid State Electrochem.*, 2014. 18: p. 1529-1536.
77. Fan, X., *et al.*, Nanocarbon-based membrane filtration integrated with electric field driving for effective membrane fouling mitigation. *Water Res.*, 2016. 88: p. 285-292.
78. Liu, J., *et al.*, Polypyrrole blending modification for PVDF conductive membrane preparing and fouling mitigation. *J. Colloid Interface Sci.*, 2017. 494: p. 124-129.
79. Song, H.-M., *et al.* Simple and Effective Preparation of Zwitterionic Anti-Fouling Poly(vinylidene fluoride) Ultrafiltration Membrane by *In Situ* Cross-Linking Polymerization Technology. 2020.
80. Vatanpour, V., *et al.*, Antifouling polyvinylidene fluoride ultrafiltration membrane fabricated from embedding polypyrrole coated multiwalled carbon nanotubes. *Mat. Sci. Eng. C-Mater.*, 2018. 89: p. 41-51.
81. Cabrol, L., *et al.*, Microbial ecology of fermentative hydrogen producing bioprocesses: useful insights for driving the ecosystem function. *FEMS Microbiol. Rev.*, 2017. 41: p. 158–181.
82. Cai, G., *et al.*, Metabolic flux analysis of hydrogen production network by Clostridium butyricum W5: Effect of pH and glucose concentrations. *Int. J. Hydrog. Energy*, 2010. 35: p. 6681-6690.
83. Islam, M.A., *et al.*, Augmentation of air cathode microbial fuel cell performance using wild type Klebsiella variicola. *RSC Adv.*, 2017. 7: p. 4798-4805.
84. Montoro, C., *et al.*, Ionic Conductivity and Potential Application for Fuel Cell of a Modified Imine-Based Covalent Organic Framework. *J. Am. Chem. Soc.*, 2017. 139(29): p. 10079-10086.
85. Wang, J., *et al.*, A Review of Recent Advances in Microbial Fuel Cells: Preparation, Operation, and Application. *BioTech*, 2022. 11.

15

Role of Smart Materials in Environmental Remediation: CO_2 Capture and CO_2 Reduction

Yogendra K. Gautam, Durvesh Gautam, Manohar Singh, Himani, Kavita Sharma*, Beer Pal Singh† and Anuj Kumar

Department of Physics, CCS University, Meerut, U.P., India

Abstract

The industrial growth and the consequential rise in energy demands in recent years have led to the widespread use of fossil fuels. This consecutively results in huge emission of greenhouse gases (GHGs) into the environment. Carbon dioxide (CO_2) is a copious gas among other GHGs and considered as the main culprit for global warming because of its increasing amount in the atmosphere. Capturing and storing of CO_2 emission needs serious attention from the research community to prevent its escalating concentration in the atmosphere. The conversion and reduction of CO_2 into other functional compounds is an alternate and potential solution to solve the environmental disturbance. The natural conversion of CO_2 to other compounds takes a lot of time and requires searching a solution to enhance this reaction. In the present chapter, the electrochemical and photochemical conversion approaches to convert CO_2 into fuel or chemical feedstock compounds are comprehensively discussed. CO_2 can be converted to carbon monoxide (CO) or CO-rich feeds, hydrocarbons, and alcohols through catalytic hydrogenation, complex metal hydrides, electrochemical reduction, photocatalysis, and biological reduction.

***Keywords*:** Global warming, green house gases, environmental remediation, CO_2 capture, CO_2 reduction, electrochemical, photocatalytic, photoelectrocatalytic process

15.1 Introduction

As the human population is increasing at an alarming rate, the need for energy to meet the industrial and humanitarian demands is growing every day. An obvious result of this is a rise in the consumption of fuels, especially traditional fossil fuel reservoirs (such as coal, oil, and natural gas), which have emerged as major energy sources since the Industrial Revolution. Global warming and depleting energy resources are the two major problems confronting the world in the 21st century. Both are interrelated to each other to a large extent. The most well-known greenhouse gas, carbon dioxide (CO_2), is released by both natural and man-made activities. Additionally, it becomes a necessity for many industrial operations

**Corresponding author*: sharmak29@gmail.com
†*Corresponding author*: drbeerpal@gmail.com

Upendra Kumar and Piyush Kumar Sonkar (eds.) Smart Materials for Science and Engineering, (305–324) © 2024 Scrivener Publishing LLC

as well as the development of all plant life on Earth [1–4]. The scientific community has acknowledged the challenge posed by the changing climate and is diligently attempting to assess and quantify this transformation. The GHGs are the main elements that are largely in charge of the environment's instability. The scientific community frequently views satellite observations as an additional or alternative tool for the detection and monitoring of anthropogenic GHG emissions such as CO_2, CH_4, NO_x, etc. because the ground-based observation networks are insufficient and because of the requirement for wide spatial and temporal coverage. The first worldwide accord on GHG emissions is the Kyoto Protocol. In this convention, developed nations committed to stabilizing or reducing GHG emissions decreased approximately by 5.2% within the commitment period from 2008 to 2012. Figure 15.1 depicts the worldwide status of the Kyoto Protocol's implementation, e.g., yellow color highlights the nations that have not reached the targets for reducing GHG emission while green color shows those that met the target of reducing GHG emission. Therefore, this protocol had to be extended until 2017 at the Durban Climate Conference in 2011 [5].

The essential need to minimize this spike in atmospheric CO_2 has driven researchers to invent methods to reduce CO_2 emissions. Emissions of carbon dioxide (CO_2) are a substantial contributor to climate change and is steadily rising due to the continuous increase in industrial activity and the usage of fossil fuels. The CO_2 level is constantly rising and has already gotten dangerously high. One estimate places the origin of CO_2 as 42.84% from the oceans, 28.56% from respiration by plants and animals, 28.56% from decomposition, and 0.03% from volcanic eruptions [6].

The average global temperature rose by 1.1°C between 1850 and 1900, according to the Intergovernmental Panel on Climate Change (IPCC), and during the next 20 years, it is expected to reach or even surpass 1.5°C. However, it goes beyond only the temperature.

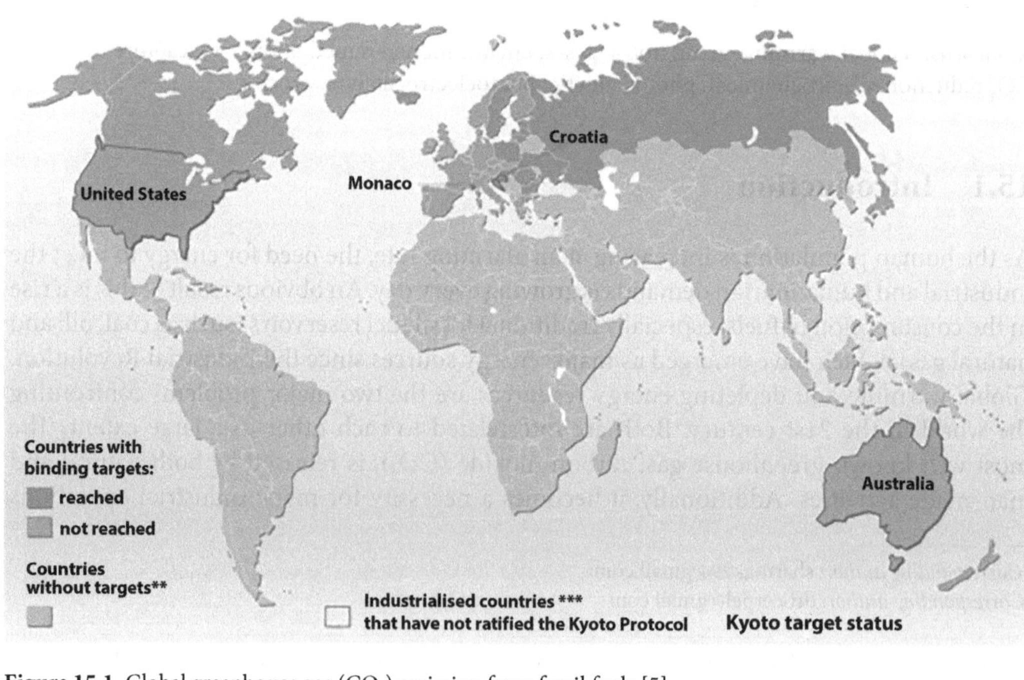

Figure 15.1 Global greenhouse gas (CO_2) emission from fossil fuels [5].

Various changes are being brought about by climate change in various locations, and these changes will all get worse as the world continues to warm up. Changes in winds, snow and ice, coastal regions, oceans, the water cycle, sea level, and many other factors are among them [7].

A plethora of options is being investigated for lowering fossil-fuel consumption and emissions of carbon dioxide. This chapter focuses on highlighting the major approaches to conversion of carbon dioxide into fuel or chemical feedstock molecules. The term "CO_2 sequestration," also referred to as "capture and storage of CO_2," is assigned to several new inventions that are aimed at trapping CO_2 produced by the industrial sector and many other energy sources. Before entering the atmosphere, CO_2 is compressed and injected into secure geological patterns to facilitate long-term storage and, at the same time, minimize the changing climatic shift [8–10]. Carbon capture and storage, which is now being explored worldwide, is a key strategy for limiting climate change. CO_2 capture techniques range from those that are now in use, such as amine-scrubbing, to others that are in their second or third generations and may have improved thermodynamics, like chemical or carbonate looping. The carbon capture techniques comprise the pumping of CO_2 into the empty oil and gas fields or unused coal mines to lock the carbon [11]. In this context, electrochemical reduction and photochemical reduction are the two main approaches to converting carbon dioxide into carbon monoxide (CO) or CO-rich feeds, hydrocarbons, and alcohols. The advantage of electrochemical CO_2 reduction is that the whole process is controllable by reaction temperature and electrode potential, and the chemical composition is minimized as the electrolytes are recyclable and might be a way to use the extra energy from intermittent renewable sources instead of large-scale energy storage [12]. Compared to other reduction methods, the electrochemical reduction of CO_2 is an effective process that can be carried out at moderate temperatures and air pressure with excellent product selectivity [13]. Innovative research was done by several researchers using transition metals, metallic oxides, alloys, and semiconductors as electrocatalysts for the electrochemical reduction of CO_2 [14].

A developing and environmentally friendly method that employs light irradiations to reduce CO_2 is CO_2 transformation by photocatalysis. This method has drawn attention since it uses low temperatures, a lot of pressure, and little input energy [15, 16].

15.2 CO_2 Reduction Techniques

There are three methods—electrochemical, photocatalytic, and photoelectrocatalytic—for CO_2 reduction, and the mechanism is shown in Figure 15.2.

15.2.1 Electrochemical CO_2 Reduction Reaction

Energy cycle and climate change are significantly impacted by electrochemical CO_2 reduction to fuel. Figure 15.3 illustrates a method for reducing carbon dioxide electrochemically using an electrocatalyst. This procedure is carried out on an electrocatalyst, such as a metal plate or an electrically conducting support layer, that is in direct contact with the cathodic current collector. An equal number of electrons are concurrently taken from the anodic reaction, which is usually the oxygen evolution reaction (OER) in most cases, and delivered

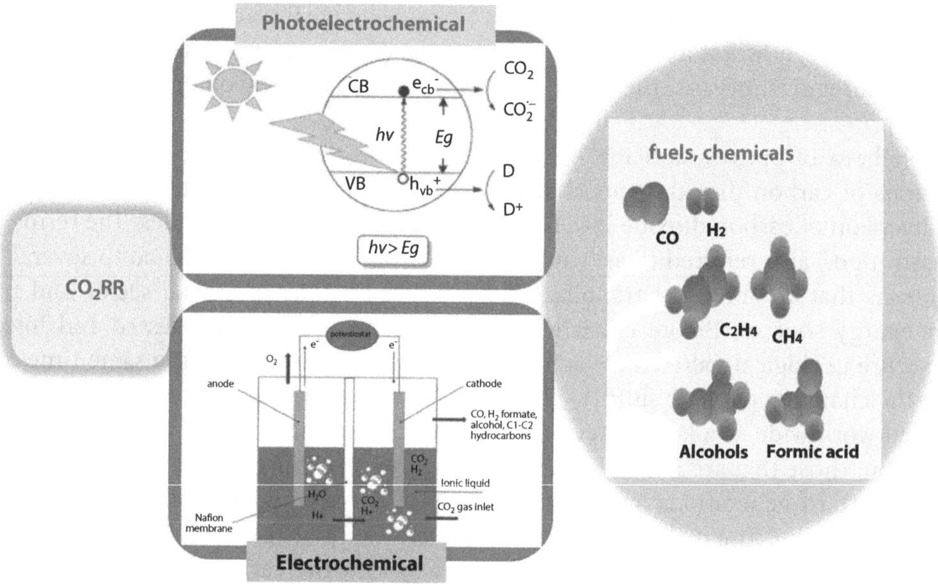

Figure 15.2 Illustration of different techniques such as electrochemical, photocatalytic, and photoelectrocatalytic, available for the reduction of CO_2.

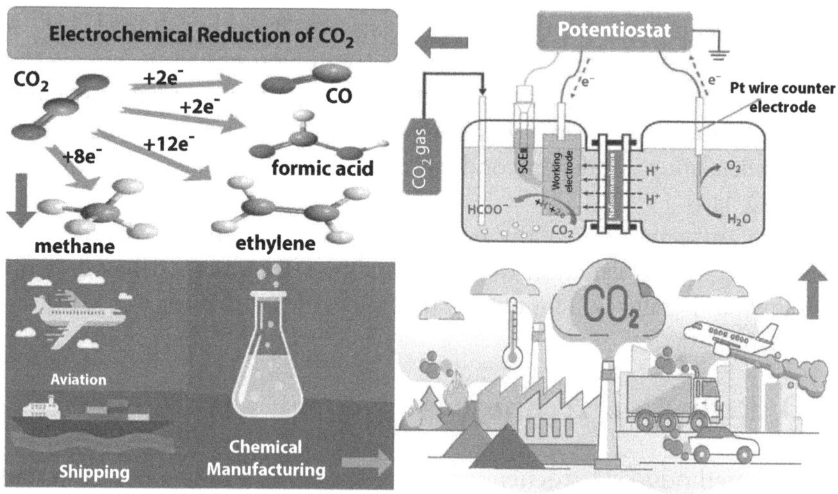

Figure 15.3 Schematic representation of process of electrochemical reduction of CO_2 into various fuel products and chemical products used in industries.

to the cathode via a type of potentiostat or galvanostat (power source) to allow the CO_2 molecules to undergo the reduction reaction. Typically, an ion exchange membrane divides the electrolyzer's cathodic and anodic chambers. The membrane serves as a barrier preventing other goods from passing through while also selectively allowing charge transfer, completing the electric circuit [17, 18].

Teeter and Ryssel Berghe developed the carbon dioxide reduction reaction (CO_2RR) in 1954 [19]. Numerous research teams predicted that CO_2RR would mostly create formate, oxalate, and carbon monoxide between the late 1960s and the early 1980s [20, 21]. According to Hori's seminal efforts from the 1980s to the early 2000s, given favorable circumstances, carbon dioxide may be electrochemically converted to valuable chemicals. The energy conversion efficiency, which is determined by dividing the free energy of products gained from electrochemical carbon dioxide reduction by the amount of energy used in the reduction, would, nevertheless, be between 30 and 40%. Such a poor efficiency would soon deter people from reducing CO_2 in practise. The focus of recent research has switched to figuring out how the structure of the electrolyzer itself impacts the rate and effectiveness of the reaction. Performance may be significantly impacted by variables such as availability of CO_2, electrolyte composition, temperature, and pH. Electrolyzers must be constructed to transport the necessary quantity of CO_2 to the cathode catalyst, which is difficult in the conventional solely aqueous systems owing to the poor solubility of CO_2 in water [22]. This is necessary to achieve high current density (CD) and selectivity for CO_2 electrolysis products. Since then, there has been an increase in the amount of research on CO_2RR, and various methodologies and materials have produced a variety of renewable chemical products, some of which have the potential to be developed further and put on the market. One of the most explored and promising CO_2 capture materials is CaO. Numerous metal-organic frameworks (MOFs), high-surface carbon-based materials, inorganic oxides, porous polymeric materials, and zeolites are examples of nano-structured amine-functionalized sorbents that have been described. In a creative design, Voskian and Hatton demonstrated electro-swing CO_2 absorption with high FE and low EC (40–90 kJ mol^{-1}) using a quinone polymerized into a solid [23].

After extensive research, two methods for CO_2RR in aqueous solutions were discovered: First, an adsorbed molecule called CO_2 undergoes surface hydrogenation and desorbs as a formate ion (CHO_2); second, CO is converted to CO and either desorbs as carbon monoxide or combines with other adsorption species to create alcohols, hydrocarbons, or organic acids [24].

The most prevalent CO_2RR products are carbon monoxide (CO), formic acid/formate (HCOOH or HCOO-), methane (CH_4), methanol (CH_3OH), ethylene (C_2H_4), acetic acid/acetate (CH_3COOH or CH_3COO-), ethanol (CH_3CH_2OH), and 1-propanol. Figure 15.4 shows the main CO_2RR products and their uses ($CH_3CH_2CH_2OH$). Table 15.1 provides an overview of the electron transfer in each of the aforementioned products as well as the primary rival process, the hydrogen evolution reaction (HER).

The discovery that methanol can be synthesized advantageously in the presence of CO_2 led to a breakthrough in methanol synthesis. The following are the primary reactions that take place when CO_2 is hydrogenated:

$$CO_2 + 3H_2 \rightarrow CH_3OH + H_2O \tag{15.1}$$

$$CO_2 + H_2 \rightarrow CO + H_2O \tag{15.2}$$

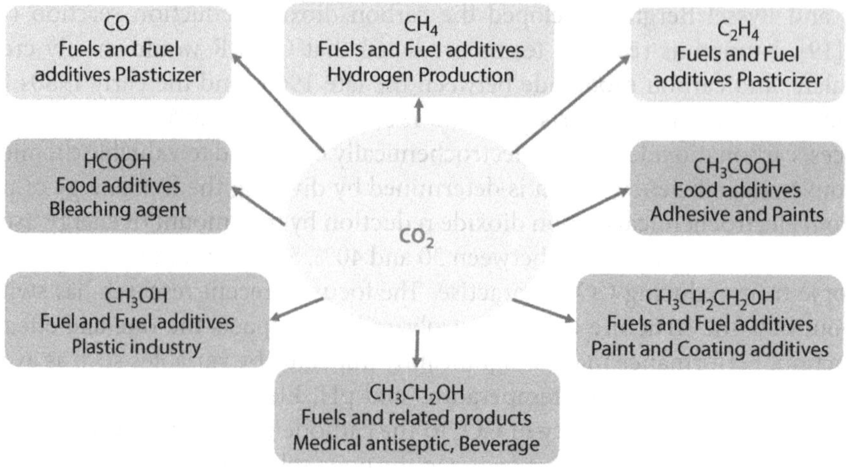

Figure 15.4 Common products from CO_2RR electrocatalysis with their applications.

Table 15.1 Summary of common CO_2RR and hydrogen evolution reaction (HER), products in alkaline solution (pH=14), standard potentials concerning the reversible hydrogen electrode [V vs. RHE] at 1 atm and 25°C, and the number of electrons transferred.

Product (phase)	Cathode reaction	E° [V vs. RHE]	Z
CO (g)	CO_2 (g) +2H_2O (l) +2e⁻= CO (g) + 2OH^-	-0.934	2
$HCOO^-$ (aq)	CO_2 (g) +2H_2O (l) +2e^- = $HCOO^-$ (aq) + OH^-	-1.078	2
CH_3OH (l)	CO_2 (g) +5H_2O (l) +6e^- = CH_3OH (l) + 6OH^-	-0.812	6
CH_4 (g)	CO_2 (g) +6H_2O (l) +8e^- = CH_4 (g) + 8OH^-	-0.659	8
$C_2O_4^{2-}$ (aq)	2CO_2 (g)+2e^- = $C_2O_4^{2-}$ (aq)	-0.590	2
CH_3COO^- (aq)	2CO_2(g)+5H_2O(l) +8e^- =CH_3COO^-(aq) +7OH^-	-0.703	8
C_2H_5OH (l)	2CO_2(g)+9H_2O(l) +12e^- = C_2H_5OH (l)+ 12OH^-	-0.744	12
C_2H_4 (g)	2CO_2(g)+8H_2O(l) +12e^- = C_2H_4 (g) + 12OH^-	-0.764	12
C_3H_7OH (l)	3CO_2(g)+13H_2O(l) +18e^- = C_3H_7OH (l)+ 12OH^-	-0.733	18
H_2 (g)	2H_2O (l) +2e^- = H_2 (g) + OH^-	-0.828	2

Methanol is directly synthesized from CO_2 in the first reaction, and reverse water gas shift one occurs in the second. Both of these reversible processes are exothermic. It should be emphasized that the cost of this CO_2 hydrogenation process is greater than that of a reaction based on a CO and CO_2 combination. However, CO_2 gases, which are often wasted in process industries, may be used in CO_2 hydrogenation for the production of methanol. Industrial catalysts for the production of methanol are available for gases that include H_2 and CO and typically have a negligible amount of CO_2 present. Therefore, it is necessary to create a novel catalyst for the synthesis of methanol that can handle input that is high in CO_2.

The ammonia plant produces ammonia (NH_3) and carbon dioxide (CO_2), which react as shown below to make urea and water.

$$2NH_3 + CO_2 \rightarrow H_2NCONH_2 + H_2O \tag{15.3}$$

Carbamate, unreacted ammonia, and CO_2 are among the intermediate products that are separated and used as new raw materials. When condensing urea [25], the water produced as a by-product is eliminated.

Another potential carbon capture method is electrochemical CO_2 capture (ECC), which employs electro-swing rather than pressure or temperature swing to absorb and release CO_2. The fundamental draw of ECC approaches is their ability to modulate CO_2 absorption/desorption in a potential-dependent manner by modifying the nucleophilicity of intermediate species in an ambient environment. Additionally, ECC techniques show greater adaptability to deal with decentralized emissions and provide reduced theoretical energy penalties of capture/release processes [26]. They are also compatible with electrified industry. While indirect ECC uses a redox mediator that competes with the redox-inactive sorbent for CO_2 binding, direct ECC is based on changing the nucleophilicity of a redox-active sorbent. Direct ECC [27] has been investigated for organic redox-active sorbents of quinones, bipyridines, and disulfides as well as transition metal complexes with macrocyclic amino and pyridine ligands.

It is not straightforward to convert carbon dioxide via electrocatalysis into clean fuels. For this goal, various research groups have examined only metallic electrodes over the last two decades and have divided them depending on the by-products of the electrocatalytic CO_2 reduction process (e-CO2RR). The heterogeneous catalytic conversion of CO_2 on the electrode surface involves three main steps: (i) the initial linear CO_2 molecule is adsorbed on an electrode material with the loss of absolute linearity; (ii) the C–O bonds are broken by electron transfer and/or proton migration processes to form C–H bonds; and (iii) the product is desorption, which is preceded by configurational rearrangement and followed by diffusion into the electrolyte [28]. It is also discovered that both e-CO2RR include numerous phases with the overall two-, four-, six-, or eight-electron transfer processes (Figure 15.4) [29]. Naturally, a wide variety of products may result from such operations. These include methane, ethylene, formaldehyde, methanol, formic acid, formate, and carbon monoxide. The half-cell reactions involved in these processes have previously been detailed below, together with their typical electrode potentials.

15.2.2 The Various Rudiments of an e-CO_2RR Experiment

Electrocatalysts may be designed for effective performance and excellent selectivity based on the mechanistic pathways of the aforementioned activities. Electrochemical methods including linear sweep voltammetry (LSV), cyclic voltammetry (CV), and electrochemical impedance spectroscopy may be used to evaluate the performance of different electrocatalysts (EIS). While the kinetics may be understood using the Tafel slopes that can be determined from the pertinent voltammograms and/or the impedance spectra, the overpotentials for distinct products at varying current densities can be estimated directly from the LSV and CV profiles. A two-compartment H-shaped electrochemical cell with a gap between the cathode and the anode should be used for all studies. A thin frit glass membrane or a semipermeable conducting membrane like Nafion may be used to accomplish this separation. Pander *et al.* have developed a thorough setup design that is gas leak-proof. Since CO_2 is reduced during e-CO2RR, a cathode is always used as the working electrode to lessen other gaseous toxins like O_2 and N_2. Therefore, air should be taken out of the cathodic chamber before beginning the e-CO2RR process. It is thus necessary to completely fill the cathodic chamber with high-purity CO_2 in order to precisely detect the products created during e-CO2RR. Typically, $KHCO_3$ serves as the electrolyte, while graphite or another carbon electrode serves as the counter-electrode (in this instance, the anode). Researchers have shown e-CO2RR in a KOH medium in a number of instances to raise the pH and hence inhibit the reduction of protons/water (hydrogen evolution). However, when the KOH electrolyte is saturated with CO_2, K_2CO_3 and $KHCO_3$ spontaneously develop, which lowers the pH. The pH of the medium should be precisely controlled even though it is often kept between 6.8 and 7.5 in order to properly convert the potential versus reference electrode (typically a pre-calibrated Ag/AgCl or SCE) to the potential vs. reversible hydrogen electrode (RHE). The H-cell's cathode chamber is where the reference electrode is positioned. To prevent poisoning and a serious decline in the catalytic performance, the working electrode should be maintained away from any metal impurities found in the electrolyte, such as those of Fe and Zn. For this, the impurities may be removed using a sacrificial cathode after the aqueous electrolyte has been pre-electrolyzed. The current densities that are reported for electrochemical experiments such as CV, LSV, etc., should be normalized with respect to the electrochemical surface area (ECSA), which can be calculated as per an earlier report [30]. Chelex is another option to prevent metal ions deposition due to its chelating action. The aforementioned electrochemical measurements are often coupled with the chromatograms obtained from GC (Gas Chromatography) and/or HPLC (High-Performance Liquid Chromatography) of the product mixtures in order to assess the Faradaic efficiencies (FE) of different electrocatalysts. To produce enough products for further study, a bulk-electrolysis experiment should be run for approximately an hour [31]. To prevent changes in the product selectivity, it is advisable to conduct studies under potentiostatic circumstances. Potentials from galvanostatic tests may not be suited for e-CO2RR or may provide a range of results. The latter, however, could only be favored under certain circumstances when the working electrode is well researched beforehand and the applied current is exact for scaling up e-CO2RR processes [32]. Therefore, the variety of physical states of the products produced in the reaction media is what makes e-CO2RR investigations difficult. The Faradaic Selectivity or Faradaic Yield, which is defined as the percentage of FE for a specific product with respect to the total FE for all goods, should also be computed and reported on the individual assessment of FE for distinct products. The evolving gases should be directly delivered to the sampling loop of an online gas chromatograph equipped with a thermal conductivity detector (TCD), using Argon as the inert carrier gas. This is necessary for the analysis of the H2 gas, which is invariably evolved during e-CO2RR experiments due to very close potentials. Additionally, the potential carbon-containing gaseous products such CO, CH_4, C_2H_4, and C_2H_6 as mentioned in the mechanistic route should be examined using a flame ionization detector (FID) (Figure 15.5). The selectivity of the product may be assessed based on the identification of the ingredients present in the liquid (for example, HCOOH, CH_3OH, and C_2H_5OH) and gaseous mixtures. Different carbon compounds are often seen producing substantially different products than what might be anticipated from

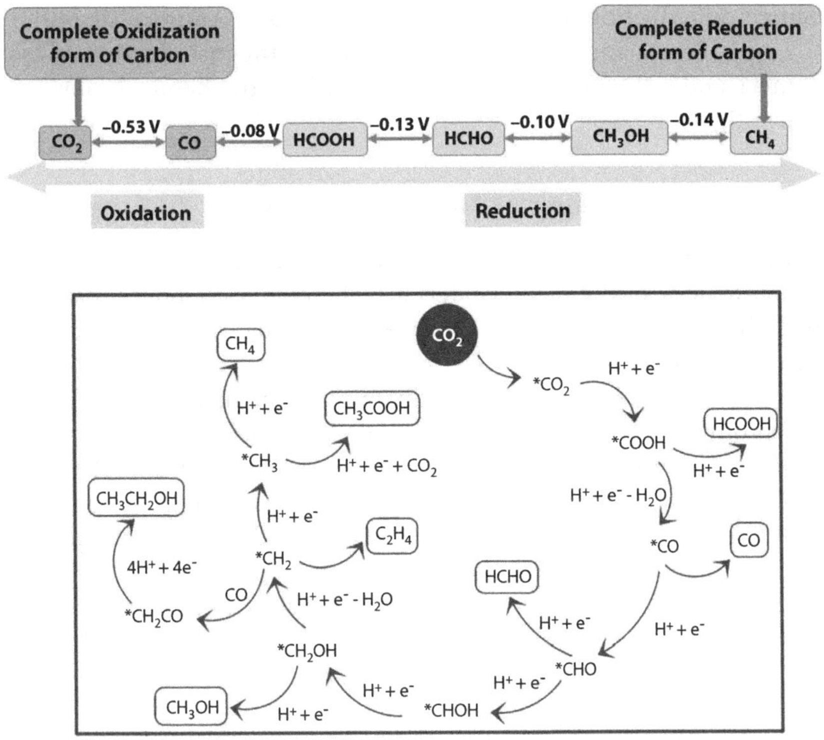

Figure 15.5 Various products formed due to multiple-step processes during electrochemical CO_2 reduction.

just metallic catalysts, even when combined with active metallic catalyst nanoparticles. It is crucial to comprehend the structure–activity correlations between active catalysts and the molecules of reactant and product. *In situ* approaches are often used [33] to conduct thorough investigations of these processes.

15.2.3 Recent Advances in CO_2 Reduction Using Metal Complex Molecular Catalysts

Molecular catalysts (homogenous or heterogeneous) in electrochemical or photochemical conditions are typically useful in the field [34, 35].

15.2.3.1 Iron-Based Molecular Catalyst

For electrocatalytic and photo electrocatalytic CO_2 reduction, iron porphyrins and metal-organic complexes have attracted a great deal of interest in the past [36–38]. Homogeneous electrocatalysts are one of the favored ways thanks to the use of a variety of contemporary techniques and modification via synthesis to increase the reactivity and product selectivity. According to Taheri *et al.*'s study [39] on "an iron electrocatalyst for selective reduction of CO_2 to formate in the water," formate is generated with a high current density and a high Faradaic efficiency (96%), even with a modest applied overpotential. "Molecular engineering for effective and discriminating iron porphyrin catalysts for the electrochemical reduction of CO_2 to CO" was studied by Ambre *et al.* [40]. Here, iron porphyrins Fe-pE, FemE, and Fe-oE were produced by the introduction of ester groups in the para, meta, and ortho positions. These catalysts' electrochemical conversion of CO_2 to CO was investigated.

15.2.3.2 Zn-Based Metal Complex Molecular Catalyst

The usage of zinc-based systems is of great interest because of the low cost, wide availability, and non-toxicity of the metal center. It is therefore astounding that so little zinc-catalyzed hydrolysis of CO_2 has been documented in the literature. These examples include two cationic, N-heterocyclic carbene (NHC) zinc complexes, a dicationic NHC-stabilized zinc hydride cluster, which includes the tris(thiopyridyl) methane-based [Tptm] ZnH, and the tris[(1-isopropylbenzimida CO_2 [41–43] and ZnH complexes catalyzed a reaction. Donovan *et al.* [44] investigated the synthesis and assessed the CO_2 stimulation and reduction potential of two novel Zn(II) complexes. A number of synthesized zeolitic imidazolate frameworks-8 (ZIF-8) nanomaterials with unusual features, such as a high surface area, uniform pore size, well-defined morphology, and strong coordination, were examined by Wang *et al.* [45] for their ability to reduce CO_2 in aqueous electrocatalysis. ZIF-8 nanocomposites are shown to be efficient during CO_2 electrochemical reduction, and they are suggested (as ZIF-8 metal nodes) as active catalysis sites.

15.2.3.3 Mn-Based Metal Complex Molecular Catalyst

The manganese compounds +2, +4, and +7 have the most distinct manganese oxidation state. The strong oxidation state of manganese, Mn+7, is commonly reduced to Mn+2. Although it is less frequent, the oxidation state of Mn+1 often occurs in manganese-based organometallic complexes. D-6 MnI tricarbonyl complexes have grown in significance in the catalysis field [46, 47]. The typical fac-[Mn(bpy) (CO)3Br] combination reveals a HOMO regulated by a Mn 3d-orbital and a diimine-based LUMO. By changing the chemical structure of the diimine ligand, it is possible to better alter the electronic and photophysical character state of similar MnI complexes and therefore adapt the LUMO energy. Reviews on transition metals and manganese-based systems for CO_2 [48–50] are numerous. The effectiveness, feature, and strategy of Mn carbonyl schemes for CO_2 electro- and photoreduction were unknown until a review (Sinopoli *et al.* [51]) was published. The first pure organometallic fac-[MnI(CO)3 (bis-MeNHC)Br] complex, according to Franco *et al.* [52], has the groundbreaking capability for selective electrocatalytic CO_2-to-CO reduction, surpassing 100 rotations of CO with exceptional FE yields (CO 95%) in anhydrous CH_3CN. In contrast, the standard bpy-based systems generally are shown to be inert without respect to specific proton source [53–56]. The novel bis-NHC catalyst also effectively and selectively adapts CO_2 to CO in an anhydrous aprotic organic solvent. The compound [MnBr(2,2'-bipyridine)(CO)3 connected to a carbon nanotube electrode over a pyrene unit was constructed, according to Reuillard *et al.* [57]. With a catalytic start overpotential of 360 mV, molecular catalyst check enables electrocatalytic CO_2 reduction in all aqueous circumstances, and restricted potential electrolysis produces turnovers at 550 mV more than 1000.

15.2.3.4 Ni-Based Molecular Catalyst Nickel

Due to its easily attained oxidation states, a non-precious metal (group VIIIB) is regarded as the best feasible substitute for expensive palladium or platinum for molecular catalysts [58, 59] (e.g., Ni0, Ni I, Ni II, Ni III, and NiIV). Nickel catalysts' recurrent ability to reduce CO_2 needed the reduction of NiII to NiI, which is related to the geometric distortion from a tetra-dentate, square planar coordination mode fitting for NiII to a tetrahedral one model for NiI [60–62]. As a result, the diverse coordination geometries' rich redox properties allow ligand-logical form scaffolds to complement nickel (II) centers, resulting in nickel complexes with precise catalyst functions [63–65] and NiFe CO dehydrogenases (CODHs) for the reversible conversion of CO_2 to CO [66]. These discoveries drove scientists to create molecular catalysts based on nickel complexes, primarily for energy conversion catalysis. Several comprehensive analyses highlighted earth-available metal complexes as CO_2 reduction catalysts [54, 67–69]. However, the literature does not yet include information on the Ni-based molecular catalyst.

As shown in Figure 15.6, the bulk of current CO_2 reduction electrocatalysts may be categorized into three categories: metallic, non-metallic, and molecular catalysts.

Due to their structural simplicity, portability, and robustness, early experiments in electrochemical CO_2 reduction based on polycrystalline metallic catalysts are appealing options for foundational research. Metallic catalysts may be further classified into numerous subgroups based on the CO_2 reduction product, including CO-selective metals (such as Au, Ag, and Zn), formate selective metals (such as Sn, In, and Pb), and hydrogen-selective metals. (e.g., Fe, Ni, and Pt). Cu stands out from the other metallic catalysts in its capacity to catalyze the production of a variety of CO_2 reduction products, such as CO, formate, ethanol, and ethylene [2, 3, 70, 71], and catalysts other than bulk metallic, such nanostructured metallics [4–7, 72, 73]. Recently, it has also been shown that ion-modified metallics [8–10, 74–76], bimetallics [11–14, 77–79], and non-metallic materials [15–20, 80–83] are potential catalysts for CO_2 reduction. They signify fresh patterns and chances for electrochemical CO_2 conversion.

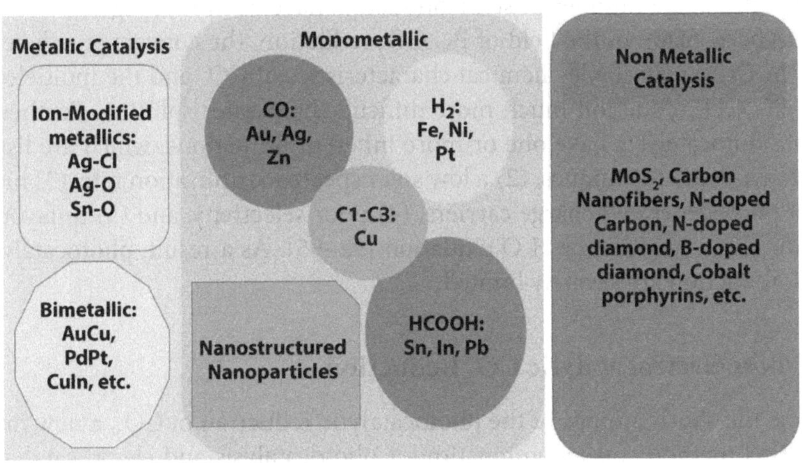

Figure 15.6 Various electrocatalysts; (i) Metallic (ii) Molecular, and (iii) Non-Metallic, used for the reduction of CO_2.

15.2.4 Photocatalytic Reduction of CO_2

Photocatalytic (PC) reduction of CO_2 is carried out in the presence of photocatalysts and water, under ambient pressure and temperature and driven by sunlight energy. In this technique, CO_2 can be converted into various fuels like HCHO, HCOOH, CH_4, CH_3OH, etc. [84]. The reduction of CO_2 by PC is first aided by Halmann in 1978. He used GaP as a photocatalyst. After that, several researches have been done using various semiconductors as photocatalysts like CdS, ZnO, CuO, WO, InP, etc. [85].

The effectiveness of a PC process for CO_2 reduction is strongly reliant on the properties like band gap, exciton binding energy, superficial area, morphological structure, etc. [86, 87] of the semiconductors used as photocatalysts. Also, every semiconductor has its intrinsic conduction band (CB) and valence band (VB), with CB being at a higher potential and VB being at a lower potential. When exposed to sunlight with an energy greater than a semiconductor's Eg, the semiconductor absorbs the photons and electrons (e-) are excited from the VB to the CB, leaving the corresponding holes (h+) in the VB. The positions of a semiconductor's CB and VB determine the reduced ability of photogenerated electrons and the oxidation ability of photogenerated holes, respectively. As a result, selecting a semiconductor with appropriate CB and VB positions is essential [88, 89]. The photocatalytic operation can only occur if the potentials of the photo-generated electron and hole are more positive or negative than the targeted reduction or oxidation reactions, respectively. The PC process can be demonstrated as follows:

$$\text{Catalyst} + h\nu \rightarrow e^- + h^+$$

$$2H_2O + 4h^+ \rightarrow O_2 + 4\,H^+$$

$$CO_2\,(aq) + n\,H^+ + m\,e^- \rightarrow \text{Hydrocabanfule} + E^0\,(V)$$

The electrochemical reduction potential values (E^0) are associated with the standard hydrogen electrode (SHE) at pH 7 [90, 91].

Even though a huge number of semiconductors have been investigated and significant progress has been made in the field of PC CO_2 reduction, the conversion efficiency is still very low. The extremely stable chemical characteristics of CO_2 and the multielectron process make PC CO_2 reduction much more difficult than water splitting. Furthermore, the majority of photocatalysts have one or more inherent limitations, which are listed below: (1) intricate synthesis techniques, (2) a low solar spectrum utilization rate, (3) high recombination of photogenerated charge carriers, (4) poor selectivity, and (5) unfavorable band positions for CO_2 reduction or H_2O oxidation [92–95]. As a result, photocatalyst options for PC CO_2 reduction are severely limited.

15.2.5 Photoelectrocatalytic CO_2 Reduction

To overcome the shortcomings of the photocatalytic reduction of CO_2, a new method was proposed. This method was a combination of photocatalysis and electrocatalysis, that is, photoelectrocatalytic. Photoelectrocatalytic (PEC) is the best technique among the others for converting CO_2 into specific gases like ethane, methane etc., and some liquid products such as ethanol, methanol, formate, etc., driven by inexhaustible solar energy, particularly for liquid products at environmental temperature and pressure [89]. Figure 15.7 schematically illustrates the PEC CO_2 conversion into fuels and value-added products upon exposure to sunlight.

A typical CO_2 reduction PEC cell is made up of a reference electrolyte and two electrodes (Figure 15.8) [96]. In most comment types of PEC, one electrode or both electrodes is/are made of semiconducting material, which is photoactive [97]. The basic principle of photocatalytic reduction of CO_2 is that photoexcitation produces electrons in the semiconductor photoelectrode, and then these photo-generated electrons travel to the electrode surface under the supervision of an external voltage to carry out catalytic CO_2 reduction [98].

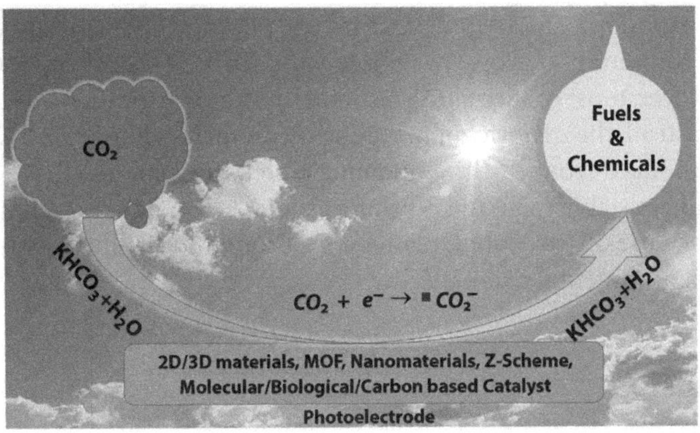

Figure 15.7 Schematic representation of CO_2 reduction into fuels and value-added products in the presence of sunlight.

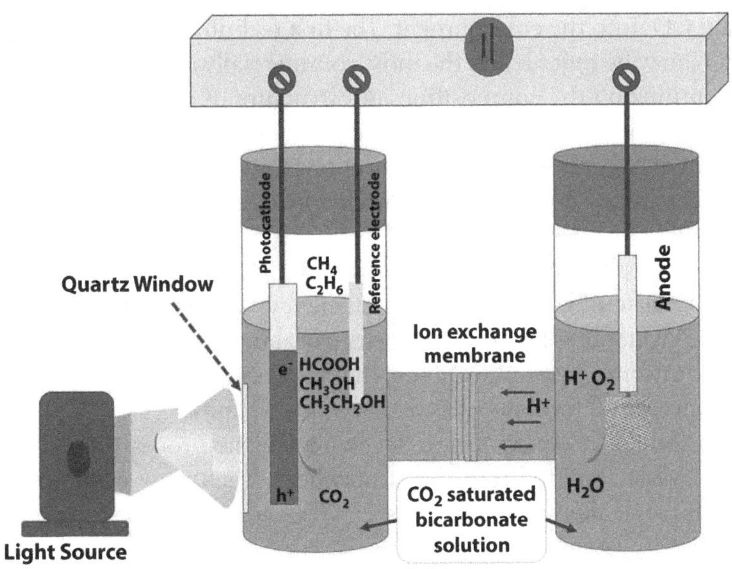

Figure 15.8 PEC cell for CO_2 reduction.

When compared to PC reduction of CO_2, in PEC, the bias imposed by the cell might produce band bending and aid in the oriented transfer of photogenerated electrons, which reduces the recombination of the electron–hole pairs created by photons [99]. The action of the external electric field is beneficial in promoting the directional transfer of photogenerated electrons and holes, hence increasing the separation efficiency of photogenerated carriers and so considerably improving the redox ability. Simultaneously, when the band position of the photocatalyst is not favorable to CO_2 reduction and H_2O oxidation, the redox potentials in the photoreduction system can be changed by applying a suitable bias voltage.

PEC is also called the artificial photosynthesis technique. The CO_2 reduction system may be categorized into the following groups based on the different semiconductor materials of the photoelectrode: (1) p-type photocathode and dark anode [100, 101], (2) dark cathode and n-type photoanode [102, 103], (3) and p-type photocathode and n-type photoanode [104]. The most commonly used p-type photocathodes are p-Cu_2O, p-CuO p-GaP, p-InP, and p-CdTe; however, most of them are poisonous, costly, and/or unstable in aqueous solutions [88, 105]. Instead, the majority of n-type semiconductors like ZnO, TiO_2, Fe2O_3, WO_3, and $BiVO_4$ are produced from earth-abundant elements and so these have low toxicity, are cost-effective, and are stable in aqueous solutions [106]. PEC has several important advantages, including the use of renewable solar energy, economic viability, control over product selectivity, and environmental compatibility.

15.3 Conclusion

Since the mid-20th century, the earth's climate has been changing for the worse. These dire changes are driven by anthropogenic activities, particularly burning of fossil fuel, which leads to increased levels of heat trapping GHGs in the earth's atmosphere. This chapter provides an insight into the existing technologies for transforming/reducing the levels of potent GHG and CO_2 into the environment. The first technique discussed, electrochemical reduction of CO_2, may be regarded as the most commercially available CO_2 reduction pathway. However, optimizing the composition and structure of catalysts is yet to be achieved for further improvement in performance of electrocatalysts. The inclusion of molecular catalysts in electrochemical or photochemical reactions for CO_2 reduction opens new avenues for yielding highly reduced products of electrochemical reactions such as carbon monoxide and formic acid.

Another technique, photochemical conversion of CO_2 reduction, is not considered as a good technique because of poor conversion efficiency. To overcome the shortcomings of the photocatalytic reduction of CO_2, a photoelectrocatalytic approach is adopted for transforming CO_2 into specific gases like ethane, methane, etc., and some liquid fuel products such as ethanol, methanol, formate, etc.

More research should be conducted to develop adequate processes to efficiently change and recycle this "waste" emitted into the atmosphere. Thus, new economic value can be generated in the form of chemicals and fuels.

References

1. Libya should stop denying scientific evidence on HIV Climate: Sawyer predicted rate of warming in 1972. *Nature,* 448, 2007–2007 (2007).
2. Ebi, K. L. *CO2 Rising: The World's Greatest Environmental Challenge* By Tyler Volk . Cambridge, MA:MIT Press, 2008. 206 pp. ISBN: 978-0-262-22083-5, $22.95. *Environ. Health Perspect.,*117, (2009).
3. Olah, G. A., Goeppert, A. & Prakash, G. K. S. Chemical recycling of carbon dioxide to methanol and dimethyl ether: From greenhouse gas to renewable, environmentally carbon neutral fuels and synthetic hydrocarbons. *J. Org. Chem.,* 74, 487–498 (2009).
4. Spinner, N. S., Vega, J. A. & Mustain, W. E. Recent progress in the electrochemical conversion and utilization of CO 2. *Catal. Sci. Technol.,* 2, 19–28 (2012).
5. Olivier, J. G. J., Janssens-Maenhout, G. & Peters, J. a H. W. Trends in global CO2 emissions: 2012 Report. (2012).
6. Goli, A. *et al.* An overview of biological processes and their potential for CO2 capture. *J. Environ. Manage.,* 183, 41–58 (2016).
7. IPCC a. Climate Change 2014 Synthesis Report Summary Chapter for Policymakers. Ipcc 31 (2014).
8. Stauffer, P. H. *et al.* Greening coal: Breakthroughs and challenges in carbon capture and storage. *Environ. Sci. Technol.,* 45, 8597–8604 (2011).
9. D'Alessandro, D. M., Smit, B. & Long, J. R. Carbon dioxide capture: Prospects for new materials. *Angew. Chem. Int. Ed.,* 49, 6058–6082 (2010).
10. Yang, H. *et al.* Progress in carbon dioxide separation and capture: A review. *J. Environ. Sci.,* 20, 14–27 (2008).
11. Kantola, I. B., Masters, M. D., Blanc-Betes, E., Gomez-Casanovas, N. & DeLucia, E. H. Long-term yields in annual and perennial bioenergy crops in the Midwestern United States. *GCB Bioenergy,* 14, 694–706 (2022).
12. Jhong, H. R. M., Ma, S. & Kenis, P. J. Electrochemical conversion of CO2 to useful chemicals: Current status, remaining challenges, and future opportunities. *Curr. Opin. Chem. Eng.,* 2, 191–199 (2013).
13. Olah, G. A., Prakash, G. K. S. & Goeppert, A. Anthropogenic chemical carbon cycle for a sustainable future. *J. Am. Chem. Soc.,* 133, 12881–12898 (2011).
14. Rao, H., Schmidt, L. C., Bonin, J. & Robert, M. Visible-light-driven methane formation from CO2 with a molecular iron catalyst. *Nature,* 548, 74–77 (2017).
15. Das, S. & Wan Daud, W. M. A. A review on advances in photocatalysts towards CO2 conversion. *RSC Adv.,* 4, 20856–20893 (2014).
16. Shehzad, N., Tahir, M., Johari, K., Murugesan, T. & Hussain, M. A critical review on TiO2 based photocatalytic CO2 reduction system: Strategies to improve efficiency. *J. CO2 Util.,* 26, 98–122 (2018).
17. Chu, S., Cui, Y. & Liu, N. The path towards sustainable energy. *Nat. Mater.* 16, 16–22 (2016).
18. Nitopi, S. *et al.* Progress and Perspectives of Electrochemical CO2 Reduction on Copper in Aqueous Electrolyte. *Chem. Rev.,* 119, 7610–7672 (2019).
19. Teeter, T. E. & Van Rysselberghe, P. Reduction of carbon dioxide on mercury cathodes. *J. Chem. Phys.,* 22, 759–760 (1954).
20. Haynes, L. V. & Sawyer, D. T. Electrochemistry of Carbon Dioxide in Dimethyl Sulfoxide at Gold and Mercury Electrodes. *Anal. Chem.,* 39, 332–338 (1967).
21. Kaiser, U. & Heitz, E. Zum Mechanismus der elektrochemischen Dimerisierung von CO2 zu Oxalsäure. *Berichte der Bunsengesellschaft für Phys. Chemie,* 77, 818–823 (1973).

22. Hori, Y., Takahashi, I., Koga, O. & Hoshi, N. Electrochemical reduction of carbon dioxide at various series of copper single crystal electrodes. *J. Mol. Catal. A Chem.*, 199, 39–47 (2003).
23. Voskian, S. & Hatton, T. A. Faradaic electro-swing reactive adsorption for CO2 capture. *Energy Environ. Sci.*12, 3530–3547 (2019).
24. **พวงผกา มะเสนา และประณต นันทิยะกุล.** No Title**การบริหารจัดการการบริการที่มีคุณภาพในโรงพยาบาลสังกัดกระทรวงสาธารณสุข. วารสารวิชาการมหาวิทยาลัยอีสเทิร์นเอเชีย**, 4, 88–100 (2557).
25. Hori, Y. *et al.* 'Deactivation of copper electrode' in electrochemical reduction of CO2. *Electrochim. Acta*, 50, 5354–5369 (2005).
26. Sharifian, R., Wagterveld, R. M., Digdaya, I. A., Xiang, C. & Vermaas, D. A. Electrochemical carbon dioxide capture to close the carbon cycle. *Energy Environ. Sci.*, 14, 781–814 (2021).
27. Electrochemical approaches toward CO2 capture and concentration Sara E. Renfrew, 1 David E. Starr, 2 and Peter Strasser 1,*.
28. Zhu, D. D., Liu, J. L. & Qiao, S. Z. Recent Advances in Inorganic Heterogeneous Electrocatalysts for Reduction of Carbon Dioxide. *Adv. Mater.*, 28, 3423–3452 (2016).
29. Porosoff, M. D., Yan, B. & Chen, J. G. Catalytic reduction of CO2 by H2 for synthesis of CO, methanol and hydrocarbons: Challenges and opportunities. *Energy Environ. Sci.*, 9, 62–73 (2016).
30. Electroanal, I. & Milano, U. Surface area measurements. *Appl. Catal. A Gen.*, 144, N3–N4 (1996).
31. Bard, A. J. & Faulkner, L. R. Double-Layer Structure and Adsorption. *Electrochemical Methods - Fundamentals and Applications*, 534–579 (2001).
32. Pander, J. E., Ren, D. & Yeo, B. S. Practices for the collection and reporting of electrocata- lytic performance and mechanistic information for the CO2 reduction reaction. *Catal. Sci. Technol.*, 7, 5820–5832 (2017).
33. Samanta, S. & Srivastava, R. Catalytic conversion of CO2 to chemicals and fuels: The collective thermocatalytic/photocatalytic/electrocatalytic approach with graphitic carbon nitride. *Mater. Adv.*, 1, 1506–1545 (2020).
34. Qiao, J., Liu, Y., Hong, F. & Zhang, J. A review of catalysts for the electroreduction of carbon dioxide to produce low-carbon fuels. *Chem. Soc. Rev.*, vol. 43 (2014).
35. Portenkirchner, E., Oppelt, K., Egbe, D. A. M., Knör, G. & Sariçiftçi, N. S. Electro- and photochemistry of rhenium and rhodium complexes for carbon dioxide and proton reduction: A mini review. *Nanomater. Energy*, 2, 134–147 (2013).
36. Lin, S. *et al.* Covalent organic frameworks comprising cobalt porphyrins for catalytic CO2 reduction in water. *Science, (80-.).*349, 1208–1213 (2015).
37 Matlachowski, C., Braun, B., Tschierlei, S. & Schwalbe, M. Photochemical CO2 reduction catalyzed by phenanthroline extended tetramesityl porphyrin complexes linked with a Rhenium(I) tricarbonyl unit. *Inorg. Chem.*, 54, 10351–10360 (2015).
38. Hod, I. *et al.* Fe-Porphyrin-Based Metal-Organic Framework Films as High-Surface Concentration, Heterogeneous Catalysts for Electrochemical Reduction of CO2. *ACS Catal.*, 5, 6302–6309 (2015).
39. Taheri, A., Thompson, E. J., Fettinger, J. C. & Berben, L. A. An Iron Electrocatalyst for Selective Reduction of CO2 to Formate in Water: Including Thermochemical Insights. *ACS Catal.*, 5, 7140–7151 (2015).
40. Ambre, R. B. *et al.* Molecular engineering for efficient and selective iron porphyrin catalysts for electrochemical reduction of CO2 to CO. *Chem. Commun.*, 52, 14478–14481 (2016).
41. Specklin, D. *et al.* Cations: Synthesis , Structures , and Use in CO2 Functionalization To cite this version: HAL Id: hal-02121106 NHC-based tri-organyl Zn alkyl cations: synthesis, structure, isomerization for improved stability and use in CO2 functionalization (2022).

42. Sattler, W. & Parkin, G. Zinc catalysts for on-demand hydrogen generation and carbon dioxide functionalization. *J. Am. Chem. Soc.*, 134, 17462–17465 (2012).
43. Specklin, D. *et al.* Accessing Two-Coordinate ZnII Organocations by NHC Coordination: Synthesis, Structure, and Use as π-Lewis Acids in Alkene, Alkyne, and CO2 Hydrosilylation. *Chem. - A Eur. J.*, 23, 15908–15912 (2017).
44. Donovan, E. S. *et al.* https://doi.org/10.1039/C5CC07318A. *Chem. Commun.*, 52, 1685–1688 (2016).
45. *ChemPhysChem* - 2017 - Wang - Zinc Imidazolate Metal Organic Frameworks ZIF-8 for Electrochemical Reduction of CO2 to CO.pdf.
46. West, A. J. Manganese, technetium and rhenium. *Annu. Reports Prog. Chem. - Sect. A*, 109, 131–141 (2013).
47. Duplais, C. & Buendia, J. <Buendia-Mn-Catalyst-cr800341a.pdf>. 1434–1476 (2009).
48. Louis, H., Akakuru, O. U., Monday, P. & Funmilayo, O. O. A review on the state-of-the-art advances for CO2 electro-chemical reduction using metal complex molecular catalysts. *Eclet. Quim.*, 44, 11–39 (2019).
49. Feng, D. M., Zhu, Y. P., Chen, P. & Ma, T. Y. Recent advances in transition-metal-mediated electrocatalytic CO2 reduction: From homogeneous to heterogeneous systems. *Catalysts*, 7, (2017).
50. Darensbourg, D. J., Rokicki, A. & Darensbourg, M. Y. Facile Reduction of Carbon Dioxide by Anionic Group 6B Metal Hydrides. Chemistry Relevant to Catalysis of the Water-Gas Shift Reaction. *J. Am. Chem. Soc.*, 103, 3217–3219 (1981).
51. Siritanaratkul, B., Eagle, C. & Cowan, A. J. Manganese Carbonyl Complexes as Selective Electrocatalysts for CO2 Reduction in Water and Organic Solvents. *Acc. Chem. Res.*, 55, 955–965 (2022).
52. *Angew. Chem. Int. Ed.* - 2018 - Franco - A Highly Active N-Heterocyclic Carbene Manganese I Complex for Selective.pdf.
53. Bourrez, M., Molton, F., Chardon-Noblat, S. & Deronzier, A. [Mn(bipyridyl)(CO)3Br]: An abundant metal carbonyl complex as efficient electrocatalyst for CO2 reduction. *Angew. Chemie - Int. Ed.*, 50, 9903–9906 (2011).
54. Smieja, J. M. *et al.* Manganese as a substitute for rhenium in CO2 reduction catalysts: The importance of acids. *Inorg. Chem.*, 52, 2484–2491 (2013).
55. Sampson, M. D. *et al.* Manganese catalysts with bulky bipyridine ligands for the electrocatalytic reduction of carbon dioxide: Eliminating dimerization and altering catalysis. *J. Am. Chem. Soc.*, 136, 5460–5471 (2014).
56. Franco, F. *et al.* Local Proton Source in Electrocatalytic CO2 Reduction with [Mn(bpy–R)(CO)3Br] Complexes. *Chem. - A Eur. J.*, 23, 4782–4793 (2017).
57. Reuillard, B. *et al.* Tuning Product Selectivity for Aqueous CO2 Reduction with a Mn(bipyridine)-pyrene Catalyst Immobilized on a Carbon Nanotube Electrode. *J. Am. Chem. Soc.*, 139, 14425–14435 (2017).
58. Keim, B. W. Keim1990.Pdf. *Angew. Chemie - Int. Ed.*, 29, 235–244 (1990).
59. Rosen, B. M. *et al.* Nickel-Catalyzed Cross-Couplings Involving Carbon - Oxygen Bonds. 1346–1416 (2011).
60. Wiese, S., Kilgore, U. J., Dubois, D. L. & Bullock, R. M. [Ni(P Me2N Ph2) 2] (BF 4) 2 as an electrocatalyst for H 2 production. *ACS Catal.*, 2, 720–727 (2012).
61. Gan, L. *et al.* A nickel phosphine complex as a fast and efficient hydrogen production catalyst. *J. Am. Chem. Soc.*, 137, 1109–1115 (2015).
62. Thoi, V. S., Kornienko, N., Margarit, C. G., Yang, P. & Chang, C. J. Visible-light photoredox catalysis: Selective reduction of carbon dioxide to carbon monoxide by a nickel N-heterocyclic carbene-isoquinoline complex. *J. Am. Chem. Soc.*, 135, 14413–14424 (2013).

63. Solis, B. H., Maher, A. G., Dogutan, D. K., Nocera, D. G. & Hammes-Schiffer, S. Nickel phlorin intermediate formed by proton-coupled electron transfer in hydrogen evolution mechanism. *Proc. Natl. Acad. Sci. U. S. A.*, 113, 485–492 (2016).
64. Tasker, S. Z., Standley, E. A. & Jamison, T. F. Recent advances in homogeneous nickel catalysis. *Nature*, 509, 299–309 (2014).
65. BELEY, M., COLLIN, J.-P., RUPPERT, R. & SAUVAGE, J.-P. ChemInform Abstract: Electrocatalytic Reduction of CO2 by Ni Cyclam2+ in Water: Study of the Factors Affecting the Efficiency and the Selectivity of the Process. *Chem. Inform.*, 18, 7461–7467 (1987).
66. Fesseler, J., Jeoung, J. H. & Dobbek, H. How the [NiFe4S4] Cluster of CO Dehydrogenase Activates CO2 and NCO-. *Angew. Chemie - Int. Ed.*, 54, 8560–8564 (2015).
67. Costentin, C., Robert, M. & Savéant, J. M. Current Issues in Molecular Catalysis Illustrated by Iron Porphyrins as Catalysts of the CO2-to-CO Electrochemical Conversion. *Acc. Chem. Res.*, 48, 2996–3006 (2015).
68. Benson, E. E., Kubiak, C. P., Sathrum, A. J. & Smieja, J. M. Electrocatalytic and homogeneous approaches to conversion of CO2 to liquid fuels. *Chem. Soc. Rev.*, 38, 89–99 (2009).
69. Luc, W. W. *et al.* Subscriber access provided by UNIV OF CALIFORNIA SAN DIEGO LIBRARIES Ag-Sn Bimetallic Catalyst with a Core-Shell Structure for CO 2 Reduction. *J. Am. Chem. Soc.*, Just Accept. Manuscr.• Publ. Date 17 (2017).
70. Hori, Y. Electrochemical CO2 Reduction on Metal Electrodes. *Mod. Asp. Electrochem.*, 89–189 (2008) doi:10.1007/978-0-387-49489-0_3.
71. Kuhl, K. P. *et al.* Electrocatalytic conversion of carbon dioxide to methane and methanol on transition metal surfaces. *J. Am. Chem. Soc.*, 136, 14107–14113 (2014).
72. Lu, Q., Rosen, J. & Jiao, F. Nanostructured metallic electrocatalysts for carbon dioxide reduction. *ChemCatChem.*, 7, 38–47 (2015).
73. Rosen, J. *et al.* Electrodeposited Zn Dendrites with Enhanced CO Selectivity for Electrocatalytic CO2 Reduction. *ACS Catal.*, 5, 4586–4591 (2015).
74. Ma, S., Lan, Y., Perez, G. M. J., Moniri, S. & Kenis, P. J. A. Silver supported on titania as an active catalyst for electrochemical carbon dioxide reduction. *ChemSusChem*, 7, 866–874 (2014).
75. Hsieh, Y. C., Senanayake, S. D., Zhang, Y., Xu, W. & Polyansky, D. E. Effect of Chloride Anions on the Synthesis and Enhanced Catalytic Activity of Silver Nanocoral Electrodes for CO2 Electroreduction. *ACS Catal.*, 5, 5349–5356 (2015).
76. Kim, D., Resasco, J., Yu, Y., Asiri, A. M. & Yang, P. Synergistic geometric and electronic effects for electrochemical reduction of carbon dioxide using gold-copper bimetallic nanoparticles. *Nat. Commun.*, 5, 1–8 (2014).
77. Kortlever, R., Peters, I., Koper, S. & Koper, M. T. M. Electrochemical CO2 Reduction to Formic Acid at Low Overpotential and with High Faradaic Efficiency on Carbon-Supported Bimetallic Pd-Pt Nanoparticles. *ACS Catal.*, 5, 3916–3923 (2015).
78. Rasul, S. *et al.* A highly selective copper-indium bimetallic electrocatalyst for the electrochemical reduction of aqueous CO2to CO. *Angew. Chemie - Int. Ed.*, 54, 2146–2150 (2015).
79. Zhang, S. *et al.* Polymer-supported CuPd nanoalloy as a synergistic catalyst for electrocatalytic reduction of carbon dioxide to methane. *Proc. Natl. Acad. Sci. U. S. A.*, 112, 15809–15814 (2015).
80. Liu, Y., Chen, S., Quan, X. & Yu, H. Efficient Electrochemical Reduction of Carbon Dioxide to Acetate on Nitrogen-Doped Nanodiamond. *J. Am. Chem. Soc.*, 137, 11631–11636 (2015).
81. Varela, A. S. *et al.* Metal-Doped Nitrogenated Carbon as an Efficient Catalyst for Direct CO2 Electroreduction to CO and Hydrocarbons. *Angew. Chemie - Int. Ed.*, 54, 10758–10762 (2015).
82. Asadi, M. *et al.* Robust carbon dioxide reduction on molybdenum disulphide edges. *Nat. Commun.*, 5, 1–8 (2014).
83. Rosen, A. *et al.* No 主観的健康感を中心とした在宅高齢者における健康関連指標 に関する共分散構造分析 Title. *Teach. Teach. Educ.*, 12, 1–17 (2015).

84. Xie, S., Zhang, Q., Liu, G. & Wang, Y. Photocatalytic and photoelectrocatalytic reduction of CO2 using heterogeneous catalysts with controlled nanostructures. *Chem. Commun.*, 52, 35–59 (2016).
85. Inoue, T., Fujishima, A., Konishi, S. & Honda, K. Photoelectrocatalytic reduction of carbon dioxide in aqueous suspensions of semiconductor powders [3]. *Nature*, vol. 277 637–638 (1979).
86. Biswas, T. & Singh, A. K. Excitonic effects in absorption spectra of carbon dioxide reduction photocatalysts. *NPJ Comput. Mater.*, 7, (2021).
87. Byzynski, G. *et al.* The interplay between morphology and photocatalytic activity in ZnO and N-doped ZnO crystals. *Mater. Des.*, 120, 363–375 (2017).
88. Kumar, B. *et al.* Photochemical and photoelectrochemical reduction of CO2. *Annu. Rev. Phys. Chem.*, 63, 541–569 (2012).
89. Wang, P., Wang, S., Wang, H., Wu, Z. & Wang, L. Recent Progress on Photo-Electrocatalytic Reduction of Carbon Dioxide. *Part. Part. Syst. Charact.*, 35, 1–25 (2018).
90. Xu, S. & Carter, E. A. Theoretical Insights into Heterogeneous (Photo)electrochemical CO2 Reduction. *Chem. Rev.*, 119, 6631–6669 (2019).
91. Pawar, A. U., Kim, C. W., Nguyen-Le, M. T. & Kang, Y. S. General Review on the Components and Parameters of Photoelectrochemical System for CO2 Reduction with *in Situ* Analysis. *ACS Sustain. Chem. Eng.*, 7, 7431–7455 (2019).
92. Wang, S., Lin, J. & Wang, X. Semiconductor-redox catalysis promoted by metal-organic frameworks for CO2 reduction. *Phys. Chem. Chem. Phys.*, 16, 14656–14660 (2014).
93. Giannakoudakis, D. A., Seredych, M., Rodríguez-Castellón, E. & Bandosz, T. J. Mesoporous Graphitic Carbon Nitride-Based Nanospheres as Visible-Light Active Chemical Warfare Agents Decontaminant. *ChemNanoMat*, 2, 268–272 (2016).
94. Song, X., Yang, Q., Yin, M., Tang, D. & Zhou, L. Highly efficient pollutant removal of graphitic carbon nitride by the synergistic effect of adsorption and photocatalytic degradation. *RSC Adv.*, 8, 7260–7268 (2018).
95. Mao, J., Li, K. & Peng, T. Recent advances in the photocatalytic CO2 reduction over semiconductors. *Catal. Sci. Technol.*, 3, 2481–2498 (2013).
96. Decker, F. & Cattarin, S. Photoelectrochemical Cells | Overview. *Encycl. Electrochem. Power Sources*, 1–9 (2009) doi:10.1016/B978-044452745-5.00035-6.
97. Bonomo, M. & Dini, D. Nanostructured p-type semiconductor electrodes and photoelectrochemistry of their reduction processes. *Energies*, 9, (2016).
98. Chen, P., Zhang, Y., Zhou, Y. & Dong, F. Photoelectrocatalytic carbon dioxide reduction: Fundamental, advances and challenges. *Nano Mater. Sci.*, 3, 344–367 (2021).
99. Maeda, K. Photocatalytic water splitting using semiconductor particles: History and recent developments. *J. Photochem. Photobiol. C Photochem. Rev.*, 12, 237–268 (2011).
100. Li, H., Zhang, X. & MacFarlane, D. R. Carbon quantum dots/Cu2O heterostructures for solar-light-driven conversion of CO2 to methanol. *Adv. Energy Mater.*, 5, 1–6 (2015).
101. Chang, X. *et al.* Stable Aqueous Photoelectrochemical CO2 Reduction by a Cu2O Dark Cathode with Improved Selectivity for Carbonaceous Products. *Angew. Chemie - Int. Ed.*, 55, 8840–8845 (2016).
102. Zhang, M., Cheng, J., Xuan, X., Zhou, J. & Cen, K. CO2 Synergistic Reduction in a Photoanode-Driven Photoelectrochemical Cell with a Pt-Modified TiO2 Nanotube Photoanode and a Pt Reduced Graphene Oxide Electrocathode. *ACS Sustain. Chem. Eng.*, 4, 6344–6354 (2016).
103. Cheng, J. *et al.* Photoelectrocatalytic reduction of CO2 into chemicals using Pt-modified reduced graphene oxide combined with Pt-modified TiO2 nanotubes. *Environ. Sci. Technol.*, 48, 7076–7084 (2014).

104. Weng, B. *et al.* Bifunctional CoP and CoN porous nanocatalysts derived from ZIF-67: *In situ* grown on nanowire photoelectrodes for efficient photoelectrochemical water splitting and CO2 reduction. *J. Mater. Chem., A*4, 15353–15360 (2016).
105. Li, P. *et al.* The photoelectric catalytic reduction of CO2 to methanol on CdSeTe NSs/TiO2 NTs. *Catal. Sci. Technol.*, 4, 1070–1077 (2014).
106. Wang, Q. *et al.* Particulate photocatalyst sheets based on carbon conductor layer for efficient Z-scheme pure-water splitting at ambient pressure. *J. Am. Chem. Soc.*, 139, 1675–1683 (2017).

16

Soft Perovskite Semiconductors for Future Optical Electronics

Rashmi Yadav[1]* and Bhoopendra Yadav[2]

[1]*Department of Physics, Deenbandhu Chhotu Ram University of Science and Technology, Murthal, India*

[2]*Department of Physics, Institute of Science, Banaras Hindu University, Varanasi, India*

Abstract

This chapter investigates the non-linear optical properties of induced stained straddled-type InGaAsP/InP semiconductor-based nanoscale heterostructures by using the GAIN software package. We observed from the evaluated results a significant sound in the case of TE mode. Moreover, the non-linear optical properties such as optical gain, refractive index, differential gain, and anti-guiding factor have been evaluated for compressive and tensile strain InGaAsP/InP single quantum well (SQW)-based lasing nano-heterostructures in TE and TM (Transverse magnetic) mode, and an optical gain increase was found as we increase the introduced strain value; the maximum gain with lasing wavelength can be controlled by varying the strain value. The maximum value of material gain has been obtained at a wavelength of 1.46 μm whenever tensile strain is introduced, while for the compressive strain, the maximum material gain value has been achieved at 1.41 μm lasing wavelength. Hence, the nano-structure studied in this work is very useful as a light source for the optical fiber-based communication system functioning in the NIR (near-infrared) region due to less attenuation.

Keywords: Perovskite, lead free, semiconductors, photovoltaic devices, optoelectronics, thin film

16.1 Introduction

Since its initial presentation in 2009, perovskite solar cells (PSCs) have received a great deal of interest [1–5]. The field has advanced quickly since 2012, when two groups reported the first metal-halide perovskite solar cell power conversion efficiency (PCE). PSCs' attractive optoelectronic characteristics, such as a tunable bandgap [6], a high absorption coefficient [7], a low recombination rate [8], and a high mobility of charge carriers [9], have contributed to the quick growth of PSC research.

The single-junction PSC's power conversion efficiency (PCE) increased from 3% to a certified value of 25.6% [10, 11] within a decade, which is the greatest figure ever attained

**Corresponding author*: rashmiiyadav@gmail.com

Upendra Kumar and Piyush Kumar Sonkar (eds.) Smart Materials for Science and Engineering, (325–336)

for thin-film photovoltaics. Moreover, by utilizing additive and interface engineering techniques, PSCs' long-term operating stability has increased to more than 1,000 hours in full sunlight [12, 13]. As the next generation of photovoltaics, PSCs consequently have a lot of potential for adoption. To achieve extremely effective and stable PSCs, compositional engineering is crucial.

Additionally, the consequence of enormous efforts, significant advancements in the field of perovskite semiconductors, and their application in optoelectronic devices have been achieved. Our knowledge of this interesting semiconductor material has greatly improved. Despite the advancement in the industry, there are still many issues that need to be resolved, and we often find that after addressing one issue, another one pops up.

The newest wave of optoelectronic applications is the focus of this special issue on perovskite semiconductors. With regard to interface engineering, device longevity, cutting-edge applications, and physical comprehension of perovskite semiconductor applications' processes, we wish to make a brief summary of some key points. In this article, we will just use a few carefully chosen examples to illustrate the field's development as well as the intriguing research problems that still need to be solved.

The physical understanding of the processes involved in perovskite semiconductors, perovskite structure, characteristics, innovative applications, current challenges, and some key features of interface engineering, compositional engineering, and bandgap engineering effects are summarized in this chapter.

16.2 Perovskite Structure and Characteristics

In 1839, Lev Perovski discovered the mineral $CaTiO_3$ and ascertained its crystalline structure. Lev Perovski identified the crystalline structure of the material $CaTiO_3$ in 1839. That specific crystal structure is known as a perovskite and it describes a group of substances with the ABX_3 structure, shown in Figure 16.1.

After examining its structural characteristics and discovering a number of uses in solar cells, photodetectors, light-emitting diodes, transistors, and lasing applications, perovskite became well-known in the field of electronics. Due to perovskite's high PCE and low cost, which changed the efficiency of solar cells, it attracted particular attention in the

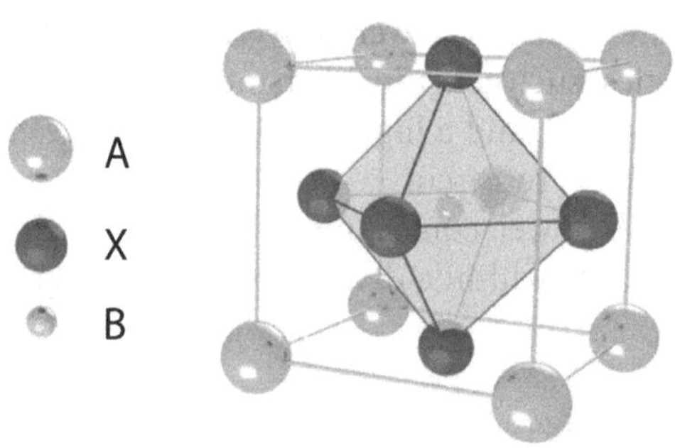

Figure 16.1 ABX_3 perovskite structure. Reproduced from Ref. [14] with permission from the Royal Society of Chemistry.

photovoltaics industry. The Goldschmidt tolerance factor t can be used to forecast the stability of a particular perovskite structure [15].

$$t = \frac{R_A + R_X}{\sqrt{2}(R_A + R_X)}$$

A is a larger monovalent cation, B is a smaller bivalent cation, and X is a monovalent anion that bonds with both A and B, where the ionic radii of the A, B, and X ions are R_A, R_B, and R_X, respectively. Structures that are hexagonal or tetragonal are produced when $t > 1$. A cubic structure is produced when $1 > t > 0.9$, an orthorhombic or rhombohedral structure is produced when $0.9 > t > 0.71$, and a perovskite crystal structure is not produced when t is 0.71 [16].

The Goldschimdt tolerance factor t makes calculations easier in the majority of situations; however, anomalies have been discovered in about 26% of situations, according to some literature. The following equation can be used to present the additional tolerance factor that Bartel *et al.* proposed [17].

$$t = \frac{R_X}{R_B} \text{-} n_A \text{-} \left(n_A \text{-} \frac{\frac{R_A}{R_B}}{ln\frac{R_A}{R_B}}\right)$$

where the ionic radii of the A, B, and X ions are R_A, R_B, and R_X, respectively, and n_A is the oxidation state of the A cation. When $t < 4.18$, the production of a perovskite structure is possible, and the likelihood of perovskite formation increases with a lower value.

The size of the individual moieties is crucial. In contrast, a much larger A would result in the formation of lower-dimensional face-sharing octahedral structures, which may not be conducive to effective solar photon harvesting due to factors like a wider bandgap. For instance, if the ionic radius of A is too small, it cannot effectively separate two B moieties, leading to an edge-sharing octahedron.

16.3 Composition Engineering Effects

It has been established that composition engineering of the perovskite is a successful strategy to improve effectiveness and stability. A simple compositional modification often results in improvements to a variety of stability problems. Conduction band minimum and valence band maximum in perovskite materials like $MAPbI_3$ are typically guided by the p orbital of Pb, the s and p orbitals of Pb, and the p orbital of I, respectively [18]. The A site cation does not contribute to the band edges, only the deep energy levels. The optical and optoelectronic characteristics are unaffected by the change in A-site. Consequently, without modifying the bandgap's effect on light-harvesting effectiveness, the A-site may be simply built to tune other physical features like thermal stability. FA is able to make more strong bonds with the PbX_6 octahedra than MA because it contains more H, which results in stronger H-bonding with PbX_6 [19]. Thus, the thermal stability increases [20]. The temperature stability of a solar cell can be increased by substituting an inorganic species, such as Cs+, for the A-site [21]. As non-perovskite phases frequently occur, $FAPbI_3$ and $CsPbI_3$'s thermodynamic instability prevents their employment in PSCs [22]. It is possible to boost the thermal stability of perovskite while maintaining its structural stability by using alloyed perovskite [23–25].

16.4 Interface Engineering Effects

As mentioned in the section on perovskite structure, halide perovskites can contain a number of distinct cation and anion species. It is well known that improving the stability and effectiveness of the perovskite through interface engineering is a successful strategy. Nonradiative recombination events, either in the perovskite material itself or at the perovskite/transport layer interface, are what primarily limit the device performance of perovskite solar cells [26]. In order to improve the film quality and prevent recombination at the perovskite interfaces, numerous interfacial engineering techniques have been used to create perovskite films [27]. Interfacial traps have been proven to occur in a variety of hole transporting materials (HTMs) by Droseros *et al.* It is interesting to note that the energy level of the various HTMs had no bearing on hole injection, which led to the conclusion that, given its speed and proximity to the thermalization timescale, hole injection must originate from hot states [28]. The high efficiency despite the interfacial trap generation is explained by spectroscopic studies in which interfacial electron trapping occurs on a considerably longer timescale than hole transfer. Another illustration is provided by Shibayama *et al.*, who demonstrated how a bathocuproine (BCP) buffer layer between the perovskite and the metal top electrode greatly affects the device characteristics in inverted perovskite solar cells. The n-type perovskite/metal contact was converted from a Schottky-type contact to an ohmic contact by this buffer layer [29]. The charge transport across interfaces and the hysteretic behavior shown in numerous current–voltage scans of perovskite solar cells are both significantly influenced by interfacial modifications. Phenylethylmmmonium (PEA) fomamidinium tin iodide, $PEA_{0.08}FA_{0.92}SnI_3$, is a 2-dimensional interface layer that has been described by Horn *et al.* as appearing to inhibition migration at the interface [30].

16.5 Bandgap Engineering Effects

Halide perovskites, having the structural formula ABX_3, can include a wide range of cation and anion species. The bandgap Eg can range from 1.2 to 3.6 eV depending on this composition, as seen in Figure 16.2a, for all first-generation lead and tin compounds [31]. For the area of optoelectronics, this variation in Eg is quite exciting since the bandgap may be adjusted to match the needs of each individual. For single-absorber solar cells, an ideal bandgap is about 1.3 eV; however, significantly greater bandgaps are preferred for wide gap cells used in tandem applications. The acceptable bandgaps for display applications for light emission range from 2 eV for red emission to 2.8 eV for blue emission (see Secs. V B and V C). The differences in the valence band (VB) and conduction band (CB) position, which are influenced by the positions of the atomistic levels, the size of the lattice, and lattice distortion, are what causes the changes in bandgap. According to band structure calculations [32], the hybridized metal s (Ms) and halide p (Xp) states are what make up the top of the VB, as seen in Figure 16.2b. The energetic positions of the atomic Ms and Xp orbitals, which are associated with the element's electronegativity, thus have a direct impact on the VB [with (Pb) (Sn) and (I) (Br) (Cl)]. Hence, a change in metal or halide component will lead to a shift in the VB, as depicted in Figure 16.2b-ii and b-iii. The bottom of the CB consists of metal p (Mp) and halide s (Xs) orbitals. It was discovered that the atomic p states of Sn (i.e., 4p) and Pb (5p) are rather similar in energy (Figure 16.2b-ii), suggesting that the orbital overlap is smaller for the conduction band and, as a result, hybridization is not as noticeable. This results in a more ionic bond in which the metal state dominates the CB. The lattice size, which is governed by the ionic radii of the various anions and cations, also influences the onsets of CB and VB. A smaller lattice will cause confinement effects, pushing the atomic levels slightly upward, affecting both VB and CB, as shown in Figure 16.2b-iv. Additionally, hybridization will rise, which primarily influences and raises the VB. The degree of hybridization will also be influenced by the bond angles. A cubic perovskite structure achieves the maximum orbital overlap, resulting in a narrow bandgap. The lattice

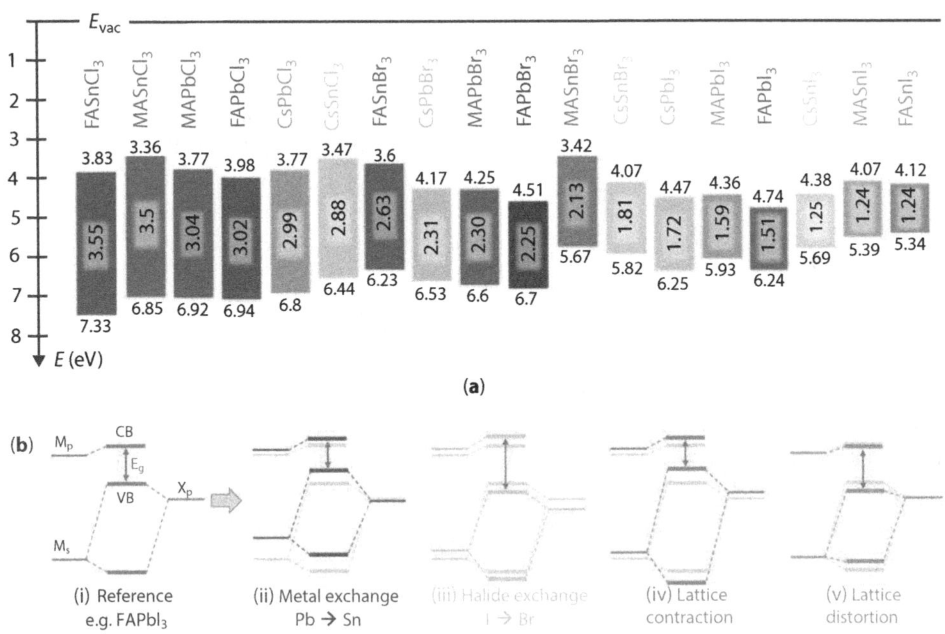

Figure 16.2 Changes in bandgap in the halide perovskite. (a) Measured changes in bandgap as a result of changes in perovskite chemistry. Reproduced with permission from Tao *et al.*, *Nat. Commun.* 10, 2560 (2019). Copyright 2019 Nature Publishing Group [31]. (b) Schematic depictions ((i), (ii) and (iii)) of the energy levels that depict the formation and ((iv) and (v)) show changes in composition, lattice size, or lattice distortion influence these levels of perovskite VB and CB from the atomic metal and halide states, respectively. Reproduced with permission from ref. [34]. Copyright © 2021, Author(s). This article is distributed under a Creative Commons Attribution (CC BY) license.

becomes deformed and the degree of hybridization decreases during a phase transition, such as when an ion with a large radius mismatch is introduced or the temperature is lowered to produce a tetragonal or orthorhombic structure. Moreover, the CB is presumably less impacted because of its stronger ionic character. This will result in a drop in the VB, as illustrated in Figure 16.2b-v, which will increase the bandgap. Naturally, pure ABX_3 perovskite structures can also be produced by combining different ions on the A, B, or X site and the bandgap can be gradually controlled by cooling melts containing the right amount of precursors [33].

16.6 Stability and Degradation Mechanism in Perovskite Solar Cells (PSCs)

For perovskite solar cells (PSCs) to be commercialized, their long-term stability is crucial [35, 36]. However, various factors contribute to the degradation of perovskite and other PSC components. To develop methods to stop such processes from occurring in the first place, it is crucial to study different decomposition mechanisms. Degradation mechanisms can be grasped by reconciling different experiments with theoretical calculations. As an example, the decomposition of perovskite by a metallic catalyst can result in the degradation of the other PSC component. The stability issues can be divided into two types: Issues with intrinsic stability—stability only resulting from the perovskite's molecular and crystallographic structure (ideally MAPbI3 is discussed).

Li *et al.* [37] identified four main external components as being in charge of the degradation mechanisms in perovskite solar cells. They are oxygen, heat, UV light, and moisture [38].

16.6.1 Moisture

It was shown that moisture affects perovskite solar cells in both favorable and unfavorable ways. Large crystals form as a result of the building-up process in film production and grain boundary movement under controlled humidity conditions, which also reduces the number of pinholes in the film [39]. Reconstruction of the film speeds up perovskite phase formation and crystallization [40]. A small amount of water aids in making the perovskite films dense and smooth [41]. Niu *et al.* measured the XRD (X-ray diffraction) and UV-VIS (ultraviolet–visible) absorption of the $CH_3NH_3PbI_3$ films [38, 42]. The beginning of the perovskite's disintegration under an environment of 80% humidity was attributed to the degradation process, which was studied using grazing incidence X-ray diffraction (GIXRD). A novel intermediate crystalline phase was formed as evidenced by the emergence of a new phase that was ascribed to neither $CH_3NH_3PbI_3$ nor PbI_2 [43]. Additionally, it was established that the water-induced breakdown of perovskite results in the formation of a new intermediate crystalline phase [44]. The interface between water molecules and perovskite materials was examined using *ab initio* molecular dynamic simulations, and it was discovered that surfaces with PbI_2 terminations are less vulnerable to water while those with CH_3NH_3I terminations tend to be soluble [45].

16.6.2 Oxygen

Oxygen molecules deteriorate perovskite materials in the work under illumination and a dry atmosphere, which caused $CH_3NH_3PbI_3$ to photoexcite electrons, which then reacted with oxygen molecules to generate superoxide (O_2^-). The disintegration of the perovskite material, which produced the water molecule, iodine, and PbI_2 in the process, is what causes the reaction between superoxide and the methylammonium cation [46].

16.6.3 Temperature

The perovskite layer deteriorates as a result of exposure to high temperatures. It is essential to comprehend this point because the perovskite layer requires an annealing step and because solar modules will eventually operate at high temperatures. The effects of high temperatures on $CH_3NH_3PbI_3$ and $CH_3NH_3PbI_{3x}Cl_x$ films were examined by Phillippe *et al.* [47].

Instead of using XRD to explain the film breakup, the scientists employed hard X-ray photoelectron spectroscopy (PES), which enables the chemical content to be determined regardless of the sample's crystallinity.

The performance of perovskite solar cells is impacted by temperature. In order to generate perovskite films, heat treatment that improved an obvious structural change at 85°C in $CH_3NH_3PbI_3$ is essential [48]. Investigation of perovskite characteristics under N_2, O_2, and 50% relative humidity in the dark for 24 hours revealed that the loss of the morphology of films handled under ambient conditions made them difficult to distinguish from pristine films. It was discovered that the samples treated with N_2 and O_2 had degraded as a result of morphological alterations. The thermal stability of perovskite material was established by Phillipe [49] and colleagues [50, 51].

16.6.4 UV Light or Light Stability

The device's stability as well as the stability of the light harvester material are both important factors in determining overall stability. By implementing specific tactics, factors like oxygen, moisture, and temperature can be decreased or eliminated, but solar illumination is obvious. UV sunlight can lead to degradation in perovskite solar cells, as it has in a number of other solar cell technologies [52–54]. According to numerous reports, $MAPbI_3$ does not degrade even after being illuminated for hundreds of hours. This is also supported by absorption spectra and other characterizations [55–57]. Even the photo-induced modifications in the perovskite can be advantageous for the PV characteristics; for instance, when illuminated by visible light, the ionic conductivity increases by several orders of magnitude [58]. Light-induced perovskite lattice growth improves device PEC [59]. Early reports of highly efficient perovskite solar cells were supported by the development of an electron transport layer made of TiO_2. Device architectures using this scaffold were developed by DSSC [60–62]. For perovskite solar cells, Leijtens *et al.* showed that this TiO_2 layer is vulnerable to UV-induced deterioration [63]. This was examined by measuring a 5-h efficiency decay curve for devices with and without encapsulation and a UV filter under 1 sun AM 1.5G illumination. Remarkably, the data revealed that the encapsulated devices age more rapidly than the non-encapsulated device.

16.7 Novel Applications

The rapid advancement of photovoltaic and light-emitting diode technology has generated a great deal of interest in metal halide perovskites in optoelectronic applications. The outstanding optoelectronic capabilities of metal halide perovskites are important for many applications; one intriguing property is the tunability of the bandgap via compositional engineering, and recombination and interfacial engineering are other crucial issues for perovskite light-emitting diodes (LEDs), in addition to their low-cost solution processability, extensive wavelength tunability, and broad substrate compatibility, which have made them a highly sought-after class of emerging semiconducting materials [64, 65]. These materials have recently been shown to be ideal for laser applications because of their high photoluminescence quantum yield, high photoluminescence quantum yield, low Auger recombination rate, large absorption coefficient, low defect density, high charge carrier mobility, long carrier diffusion length, and low Auger recombination rate. The development of perovskite lasers is still in its infancy, and the creation of electrically pumped lasers has not yet been achieved, despite the significant advancements in laser applications. In high-performance light-emitting applications like light-emitting diodes, organic–inorganic perovskites should also be used due to their outstanding material quality and very high photoluminescence quantum efficiencies (LEDs). Prior to being utilized as absorbers [66] in DSSC device architectures, organic–inorganic lead-based perovskites were originally studied for light-emitting applications. Certainly, additional applications are also of interest. Bruzzi *et al.* provide an intriguing application as proof of concept for the use of inorganic perovskites as clinical radiation dosimeters [67]. In addition to solar cells, a wide range of optoelectronic applications have been made possible by the exceptional optical characteristics of metal-halide perovskites. The use of perovskites for light-emitting applications is expected to attract a lot of research attention in the near future, opening up a new field of study for perovskites.

16.8 Conclusion

For laboratory-scale devices, the PSCs' efficiency quickly exceeded 25%. However, in order to be commercialized, stability and dependability must be established rapidly. In reality, the development of highly efficient, large-area devices that are stable over the long period is necessary for commercialization. Understanding the mechanisms for degradation is crucial for preventing it. PSCs have shown promising improvement in terms of their efficiency in power conversion, creating a competition to find ways to improve efficiencies, but until now, it has not been possible to fully understand all the pathways that lead to degradation. To further understand degradation mechanisms, more research should be conducted. The adoption of strategies to further increase the PSCs' efficiency and stability will be tremendously aided by this knowledge. To comprehend the mechanisms of PSC degradation, a variety of characterization techniques, including spectroscopy, electrochemistry, and electron microscopy, are used. For a better understanding of degradation, these techniques can be combined with appropriate theoretical support. In order to combat the instabilities, several techniques are used. It has been demonstrated that many strategies can improve multiple factors with just one strategy or one additional information.

References

1. Kojima, A., Teshima, K., Shirai, Y. & Miyasaka, T. Organometal halide perovskites as visible-light sensitizers for photovoltaic cells. *J. Am. Chem. Soc.*, 131, 6050–6051 (2009).
2. Grätzel, M. The light and shade of perovskite solar cells. *Nat. Mater.*, 13, 838–842 (2014).
3. Park, N.-G. *et al.* Towards stable and commercially available perovskite solar cells. *Nat. Energy,* 1, 16152 (2016).
4. Correa-Baena, J. P. *et al.* Promises and challenges of perovskite solar cells. *Science,* 358, 739–744 (2017).
5. Lu, H., Krishna, A., Zakeeruddin, S. M., Grätzel, M. & Hagfeldt, A. Compositional and interface engineering of organic-inorganic lead halide perovskite solar cells. *iScience, 23,* 101359 (2020).
6. Eperon, G. E. *et al.* Formamidinium lead trihalide: A broadly tunable perovskite for efficient planar heterojunction solar cells. *Energy Environ. Sci.,* 7, 982–988 (2014).
7. De Wolf, S. *et al.* Organometallic halide perovskites: Sharp optical absorption edge and its relation to photovoltaic performance. *J. Phys. Chem. Lett.,* 5, 1035–1039 (2014).
8. Stranks, S. D. *et al.* Electron-hole diffusion lengths exceeding 1 micrometer in an organometal trihalide perovskite absorber. *Science, 342,* 341–344 (2013).
9. Herz, L. M. *et al.* Charge-carrier mobilities in metal halide perovskites: Fundamental mechanisms and limits. *ACS Energy Lett.,* 2, 1539–1548 (2017).
10. NREL. Best Research-Cell Efficiency Chart https://www.nrel.gov/pv/cell-efficiency.html, (2021).
11. Jeong, J., Kim, M., Seo, J., Lu H., Ahlawat P., Mishra A., Yang Y., Hope A. M., Eickemeyer T. F., Kim M., Yoon Jin y., Choi W. I., Darwich P.B., Choi Ju S., Jo Y., Lee H. J., Walker B., Zakeeruddin M. S., Emsley L., Rothlisberger U., Hagfeldt A., Kim S. D., Gratzel M., Kim Y. J., Pseudo-halide anion engineering for α-FAPbI3 perovskite solar cells. *Nature,* 592, 381–385 (2021).
12. Zheng, X. *et al.* Managing grains and interfaces via ligand anchoring enables 22.3%-efficiency-inverted perovskite solar cells. *Nat. Energy,* 5, 131–140 (2020).
13. Liu, Z. *et al.* A holistic approach to interface stabilization for efficient perovskite solar modules with over 2,000-hour operational stability. *Nat. Energy,* 5, 596–604 (2020).
14. Y. Chen, L. Zhang, Y. Zhang, H. Gao, and H. Yan, "Large-area perovskite solar cells – a review of recent progress and issues." *RSC Adv.*, vol. 8, no. 19, pp. 10489–10508, (2018).
15. Li, Z., Yang, M., Park, J.-S., Wei, S.-H., Berry, J. J., and Zhu, K., Stabilizing Perovskite Structures by Tuning Tolerance Factor: Formation of Formamidinium and Cesium Lead Iodide Solid- State Alloys. *Chem. Mater.*, 28, 284–292 (2016). doi: 10.1021/acs.chemmater.5b04107.
16. Ma, X., Yang, L., Lei, K., Zheng, S., Chen, C., and Song, H., Doping in Inorganic Perovskite for Photovoltaic Application. *Nano Energy,* 78, 105354 (2020). doi: 10.1016/j.nanoen.2020.105354.
17. Bartel, C. J., Sutton, C., Goldsmith, B. R., Ouyang, R., Musgrave, C. B., Ghiringhelli, L. M., *et al.*, "New Tolerance Factor to Predict the Stability of Perovskite Oxides and Halides". *Sci. Adv.*, 5, (2019). doi:10.1126/sciadv. aav0693.
18. Yin, W.-J., Shi, T., and Yan, Y. "Unusual defect physics in CH3NH3PbI3 perovskite solar cell absorber". *Appl. Phys. Lett.*, 104, 063903 (2014). doi:10.1063/ 1.4864778.
19. Amat, A., Mosconi, E., Ronca, E., Quarti, C., Umari, P., Nazeeruddin, M. K., *et al.*, "Cation-Induced Band-Gap Tuning in Organohalide Perovskites: Interplay of Spin-Orbit Coupling and Octahedra Tilting". *Nano Lett.*, 14, 3608–3616(2014). doi:10.1021/nl5012992.
20. Pering, S. R., Deng, W., Troughton, J. R., Kubiak, P. S., Ghosh, D., Niemann, R. G., *et al.*, "Azetidinium lead Iodide for Perovskite Solar Cells". *J. Mater. Chem. A.*, 5, 20658–20665 (2017). doi:10.1039/c7ta07545f.

21. Ma, S., Bai, Y., Wang, H., Zai, H., Wu, J., Li, L., *et al.*, "1000 H Operational Lifetime Perovskite Solar Cells by Ambient Melting Encapsulation". *Adv. Energ. Mater.,* 10, 1902472 (2020). doi:10.1002/aenm.201902472.
22. Lin, J., Lai, M., Dou, L., Kley, C. S., Chen, H., Peng, F., *et al.,*"Thermochromic Halide Perovskite Solar Cells". *Nat. Mater.,* 17, 261–267 (2018). doi:10.1038/s41563-017- 0006-0.
23. Hou, F., Han, C., Isabella, O., Yan, L., Shi, B., Chen, J., *et al.*, "Inverted Pyramidally-Textured PDMS Antireflective Foils for Perovskite/silicon Tandem Solar Cells with Flat Top Cell". *Nano Energy,* 56, 234–240 (2019). doi: 10.1016/j.nanoen.2018.11.018.
24. Fuhua Hou, Biao Shi, Tiantian Li, Chenguang Xin, Yi Ding, Changchun Wei, Guangcai Wang, Yuelong Li, Ying Zhao, and Xiaodan Zhang, "Efficient and Stable Perovskite Solar Cell Achieved with Bifunctional Interfacial Layers". *ACS Appl. Mater. Interfaces, 11* (28), 25218-25226 (2019). DOI: 10.1021/acsami.9b06424.
25. Hou, M., Xu, Y., Zhou, B., Tian, Y., Wu, Y., Zhang, D., *et al.* Aryl Diammonium Iodide Passivation for Efficient and Stable Hybrid Organic Inorganic Perovskite Solar Cells. *Adv. Funct. Mater.*, 30, 2002366 (2020). doi:10.1002/adfm.202002366.
26. M. Stolterfoht *et al.*, "Visualization and suppression of interfacial recombination for high-efficiency large-area pin perovskite solar cells." *Nat. Energy,* 3, 847 (2018).
27. A. Fakharuddin, L. Schmidt-Mende, G. Garcia-Belmonte, R. Jose, and I. MoraSero, "Interfaces in perovskite solar cells." *Adv. Energy Mater.*, 7, 1700623 (2017).
28. N. Droseros, B. Dänekamp, D. Tsokkou, P. P. Boix, and N. Banerji, "Charge injection and trapping at perovskite interfaces with organic hole transporting materials of different ionization energies." *APL Mater.*, 7, 041115 (2019).
29. N. Shibayama, H. Kanda, T. W. Kim, H. Segawa, and S. Ito, "Design of BCP buffer layer for inverted perovskite solar cells using ideal factor." *APL Mater.,* 7, 031117 (2019).
30. J. Horn, M. Scholz, K. Oum, T. Lenzer, and D. Schlettwein, "Influence of phenylethylammonium iodide as additive in the formamidinium tin iodide perovskite on interfacial characteristics and charge carrier dynamics." *APL Mater.,* 7, 031112 (2019).
31. S. Tao, I. Schmidt, G. Brocks, J. Jiang, I. Tranca, K. Meerholz, and S. Olthof, "Absolute energy level positions in tin- and lead-based halide perovskites". *Nat. Commun.,* 10, 2560 (2019).
32. T. Umebayashi, K. Asai, T. Kondo, and A. Nakao, "Electronic structures of lead iodide based low-dimensional crystals". *Phys. Rev. B* 67, 155405 (2003).
33. J. Barrett, S. R. A. Bird, J. D. Donaldson, and J. Silver, "The Mössbauer effect in tin(II) compounds. Part XI. The spectra of cubic trihalogenostannates(II)". *J. Chem. Soc. A*, 3105 (1971).
34. L. Schmidt-Mende *et al.*, "Roadmap on organic-inorganic hybrid perovskite semiconductors and devices." *APL Mater.*, vol. 9, no. 10, pp. 1–82, 2021, doi: 10.1063/5.0047616.
35. Wang, J., Zhang, J., Zhou, Y., Liu, H., Xue, Q., Li, X., *et al.*, "Highly efficient all-inorganic perovskite solar cells with suppressed non-radiative recombination by a Lewis base". *Nat. Commun.*, 11, 177 (2020). doi:10.1038/s41467-019-13909-5.
36. Yang, Z., Wang, H., Huang, M., Liu, Y., Lv, Q., Lv, F., *et al.*, Hot Debate on Perovskite Solar Cells: Stability, Toxicity, High-Efficiency and Low Cost. *J. Energ. Chem.,* 53, 407–411 (2021). doi: 10.1016/j.jechem.2020.04.028.
37. Li, B., Li, Y., Zheng, C., Gao, D., Huang, W. Advancements in the stability of perovskite solar cells: Degradation mechanisms and improvement approaches. *RSC Adv.,* 6, 38079 – 38091(2016).
38. Berhe, T. A., Su, W-N., Chen, C-H., Pan, C-J., Cheng, J. H., Chen, H-M., Tsai, M-C., Chen, L-Y., Amare Aregahegn Dubale, A. A and Hwang, B-J, Organometal halide perovskite solar cells: Degradation and stability. *Energy Environ. Sci.*, 9, 323-356 (2016).
39. Jingbi You, Yang (Michael) Yang, Ziruo Hong, Tze-Bin Song, Lei Meng, Yongsheng Liu, Chengyang Jiang, Huanping Zhou, Wei-Hsuan Chang, Gang Li, and Yang Yang, "Moisture

assisted perovskite film growth for high performance solar cells". *Appl. Phys. Lett.*, 105, 183902 (2014).
40. Bass, K. K., Mcanally, R. E., Zhou, S., Djurovich, P.I., Thompson, M. E., Melot, B.C., Influence of moisture on the preparation, crystal structure, and photophysical properties of organohalide perovskites. *Chem. Commun.*, 50, 15819 – 15822 (2014).
41. C.-G. Wu, C.-H. Chiang, Z.-L. Tseng, M. K. Nazeeruddin, A. Hagfeldt, and M. Grätzel, "High efficiency stable inverted perovskite solar cells without current hysteresis." *Energy Environ. Sci.*, vol. 8, no. 9, pp. 2725–2733 (2015).
42. Niu, G. D., Guo, X. D., Wang, L.D. Review of recent progress in chemical stability of perovskite solar cells. *J. Mater. Chem. A.*, 3, 8970 – 8980 (2015).
43. Niu, G. D., Li, W. Z., Meng, F. Q., Wang, L. D., Dong, H. P., Qiu, Y. Study on the stability of CH3NH3PbI3 films and the effect of post modification by aluminium oxide in all solid – state hybrid solar cells. *J. Mater. Chem. A*, 2, 705 – 710, 2014.
44. Yang, W.S., Noh, J.H., Jeon, N.J., Kim, Y.C., Ryu, S., Seo, J., Seok, S.I. High – performance photovoltaic perovskite layers fabricated through intramolecular exchange. *Science*, 348(6240), 1234-1237 (2015).
45. Christians, J. A., Miranda Herera, P. A., Kamat, P. V. Transformation of the excited state and photovoltaic efficiency of CH3NH3PbI3 perovskite upon controlled exposure to humidified air. *J. Am. Chem. Soc.*, 137, 1530 – 1538 (2015).
46. Mosconi, E., Azpiroz, J. M., Angelis, F. De. Ab initio molecular dynamic simulation of methylam- monium lead iodide perovskite degradation by water. *Chem. Mater.*, 27 (13), 4885 – 4892 (2015).
47. Aristidou, N., Sanchez – Molina, I., Chotchuangchutchaval, T., Brown, M., Martinez, L., Rath, T., Haque, S. A. The Role of Oxygen in the Degradation of Methylammonium Lead Trihalide Perovskite Photoactive Layers. *Angew. Chem. Int. Ed.*, 54, 8208 – 8212, (2015).
48. B. Philippe, B.-W. Park, R. Lindblad, J. Oscarsson, S. Ahmadi, E.M.J. Johansson, H. Rensmo, Chemical and electronic structure characterization of lead halide perovskites and stability behavior under different exposures—a photoelectron spectroscopy investigation. *Chem. Mater.*, 27, 172 (2015).
49. Conings, B., Dijkoningen, J., Gauquelin, N., Babayigit, A., D'Haen, J., D'Olieslaeger, L., Ethirajan, A., Verbeeck, J., Manca, J., Mosconi, E., Angelis, F.D., Boyen, H. – G. Intrinsic Thermal Instability of Methylammonium Lead Trihalide Perovskite. *Adv. Energy Mater.*, 5(15), 1500477 (2015).
50. Philippe, B.; Park, B.-W.; Lindblad, R.; Oscarsson, J.; Ahmadi, S.; Johansson, E.M.J.; Rensmo, H. Chemical and Electronic Structure Characterization of Lead Halide Perovskites and Stability Behavior under Different Exposures—A Photoelectron Spectroscopy Investigation. *Chem. Mater.*, 27(5), 1720–1731 (2015).
51. M. Jørgensen, K. Norrman, F.C. Krebs, Stability/degradation of polymer solar cells. *Sol. Energy Mater. Sol. Cells*, 92, 686–714 (2008).
52. A. Hinsch, J.M. Kroon, R. Kern, I. Uhlendorf, J. Holzbock, A. Meyer, J. Ferber, Long-term stability of dye-sensitised solar cells. *Prog. Photovolt. Res. Appl.*, 9, 425–438 (2001).
53. C.R. Osterwald, A. Anderberg, S. Rummel, L. Ottoson, Degradation analysis of weathered crystalline-silicon PV modules, in: *Proceedings of the Photovoltaic Specialists Conference Record of the Twenty-Ninth IEEE.* 1392–1395 (2002).
54. Mei, A., Sheng, Y., Ming, Y., Hu, Y., Rong, Y., Zhang, W., *et al.*, Stabilizing Perovskite Solar Cells to IEC61215:2016 Standards with over 9,000-h Operational Tracking. *Joule*, 4, 2646–2660 (2020). doi: 10.1016/j.joule.2020.09.010.

55. Mei, A., Li, X., Liu, L., Ku, Z., Liu, T., Rong, Y., *et al.*,A Hole-conductor-free, Fully Printable Mesoscopic Perovskite Solar Cell with High Stability. *Science,* 345, 295–298 (2014). doi:10.1126/science.1254763.
56. Seo, S., Jeong, S., Bae, C., Park, N.-G., and Shin, H., Perovskite Solar Cells with Inorganic Electron- and Hole-Transport Layers Exhibiting Long-Term (≈500 H) Stability at 85 °C under Continuous 1 Sun Illumination in Ambient Air. *Adv. Mater.*, 30, 1801010 (2018). doi:10.1002/adma.201801010.
57. Kim, G. Y., Senocrate, A., Yang, T.-Y., Gregori, G., Grätzel, M., and Maier, J., Large Tunable Photoeffect on Ion Conduction in Halide Perovskites and Implications for Photodecomposition. *Nat. Mater.,* 17, 445–449 (2018). doi:10.1038/ s41563-018-0038-0.
58. Tsai, H., Asadpour, R., Blancon, J.-C., Stoumpos, C. C., Durand, O., Strzalka, J. W., *et al.* Light-induced Lattice Expansion Leads to High-Efficiency Perovskite Solar Cells. *Science,* 360, 67–70 (2018). doi:10.1126/science. aap8671.
59. B. O'Regan, M. Gratzel, A low-cost, high-efficiency solar cell based on dyesensitized colloidal TiO2 films. *Nature,* 353, 737–740 (1991).
60. M. Grätzel, Recent advances in sensitized mesoscopic solar cells. *Acc. Chem. Res.*, 42, 1788–1798 (2009).
61. G.K. Mor, O.K. Varghese, M. Paulose, K. Shankar, C.A. Grimes, A review on highly ordered, vertically oriented TiO2 nanotube arrays: fabrication, material properties, and solar energy applications. *Sol. Energy Mater. Sol. Cells,* 90, 2011–2075 (2006).
62. M. Grätzel, Dye-sensitized solar cells. *J. Photochem. Photobiol. C. Photochem. Rev.*, 4, 145–153 (2003).
63. T. Leijtens, G.E. Eperon, S. Pathak, A. Abate, M.M. Lee, H.J. Snaith, Overcoming ultraviolet light instability of sensitized TiO(2) with meso-superstructured organometal tri-halide perovskite solar cells. *Nat. Commun.*, 4, 2885 (2013).
64. A. Dubey *et al.*, "A strategic review on processing routes towards highly efficient perovskite solar cells." *J. Mater. Chem. A,* 6, 2406–2431 (2018).
65. M. Stolterfoht *et al.*, "Visualization and suppression of interfacial recombination for high-efficiency large-area pin perovskite solar cells." *Nat. Energy,* 3, 847 (2018).
66. C. C. Boyd, R. Cheacharoen, T. Leijtens, and M. D. McGehee, "Understanding degradation mechanisms and improving stability of perovskite photovoltaics." *Chem. Rev.,* 119, 3418–3451 (2019).
67. M. Bruzzi, C. Talamonti, N. Calisi, S. Caporali, and A. Vinattieri, "First proof-of-principle of inorganic perovskites clinical radiotherapy dosimeters." *APL Mater.,* 7, 051101 (2019).

17

Band Gap Engineering and Nanopatterning of Muscovite Mica by Low-Energy Ion Beams Applicable for Futuristic Microelectronics

Dipak Bhowmik[1]*, Joy Mukherjee[2] and Prasanta Karmakar[2]

[1]*Department of Physics, Indian Institute of Technology Kanpur, Kanpur, India*
[2]*Ion Beam Development and Application Section, RIB Group, Variable Energy Cyclotron Centre, HBNI, 1/AF Bidhannagar, Kolkata, India*

Abstract
Muscovite mica [$KAl_2(Si_3Al)O_{10}(OH)_2$] is a naturally available insulating layered smart material having thickness of each layer 1nm and band gap ~ 4.6 eV. In this chapter, we propose an approach to engineer the surface properties and bandgap of 2D nanosheets of muscovite mica in terms of nano patterning and defect formation by low energy ion beam sputtering. The ion beams induced sputtering and mass redistribution leads to well-defined nano-ripple pattern as well as defect formation on atomically flat mica surface. The band gap tailoring and nano ripple patterning on mica surface are elaborated for several ions bombardments and it has been observed that the projectile ion mass plays a key role for both ripple patterning and band gap engineering. Also, the presence of nanopatterns and chemical defects modifies the surface energy, which alters the wetting properties of the mica surface. Band gap narrowing of mica could be potentially useful for microelectronic devices, while nanostructured and chemically modified mica surface could be beneficial for self-cleaning and photocatalytic applications.

***Keywords*:** Smart material, mica, 2D nanosheet, low energy ion beam, nanopatterning, band gap engineering, wettability

17.1 Introduction

Smart materials are materials that can be modified in a controllable and reversible way in terms of several properties by some external stimuli such as a certain mechanical stress, pressure, electric field, magnetic field, or temperature. To use a material for a particular purpose, one needs to modify the material very precisely. The layer of naturally available muscovite mica is an atomically flat important two-dimensional (2D) material, which is used as a unique substrate for biomolecule deposition, atomic force microscopy calibration, deposition of plasma glycoproteins, membrane bilayers, etc. [1]. Four types of mica

**Corresponding author*: dipakbhowmik1@gmail.com

Upendra Kumar and Piyush Kumar Sonkar (eds.) Smart Materials for Science and Engineering, (337–360) © 2024 Scrivener Publishing LLC

sheets, namely, (1) muscovite, (2) phlogopile, (3) biotile, and (4) synthetic, are available in nature. Out of them, muscovite mica is transparent and flexible, and a thin sheet of atomic thickness can be extracted from bulk mica. In addition, mica is easily available and inexpensive, which is advantageous to researchers who may play with it and explore its various properties. Each mica layer consists of positively charged K atoms interconnected with a negatively charged aluminosilicate layer.

The self-organized nanoscale periodic pattern on a large area solid surface is very promising and potentially applicable in microelectronics and optoelectronic devices [2]. The periodic type of nano ripple structure can be formed on the mica surface just by sputtering the mica atoms using low-energy ion bombardment [3]. The chemical structure of mica can also be modified by both reactive and inert ion bombardment [4, 5]. Both physical modification in terms of pattern formation and chemical modification lead to the alteration of the super-hydrophilic surface to the hydrophobic surface useful for self-cleaning [6–8]. These types of nanostructures are very useful in the field of thin film growth, plasmonic, nanoscale magnetism, and catalysts [9, 10]. There are several methods to form a nanostructure on a solid surface; however, the low energy ion beam is treated as a very useful technique due to its precise control in its dimension [11]. The ion irradiation can manage defect engineering, surface morphology control, band gap tailoring, chemical compound formation, thin layer formation, and so on [12]. Since mica is an insulator with band gap ~ 4 eV, its band gap narrowing in semiconductor level would be interesting and useful for the application of futuristic microelectronics devices. Each mica layer can be considered as a 2D nanosheet of thickness 1 nm [5]. One of the most important and challenging task is to narrow down the band gap of 2D nanosheets due to quantum confinement effect [13]. Researchers introduced several techniques such as thermo-luminescence, exfoliation of layer, and swift heavy ion irradiation for the modification of dielectric property and band gap of muscovite mica [14, 15]. Not all of these techniques provide controlled tailoring of the band gap of mica. Thus, the band gap tailoring of muscovite mica in terms of modification of the topmost layer using low-energy and ultra-low-energy ion sputtering could be an interesting finding.

In this chapter, we have elaborated the nanopattern formation and band gap tailoring of mica surface in terms of low-energy ion irradiation. The ripple-type periodic pattern is formed on mica surface by inert Ar^+ and other reactive ions (O_2^+, NO^+, N_2^+, and C^+). The growth mechanism of anisotropic ripple structure on mica surface using 12 keV Ar^+ ion bombardment has been studied. The ion bombardment also modifies the mica surface chemically along with the nanopattern formation, which alters its wettability property from hydrophilic to hydrophobic in nature. The band gap tailoring of mica by modifying its several layers and monolayer is investigated, which illustrates that the band gap depends on its modifying layer. The present study concludes that the projectile mass plays a vital role for both ripple patterning and band gap tailoring.

17.2 Experimental Details

Raw muscovite mica sheets, available in Indian mica mines, are collected from the local market. A mica sheet is cut into small (1 cm × 1 cm) pieces, and are cleaved by adhesive tape to get the clean and atomically flat surface. Each mica sample of the same thickness (1 mm) is irradiated by 12–14 keV Ar^+, O_2^+, NO^+, N_2^+, and C^+ ions with varying ion fluences from 1×10^{17} ions/cm^2 to 1×10^{18} ions/cm^2 using a 2.4-GHz ECR ion source-based low-energy ion beam system at VECC Kolkata. The total energy of a homo-nuclei molecular ion like O_2^+ and N_2^+ is equally shared between its atoms, whereas in case of hetero-nuclei NO^+ ion bombardment, the distribution of the ion energy depends on the mass of atoms (i.e., 6.53 keV N^+ and 7.47 keV O^+ for 14 keV NO^+) [16]. The details of the experimental setup for the ion beam irradiation experiment and selection of ion species are schematically shown in Figure 17.1. The low-energy ion beam system comprises an ECR ion source, accelerating system, analyzing dipole magnet, and the target chamber [11]. The Electron Cyclotron Resonance (ECR) contains a plasma chamber, axial and radial permanent magnets, a microwave source, and a gas feeding system. The plasma is formed by injecting gas into the ECR plasma chamber in terms of microwave heating. The creation of circular motion of electrons in the plasma chamber has taken place due to the Lorenz force. The frequency with which the electrons gyrate is $\omega_{ECR} = \frac{eB}{m}$, where e is the electronic charge, B is the magnetic field, and m is the mass of the electron. When the frequency of electron gyration matches with the external microwave frequency (RF), the resonance takes place in the system. In this condition, the gas transfers into plasma of singly, doubly, and triply ionized state. As the microwave generates heat in the chamber, the plasma chamber and the ECR body are connected with low conductivity water (LCW) for cooling. The ion produced in the plasma chamber is extracted by applying accelerating potential between the plasma chamber and extraction plasma electrode. An Einzel lens is used to focus the ion and transport it to the analyzing bending magnet. The analyzing bending magnet is a very crucial part of the ion beam extraction system as it filters the pure ion beam by selecting the A/q ratio. Here, A is the mass of the ion and q is the charge state of the ion. The typical mass spectrum of 5 keV ions with Ar gas plasma is shown in the inset of Figure 17.1. As pure Ar

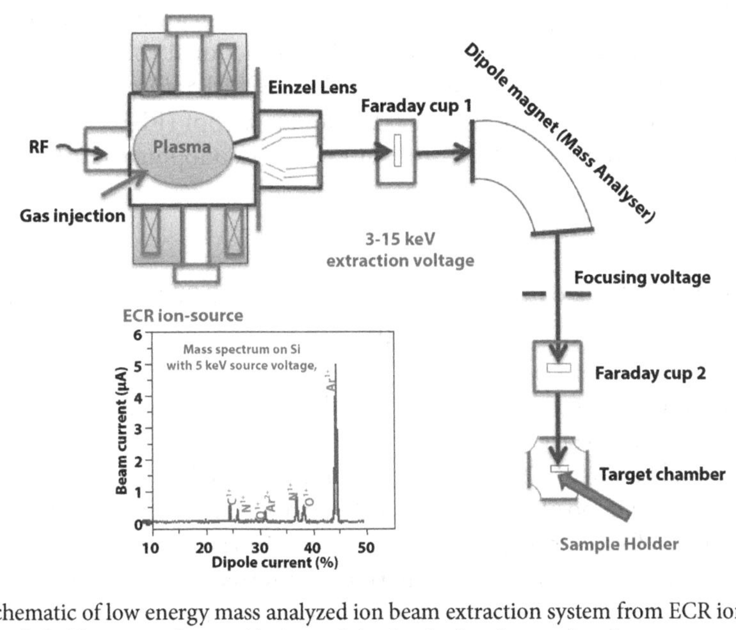

Figure 17.1 Schematic of low energy mass analyzed ion beam extraction system from ECR ion source and mass spectrum of 5 keV source voltage with Ar gas injection in the plasma chamber.

gas plasma is generated in the chamber, the current of Ar^+ ion is maximum and common impurities like C^+, O^+, and N^+ are present [17], which are filtered in the analyzing magnet.

In this chapter, several characterization techniques like atomic force microscopy (AFM) for surface morphology, X-ray photoelectron spectroscopy (XPS) for surface chemical analysis, and a UV-visible spectrophotometer for optical measurement are used. Here, the nanoscope Multi-Mode V has been used for AFM in tapping mode. An Omicron Multiprobe (Omicron Nano Technology, UK) has been used for XPS in which a monochromatic Al K_α source of X-ray is used to provide a photon energy of 1486.6 eV, and the hemispherical analyzer is used to calculate the kinetic energy of the photoelectrons [18]. Corresponding X-ray photoelectron spectroscopy data were characterized by CASA peak fit software [19] using the Marquart Levenberg optimization algorithm [20]. The entire system is maintained in high vacuum ($\sim 10^{-10}$ mbar) to avoid external impurities in the spectrum. The optical characterization of the ion-modified mica surfaces was investigated in both transmittance as well as reflectance mode throughout the wavelength region 200–1100 nm by a PerkinElmer Lambda 750 UV/VIS and Shimadzu 360 UV-IR spectrometer, respectively. Further, the local current voltage characteristics (I–V) of the mica surfaces are studied using a conducting Atomic Force Microscope (C-AFM, Bruker Multimode V).

17.2.1 Simulation Methodology

The vacancy, defect, and penetration depth of the projectile ions are estimated by Monte Carlo simulation using SRIM(BCA) [21]. The calculations are carried out using full cascade assumptions and the density of mica was considered as 2.73 gm/cm^3. The curvature-dependent erosive coefficient considering mass redistribution and sputtering for 12 keV Ar^+ ion-bombarded mica surfaces with ion incidence angles is estimated using TRIM simulation. The sputtering yield has been calculated considering the stoichiometric ratio of all the elements in mica [22].

The band structure of the pristine and defect-induced mica was estimated theoretically using first-principles calculations based on density functional theory (DFT) by employing the plane wave pseudo-potential in QUANTUM ESPRESSO code [23]. For the use of exchange co-relation function, the Perdew–Burke–Ernzerhof (PBE) version of the generalized gradient approximation (GGA) with ultra-soft pseudopotential (USPP) is taken as consideration [24].

17.3 Nanopattern Formation on Mica Surface and Its Wettability Property by Low-Energy Ion

Low-energy ion beam sputtering (IBS) is one of the important tools for the pattern formation on large-area solid surface at the nanometer scale. This type of periodic nanostructure could be a potential candidate in the field of thin film growth, magnetic anisotropy study, substrate for biological deposition, etc. [8, 10, 25]. Navez and co-workers first observed the periodic nanopattern formation on the glass surface in 1962 by 4 keV air ions [26]. However, there was no clear understanding of those periodic patterns among the community. Bradley and Harper [27] proposed a breakthrough theoretical model for understanding the pattern formation, which advances the field of nanopattern formation. This section is divided into three parts. First, the pattern formation and growth mechanism on mica surface will be discussed. Second, the ion-induced wettability modification will be emphasized. Finally, the role of projectile mass in pattern formation will be highlighted.

17.3.1 Ripple Pattern Formation and Its Growth Mechanism by 12 keV Ar^+ Ion Sputtering on Mica Surface

Figure 17.2 represents the surface morphologies of pristine and 12 keV Ar^+ ion-bombarded mica surface showing the ripple-like periodic structure on mica surface with varying ion fluences (2×10^{17} to 1×10^{18} ions/cm^2) at oblique angle incidence (60°). The pristine mica surface is nearly atomically flat with nominal rms roughness 0.2 nm (Figure 17.2a). The surface morphologies of ion-bombarded mica surfaces show that there is no such well-defined periodic pattern formation up to ion fluence 2×10^{17} ions/cm^2. However, well-defined alternative descending and ascending step ripples are formed at ion fluence 3×10^{17} ions/cm^2, which may be considered as the threshold ion fluence for the ripple initiation. The alignment of the ripples' wave vector is parallel to the ion beam incidence direction, which is called parallel mode ripples. The Fast Fourier Transform (FFT) of each AFM image is also shown in the inset of each image, which indicates parallel mode ripple. We have also performed the experiment with other ion incident angles like 20°, 30°, 50°, and 70°, but well-defined ripples are not found to form (the AFM images of those mica surfaces are not shown). To explain the ripple pattern formation around 60° for multi-elemental mica surface, we will introduce the Bradley and Harper (BH) theory [27]. According to the BH theory, the energetic ions sputter the surface atoms depending on local curvature. The curvature-dependent nonuniform sputtering of elements of the substrate generates surface instability.

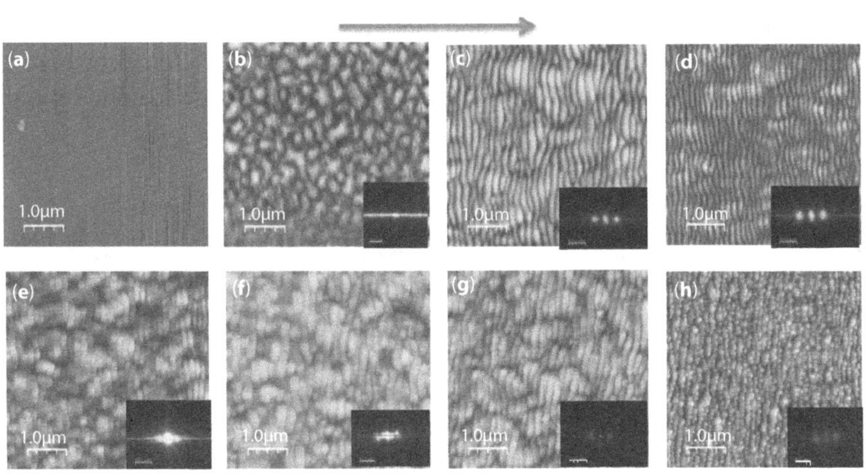

Figure 17.2 AFM images of (a) pristine and 12 keV Ar^+ bombarded mica surfaces at oblique incidence with fluence (b) 2×10^{17}, (c) 3×10^{17}, (d) 5×10^{17}, (e) 6×10^{17}, (f) 8×10^{17}, (g) 9×10^{17}, and (h) 1×10^{18} ions/cm^2. The Fast Fourier Transform (20 µm^{-1} × 20 µm^{-1}) spectra are shown in the corner of each AFM images. The ion beam direction is indicated by arrows.

On the other hand, re-arrangement of the surface atoms takes place, simultaneously, due to the momentum transfer from projectile ions to the target atoms. This leads to smoothening of the surface. The combined effects of sputtering and smoothening develop a ripple-like structure on the surface. The formation of a ripple pattern can be well understood from the height evolution equation as given below [28]:

$$\frac{\partial h}{\partial t} = -\vartheta_0 + \frac{\partial \vartheta_0}{\partial \theta}\frac{\partial h}{\partial x} + S_{11}\frac{\partial^2 h}{\partial x^2} + S_{22}\frac{\partial^2 h}{\partial y^2} - B\nabla^4 h \quad (17.1)$$

where ϑ_0 is the erosion rate and S_{11} and S_{22} are the curvature-dependent coefficient. The above equation can be solved in Fourier space with wave vectors k_x and k_y. For smooth surface having an initial amplitude $A_0(k_x, k_y)$, the evolution of h can be written as

$$h(x,y,t) = -\vartheta_0 t + \sum_{k_x,k_y} A_0(k_x,k_y)exp\left\{i\left[k_x(x+\gamma t)+k_y y\right]+rt\right\} \quad (17.2)$$

where $\gamma = \frac{\partial \vartheta_0}{\partial \theta}$ and r is the amplification factor as given by

$$r\left(k_x,k_y\right) = -S_{11}k_x^2 - S_{22}k_y^2 - B(k_x^2+k_y^2)^2 \quad (17.3)$$

From Equation 17.3, the amplitude of Fourier components with $r>0$ will grow exponentially with time. As B is always +ve, the amplification factor will be positive over some range where either S_{11} or S_{22} is negative, which leads to the growth of those wave vectors in a particular direction. S_{11} or S_{22} consists of three contributions like sputtering, mass redistribution, and ion implantation. However, the contribution of ion implantation is very small and hence neglected [29]. Thus, in the present case, S_{11} and S_{22} are calculated considering sputtering and mass redistribution effect and plotted with different angles and are shown in Figure 17.3a and b. The details of calculation of S_{11} or S_{22} can be found elsewhere [25, 29]. It is clear from Figure 17.3 that S_{11} is more negative around 60° ion incidence and the wave

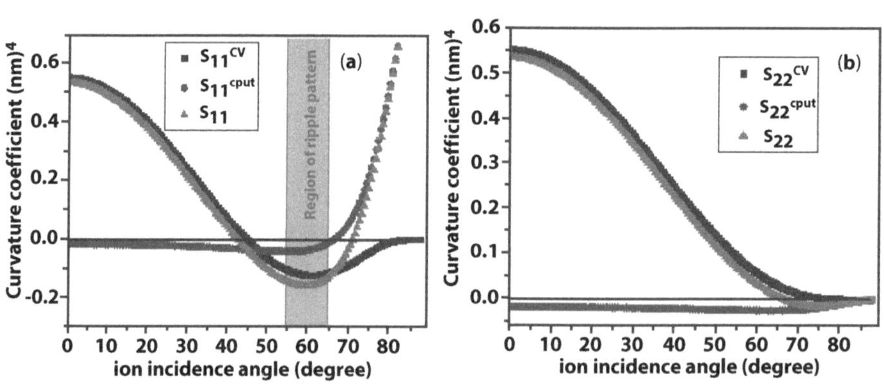

Figure 17.3 Curvature-dependent coefficient S_{11} and S_{22} with ion incidence angle showing the region of ripple pattern formation.

vectors along the x-direction will grow, resulting in the x-directional ripple pattern formation as observed in Figure 17.2.

To study the in-depth ripple growth mechanism, the dynamics of the ripple is investigated in terms of scaling as an alternative tool for explaining the nanopattern dynamics. The nanopattern evolution is the effect of random fluctuations of height as well as smoothening or roughening on the surface. The height fluctuation $(h(\vec{l},t))$ at the position $\vec{l}$ is characterized by the temporal evolution of interface width, which is rms roughness. It is defined as $W(L,t) = \langle [h(\vec{l},t) - \langle h(\vec{l},t) \rangle]^2 \rangle^{1/2}$, where the angular brackets symbolize averaging over all the length scale l within a system size L. According to the Family–Vicsek (FV) equation [30], W scales with time as t^{β} at the length scale after correlation length (ξ), where β is the growth exponent. The correlation length (ξ), calculated from the width of autocorrelation function $C(l) = \langle h(l)h(0) \rangle$, is the length at which $C(\xi) = C(0)/e$ [31]. ξ scales with time as $t^{1/z}$, where $1/z = \beta/\alpha$ is termed as dynamic exponent. The roughness exponent (α) can be calculated from the structure factor as extracted from the power spectral density (PSD) function, which is defined as $S(k,t) = \langle \tilde{h}(k,t)\tilde{h}(-k,t) \rangle$. Here, $\tilde{h}(k,t)$ denotes the Fourier transform of $h(l,t)$ and k denotes the spatial frequency. According to the Family–Vicsek relation, the structure factor scales as

$$S(k,t) = k^{-(2\alpha+1)} s(kt^{1/z})$$

with the scaling function, $s(u) \sim \begin{cases} \text{const.} & \textit{if } u \gg 1, \\ u^{(2\alpha+1)} & \textit{if } u \ll 1. \end{cases}$

The roughness exponents along the parallel (x) and normal (y) direction to the ion beam projection are extracted from the one-dimensional structure factor $S(k_{x,y}) \sim k_{x,y}^{-(2\alpha_{x,y}+1)}$ in the corresponding direction [32]. From the above scaling equation, one can estimate the values of scaling exponents (α, β, z) from the analysis of AFM images for a particular system of surface growth.

For the deeper understanding of growth mechanism, the dimension of the nanostructure, i.e., amplitude and wavelength of the ripple, is very important. The wavelength of the ripple structure is calculated from the height profile of AFM images and plotted with ion fluence (time of irradiation) in Figure 17.4a. It is observed that the wavelength increases slowly first up to a certain fluence, and after that, it increases rapidly with exponent $n = 0.43 \pm 0.04$. This phenomenon is called coarsening of ripple growth, which is associated with non-linear effect as described by Garcia *et al.* [33]. The wavelength of the ripple can also be calculated from the peak of 1D power spectral density function (as described later in Figure 17.5). To study the dynamic scaling behavior of the ripple pattern, the rms roughness (W), correlation length along the ripple wave vector direction (ξ_x), and perpendicular to the ripple wave vector (ξ_y) are extracted from corresponding AFM images. The rms roughness is plotted with ion fluence and is shown in Figure 17.4b. The growth exponent (β) is calculated from linear fitting of the plot, which is 0.51 ± 0.04. This indicates the roughening behavior of the ripple structure with the ion fluences. The correlation lengths along the ripple direction and perpendicular to the ripple direction are plotted with ion fluence as shown in Figure 17.4c and d, respectively. The correlation length along the ripple direction illustrates a monotonic increase with the dynamic exponent $(1/z_x) = 0.42 \pm 0.08$, which indicates the lateral growth

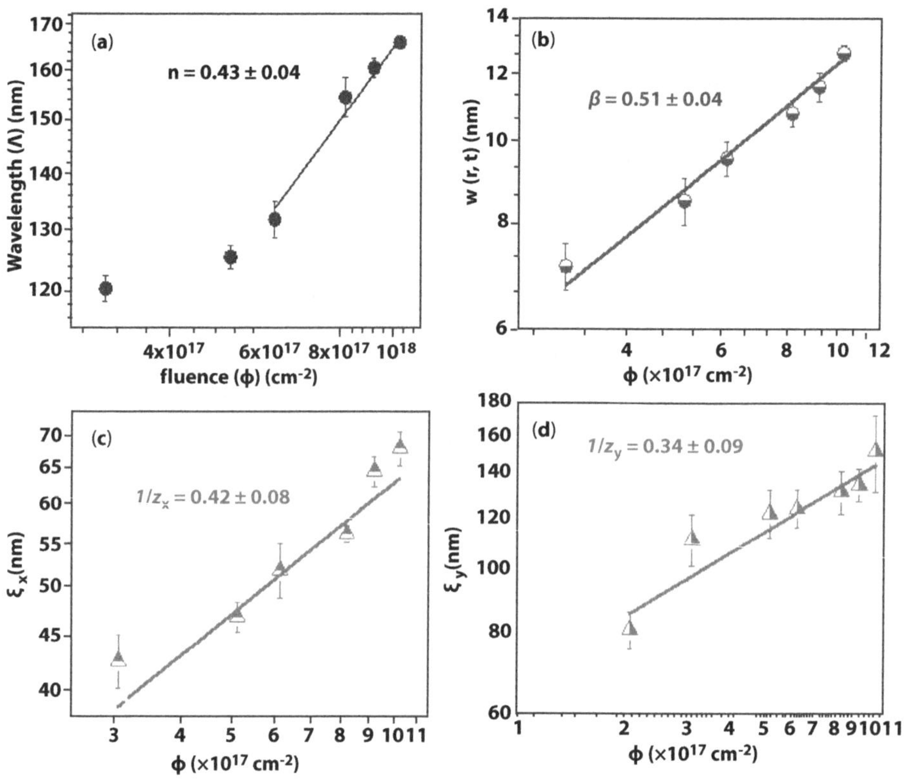

Figure 17.4 Variation of (a) ripple wavelength and (b) surface rms roughness with ion fluence. (c) Lateral and (d) longitudinal correlation length for the mica surface with well-defined ripple pattern. Reprinted from [25], Copyright (2019), with permission from Elsevier.

of the ripple along the wave vector direction. On the other hand, the correlation length perpendicular to the ripple direction also increases with ion fluence but slightly slower compared to parallel correlation length with dynamic exponent $(1/z_y) = 0.34 \pm 0.09$. These exponents give us an idea about the dynamic growth of the system.

To further extract the roughness exponent, the power spectral density function along ripple direction ($S_x(k_x)$) and perpendicular to the ripple direction ($S_y(k_y)$) is plotted with wave vector (Figure 17.5a and b). The structure factor $S_x(k_x)$ shows a peak in the curve at k_x^* corresponding to the ripple wavelength. At the large value of k_x, all the curves tend to coincide with power law behavior $S(k_x) = k_x^{-(2\alpha_x+1)}$ having the same slope $m_x = 3.54 \pm 0.42$. However, for the perpendicular structure factor, the curves coincide at a slightly larger value of wave vector and follow the same power law behavior with slope $m_y = 3.56 \pm 0.2$. The rescaled structure factors $S(k,t)k^{(2\alpha+1)}$ versus $kt^{1/z}$ for the different fluences are plotted to confirm the same power law behavior as shown in Figure 17.5c and d. All the curves collapse in a single line and the average slope of the collapsed structure factors is found to be $m_x = 3.5 \pm 0.15$ and $m_y = 3.4 \pm 0.16$ along the ripple direction and perpendicular to the ripple direction, respectively. The roughness exponents are calculated from these slopes as $(m = 2\alpha + 1)\, \alpha_x = 1.25 \pm 0.07$ and $\alpha_y = 1.20 \pm 0.08$. This indicates that the structure is anisotropy. The degree of anisotropy (ζ) was introduced by Vivo *et al.* [34] determined from

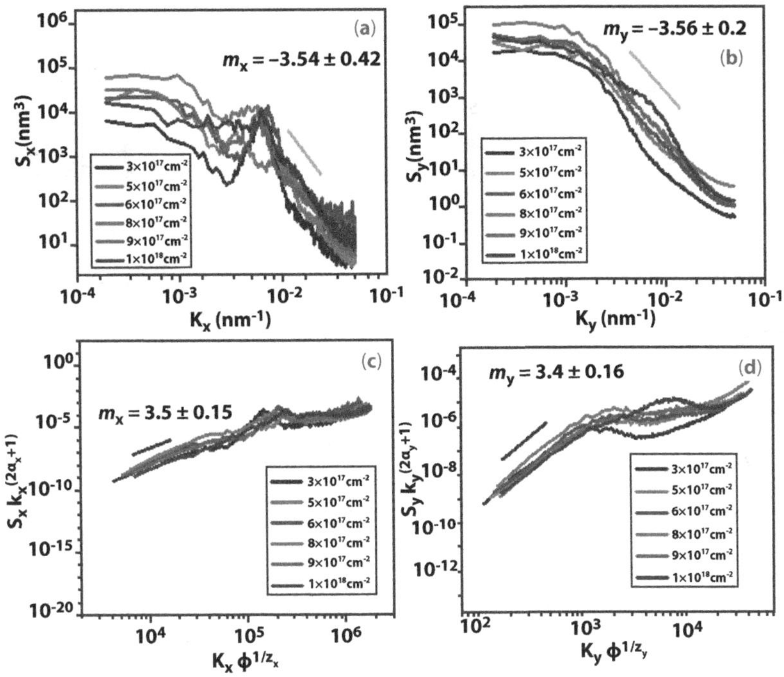

Figure 17.5 Structure factors S_{xy} along (a) ripple direction and (b) normal to the ripple direction. (c and d) Collapsed structure factors along parallel and normal to ripple direction calculated from scaling coefficients. Reprinted from [25], Copyright (2019), with permission from Elsevier.

the ratio of two roughness exponents α_x and α_y. In the present case, ζ is found marginally greater than 1, which indicates the weak anisotropy of ripples. The value of $\frac{\alpha_x}{z_x}$ is found to be 0.52, which is close to the β value 0.51 within the error limit. However, the value of $\frac{\alpha_y}{z_y}$ (0.40) is far close to β for the perpendicular direction, which may be due to the ripple growth along the parallel direction and the inconsistent correlation length along the perpendicular direction.

The main reason for the ripple pattern formation on a solid surface is the generation of surface instabilities during ion bombardment by several phenomena like initial perturbation on the surface [35], presence of contamination [36], unequal sputtering [16, 37], reactive impurities with ion beam [17], redistribution of surface atoms [38], and local curvature-dependent sputtering [27]. To investigate the reason for the ripple pattern formation for the present case, the surface chemistry and sputtering yield of constituent elements are studied in terms of X-ray photoelectron spectroscopy and SRIM simulation. Figure 17.6a shows the layered structure of the crystalline muscovite mica surface, which indicates that positively charged K atoms are attached to a negatively charged aluminosilicate layer. The XPS survey scan of pristine and ion-modified mica with ion fluence 5×10^{17} and 1×10^{18} ions/cm^2 is shown in Figure 17.6b. The survey scan indicates the presence of all the elements in mica. The upper K atoms are sputtered most as obvious from XPS spectra, and ion-modified mica surfaces adsorb environmental C due to the chemical exposure of the surface. This

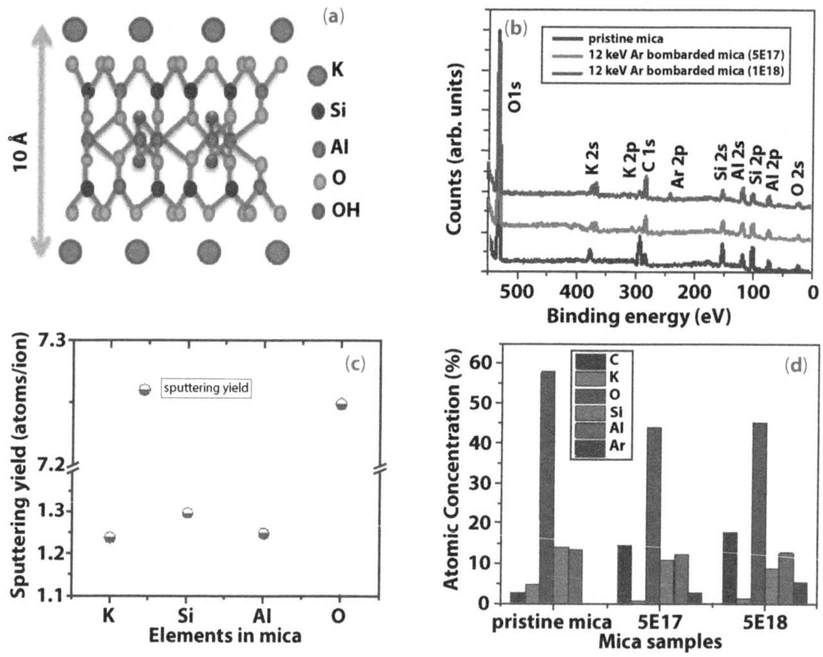

Figure 17.6 (a) Schematic diagram of muscovite mica showing layer nanosheet. (b) XPS survey scan of pristine and ion-bombarded mica showing all elements of mica. (c) Sputtering yield of mica elements calculated from SRIM simulation and (d) percentage of atomic concentration of mica elements for pristine and ion-bombarded surfaces calculated from XPS analysis.

type C adsorption was also observed earlier after ion bombardment [5, 39]. The intensities of several elements of mica in XPS spectra indicate the unequal sputtering during ion bombardment. To justify this observation, the sputtering yield of several elements of mica is calculated from SRIM simulation and is shown in Figure 17.6c. This shows the different sputtering yield of several elements due to their different mass. The percentage of atomic concentration of several elements of mica (Figure 17.6d) as calculated from XPS spectra also indicates the unequal sputtering. As mica is a multi-elemental compound material, the elements in mica are sputtered at different rates during ion bombardment. Due to preferential sputtering of mica elements and rearrangement of atoms using ion bombardment, surface instability is generated, resulting in the formation of a well-defined ripple pattern after long-term ion bombardment.

The morphological evolution of mica ripples has been investigated in terms of the dynamic scaling parameters as well as surface chemical changes. A well-defined parallel mode ripple pattern is formed on multi-elemental mica surface after certain fluence due to preferential sputtering of several elements of mica. The surface rms roughness and ripple wavelength both increase with ion fluence. The correlation length along both directions also increases with ion fluence. However, the faster growth of correlation length along the ripple direction is observed. The scaling exponents α, β, and z are calculated, which gives the quantitative growth dynamics of the system. The curvature-dependent coefficient considering mass redistribution, sputtering, and ion implantation is estimated for the present system, and it is observed that mass redistribution plays a major role for ripple patterning.

17.3.2 Surface Wettability Property of Ion-Bombarded Mica Surface

Surface wettability is measured in terms of contact angle, and it depends on surface chemical changes and nano- or microstructure formation on that surface. Interestingly, low-energy ion beams could modify the wetting properties of mica surface due to its physicochemical modification. The wetting property of ion-modified mica is studied in terms of water contact angles. The schematic of contact angle measurement is shown in Figure 17.7a and the droplet of water on pristine and ion-modified mica is shown in Figure 17.7b and c. The contact angles are measured after they reach equilibrium as shown in Figure 17.7d. The contact angles of pristine and 12 keV Ar^+ bombarded mica surfaces with different ion fluences are measured and plotted in Figure 17.7e. It is observed that the water contact angle of pristine mica is ~5°, i.e., it is super hydrophilic in nature. However, it increases for ion-modified surfaces and the contact angles of ion-modified mica surface are greater than 90°, which are considered as hydrophobic. The contact angles of all the ion-bombarded mica surfaces are from 100° to 110° within the error bar. The ion beam-induced chemical and physical modification leads to a decrease in the surface energy of the surface, resulting in the transformation of hydrophilic to hydrophobic surface. It is observed from the XPS analysis that the ion-modified surfaces adsorb carbon from the environment, which is mainly responsible for the decrease in surface energy [6]. The ripple-like structure on the surface also influences the contact angle. The previous study shows that the surface roughness of the microstructure may influence the wettability property [40]. However, the nanoscale-level structure is less effective than surface chemical changes for the modification of wettability property. In the present case, the adsorption of C on the surface after ion bombardment as observed from XPS study leads to a decrease in surface energy. The contact angle increases with increasing ion fluence as the adsorption of C in the surface is greater for larger fluence

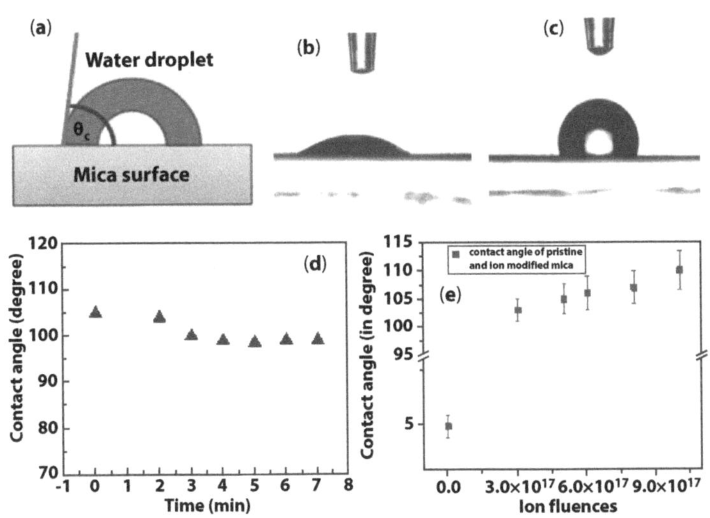

Figure 17.7 (a) Schematic of the contact angle measurement. Images of droplets of water on (b) pristine mica and (c) 12 keV Ar^+ irradiated mica with fluence 5×10^{17} ions/cm^2. (d) Plot of equilibrium contact angle for 12 keV Ar^+ bombarded mica surface (fluence 5×10^{17} ions/cm^2). (e) Variation of contact angle with ion fluence for ion-bombarded mica surfaces. Panels (a) to (d) are reprinted from [8], Copyright (2018), with permission from Elsevier.

as indicated in Figure 17.6d. This type of transformation of hydrophilicity to hydrophobicity is very important for self-cleaning surface.

17.3.3 Projectile Ion Mass-Dependent Nano Ripple Patterning on Mica Surface

The nanopatterning on solid surface by a low-energy ion beam depends on several parameters like projectile ion, ion energy, ion mass, incident angle, substrate, etc. Out of them, projectile ion mass plays an important role for ripple patterning on solid surface. In this sub-section, we have addressed the role of projectile mass in nanopatterning on mica surface. The clean mica surfaces are irradiated by 14 keV C^+, N_2^+, NO^+, O_2^+, and Ar^+ ion with increasing mass to observe the mass-dependent ripple patterning effect. The AFM images of ion-bombarded mica surface with 14 keV C^+, N_2^+, NO^+, O_2^+, and Ar^+ ion are shown in Figure 17.8a–e at ion fluence 7×10^{17} ions/cm^2 and ion incidence angle 60°. It is observed that C^+ is unable to form a well-defined ripple structure on the mica surface. The well-defined ripple structure starts from N_2^+ ion and the pattern becomes prominent with increasing ion mass. The parallel mode ripple structure is confirmed from Fast Fourier Transform (FFT) images as shown in the inset (right corner) of each AFM image. The quantitative analysis is carried out to understand the ripple dimension. We have extracted the rms roughness value and ripple wavelength from AFM images for different ion species. As we know that the ion loses its energy in terms of nuclear and electronic energy loss and for low energy ion bombardment, nuclear energy loss is mainly dominated, we have calculated nuclear energy loss for several ions from SRIM simulation. The XPS analysis of ion-modified mica surfaces is also done to observe the post-bombardment effect. The rms roughness of ion-modified

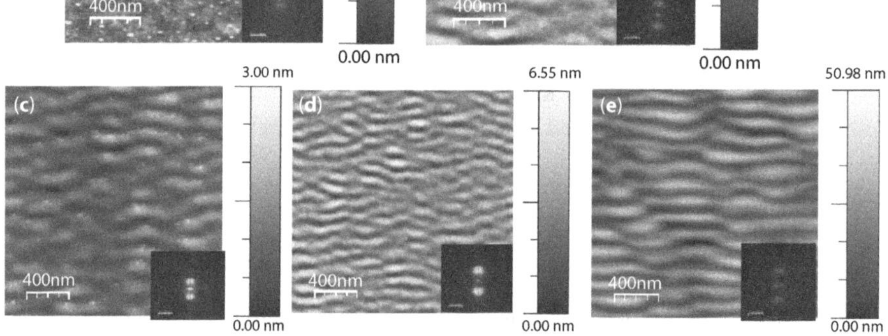

Figure 17.8 AFM images of 14 keV (a) C^+, (b) N_2^+, (c) NO^+, (d) O_2^+, and (e) Ar^+ ion-bombarded mica at angle 60° with ion fluence 7×10^{17} ions/cm^2. The arrow indicates the ion beam direction. Reprinted from [42], Copyright (2021), with permission from Elsevier.

surfaces is shown in Figure 17.9a with nuclear energy loss for different ion mass. The rms roughness increases with ion mass and the nuclear energy loss is also greater for heavier ions as observed from the figure. The increase of rms roughness for heavier ions is due to the higher sputtering yield for heavier ion mass. The ripple wavelength, calculated from AFM images, is found to be proportional to the range of ion as shown in Figure 17.9b. This type of phenomena was also observed earlier [41]. The ion penetration depth (range) can be tuned by changing the energy or mass of the projectile ion.

Figure 17.9c, the XPS spectra of pristine and ion-modified mica, shows the presence of all the elements (except H). Interestingly, the upper K atoms are sputtered most as discussed in the previous section. We have calculated the percentage of K concentration after ion bombardment quantitatively and plotted it with sputtering yield (from SRIM simulation) in Figure 17.9d. The depletion of K atoms is greater for heavier ions and the percentage of K atoms is nearly negligible for Ar^+ ion. This indicates the depletion of almost all the K atoms from the mica surface. As the mass of Ar^+ ion is comparable with the mass of K, the transfer of energy by Ar^+ ion to K atom during bombardment is the highest, resulting in the sputtering of almost all the K atoms.

The preferential sputtering of different elements in the mica surface generates surface instability, which initiates the development of ripple-like morphology. The growth rate of the ripple wave vector is maximum at oblique ion incidence. As a result, a periodic ripple pattern is formed for all the ions except C^+. However, the preferential sputtering is the cause

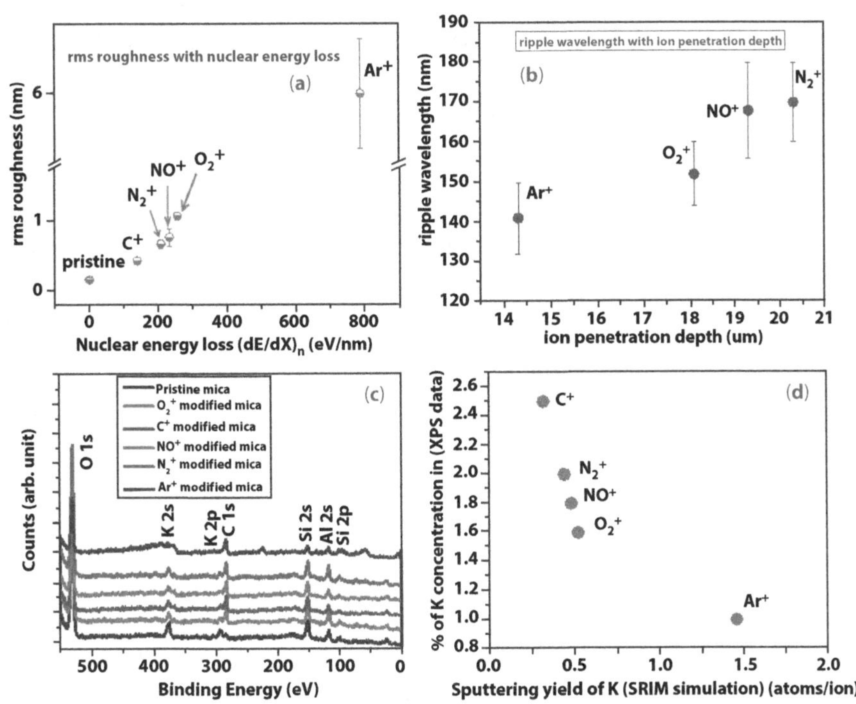

Figure 17.9 (a) Variation of rms roughness with nuclear energy loss and (b) ripple wavelength with ion penetration depth. (c) Survey scan of XPS of pristine and ion-modified mica surfaces with different ions and (d) calculated K concentration from XPS with sputtering yield of K. Reprinted from [42], Copyright (2021), with permission from Elsevier.

of ripple pattern formation for mica surface, and a minimum sputtering yield is essential for initiation of ripple structure [8]. The carbon is the lightest mass among all these projectile ions. The sputtering yield of different constituent elements of mica is different, and it is less for carbon ion bombardment compared to other ions (Figure 17.9d). However, due to the small amount of sputtering of Al, Si, and H in mica, irregular dot-like patterns are observed for C^+ bombardment. This type of dot was also observed at oblique ion incidence on the InP surface due to enrichment of elements (In or P) [43]. On the other hand, a well-defined ripple pattern is developed on the mica surface due to 14 keV Ar^+, O_2^+, N_2^+, and NO^+ ion bombardment. In the case of heavier ion bombardment like Ar^+, O^+, etc., the energy momentum transfer to the target atoms of mica is larger compared to lighter ions. This directly implies that the formation of vacancy or defect by the rate of sputtering of the constituent elements in mica is larger. This is also consistent with the concentration and sputtering yield of K atoms as indicated in Figure 17.9d. Further, the mass of the Ar^+ ion is nearly comparable to that of the potassium; hence, the potassium vacancy formation is the largest due to Ar^+ ion bombardment. Therefore, the mass of the ion plays an important role on controlling the surface instability. Hence, the Ar^+ ion having a higher mass is favorable for the well-defined ripple pattern formation on mica. Ziberi *et al.* also observed that the mass of ion plays a vital role on ripple pattern formation in case of ultra-low energy (eV) inert ion bombardment on the Si surface [44].

17.4 Band Gap Engineering of Muscovite Mica by Low-Energy Ion Beam via Few-Layer and Monolayer Modification

The energetic ion loses its energy due to energy and momentum transfer to the target atoms during ion bombardment on solid surface, which leads to the formation of defect/damage on the substance. In general, for the case of low-energy ion bombardment on solid surface, the formation of defect/damage on the substrate occurs as a consequence of sputtering and nuclear energy loss by the projectile ion. This energy loss depends not only on the incident ion energy but also on the projectile mass, resulting in the difference of penetration depth in the material for the ions of different mass. In case of mica, the penetration depth is directly related to the modified number of layers as each layer is 1 nm thick. The ion beam-induced modification by means of defect formation changes the electronic configuration of the material, which results in the change of density of state of the material, and hence the band gap is modified.

In order to investigate the change in band gap due to the defect formation by these projectile ions on mica, the optical response of the pristine and ion-modified muscovite mica is recorded by means of UV-IR absorbance spectra. Figure 17.10a shows the UV-Visible spectroscopy of pristine and 14 keV several ion-modified muscovite mica. The pristine mica is UV sensitive since the absorption is larger near the UV region. The ion bombardment leads to a change in UV absorption. The optical absorption is less at higher wavelength, which increases at higher wavelength. The light absorption of a material depends on several parameters like the band structure, density of material, band gap, etc. [45]. The introduction of defect changes the band structure and the band gap of the material, resulting in the change of absorption. However, the energetic ion beam irradiation leads to chemical

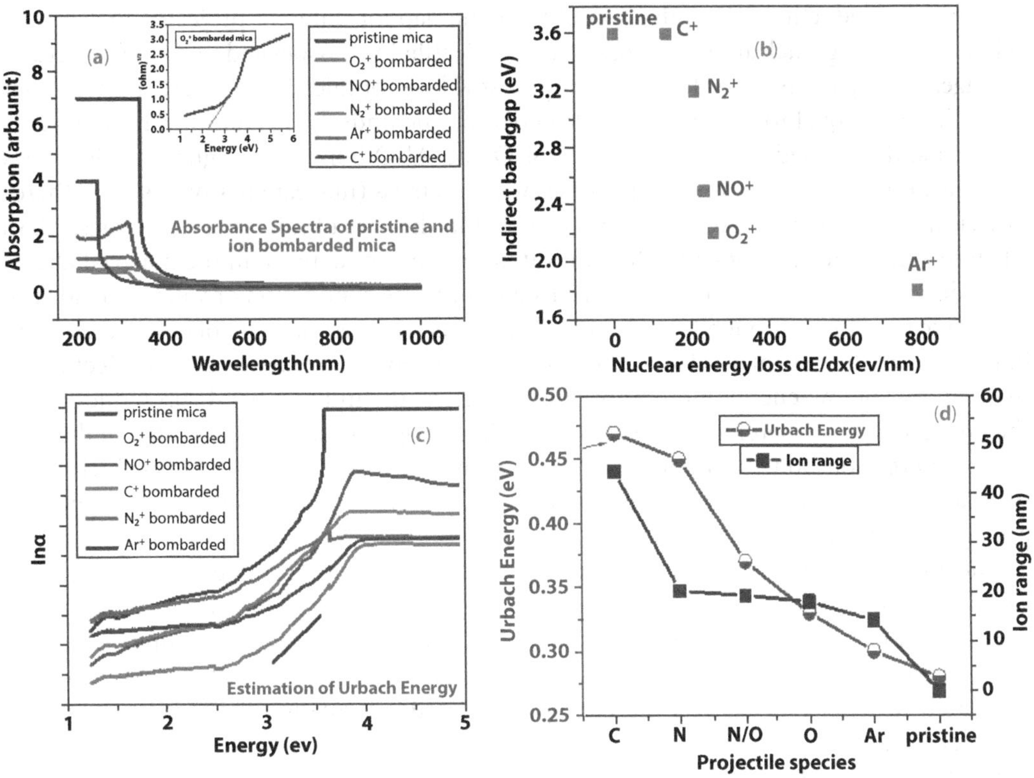

Figure 17.10 (a) UV-VIS-IR absorption spectra of pristine and ion-modified mica surface. The inset of panel (a) is the Tauc's plot showing the band gap estimation. (b) Plot of ln α vs energy. (c) Variation of Urbach energy and ion range (from SRIM) with different projectile ions. (d) Calculated indirect band gap for ion-modified mica with nuclear energy loss. Reprinted from [42], Copyright (2021), with permission from Elsevier.

modification of the surface, but in this case, the possibility of such modification is very low. Therefore, the band gap modification is mainly affected by defect formation. The ions create several types of defects or damages in mica surface; as a result, it modifies the electronic and chemical configuration of the surface. To ensure that, the band gap of ion-modified mica surfaces is calculated from the absorption spectra. The estimation of band gap from Tauc's plot is shown in the inset of Figure 17.10a. The corresponding absorption data are used to generate the plot $\alpha(hv) = B\,(hv - Eg)^{n}$ where 'B' is the correction coefficient and 'E_g' is the optical band gap. The value of n is equal to ½ and 2 for direct and indirect bang gap, respectively. The absorption coefficient can be obtained from the absorption intensity using the equation $\alpha(hv) = 2.303A/t$, where 't' is the thickness of the sample, A is the absorbance and h is the Planck constant. The optical band gap is estimated by extrapolating the curve to the $\alpha(hv) = 0$ axis. The pristine mica exhibits a direct band gap of 3.55 eV and an indirect band gap of 3.6 eV. The ion bombardment does not lead to a significant change in direct optical band gap, but the indirect band gap changes significantly. The direct and indirect band gaps as estimated from Tauc's plot are given in Table 17.1. The penetration depths are also provided to establish a relation between the numbers of modified layers with the amount of the band gap reduction. The reduction of the band gap for different projectile ions can be enlightened using nuclear energy loss of the projectile ions associated with the structural damage. During ion bombardment, the topmost K layer of mica is massively sputtered out up to the layer equal to the ion penetration depth. The underlying aluminosilicate layer is exposed and modified by the breaking of Si-O and Al-O bonds [5]. Figure 17.10b shows that the indirect band gap of mica decreases with ion mass (nuclear energy loss). It is therefore clear from Figure 17.10b and Table 17.1 that in the case of mica, projectile mass plays a key role in band gap tailoring. The introduction of defect/damage in the material affects the indirect band gap by creating localized states near the conduction or valence band. The electronic transition through these localized states requires less amount of energy, resulting in lowering of the optical band gap. The nuclear energy loss increases with the projectile ion mass in case of low-energy ion bombardment; hence, the indirect band gap reduces with the projectile mass. Therefore, a precise control in damage formation is expected in the case of low-energy ion bombardment. As mica is a layered material with a 1-nm-thick layer, several numbers of such layers can be modified in a single-step irradiation by changing the projectile ion mass.

The quantification of the damage or disorder generation on mica due to low energy ion irradiation can be explored by means of estimating Urbach energy (*Eu*). The Urbach energy is estimated using the Urbach equation $ln\alpha = ln\alpha_0 + h\nu/E_u$, as shown in Figure 17.10c and listed in Table 17.1. Table 17.1 indicates that the penetration depth of the C^+ ion is higher compared to other ions; hence, the damage or disorder formation by C^+ over the entire penetration depth is also greater. Thus, the Urbach energy increases with the depth of the implanted ion, i.e., the number of modified layers (Figure 17.10d). Thus, the indirect band

Table 17.1 Band gap values of ion-modified mica surfaces with ion ranges and energy loss. Reprinted from [42], Copyright (2021), with permission from Elsevier.

Ion species	Range (nm)	$(dE/dx)_{nuclear}$ (eV/nm)	Direct band gap energy (eV)	Indirect band gap energy (eV)	Urbach energy (eV)
Pristine			3.55	3.6	0.28
14 keV C+	44.5	137.9	3.27	3.6	0.47
14 keV N_2^+	20.3	207.9	3.1	3.2	0.45
(7 keV N^+)					
14 keV NO^+	19.3	232.4	3.6	2.5	0.37
(7.5keV O^+& 6.5keV N^+					
14 keV O_2^+	18.1	255.4	3.3	2.2	0.33
(7 keV O^+)					
14 keV Ar^+	14.3	789.3	3.2	1.8	0.30

gap of ion-modified mica decreases with the increase of nuclear energy loss by the projectile ions, whereas the Urbach energy is proportional to the ion penetration depth.

Therefore, it is established that the band gap modification of muscovite mica depends on the number of layers in which the defects are introduced. Thus, it is very interesting to investigate the band gap reduction due to the modification of the top layer (mono layer) of mica. The Monte Carlo simulations (SRIM) are carried out to estimate the ion beam parameters (energy and incident angle), which ensure the damage formation into a single layer of mica. The first layer of mica surface can be modified by selecting the Ar^+ ion beam of energy 3 keV at near-grazing incidence (85°). The vacancy profile for 3 keV Ar^+ ion in mica surface is shown in Figure 17.11a, which indicates that the maximum vacancy is created up to 1 nm depth, i.e., the first monolayer. The experimental verification of defect formation or vacancy creation is investigated by measuring the change in atomic concentration of constituent element of mica. Figure 17.11b indicates the XPS of pristine and 3 keV Ar^+ ion-bombarded mica at 85°. The change in atomic concentrations of each constituent element before and after irradiation is estimated from high-resolution spectra and is presented in Figure 17.11c. It is evident that ion- induced sputtering removes all the constituent elements, resulting in the decrease in atomic concentration. The arrival of an additional C 1s peak in all the surfaces is due to adsorption of carbon from the environment during air exposure before XPS

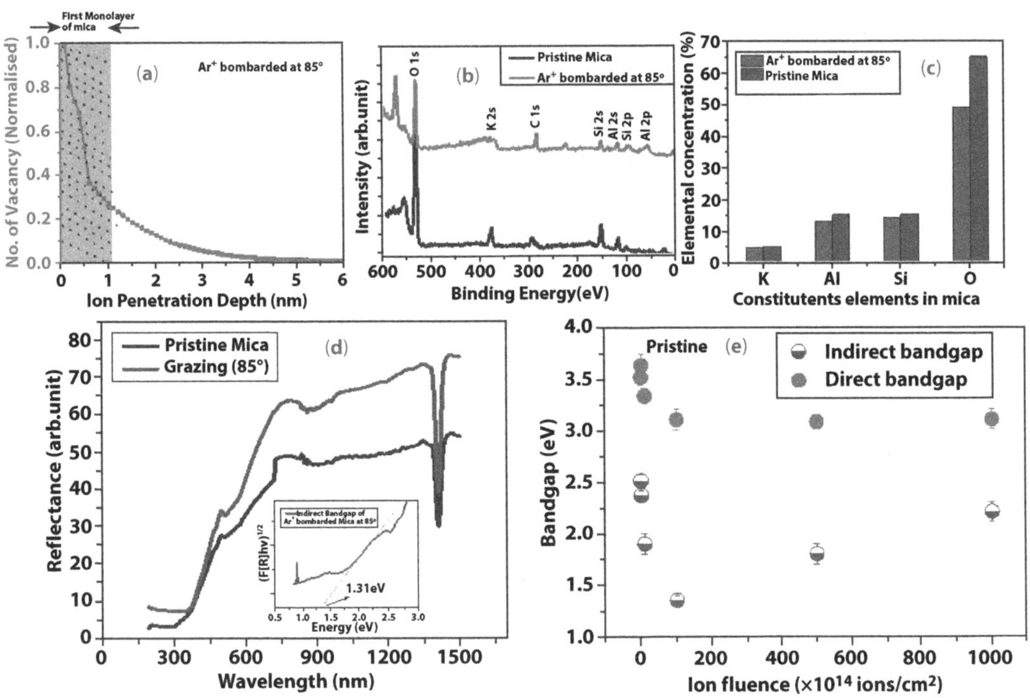

Figure 17.11 (a) XPS survey scan of pristine and Ar^+ ion (3 keV) bombarded mica surface at grazing incidence (85°), (b) elemental composition of mica elements after ion bombardment, and (c) vacancy profile for 3 keV Ar ion in mica surface by SRIM simulation. (d) Reflectance of pristine and ion-modified mica from UV-Vis spectra. The Kubelka–Munk's plot for estimation of direct band gap is shown in the inset of panel (d). (e) Change in indirect band gap with increase of Ar^+ ion fluences indicates maximum band gap reduction at fluence 1×10^{16} ions/cm². Reprinted from [46], Copyright (2022), with permission from Elsevier.

measurement. The intensity of the C 1s peak is higher for ion-bombarded mica, which indicates hydrocarbon absorption of mica surfaces as discussed in the previous section.

In order to estimate the change in band gap of pristine as well as defect-induced mica surface, UV-Vis-IR reflectance spectra of the pristine and defect-induced mica surfaces are presented. The band gaps are estimated using Kubelka–Munk's plot. Figure 17.11d shows the UV-IR reflectance spectra for pristine and ion-modified mica samples, indicating the change in reflectivity due to ion bombardment. The Kubelka–Munk's (K-M) plot is generated using these reflectance spectra as shown in the inset of Figure 17.11d. The linear portion of the plot $[F(R)h\nu]^{1/n} = B\,(h\nu - E_g)]$ is extrapolated (from the point where $d(F(R)h\nu)^{1/n}/d(h\nu) = 0$) in energy axis to determine the optical band gap of the samples [47]. Here, the Kubelka-Munk's function (F(R)) is defined as $F(R) = (1- R)^2/2R$, where E_g is the optical band gap, B is a constant [48] and R is the reflectance. The extrapolation of band gap from the K-M plot is given in the inset of Figure 17.11d for 3 keV Ar^+ ion-bombarded mica at an oblique ion incidence of 85° with an ion fluence of 1×10^{16} ions/cm^2. It is clear from the K-M plot that the indirect band gap of pristine mica (around 3.6 eV) reduces to 1.3 eV due to the introduction of defect in the topmost layer. The change in direct band gap is very nominal, since the mica is not a direct band gap material. The band gap is also studied for several ion fluences starting from 1×10^{14} ions/cm^2 to 1×10^{17} ions/cm^2 as given in Figure 17.11e. The variation of indirect band gap with ion fluence indicates that the band gap reduction is minimum at bombarded ion fluence 1×10^{16} ions/cm^2. This indicates that at fluence 1×10^{16} ions/cm^2, the defect formation is sufficient enough. However, beyond fluence 1×10^{16} ions/cm^2, there is slight increment in band gap due to sputtering of upper aluminosilicate layer, and exposure of the underlying layer.

The measurement of local current voltage characteristics by Conductive Atomic Force Microscopy (C-AFM) is used to confirm the change in band gap due to defect formation on the top layer as observed in a previous method. Considering Wentzel-Kramer's Brillouin (WKB) approximation, the plot of dI/dV as a function of tip-sample bias voltage (V) is directly related to the density of states (DOS) of the sample, with the assumption that the bias voltage does not exceed the work function of the sample [49]. The band gap is calculated from the difference of bias voltage as $E_g = |V_{+bias}| + |V_{-bias}|$. Figure 17.12 represents the conducting AFM characteristics of mica before and after the defect was introduced. The two-dimensional current images from C-AFM are shown in Figure 17.12a and c for the pristine and 3 keV Ar^+ bombarded mica surface. The current–voltage measurements are also shown in the inset of each image. Since the C-AFM measurements involve only the top surface of the mica, the information are mainly gathered from the topmost layer. The band gap is estimated as the length of the flat portion of the dI/dV –V curve (Figure 17.12b and d) [15, 50]. The pristine mica exhibits a band gap of 3.5 ± 0.1 eV approximately, whereas the same reduces to 1.2 ± 0.1 eV after introduction of defect in the topmost layer of mica. Thus, the entire results measured by C-AFM is consistent with the previous observation from UV-IR reflectance using the KM plot. Though these two techniques are used to determine the optical and electrical band gap, they are comparable due to the less exitonic binding energy [51]. Therefore, from both the theoretical and experimental data, the band gap of muscovite mica reduction is obvious.

The band gap reduction of muscovite mica due to defect formation is also supported theoretically from first-principles simulation of density functional theory. The band structure

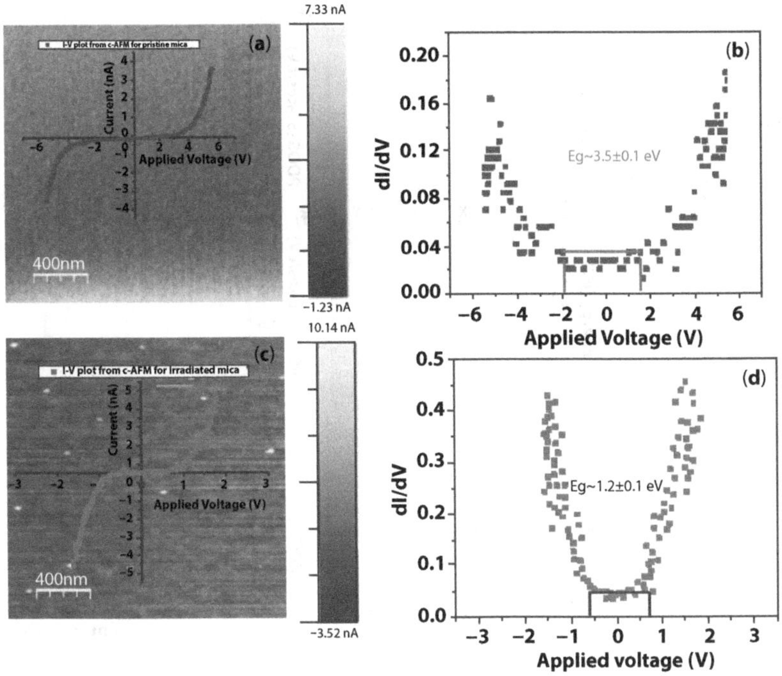

Figure 17.12 (a and c) Variation of current (I) measured from conducting AFM with applied voltage for pristine and ion-bombarded (85°) mica. (b and d) The local electronic density of state (dI/dV) as a function of applied voltage for pristine and ion-bombarded mica, showing the band gap. Reprinted from [46], Copyright (2022), with permission from Elsevier.

and the electronic density of states of a single layer of the pristine and defect-induced mica are presented in Figure 17.13. The pristine mica contains potassium (K), aluminum (Al), silicon (Si), oxygen (O), and hydrogen (H) atoms. The band gap of pristine mica is 4.63 eV along Z-A-X-Γ-Z-D-Y-Γ high symmetry points, consistent with the previously reported studies [52]. The projected electronic density of states of pristine mica (Figure 17.13c) indicates that 2p electrons of the O atoms have a major contribution in the valence band (VB), whereas the s and p orbitals of K atoms contribute mainly in the conduction band. Further, a change in band structure due to the introduction of two K, one Al, two Si, four O, and two H atoms with defect (Figure 17.13d) indicates a significant change in band gap from 4.63 eV (without defect) to 1.32 eV (all defect). Here, the VB is mainly composed of s orbitals of Si and p orbitals of O atom, respectively, while the s and p orbitals of K atoms mainly contribute in the conduction band. The above studies clearly suggest that the O atoms play a major role in the band structure of mica. It is therefore clear that the introduction of defect on the topmost layer of mica reduces the band gap, which is consistent with the experimental result. Further, we have also determined that the band gap reduction is caused by potassium removal. The band gap reduces to 2.2 eV due to such K removal. In this case, the 2p orbital of the O atoms contributes more in the valence band whereas the 2p and 2s of the potassium contributes highly in the conduction band. Therefore, it can be concluded that, in the case of muscovite mica, the maximum band gap reduction depends on defect formation on the topmost layer.

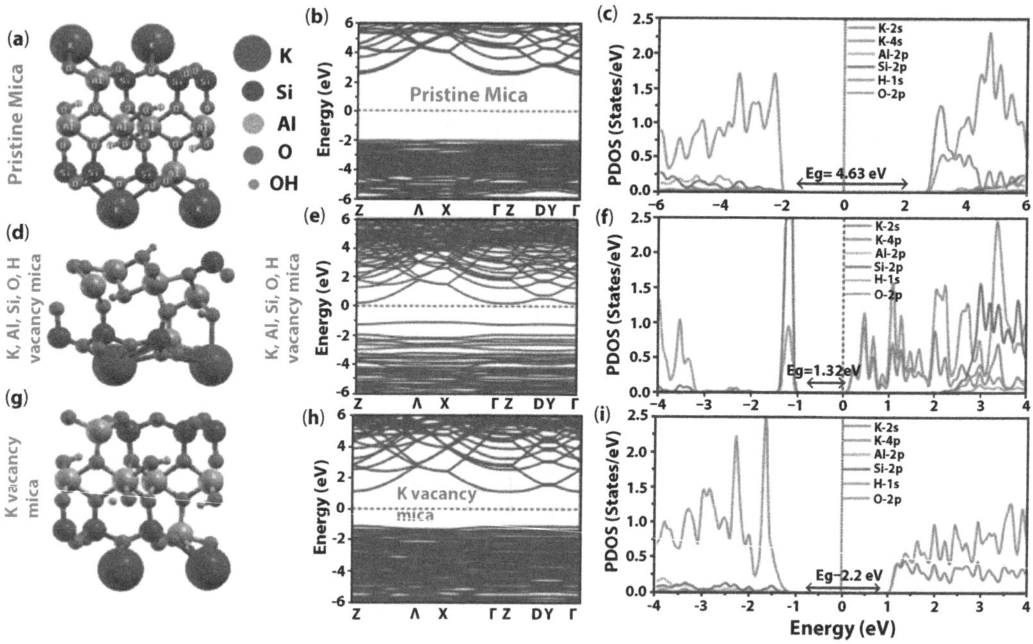

Figure 17.13 First-principles calculation for the estimation of band gap: 3D structure of Mica, band structure, and partial density of state for (a–c) pristine mica, (d–f) mica with K, Al, Si, O, H vacancies, and (g–i) mica with only K vacancy. Reprinted from [46], Copyright (2022), with permission from Elsevier.

17.5 Conclusion

The present work describes the engineering of several properties of muscovite mica, a naturally available important material, by low-energy ion beam irradiation. The periodic ripple-like nanostructure formation and the growth mechanisms are elaborated in this chapter. Since a patterned solid surface has several applications, it has been developed by ion bombardment on a large area, and their size and shape have been controlled by ion beam parameters like ion energy, species, incident angle, etc. As mica can be considered as a 2D nanosheet having atomic thickness, the band gap engineering of mica surface could be potentially applicable in microelectronics and optoelectronic devices. The band gap engineering of this type of nanosheet becomes sometimes challenging due to the quantum confinement effects [13]. However, the ion beam technique can manage this limitation by controlling its parameters. By changing the bombarding species, energy, and angle of incidence, the dimension of the modified region can be controlled. Due to differential sputtering of the elements of mica and defect formation during ion bombardment, the electronic structure of the mica nanosheet is modified; as a result, the density of states of mica is altered. It was observed by Kim *et al.* [15] using DFT calculation that the density of states of oxygen atoms in the valence band and other elements (K, Al, Si) in the conduction band changes, which modifies the band gap of mica. In our DFT results, the partial density of states of each element changes, leading to band gap modification.

In summary, low-energy ion bombardment is a unique tool to develop nanostructure and to modify the band gap of 2D materials like muscovite mica in a controlled way. Both theoretical and experimental studies of defect-induced band gap modification of muscovite mica evidence a good agreement. The depthwise introduction of defects on insulating muscovite mica changes the electronic band structure; as a result, the optical properties are also changed. This is useful for different applications such as photocatalytic and biological aspects. Therefore, ion bombardment-modified mica with a patterned surface and a narrow band gap bears promise and could be a potential smart material for future micro- and optoelectronic device applications.

Acknowledgments

We thank Ms. Madhavi and Prof. P. K. Jha for DFT computational support. We acknowledge Mr. Goutam Sarkar and Prof. M. Mukherjee for providing the XPS facility. We also thank RRCAT synchrotron XPS beam line members for their cooperation. Smt. Soma Roy and Prof. P.M.G Nambisan are acknowledged for providing the UV-Vis facility. The authors thank the Department of Atomic Energy for financial support to carry out this work. One (DB) of the authors also thanks IIT Kanpur for providing Institute Postdoctoral Fellowship.

References

1. Singh, S.; Keller, D. J. *Biophys. J.*, 1991, *60*, 1401. Thundat, T.; Allison, D. P.; Warmack, R. J.; Brown, G. M.; Jacobson, K. B.; Schrick, J. J.; Ferrell, T. L.; Henderson, E.; Heckl, W.; Lindsay, S. M. S. M. *Scanning Microsc.*, 1992, *6*, 911. Eaton, P.; West, W. Substrates for AFM, in: *Atomic Force Microscopy*; Oxford University Press, 2010. Hansma, P. K.; Cleveland, J. P.; Radmacher, M.; Walters, D. A.; Hillner, P. E.; Bezanilla, M.; Fritz, M.; Vie, D.; Hansma, H. G.; Prater, C. B.; *et al. Appl. Phys. Lett.*, 1994, *64*, 1738. Cai, L.; Tabata, H.; Kawai, T. *Appl. Phys. Lett.*, 2000, *77*, 3105.
2. Osada, M.; Sasaki, T. *Adv. Mater.*, 2011, *24*, 210. Geim, A. K.; Novoselov, K. S. *Nat. Mater.*, 2007, *6*, 183. Novoselov, K. S.; Geim, A. K.; Morozov, S. V.; Jiang, D.; Zhang, Y.; Dubonos, S. V.; Grigorieva, I. V.; Firsov, A. A. *Science*, 2004, *306*, 666.
3. Metya, A.; Ghose, D. *Appl Phys Lett.*, 2013, *103*, 161602.
4. Metya, A.; Ghose, D.; Mollick, S. A.; Majumdar, A. *J. Appl. Phys.*, 2012, *111*, 074306.
5. Bhowmik, D.; Karmakar, P. *Surf. Interface Anal.*, 2019, *51*, 667.
6. Keller, A.; Fritzsche, M.; Ogaki, R.; Bald, I.; Facsko, S.; Dong, M.; Kingshott, P.; Besenbacher, F. *J. Chem. Phys.*, 2011, *134*, 104705.
7. Metya, A.; Ghose, D.; Ray, N. R. *Appl. Surf. Sci.*, 2014, *293*, 18.
8. D. Bhowmik; P. Karmakar. *Nucl. Instrm. Methd. Sect. B*, 2018, *422*, 41.
9. Oates, T.; Keller, A.; Facsko, S.; Mücklich, A. *Plasmonics*, 2007, *2*, 47. Liedke, M.; Körner, M.; Lenz, K.; Grossmann, F.; Facsko, S. *Appl Phys Lett.*, 2012, *100*, 242405.
10. Fassbender, J.; Strache, T.; Liedke, M.; Marko, D.; Wintz, S.; Lenz, K.; Keller, A.; Facsko, S.; Monch, I.; McCord, J. *New J. Phys.*, 2009, *11*, 125002.
11. Bhowmik, D. Surface and interface modification by low energy ion beams. Homi Bhabha National Institute, Mumbai, 2019. http://www.hbni.ac.in/phdthesis/phys/PHYS04201404008.pdf.

12. Wang, X.; Wan, W.; Shen, S.; Wu, H.; Zhong, H.; Jiang, C.; Ren, F. *Appl. Phys. Rev.*, 2020, *7*, 041303.
13. Mak, K. F.; Lee, C.; Hone, J.; Shan, J.; Heinz, T. F. *Phys. Rev. Lett.*, 2010, *105*, 136805.
14. Wary, G.; Kalita, J. M. *Nanoscale*, 2016, *22*, 3831. Kaur, S.; Singh, S.; Singh, L. *Appl. Radiat. Isot.*, 2017, *121*, 116.
15. Kim, S. S.; Khai, T. V.; Kulish, V.; Kim, Y.-H.; Na, H. G.; Katoch, A.; Osada, M.; Wu, P.; Kim, H. W. *Chem. Mater.*, 2015, *27*, 4222.
16. Mukherjee, J.; Bhowmik, D.; Mukherjee, M.; Satpati, B.; Karmakar, P. *J. Appl. Phys.*, 2020, *127*, 145302.
17. Bhowmik, D.; Mukherjee, M.; Karmakar, P. *Nucl. Instr. Meth. B*, 2019, *444*, 54.
18. Mandal, S.; Mukherjee, M.; Hazra, S. *ACS Appl. Mater. Interfaces*, 2020, *12*, 45564.
19. Levenberg, K. *Quart. Appl. Math.*, 1944, *2*, 164.
20. Beamson, G.; Briggs, D. *J. Chem. Educ.*, 1993, *70*, A25.
21. Ziegler, J. F.; Ziegler, M. D.; Biersack, J. P. *Nucl. Instrum. Meth. Phys. Res. B*, 2010, *268*, 1818.
22. Bhowmik, D.; Mukherjee, J. *Mater. Today: Proc.*, 2021, *2369*, 020038.
23. Kohn, W.; Becke, A. D.; Parr, R. G. *J. Phys. Chem.*, 1996, *100*, 12974. Giannozzi, P.; Baroni, S.; Bonini, N.; Calandra, M.; Car, R.; Cavazzoni, C.; Wentzcovitch, R. M. *J. Phys. Cond. Matter*, 2009, *21*, 395502.
24. Perdew, J. P.; Burke, K.; Ernzerhof, M. *Phys. Rev. Lett.*, 1996, *77*, 3865.
25. Bhowmik, D.; Chowdhury, D.; Karmakar, P. *Surf. Sci.*, 2019, *679*, 86.
26. Navez, M.; Sella, C.; Chaperot, D. *C.R. Acad. Sci.*, 1962, *254*, 260.
27. Bradley, R. M.; Harper, J. M. E. *J. Vac. Sci. Technol. A*, 1988, *6*, 2390.
28. Chan, W. L.; Chason, E. *J. Appl. Phys.*, 2007, *101*, 121301.
29. Garcia, M. A.; Rickards, J.; Cuerno, R.; Trejo-Luna, R.; Cañetas-Ortega, J.; Vega, L. R. d. l.; Rodríguez-Fernández, L. *Phy. Rev. App.*, 2017, *8*, 064027.
30. Keller, A.; Facsko, S. *Materials*, 2010, *3*, 4811.
31. Zhao, Y.-P.; Drotar, J. T.; Wang, G.-C.; Lu, T.-M. *Phys. Rev. Lett.*, 1999, *82*, 4882.
32. Schmittmann, B.; Pruessner, G.; Janssen, H.-K. *Phys. Rev. E*, 2006, *73*, 051603.
33. Munoz-Garcia, J.; Castro, M.; Cuerno, R. *Phys. Rev. Lett.*, 2006, *96*, 086101.
34. Vivo, E.; Nicoli, M.; Engler, M.; Michely, h.; V ́azquez, L.; Cuerno, R. *Phys, Rev. B*, 2012, *86*, 245427.
35. Karmakar, P.; Mollick, S. A.; Ghose, D.; Chakrabarti, A. *Appl. Phys. Lett.*, 2008, *93*, 103102.
36. Hofsäss, H.; Zhanga, K. *Nucl. Instr. Meth. B*, 2009, *267*, 2731.
37. Karmakar, P.; Liu, G. F.; Yarmoff, J. A. *Phys. Rev. B*, 2007, *76*, 193410.
38. Carter, G.; Vishnyakov, V. *Phys. Rev. B*, 1996, *54*, 17647. Madi, C. S.; Anzenberg, E.; Ludwig, K. F.; Aziz, M. J. *Phys. Rev. Lett.*, 2011, *106*, 066101.
39. Bhattacharyya, K. G. *J. Electr. Spectrosc. Related Phenomena*, 1993, *63*, 289. Buzio, R.; Toma, A.; Chincarini, A.; Mongeot, F. B. D.; Boragno, C.; Valbusa, U. *Surf. Sci.*, 2007, *601*, 2735.
40. Neinhuis, C.; Barthlott, W. *Annals of Botany*, 1997, *79*, 667.
41. Makeev, M. A.; Cuerno, R.; Barabási, A.-L. *Nucl. Instrum. Methods Phys. Res. B*, 2002, *197*, 185. Kahng, B.; Jeong, H.; Barabasi, A. L. *Appl. Phys. Lett.*, 2001, *78*, 805. Bhowmik, D.; Karmakar, P. *AIP Conference Proceedings*, 2017, *1832*, 080019.
42. Bhowmik, D.; Mukherjee, J.; Karmakar, P. *Radiat. Phys. Chem.*, 2021, *187*, 109568.
43. Som, T.; Chini, T. K.; Katharia, Y. S.; Tripathy, S.; Kanjilal, D. *Appl. Surf. Sci.*, 2009, *256*, 562.
44. Ziberi, B.; Frost, F.; Höche, T.; Rauschenbach, B. *Phys. Rev. B*, 2005, *72*, 235310.
45. Mattheis, J.; Rau, U.; Werner, J. H. *J. Appl. Phys.*, 2007, *101*, 113519.
46. Mukherjee, J.; Dalsaniya, M. H.; Bhowmick, S.; Bhowmik, D.; Jha, P. K.; Karmakar, P. *Surf. Interfaces*, 2022, *33*, 102283.
47. Zanatta, A. R. *Sci. Rep.*, 2019, *9*, 11225.

48. Gesesse, G. D.; Gomis-Berenguer, A.; Barthe, M.-F.; Ania, C. O. *J. Photochem. Photobiol. A*, 2020, *398*, 112622.
49. Wiesendanger, R. *Scanning Probe Microscopy and Spectroscopy: Methods and Applications*; Cambridge University Press, 1994. Hamers, R. J.; Padowitz, D. F.; Bonnell, D. A. Methods of Tunneling Spectroscopy with the STM," from *Scanning Probe Microscopy and Spectroscopy: Theory, Techniques, and Applications*; Wiley-VCH, 2001.
50. Bhowmik, D.; Bhattacharjee, S. *Physica B Condens. Matter.*, 2021, *623*, 413377.
51. Duman, S.; Gurbulak, B.; Dogan, S.; Ozcelik, F. S. *J. Phys. Conf. Ser.*, 2016, *707*, 012027.
52. Rudenko, A. N.; Keil, F. J.; Katsnelson, M. I.; Lichtenstein, A. I. *Phys. Rev. B*, 2011, *83*, 045409. Wu, S.; Lee, W. P. C.; Wu, P. *Sci. Rep.*, 2022, *12*, 2868.

About the Editors

Upendra Kumar, PhD has been an assistant professor in the Department of Applied Science at the Indian Institute of Information Technology, Allahabad, Uttar Pradesh, India since 2021. His research has been published in prestigious international Science Citation Index physics and materials science publications and he has received numerous awards in the field. He has also made an impact by attending numerous scientific conferences, seminars, and workshops and serving on a number of national committees and professional bodies and associations.

Piyush K. Sonkar, PhD is an assistant professor in the Department of Chemistry, Banaras Hindu University, Varanasi-India. His research interests include nanomaterials, nanocomposites, fuel cells, electrochemical devices, supercapacitors, bio-sensors, chemical sensors, and new materials. He has published more than 38 international and national research papers in various reputed journals.

Index

Also of Interest

Check out these other Related Titles

MATERIALS FOR SOLAR ENERGY CONVERSION: Materials, Methods and Applications, Edited by R. Rajasekar, C. Moganapriya and A. Mohankumar, ISBN: 9781119750604. This book provides professionals and students with a resource on the basic principles and applications of solar energy materials and processes, as well as practicing engineers who want to understand how functional materials operate in solar energy conversion systems.

BIODEGRADABLE MATERIALS AND THEIR APPLICATIONS, Edited by Inamuddin and Tariq Altalhi, ISBN: 9781119904908. Biodegradable materials have ascended in importance in recent years and this book comprehensively discusses all facets and applications in 29 chapters making it a one-stop shop.

GREEN AND SUSTAINABLE ADVANCED MATERIALS, VOLUME 1: Processing and Characterization, Edited by Shakel Ahmed and Chaudhery Mustansar Hussain, ISBN: 9781119407041. This book is industry-oriented and shows current challenges for scaling up green and sustainable advanced materials design and manufacturing technology.

GREEN AND SUSTAINABLE ADVANCED MATERIALS, VOLUME 2: Applications, Edited by Shakeel Ahmed and Chaudhery Mustansar Hussain, ISBN: 9781119528364. This book is industry-oriented and shows current challenges for scaling up green and sustainable advanced materials design and manufacturing technology.

HANDBOOK OF COMPOSITES FROM RENEWABLE MATERIALS, VOLUME 1: Fundamentals, Synthesis and Methods, Edited by Vijay Kumar Thakur, Manju Kumari Thakur and Michael R. Kessler, ISBN: 9781119223627. This unique multidisciplinary 8-volume set focuses on the emerging issues concerning synthesis, characterization, design, manufacturing and various other aspects of composite materials from renewable materials and provides a shared platform for both researcher and industry.

HANDBOOK OF COMPOSITES FROM RENEWABLE MATERIALS, VOLUME 2: Structure and Chemistry, Edited by Vijay Kumar Thakur, Manju Kumari Thakur and Michael R. Kessler, ISBN: 9781119223658. This unique multidisciplinary 8-volume set focuses on the emerging issues concerning synthesis, characterization, design, manufacturing and various other aspects of composite materials from renewable materials and provides a shared platform for both researcher and industry.

Printed in the USA/Agawam, MA
September 17, 2026

910661.172